PRINCIPLES OF LIGHT VEHICLE OPERATIONS

Level 1

Graham Stoakes

Eric Sykes

Catherine Whittaker

babcock
trusted to deliver™

Heinemann
Part of Pearson

Heinemann is an imprint of Pearson Education Limited, Edinburgh Gate, Harlow, Essex, CM20 2JE.

www.pearsonschoolsandfecolleges.co.uk

Heinemann is a registered trademark of Pearson Education Limited

Text © Babcock International Group and Graham Stoakes 2011
Edited by Liz Cartmell, Caroline Low and Liz Evans
Designed by Pearson Education Limited
Typeset by Tek-Art, Crawley Down, West Sussex
Original illustrations © Pearson Education Limited 2011
Illustrated by KJA Artists
Cover design by Woodenark Studios
Cover photo © Front: **Car Photo Library:** David Kimber

The rights of Babcock International Group and Graham Stoakes to be identified as authors of this work have been asserted by them in accordance with the Copyright, Designs and Patents Act 1988.

First published 2011

15 14
10 9 8 7 6 5

British Library Cataloguing in Publication Data
A catalogue record for this book is available from the British Library

ISBN 978 0 435 04815 0

Printed in Slovakia by Neografia

Acknowledgements

Pearson Education Limited would like to thank the following people for providing technical feedback: Ian Gillgrass and Beverley Lilley of the Institute of the Motor Industry (IMI) and Stephen Mitchell of Whitstable Community College for his invaluable help in the development of this title.

Babcock International Group would like to thank Eric Sykes for his technical direction and Catherine Whittaker, Sarah Harrison and Christine Potts of Babcock International Group for their invaluable help in the development of this material. We would also like to thank Mike Davies and Gary Hurst for assistance in the development of the material, Derrick Insley for photography direction and Zoe Fielding and Jake Davenport for modelling during the photo shoot.

Graham Stoakes would like to thank Stella Mbubaegbu CBE, Highbury College Principal and Chief Executive, for the use of the college workshops during the photo shoot. Graham Stoakes would also like to thank Holly Stoakes and Jonny Walker for modelling during the photo shoot.

The author and publisher would like to thank the following individuals and organisations for permission to reproduce photographs:

The publisher would like to thank the following for their kind permission to reproduce their photographs:

(Key: b-bottom; c-centre; l-left; r-right; t-top)

Alamy Images: Alexander Podshivalov 291, Arunas Gabalis 299, Carpe Diem - UK 58, Chris Howes / Wild Places Photography 57, David J. Green 31 (4), Flake 29bl (2), 32 (3), 239bl (2), Hympi 61tr, 64cl, Johner Images 185 (1), Justin Kase 62, kpzfoto 61cr, metalpix 33tr, NUAGE 306, 307, Oleg Shpak 186, Peter Huggins 60 (4), 60tl, Prisma Bildagentur AG 235, RT Images 9tl, Transtock Inc. 55, Vibe Images 226; **Corbis:** Fancy 158cr, TongRo Image Stock 27; **Getty Images:** PhotoDisc 35 (5); **Glow Images:** Felbert & Eickenberg / STOCK4B 115; **iStockphoto:** Alistair Forrester Shankie 17 (2), 37 (3), Brian Sullivan 123cl; **Masterfile UK Ltd:** Ron Stroud 273, Transtock 175; **NHPA Ltd / Photoshot Holdings:** A.N.T. Photo Library 278; **Pearson Education Ltd:** David Sanderson 5 (No smoking sign), 5 (Safety helmet sign), 20tl, 21, 22tl, 22cl, 22bl, 23tr, 23cr, Gareth Boden 16, 29bl (3), 31 (1), 31 (2), 31 (3), 31 (6), 32 (2), 33cl, 33bl, 35 (2), 35 (3), 39 (2), 119cr, 239tr, 239bl (3), 277tl, Naki Photography 36 (1), Stuart Cox 19, Trevor Clifford 36 (3), 39 (1); **Rex Features:** Action Press 71, David Hartley 63tr; **Shutterstock.com:** Adrin Shamsudin 34tr, Alaettin YILDRIM 29bl (4), 32 (5), 239bl (4), Alex Kosev 1, ching 23bc, chrisbrignell 39 (4), Dzarck 195tr, Four Oaks 15cr, 17 (1), 36 (2), John Schiffer 37 (4), Kromkrathog 185 (2), Monkey Business Images 9bl, Quanta Photography 31 (5), Rafa Irusta 17 (3), Ramira 30 (4), Shi Yali 47, Steveball 9tr, StockHouse 23bl, TerryM 37 (1), TsR 63cr; **SuperStock:** imagebroker.net 185 (3)

All other images © **Pearson Education Ltd:** Clark Wiseman, Studio 8

Every effort has been made to trace the copyright holders and we apologise in advance for any unintentional omissions. We would be pleased to insert the appropriate acknowledgement in any subsequent edition of this publication.

Contents

Introduction

Welcome to Principles of Light Vehicle Maintenance & Repair!

Working in the automotive industry is going to give you a great opportunity to work here in the UK or overseas: it's a challenging, stimulating and fulfilling career.

Working in this sector combines many different practical skills with a knowledge of specialised materials and techniques. It also requires good people skills: customers are as much a part of the sector as the light vehicles. This book will introduce you to the automotive industry and all the important systems of light vehicles from engines, chassis, transmission and electrics through to inspection and servicing.

About this book

This book has been produced to help you build a sound knowledge and understanding of all aspects of the Certificate and NVQ requirements associated with the automotive industry.

The topics in this book cover all the information you will need to attain your Level 1 qualification in Light Vehicle Maintenance and Repair Principles. Each chapter of the book relates to particular units of the Certificate and provides the information needed to form the required knowledge and understanding of that area. It can be used for any awarding organisations qualification including IMIAL and City & Guilds.

This book is just one part of a series of publications aimed at light vehicle maintenance and repair learners. This Level 1 book is covers an introduction to the industry and how the light vehicle is constructed and maintained. The second book in the series covers the knowledge and skills required for the learner to service and replace vehicle systems in order to progress to Level 3. The Level 3 book (published in Spring 2012) covers in depth diagnosis and repair for current and new vehicle technology.

Although there are many routes into a career in the automotive industry, most will involve undertaking an apprenticeship. An apprenticeship means you will probably be employed by a garage and study at a college one to two days a week. At college you will learn the theory needed to enable you to gain a 'technical certificate' which includes the knowledge and skills requirements of your qualification. At work you will practice the skills and gather evidence to show that you are able to undertake the required practical tasks competently. This will enable you to gain the NVQ or SVQ part of your apprenticeship qualification.

This book has been written by experienced trainers who have many years of experience within the sector. They believe in providing you with all the necessary information you need to support your studies and ensuring it is presented in a style which is both manageable and relevant.

This book will also be a useful reference tool for you in your professional life once you have gained your qualifications and are working in the sector.

About the Automotive industry

75 per cent of households in the UK have access to at least one car and 30 per cent own two or more cars. This means that there are more than 33 million cars licensed for use in the UK.

The motor industry has a wide variety of job opportunities ranging from design, manufacture, sales maintenance and repair. All of these jobs need specialist knowledge and bring good career prospects. In the UK, the automotive industry employs more than 600,000 people directly and supports a further 1.9 million jobs (around 6 per cent of the UK workforce). Your career doesn't have to end in the UK either – there will always be cars that need fixing!

Qualifications for the automotive industry

There are many ways of entering the automotive industry, but the most common method is as an apprentice.

Apprenticeships

You can become an apprentice by being employed, usually be working for a garage or a vehicle manufacturer.

The Institute of the Motor Industry (IMI) operates the sector skills council which is responsible for setting the standards of automotive training in the UK.

The framework of an apprenticeship is based around an NVQ (or SVQ in Scotland). These qualifications are developed and approved by industry experts and will measure your practical skills and job knowledge on-site.

You will also need to achieve:

- a technical certificate
- the appropriate level of Functional skills assessment
- an Employees Rights and Responsibilities briefing.

Diploma

The Level 1 Diploma in Light Vehicle Maintenance and Repair Principles VRQ (Vocational Related Qualification) provides the basic knowledge and skills requirement that will allow learners to progress towards higher level Maintenance and Repair qualifications. The assessment of this VRQ is made up of three components:

- practical tasks
- online testing
- written assessments.

The Level 1 Diploma meets the requirements of the new Qualifications and Credit Framework (QCF) which bases a qualification on the number of credits.

In order to pass the qualification, you must achieve a minimum of 49 credits taken from mandatory and optional units.

Another important component of this qualification is the development of your functional skills to include: Numeracy, Literacy and ICT.

Authors

Graham Stoakes MIMI is a lecturer in automotive engineering for light vehicles and motorcycles at a large college of further education. With his background as a qualified master technician, senior automotive manager and specialist diagnostic trainer, he brings 28 years of technical industry experience to this title.

Eric Sykes B.Ed, AAE, MIMI, MSOE, MIFL, Eng. Tech – Automotive Quality Manager for Babcock International Group – has 28 years of experience in the motor industry from delivery of training through to management roles both nationally and internationally and has brought his wealth of technical knowledge and experience to this title.

Catherine Whittaker BA (Hons) – Babcock International Group – used her 12 years experience in learning resource development, specifically in the area of vocational training, to assist in the development of the title.

About the Institute of the Motor Industry (IMI)

The IMI is here to help you progress throughout your career and give you the recognition you deserve.

If you're looking to start a career in the sector, **1st Gear** is here to help you do just that. With free advice from industry experts, 1st Gear aims to help point you in the right direction to get you started. To join for free, visit www.1stgear.org.uk. For those of you just starting your career in the sector, **Accelerate** is the IMI's fast-paced online community for students in training who are at the beginning of their career journey into the automotive sector. It's where, from the start of your career, you can get help and information to assist you in your formal training. Accelerate is here to plug you in to up to date information and guidance. It hooks you up with fellow industry apprentices and trainees. It provides the support that will help you achieve your goals and make your career a success.

Not only does Accelerate help your career, it also gives your toolbox and social life a real boost. To take a look at the benefits and join, visit www.motor.org/accelerate.

For more information on the work of the IMI, visit www.motor.org.uk.

Qualification mapping grid

Chapter	QCF unit reference	IMIAL	City & Guilds
1 Health and safety practices in light vehicle maintenance	T/502/4654 A/600/3296	EL01 L101	001 051
2 Tools, equipment & materials	J/502/4657 F/600/3297	EL03 L102	026/076
3 The automotive industry	A/502/4655	EL02	081
4 Light vehicle construction & maintenance	F/502/4673 J/600/3303	EL13 L108	027/077
5 Light vehicle engine systems	J/600/3298 L/600/3299 T/600/3300 R/600/3305 Y/602/0008 D/602/0009 D/600/3307	L103 L104 L105 L110 L119 L120 L112	152 701 702 703 704
6 Light vehicle chassis systems	A/600/3301 F/600/3302 L/600/3304	L106 L107 L109	706 708 709
7 Light vehicle drivelines	Y/600/3306	L111	707
8 Light vehicle electrical systems	H/600/3308 K/600/3309	L113 L114	705

Features of this book

This book has been fully illustrated with artworks and photographs. These will help to give you more information about a concept or a procedure, as well as helping you to follow a step by step technical skill procedure or identify a particular tool or material.

This book also contains a number of different features to help your learning and development.

Working practice pages

These pages pick up the key health and safety areas you need to be aware of as you work on particular light vehicle systems.

Technical skills

Throughout the book you will find step by step procedures to help you practice the technical skills you need to complete be successful in your studies (see the example to the right).

Safety tips

These four features give you guidance for working safely on cars and in the workshop.

Emergency
Green safety tips provide useful information about SAFE conditions in emergency situations and your personal safety.

Safe working
Red safety tips indicate a PROHIBITION (something you **must not** do).

Safe working
Blue safety tips indicate a MANDATORY instruction (something that you must do).

Safe working
Yellow safety tips indicate a WARNING (hazard or danger).

Other features

Key term

These are new or difficult words. They are picked out in **bold** in the text and then defined in the margin.

Did you know?

This feature gives you interesting facts about the automotive industry.

Find out

These are short activities and research opportunities, designed to help you gain further information about, and understanding of, a topic area.

Working life

This feature gives you a chance to read about and debate a real life work scenario or problem. Why has the situation occurred? What would you do?

CHECK YOUR PROGRESS

These are a few questions at the end of each section, usually related to a learning outcome to see how you are getting along.

FINAL CHECK

This is a series of multiple choice questions at the end of each chapter, in the style of the end of unit tests.

GETTING READY FOR ASSESSMENT

This feature provides guidance for preparing for the practical assessment. It will give you advice on using the theory you have learnt about in a practical way.

1 Health & safety practices in light vehicle maintenance

Working in the light vehicle industry can be very rewarding. The skills you learn could potentially take you anywhere in the world. However, there are many dangers which lurk in the workshop. For example, you might get an electric shock from faulty wiring or simply trip over a cable that someone forgot to tidy away.

You need to make sure you work safely. Remember that there are other people in your workshop too. What you do, and how you behave, has an impact on your own safety and that of other people.

You may not have realised that health and safety in your workshop has anything to do with you. You are about to find out that it is *everything* to do with you.

In this chapter you will cover all you need to know at this level in order to work safely in the light vehicle industry.

This chapter covers:

- Health and safety requirements and information in motor vehicle workshops
- Health and safety practices and equipment
- Safe manual handling procedures
- COSHH procedures
- Fire prevention and emergency procedures

Health and safety requirements and information in motor vehicle workshops

Your responsibilities for health and safety in the workshop

Health and safety in the workshop isn't just something your employer needs to worry about. It is both your employer's and your responsibility to make sure everyone is safe in the workplace.

There are specific laws which both you and your employer must follow. They are there to make sure everyone, including your workmates and customers, are safe. The law which covers you and others in your workplace or workshop is the Health and Safety at Work Act 1974 (HASAWA).

Who is responsible for what under the Health and Safety at Work Act 1974?

You are responsible for:

- taking care that you do not cause danger to yourself or others who may be affected by your work
- following the health and safety polices and procedures at your workplace
- not damaging any machinery or PPE provided for your safety
- taking care of machine guards and reporting any problems.

Your employer is responsible for:

- providing you with safe equipment and safe ways of carrying out work tasks
- making sure that the equipment you use when handling, storing and transporting materials and substances is safe and does not cause health risks
- providing information, instruction, training and supervision to guarantee health and safety
- maintaining the workplace, including means of entry and exit, in a safe condition
- providing a safe and healthy working environment with adequate facilities and arrangements for employees' welfare
- providing for employees a written statement of policy, organisation and arrangements for health and safety. Employers with fewer than five employees are exempt from this.

The organisation which makes sure health and safety laws are followed correctly is called the Health and Safety Executive (HSE).

Using Personal Protective Equipment (PPE)

Personal protective equipment (PPE) is equipment that you wear or hold at work to protect you against risks to health and safety, for example approved safety glasses or goggles and boots. It is your responsibility to make sure you are wearing your PPE when at work.

> **① Safe working**
>
> It is everyone's responsibility to report all hazards and potentially dangerous incidents.

> **Did you know?**
>
> Apart from your own behaviour, you need to understand that other people's actions also have an impact on your safety when at work. So always keep an eye on what is happening around you and report any dangerous behaviour to your manager.

> **Did you know?**
>
> PPE is covered by the Personal Protective Equipment at Work Regulations 1992.

Tom is tired today because he had a late night last night. Normally he would change into his overalls before he starts work. Today he can't be bothered and his manager isn't there anyway. He has been asked to clean the workshop. After he has finished he sits down in reception and falls asleep. He is woken when a customer comes into reception to book their car in for a tyre to be replaced. He is asked to take the keys to his **mentor**. He reluctantly picks up the keys, yawns, and takes the keys to the workshop. The customer sits down in the reception area and waits for his car to be repaired.

1   How do you think Tom should have behaved?

The following table shows the most common PPE and what it is used for.

Table 1.1 PPE and its uses

PPE	Protect from	Example of workshop task
Overalls 	Used everyday to protect you and your clothing from grease and oil	You should not go into the workshop without wearing overalls. You need to wear your overalls for everything you do
Gloves 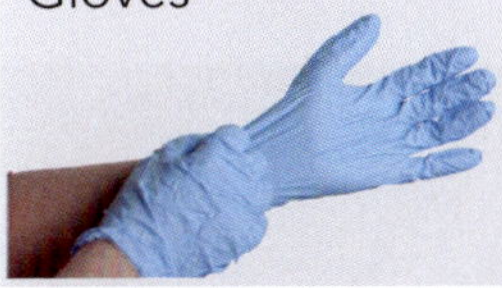	To protect your hands from harmful **chemicals** and **substances**	Changing an oil filter
Apron	To protect your clothes and overalls from harmful chemicals and substances	Cleaning greasy parts
Goggles 	Used to protect your eyes from dust and waste metal particles	Grinding
Helmet 	To protect your head from injury	Working under a vehicle
Safety footwear 	Used every day to protect your feet from injury	You should not go into the workshop without wearing boots. You need to wear your boots for everything you do

Figure 1.1 You must always wear the correct PPE for the task when in a workshop

Did you know?

By wearing the correct PPE you are protecting yourself from injury, but it is just as important to work safely with the tools and equipment that are appropriate to carry out the job. Never rely on PPE to be the only method used to protect you.

Key terms

Mentor – someone who you can look up to and learn from.

Chemical – a single fluid, gas or solid in its natural state.

Substance – a mixture of chemicals.

Figure 1.2 This technician is about to use a grinder so he is wearing his safety goggles

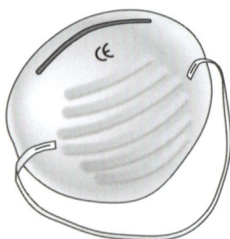

Figure 1.3 CE mark

Look after your PPE

Make sure you look after your PPE. If PPE is damaged, it will not protect you.

Failing to use the correct PPE

Table 1.2 The consequences of not wearing PPE

Work carried out	PPE not worn	Damage
Balancing a wheel	Goggles	Stones embedded in tyres could go into your eyes
Changing brake pads	Dust mask	Long-term breathing problems
Changing oil	Latex gloves	Long-term skin problems

Using vehicle protective equipment (VPE) responsibly, safely and appropriately

When you work on a vehicle you will, of course, be wearing your PPE to protect yourself from things such as oil and grease. This is a good start, but you also need to remember that the oil and grease you have on your gloves, boots and overalls can dirty the vehicle. The following vehicle protective equipment (VPE) must be used to protect the vehicle.

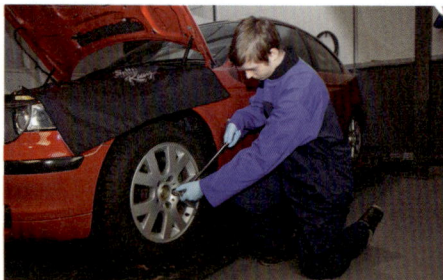

Figure 1.4 A wing cover will protect the vehicle body from scratches when you are leaning on the side of the vehicle

Figure 1.5 Use seat covers to prevent oil and dirt being left on the seats

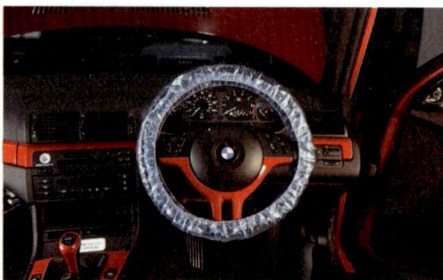

Figure 1.6 Protect the steering wheel from oil and dirt by using a steering wheel cover

Figure 1.7 Use foot mat covers to prevent oil and dirt being left on the foot mats

Working life

Chesney has been asked to help his mentor check the lights on a customer's vehicle. He notices that no VPE has been added to the vehicle. He understands the importance of keeping the vehicle clean.

1   What do you think Chesney will do to make sure the vehicle is kept clean?

Health and safety notices and instructions

In your workshop you will find many notices that make it clear what you have to do to work safely. It is important that you are aware of them and that you follow the instructions to keep yourself and your colleagues safe.

Safety signs are colour-coded so that you know at a glance the type of warning you are being given.

Table 1.3 Types of colour-coded safety signs found in a workshop

Safety sign	Colour code and what it means
 Safety helmets must be worn at all times on this site	Blue and white This is a **mandatory** sign.
 No smoking — It is against the law to smoke in these premises except in a designated area	Red and white This shows that something is **prohibited**.
 Corrosive	Yellow and black This is a **warning** sign.
 Fire exit	Green and white This is a **safe condition** sign.

Did you know?

A scratch or cut which might need a plaster or a few stitches is referred to as a small or minor injury. A serious injury could be a broken finger which could mean you need to take time off work. If the serious injury is really bad you might not be able to return to work.

Key terms

Mandatory – must be followed.

Prohibited – not allowed.

Find out

Name the following signs and describe what they mean.

Look for boxes with this symbol throughout the book for SAFE conditions you should be aware of:

+ Emergency

If the low battery symbol appears on a digital multimeter screen, replace the battery straight away. Otherwise, you might get inaccurate readings that could lead to electric shock or personal injury.

Look for boxes with this symbol throughout the book for PROHIBITIONS you should be aware of:

Safe working

Do not measure the voltages of a hybrid drive system unless you have been specifically trained. Hybrid drives operate with high voltages that can cause electric shock and death.

Look for boxes with this symbol throughout the book for MANDATORY instructions you should be aware of:

! Safe working

Following the replacement of any engine components, the car should be road tested to ensure correct function and operation.

Look for boxes with this symbol throughout the book for WARNINGS you should be aware of:

! Safe working

When checking for an electrical short circuit, only bypass the fuse with an electrical consumer like a bulb. Using other electrical components could cause a sudden discharge of electricity that may burn you.

Warning

Engines must not be started whilst hoist is raised

Safety signs of how to use equipment

Notices found in your workshop give you specific details of what to do. For example, a wet floor sign instructs that the floor is wet and that you should walk and not run. The main aim of this sign is to reduce the possibility of slipping.

Figure 1.8 shows an example of a safety notice you might see in a workshop.

Find out

Do you have a notice like the one shown in Figure 1.8 in your workshop? Is it the same or a bit different? Discuss any differences or similarities with your workmates and see if you can come up with a notice that fits your own working environment.

Figure 1.8 Workshop safety notice

Find out

Look around your workshop and locate the following health and safety information:

- fire and emergency exit signs
- equipment safety notices
- PPE instruction signs
- hazard warning signs.

Key term

Procedures – a list of operations to be performed.

Following your employer's instructions and procedures

Before you begin work you will be given a company induction. This is an introduction to your new company and will explain what your employer expects from you and in return what you can expect from your employer. A typical induction may include:

- being shown around your workshop
- general health and safety practices
- answering questions on your general health
- being told who to go to in the event of an emergency or if you have a problem.

If you work for a national company you may be required to go on an induction course. Your employer will expect you to follow their instructions and **procedures** no matter what size business you work for. They will also expect you to behave appropriately at all times.

An example of an induction checklist is shown opposite. This lists what may be covered during your induction. It would be completed after your induction with you present.

Name:	Job title:
Location:	Date commence employment:

Personnel documentation and checks completed
- ☐ National Insurance number
- ☐ Evidence of residency or right to work
- ☐ Licences and qualifications
- ☐ Medical insurance details
- ☐ Bank details (if needed for payment of salary/wages)
- ☐ Emergency contact details

Terms and conditions of employment explained
- ☐ Contract of employment or letter of confirmation issued and signed
- ☐ Probationary period complete
- ☐ Hours of work and work breaks
- ☐ Clocking on/flexi-time procedures
- ☐ Pay and payment procedures
- ☐ Holiday and sick leave entitlements and procedures
- ☐ Other leave
- ☐ Trade union membership
- ☐ Reviewing work performance and counselling
- ☐ Grievance and disciplinary procedures
- ☐ Termination procedures

Occupational health and safety
- ☐ Awareness of hazards
- ☐ Safety rules
- ☐ Emergency procedures
- ☐ Location of exits
- ☐ Clear gangways, exits
- ☐ Dangerous substances or processes
- ☐ Reporting of accidents
- ☐ Reporting of hazards
- ☐ First aid
- ☐ Personal and workplace hygiene
- ☐ Identified introductory training courses

Regulatory procedures
- ☐ Handling chemicals – COSHH training received?

Figure 1.9 An induction checklist

Behaving responsibly

The way in which you act at work has a direct effect on everything and everyone. If you work for a large organisation they will have a 'code of conduct' procedure which you will be expected to follow. This will contain specific details ranging from what you should wear to how you answer the telephone.

You might also find notices, such as the ones in Figure 1.10, which are all about what is expected of you. These notices might be about:

- not smoking
- not using your mobile
- putting equipment away in a particular order
- putting tools back in the right place
- wearing the correct PPE
- using guards on machinery.

Wear protective gloves

Guards must be in position before starting

Figure 1.10 More safety notices

Being aware of other's safety and actions

A workshop can be a busy place with several repairs and services happening at the same time. This means there is always the risk that an accident may happen. To prevent accidents occurring, follow all instructions given and pay attention to what is going on around you.

Other people in your workshop who you should be aware of may include:

- customers
- suppliers
- **subcontractors**
- workmates.

Awareness of others

Working life

It was Sylvia's 21st birthday party and at lunchtime everyone went to the pub to celebrate. The majority of staff had non-alcoholic drinks but the technician, Richard, had a few pints of beer. Everyone was in a great mood when they went back to work after lunch. In the middle of the afternoon, the service manager saw Richard about to get into a customer's car to drive it into the workshop.

1   What should the service manager do?

Working life

Ragav is an apprentice and is taking his morning break. His employer has set aside an area away from the workshop where staff and customers can smoke cigarettes. Ragav notices that some of his workmates smoke just outside the workshop and they ask him where he is going. He tells them and continues to walk over to the smoking area. He is laughed at and called names.

Ragav has followed his company procedures and used the correct area for smoking.

1   Are Ragav's workmates doing the same?

2   What would you do if your workmates teased you for following correct company procedure?

Working life

Isaac is a receptionist. He has been called into the office by his manager to discuss a customer issue. Yesterday afternoon at 5.25 p.m. a customer came to pick up his car. The customer was greeted by Isaac turning the sign on the door to closed. On knocking on the window, the customer said 'I'm here to collect my car.' Isaac said nothing and pointed to the closed sign, turned his back and walked away.

1   Why do you think that Isaac has been called into the office?

Working life

Ron is an apprentice technician. He loves his job and takes a pride in everything he does. When asked to complete a task he prepares his work area, sets out all the tools required, sorts out the parts for the job, wears the correct PPE and adds the VPE needed. When he finishes his task he tidies away.

One day when he is using a tyre balancer without a guard, a stone from the tyre tread heads towards his eyes. It bounces off his safety goggles and hits him on the nose. Ron has minor injuries which are treated immediately by the first-aider.

1   What could have happened if Ron wasn't wearing the correct PPE?

Common risks and hazards in a workshop

You need to be aware of the **hazards** you might come across and deal with any **risks** linked to them.

Some hazards and risks you might come across are shown in Table 1.4.

Table 1.4 Hazards and risks and how to prevent them

Hazard	Risk	How to prevent it
Trailing wire	Tripping and injury	Tape all wires and store them safely
Electric equipment	Electrical burn	• Use safely • Dry your hands • Check all connections • Check wire condition • Use correct fuse
Air-powered tools and air lines	Air enters the body	Do not point air tools at people
Vehicle fuels	Inhalation or fire	• **Ventilation** • No smoking • Store safely
Hazardous substances, for example chemicals, brake fluid	• Burns to skin • Burns after swallowing • Environmental damage	• Store safely • Label correctly • Dispose of safely
Movement of vehicles	• Collision • Crash injuries	• Take care when moving vehicles around • Be alert and observant • Use the handbrake • **Immobilise** vehicles
Waste materials, for example used oil filters	Contamination of the environment	Dispose of safely
Loose tools and equipment	• Tripping and injury • Damage	Store in the right place
Lifting, jacking and supporting vehicles	• Crush injury • Injury to person working underneath	• Use axle stands • Check jack for oil leaks

Find out

Look at Table 1.4. In your work today have you followed procedures to make sure you and your workmates are safe? Did you cut corners to save time? If so, did anything happen that you could have prevented and did you put anyone at risk?

CHECK YOUR PROGRESS

1 State two employee responsibilities under the Health and Safety at Work Act.
2 Name five items of PPE.
3 How would you protect your head when working under a ramp?
4 What is the difference between a risk and a hazard?

Health and safety practices and equipment

There are common tasks which you will need to carry out in your everyday working life. It is very important that they are done safely. Your employer will have health and safety procedures for tasks you need to carry out. Following a safety procedure will make sure you work safely and reduce the risk of a potentially serious accident. Whether it's a written checklist or something you just know off the top of your head, there are many benefits to following a procedure. A procedure:

* ensures you follow health and safety rules and do not miss any vital steps
* reduces the risk of accident
* limits waste
* saves you time
* helps you work out in advance the time needed for a job
* ensures a consistent standard of work is carried out.

The procedure for preparing a car for work to be completed is shown below.

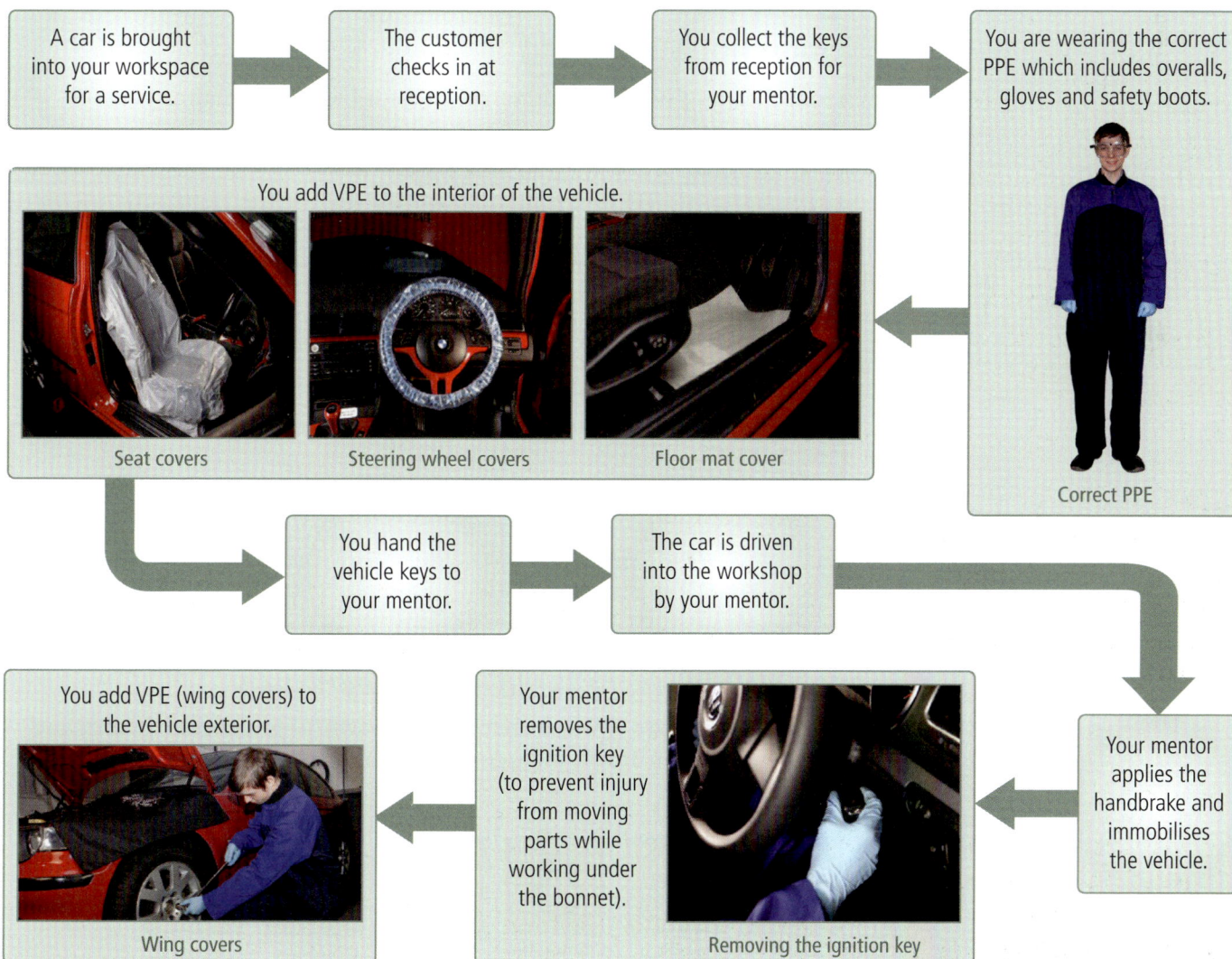

Find out

Think of a task you have to complete regularly in a workshop. Write down a procedure on how to carry it out using the flow chart below to help you.

! Safe working

Always use guards and PPE when using machinery.

A car is brought into your workspace for a service.	The customer checks in at reception.	You collect the keys from reception for your mentor.	You are wearing the correct PPE which includes overalls, gloves and safety boots.

You add VPE to the interior of the vehicle.

Seat covers

Steering wheel covers

Floor mat cover

Correct PPE

You hand the vehicle keys to your mentor.

The car is driven into the workshop by your mentor.

You add VPE (wing covers) to the vehicle exterior.

Your mentor removes the ignition key (to prevent injury from moving parts while working under the bonnet).

Your mentor applies the handbrake and immobilises the vehicle.

Wing covers

Removing the ignition key

Figure 1.11 The correct procedure to follow when a car comes into a workshop

Billy has to change the brake pads on a customer's car.

Billy reads the brake cleaner label. He also checks where the fire extinguishers are.

He completes the following steps:

- He removes the ignition key.
- He checks the handbrake is on before the car is lifted.
- He jacks up the car.
- He puts the axle stands under the car for support.
- He removes the brake fluid cap from under the bonnet.
- He removes the wheel.
- He puts on a dust mask (to prevent inhaling brake dust).
- He cleans off the brakes.
- He removes the brake pads.

1   Why does Billy remove the ignition key?

2   Find a can of brake cleaner in your workshop. Read the label to find out why Billy read the brake cleaner label and checked where the fire extinguishers are.

> PPE – Billy wears boots, overalls, dust mask and latex gloves
>
> ↓
>
> VPE – he protects the vehicle with vehicle wing, steering wheel, seat and mat covers
>
> ↓
>
> Materials – he uses brake cleaner
>
> ↓
>
> Tools – he uses a brake pad removal tool

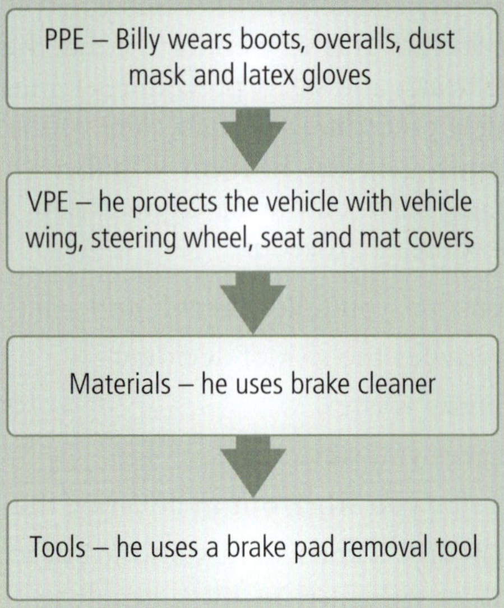

Good housekeeping

Housekeeping is the simple term used for cleaning up after yourself. This will ensure your work area is clean and tidy.

Here are some good housekeeping tips.

- Work tidily to reduce the risk of you or someone else getting hurt.
- Make sure you throw away any food waste in the correct bin.
- Store your PPE correctly.
- Don't block fire exits.

Annette is the parts receptionist. She has a message from a parts supplier and walks across the workshop to speak with the technician. As she does so she trips over a cable which someone has left on the floor. Annette sprains her ankle and has to go to hospital.

1   A trailing cable might seem a small thing but how has this affected the workshop? How could it have been avoided?

Hazardous waste found in your workshop

Hazardous waste is waste that may be harmful to people's health and the environment. It can include materials that are **flammable** and **corrosive**.

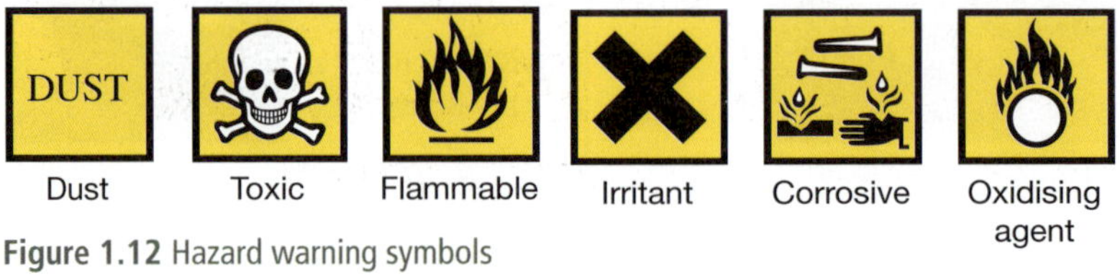

| Dust | Toxic | Flammable | Irritant | Corrosive | Oxidising agent |

Figure 1.12 Hazard warning symbols

Examples of hazardous waste in a workshop include:

- car lead acid batteries (also known as wet cell batteries)
- contaminated rags
- used oil or fuel filters
- aerosols
- antifreeze
- brake fluids
- waste vehicle components
- fluorescent tubes
- energy-saving light bulbs
- sodium lamps
- toner and ink jet cartridges
- old computer monitors
- asbestos
- cleaning materials.

Disposal of waste

All waste must be **disposed** of correctly otherwise it is dangerous to your health and the environment. This includes:

- waste oils and filters
- old units and components
- cleaning materials
- volatile materials such as petrol filters and petrol engine components.

Figure 1.13 When using an oil can, take care to clean up any spills immediately

Key terms

Hazardous – involving risk or danger.

Flammable – can be set alight easily and burn quickly.

Corrosive – causes damage to any part of the body on contact. Disposal – the action or process of throwing away or getting rid of something.

Did you know?

Used motor oil contains carcinogens. These are chemicals which may cause skin cancer. It also contains heavy metals which can cause nerve and kidney damage.

Key terms

Controlled waste – any waste which cannot be disposed of to landfill. This includes liquids, asbestos, tyres and waste that has been decontaminated. There are three types of controlled waste listed under the Environmental Protection (Controlled Waste) Regulations 2004. These are household, industrial and commercial waste.

Ecotoxic – damaging to the environment.

When a vehicle comes into the workshop for repair, and components are replaced, you must follow your company's procedures for disposal. This will include contacting a waste company to take away old oil and used filters.

Corrosive

ECOTOXIC

Figure 1.14 Symbol for corrosive substances   **Figure 1.15** Symbol for **ecotoxic** substances

Hazardous waste should be disposed of with minimal environmental impact. It is your employer's responsibility to register your workshop with the Environment Agency (EA). They should also keep records of all hazardous waste it disposes of. The actual disposal of waste will often be done by a specialist company.

Checking and using tools correctly

There are certain tools used for specific tasks. In the next chapter on pages 30–45 you will find all the information you need on choosing and using your tools correctly.

Find out

Find out how your employer gets rid of waste oil.

CHECK YOUR PROGRESS

1 Name five items covered on a work induction.
2 What is meant by 'immobilising' the vehicle?
3 Why do you remove the ignition key before you start working on a car?

Safe manual handling procedures

Manual handling means lifting and moving a piece of equipment or material from one place to another without using machinery. Lifting and moving loads by hand is one of the most common causes of injury at work. Most injuries caused by manual handling result from years of lifting items that are too heavy or are awkward shapes or sizes, or from using the wrong technique.

Correct lifting technique

The first and most important thing you can do to avoid injury from lifting is to receive proper manual handling training. **Kinetic lifting** is a way of lifting objects that reduces the chance of injury (see next page).

Before you lift anything you should ask yourself the simple questions listed in Table 1.5.

> **Key term**
>
> **Kinetic lifting** – a way of lifting objects that reduces the risk of injury to the lifter.

Table 1.5 Before you lift an object, ask yourself…

Question	Further information
Does the object need to be moved?	Can other objects be moved instead (for example to clear a walkway)?
Can I use something to help me lift the object?	A mechanical aid such as a scissor jack or a hydraulic aid such as an engine hoist may be more appropriate than a person.
Can I reduce the weight by breaking down the load?	Breaking down a load into smaller and more manageable weights may mean that you need to make more journeys, but it will also reduce the risk of injury.
Do I need help?	Asking for help to lift a load is not a sign of weakness, and team lifting will greatly reduce the risk of injury.
How much can I lift safely?	The recommended maximum weight an adult can lift is 25 kg for males and 16 kg for females. But remember that this is only an average weight, as each person is different. The amount that a person can lift will depend on their physique, age and experience.
Where is the object going?	Make sure that any obstacles in your path are out of the way before you lift. You also need to make sure there is somewhere to put the object when you get there.
Am I trained to lift?	The quickest way to receive a manual handling injury is to use the wrong lifting technique.

Figure 1.16 A scissor jack

Figure 1.17 A vehicle hoist

Kinetic lifting

It is important to keep the correct posture and to use the correct technique when lifting any load.

Lifting safely

1. Think before you lift. Where are you moving the load? Is there anything in your way?

2. Get into the correct posture before lifting.

   a. Stand with feet shoulder width apart, with one foot slightly in front of the other.
   b. Knees should be bent.
   c. Back must be straight.
   d. Arms should be as close to the body as possible.
   e. Your grip must be firm. Use the whole hand and not just the fingers.

3. Get a good grip on the load.

   a. Approach the load squarely, facing the direction of travel.
   b. Adopt the correct posture (as above).
   c. Place your hands under the load and pull the load close to your body.
   d. Lift the load using your legs and not your back.

4. Adopt the correct posture when lifting.

5. Move smoothly with the load.

6. Get into the correct posture and technique when lowering.

   a. Bend at the knees, not the back.
   b. Adjust the load to avoid trapping your fingers.
   c. Release the load.

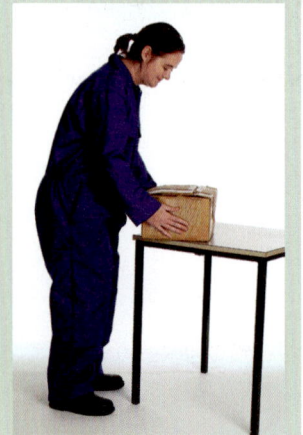

Moving loads using lifting equipment

A load is any object that needs to be moved manually or by mechanical lifting equipment. Examples of items that may need to be moved mechanically include:

- differential
- door or bonnet
- suspension assembly
- gearbox.

Table 1.6 gives some examples of lifting devices used in an automotive workshop.

Table 1.6 Lifting devices used in a workshop

Type	Source of power	Used for
Scissor jack 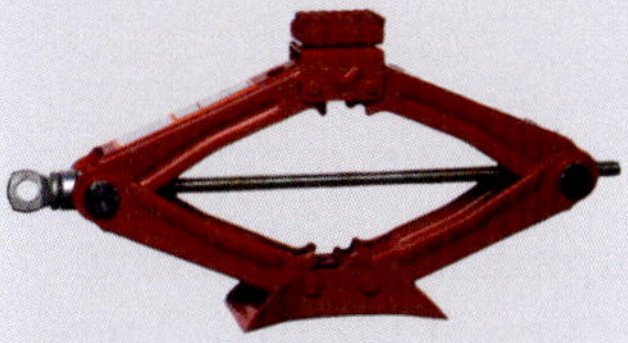	Manual/mechanical	Lifting the vehicle for repairing roadside punctures
Crane 	Manual/hydraulic	Lifting engines, gearboxes and subframe assemblies
Hydraulic jack 	Manual/hydraulic	Lifting vehicles distances above the ground
Hoist 	Manual/electric motor	Lifting vehicles to a suitable working height
Slings	Used with manual/hydraulic	Attached to the load and the lifting device
Chains	Used with manual/hydraulic	Attached to the load and the lifting device
Wire ropes	Used with manual/hydraulic	Attached to the load and the lifting device

When you are using lifting devices in a workshop, always follow the rules below.

1. Make sure you have a clear route.
2. Make sure the surface is flat otherwise an uneven surface may make the load become unsteady.
3. The maximum **safe working load (SWL)** must never be exceeded. This will be indicated on the equipment and the accessories used for lifting.
4. Never **shock load** the lifting equipment.
5. Always maintain an even balance with the load.
6. Do not push or pull the load to adjust the balance.
7. Never transport loads over the heads of people and never walk under a load.
8. Do not leave a load hanging without support.
9. Do not try to keep a load steady manually if it is about to fall.

The law which covers lifting operations and equipment is the Lifting Operations and Lifting Equipment Regulations 1988 (LOLER). These regulations state that any lifting equipment used at work for lifting or lowering loads must be:

- strong and stable enough for its particular use
- marked to indicate safe working loads (SWL)
- positioned and installed to minimise any risks
- used safely – the work is planned, organised and performed by competent people
- regularly given a thorough examination and, where appropriate, inspected by competent people.

Key terms

Safe working load (SWL) – the maximum load which a lifting device, such as a crane, cherry picker or lifting arrangement, can safely lift, suspend or lower.

Shock load – a load forced suddenly on the lifting equipment.

CHECK YOUR PROGRESS

1 State the six steps for safe manual lifting.
2 What does SWL stand for?
3 What source of power does a scissor jack use?

COSHH procedures

Not all the substances used in your workshop have the potential to be hazardous to health, but a large number of them do.

Before you start working in a workshop, your employer will take you through your COSHH training. They will make sure that you:

- understand that any substance you come into contact with can be dangerous
- understand what hazards and risks are in the workshop

Did you know?

The legislation which you and your employer must follow when using hazardous substances in the workshop is the Control of Substances Hazardous to Health Regulations 2002 (COSHH).

- are aware that the work you carry out can expose you and others to substances hazardous to health
- have the knowledge and skills to make decisions on how to control exposure
- know who to ask if you have any questions or are not sure what to do.

Hazardous substances may be:

- liquid, e.g. used engine oil
- solid, e.g. brake linings
- gas, e.g. exhaust fumes.

The hazard for each substance is shown by the hazard symbol found on the label.

Figure 1.18 Hazard warning symbols

Cleaning

As well as following the good housekeeping rules for cleaning up mess, always clean up anything that gets spilt straightaway. You should also move any objects that might trip someone up.

Table 1.7 Dealing with spillages and leaks

	PPE*	Absorbant granules	Brush and shovel	Mop and bucket	Water solvent	Extra precautions
Oil spillage	✓	✓	✓	✓	✓	Use a sign to warn people of the spillage
Fuel	✓	✓	✓	✓	✓	Ventilate the area well
Coolant	✓	✓	✓	✓	✓	Use a sign to warn people of the spillage
Nuts and bolts	✓	–	✓	–	–	–

Note: PPE worn will be latex gloves, overalls and steel toe-capped boots

Safe working

Never use any hazardous substance unless you have received COSHH training. Always follow the procedures on how to use a hazardous substance.

Figure 1.19 A trip hazards sign

Safe working

Always wash your hands after using chemicals and hazardous substances. You must also wash them before and after eating, drinking, smoking and using the lavatory.

Safe working

Never clean your hands with concentrated cleaning products or solvents.

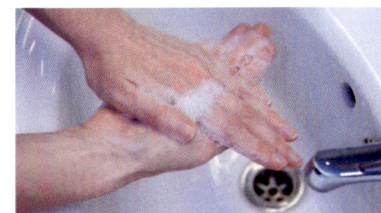

Figure 1.20 It is important to wash your hands thoroughly

CHECK YOUR PROGRESS

1   What is meant by 'corrosive'?
2   Name five hazardous substances found in a workshop.
3   What does COSHH stand for?

Figure 1.21 A sign showing instructions on what to do if there is a fire

Figure 1.22 A fire exit sign

Fire prevention and emergency procedures

Your workshop will have procedures in place to follow in case of emergency. A poster similar to the one in Figure 1.19 will be displayed in your workshop. This poster gives instructions on what to do in the event of a fire.

In the event of an emergency, flashing lights or similar visual alarm systems may be used. Visual alarm systems will be used where ear protectors are worn or where there are individuals with hearing problems. Fire exits should all be clearly signed.

Flammable liquids

There are many flammable liquids found in a workshop including:

* petrol – used as fuel for vehicles
* diesel – used as fuel for vehicles
* thinners – used to clean spray paint guns
* paraffin – used for cleaning oil off vehicle components
* paint – used for painting vehicles
* brake cleaner – used to damp down brake dust and clean brake linings.

What causes a fire?

There are three elements that a fire needs to burn:

* fuel
* oxygen
* heat.

These three elements are referred to as the 'fire triangle' (see Figure 1.23).

The purpose of firefighting is to remove elements of the fire triangle by:

* cooling – removes or reduces heat
* smothering – removes air (oxygen)
* starving – removes the fuel.

Figure 1.23 The fire triangle

Table 1.8 The fire triangle elements and how to control them

Fire triangle element	What is it?	Control
Fuel	Paper, wood, waste, rubbish and flammable substances.	• Remove waste and rubbish regularly. • Avoid using flammable substances or keep their use to a minimum. • Store flammable substances away from sources of ignition (for example hot machine parts) and preferably in fireproof stores.
Oxygen	Approximately 20% of the atmosphere is made up of oxygen.	• It is not normally possible to control the oxygen in the air. However, fires can be put out by depriving them of oxygen – by smothering the flames.
Heat	There has to be a source of ignition to create a fire. This can be produced by friction in machines, naked flames, hot surfaces and smoking.	• Machines should be checked and serviced regularly. • Smoking should be banned anywhere near flammable substances. • Open flames should be avoided.

Steps to follow in the event an emergency evacuation or fire

At your induction you will be told about the procedures that are in place in your workshop in case of emergency. You will be told where the fire exits are, what to do, where to assemble in the event of a fire and who the fire marshals are. Fire marshals are the people responsible for making sure everyone is accounted for or who you should report to.

Example of a fire evacuation drill

> • On hearing the fire alarm:
>   o stay calm
>   o stop what you are doing
>   o do not stop to gather any belongings
>   o do not run
>   o go to the assembly point and make the marshal aware you are there.
> • If you discover the fire, sound the alarm. Only tackle it if it is preventing you leaving the building or if it is small. Only re-enter the workshop after you have been told to do so by the fire marshal.

You may also be required to **evacuate** the workshop for other security reasons such as:

•   a bomb scare

•   the building has become structurally unsafe

•   an explosion

•   a leak of dangerous chemicals or gas.

The evacuation procedure will be the same for a fire evacuation, using the same exits and assembly point.

Figure 1.24 A fire assembly point sign

Safe working

Every workplace will have its own fire evacuation drill. Make sure you know yours.

Emergency

In the event of a fire keep internal fire doors closed. This cuts off the oxygen supply to the fire and helps prevent smoke and flames from spreading.

Do not use any lifts. They could fail if the electricity supply was disconnected and leave you trapped inside.

Key term

Evacuate – leave.

Figure 1.25 A fire blanket

Which fire extinguisher?

The main thing to do in the event of fire is to evacuate the building. The only reason for having firefighting equipment is to allow people to escape from a burning building. However, if you do think it is safe to put out a fire, make sure that you are close to a fire exit if the fire gets worse.

It is important that any fire extinguisher used is the correct type. You should only use an extinguisher if you have been trained to do so. You must never put yourself or others at risk.

Firefighting equipment

There are several types of firefighting equipment found in the workplace, as shown in Figures 1.25 to 1.31.

Fire blankets

Staff should be trained in the correct, safe use of fire blankets. These can be used to smother small oil fires.

Red label extinguisher – water

- This is used for fires involving solid materials such as wood, cloth, paper, plastics and coal.
- Do not use on petrol, oil or on electrical appliances.
- Use it by pointing the jet at the base of the flames and keeping it moving across the area of the fire.
- Make sure that all areas of the fire are out.
- It works mainly by cooling the burning material.

Figure 1.26 A water fire extinguisher

Black label extinguisher – CO_2 (carbon dioxide)

- This is used on liquids such as grease, fats, oil, paint and petrol. It can also be used on electrical fires.
- Do not use on molten metal fires.
- This type of extinguisher does not cool the fire very well and you need to watch that the fire does not start up again.
- The fumes can be harmful if used in confined spaces – ventilate the area as soon as the fire has been controlled.
- Use it by pointing the jet at the base of the flames. Keep the jet moving across the area of the fire.
- It works because carbon dioxide gas smothers the flames by replacing oxygen in the air.

Figure 1.27 A carbon dioxide fire extinguisher

Cream label extinguisher – foam

- This is used on fires involving solids such as wood, cloth, paper, plastics and coal.
- It is also used for liquids such as grease, fats, oil, paint and petrol.
- Do not use on electrical fires.
- For fire involving liquids, do not aim the jet straight into the liquid.
- Where the liquid on fire is in a container, point the jet at the inside edge of the container or on a nearby surface above the burning liquid.
- Allow the foam to build up and flow across the liquid. It works by forming a fire-extinguishing film on the surface of a burning liquid.

Blue label extinguisher – dry powder

- This is for fires involving solids such as wood, cloth, paper, plastics and coal.
- It can also be used on liquids such as grease, paints and petrol.
- Do not use on molten metal fires.
- Use by pointing the jet or discharge horn at the base of the flames. With a rapid sweeping motion, drive the fire towards the far edge until all the flames are out.
- If the extinguisher has a shut-off control, wait until the air clears and, if you can still see the flames, attack the fire again.
- It works by melting to form a skin smothering the fire and provides a cooling effect.

Sprinkler systems

These are automatic. They detect and control a fire in the early stages.

Hose reels

These are provided for the use of the fire service. They must be easily accessible.

Figure 1.28 A foam fire extinguisher

Figure 1.29 A powder fire extinguisher

🚫 **Safe working**

A water fire extinguisher should never be used to put out an electrical or burning oil fire. This is because the electrical current can carry along the water back to the person holding the extinguisher and electrocute them. Putting water on an oil fire will make it explode and potentially cause serious injury.

⚠️ **Safe working**

It is important that any fire extinguisher used is the correct type. You should only use an extinguisher if you have been trained to do so. You must never put yourself or others at risk.

Figure 1.30 A ceiling sprinkler system

Figure 1.31 A fire hose reel

Figure 1.32 Electrical hazard symbol

Safe working

Remember that you will need to seek professional help in all emergencies.

What to do in the event of electric shock

When someone receives an electric shock the electricity passes through their body. This is because of the large water content within the body. This may cause the skin to burn or look pale or bluish and the person may not have a pulse. In this instance, the person may not be breathing and their heart may have stopped.

In the event of an electrical shock:

- seek help
- DO NOT touch the casualty (the electricity may pass into your body and give you a shock)
- switch off the current source
- stand on dry insulating material (wood, rubber or lino) and isolate the casualty by using material that does not conduct electricity such as wood or plastic, for example a wooden broom handle
- follow your training for dealing with electric shock if you are a qualified first-aider
- obtain emergency medical assistance for the casualty.

Did you know?

It is your employer's responsibility under the Reporting of Injuries, Diseases and Dangerous Occurrences Regulations 1995 (RIDDOR) to report:

- work-related deaths
- major injuries or injuries lasting three days or more
- work-related diseases
- dangerous occurrences (near miss accidents).

CHECK YOUR PROGRESS

1 Name the three elements of fire.
2 What type of fire could be extinguished with a carbon dioxide extinguisher?
3 What colour is a dry powder extinguisher?

FINAL CHECK

1 Under the Health and Safety at Work Act, safety in the workplace is the responsibility of:

   a the employee alone

   b the employer and the employee

   c an inspector from the Health and Safety Executive alone

   d the employer alone

2 It is every person's responsibility to report:

   a only serious accidents

   b all hazards and potentially dangerous incidents

   c only incidents which affect them personally

   d only incidents which affect the general public

3 The main aim of a wet floor sign is to:

   a reduce the possibility of slipping

   b avoid distracting other operators

   c reduce congestion at fire exits

   d meet the requirements of the Factories Act

4 When using machinery and equipment it is necessary for a worker to:

   a ensure that other people take the responsibility if possible

   b avoid involvement with safety measures

   c always ensure that the safety officer is present

   d use guards and PPE at all times

5 Approved safety glasses/goggles should be worn:

   a only when machining

   b only when welding

   c whenever there is danger to the eyes

   d when working in bad light

6 Latex gloves are mainly used to:

   a keep your skin soft

   b prevent long-term skin problems

   c avoid having to use soap when washing

   d act as an insulator and prevent electric shock

7 Which one of the following fire extinguishers should be used on electrical fires?

   a water

   b foam

   c carbon dioxide

   d fire blanket

8 What colour is the label on the carbon dioxide extinguisher?

   a black

   b blue

   c green

   d red

9 Which one of the following is the correct way of lifting a weight from the ground?

   a back bent, legs straight

   b back bent, knees on ground

   c back straight, knees bent

   d back straight, knees on the ground

10 What does COSHH stand for?

   a Centre of Study for Health and Hazard

   b Contractors of Substances Hazardous to Health

   c Control of Substances Hazardous to Health

   d Control of Substances Harmful to Health

GETTING READY FOR ASSESSMENT

The information contained in this chapter, as well as continued practical assignments in your centre or workplace, will have helped you to prepare for both the end-of-unit tests and diploma multiple-choice tests.

Through reading and completing the chapter you have gained the knowledge and skills you need to work safely within the workplace. You have learned the importance of following correct health and safety practices and procedures, including what to do in the event of a fire or emergency evacuation.

Throughout your working you will have gained an understanding of the hazards and risks involved in your working environment and the importance of effective communication.

You will be assessed on the following topics:

- Health and safety requirements and information in motor vehicle workshops
- Health and safety practices and equipment
- Safe manual handling procedures
- COSHH procedures
- Fire prevention and emergency procedures

You now need to apply the knowledge you have gained in this chapter in your day-to-day working activities. For example, you are required to have knowledge and understanding of health and safety practices and equipment. The chapter has shown you how to do this by making you aware of your and your employer's responsibility for health and safety in the workshop. You have also been made aware of the importance of wearing PPE and the colour-coding system for warning signs, along with the common risks and hazards you might come across and how to deal with them. You have also gained an awareness of the types of hazardous substances found in your workshop.

You should now use this knowledge in your workplace/centre by working safely, observing set down processes and regulations and communicating effectively.

This chapter has given you an introduction and overview to general health and safety for light vehicles, providing the basic knowledge that will help you with both theory and practical assessments.

Before you try a theory end-of-unit test or multiple-choice test, make sure you have reviewed and revised any key terms and read all the questions carefully. Take time to take in the information so that you are confident about what the question is asking you. With multiple-choice tests, it is very important that you read all of the answers carefully, as it is common for two of the answers to be very similar, which may lead to confusion.

For practical assessment tasks, it is important that you have had enough practice and that you feel that you are capable of passing. Before you begin a task make sure you have the correct PPE, VPE, tools and equipment to hand and that you have a plan to follow, along with the equipment and information to complete the task. You could, for example, be asked to help your mentor lift a vehicle a small distance off the ground. For this you would need to wear boots, overalls, a dust mask and latex gloves. To carry out the task, you would need vehicle chocks, a hydraulic jack and axle stands.

Read the chapter again to confirm you have completed and understood any tasks. This will help you to be sure that you are working correctly and to avoid problems developing as you work.

When you are doing any practical assessment, always work safely throughout the task. Make sure that you observe all health and safety requirements and that you use the recommended personal protective equipment (PPE) and vehicle protection equipment (VPE). When using tools, take care to use them correctly and safely.

Good luck!

2 Tools, equipment & materials

Throughout your career in the light vehicle industry you will be required to carry out many types of vehicle maintenance.

In this chapter you will find out about the different tools you will use and how to select the correct tool for the job. You will also learn how to check if your tools are damaged and how to use tools correctly and safely.

This chapter covers:

- Safe working with tools, equipment and materials

- The correct identification and selection of hand tools and equipment for vehicle maintenance

- The correct preparation, use and maintenance of hand tools and equipment for vehicle maintenance

- The identification, selection and use of materials and consumables for vehicle maintenance

WORKING PRACTICE

You need to be safe when you work with hand tools and equipment. There are many risks to using tools and equipment which can result in small or more serious injuries. Always use tools correctly so that you don't cause harm to yourself or others.

You must always use personal protective equipment (PPE) and you also need to think about the possibility of crush or bump injury. You will also come into contact with chemicals such as lubrication oils, grease, coolant and cleaners containing solvents. Make sure that your selection of PPE will protect you from these hazards.

Personal Protective Equipment (PPE)

Safety goggles/glasses reduce the risk of small objects or chemicals coming into contact with the eyes.

Overalls provide protection from coming into contact with oils and chemicals.

Barrier cream protects the skin from old engine oil, which can cause dermatitis and may be carcinogenic (a substance that can cause cancer).

Safety gloves provide protection from oils and chemicals. They also protect the hands when handling objects with sharp edges.

Safety helmet protects the head from bump injuries when working under cars.

Safety boots protect the feet from a crush injury and often have oil- and chemical-resistant soles. Safety boots should have a steel toe-cap and steel mid-sole.

Tools

Torque wrench

Spanners

Socket set

Hammers

Air ratchet

Electric drill

Safe working

- Always use the correct tool for the job and protect yourself where there is a possibility of danger.
- Always wear safety goggles when you are grinding and drilling.
- Always wear ear defenders when you are grinding and sanding.
- Always wear a particle mask when you are working on brakes and sanding.
- Always wear head protection when you are under a lift.
- Always wear protective gloves when you are using chemicals.
- Always wear steel toe-capped boots and overalls at all times in the workshop.

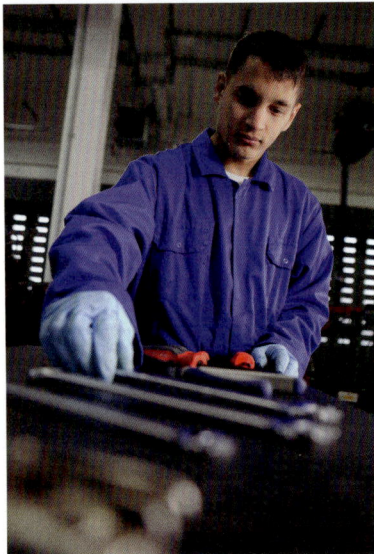

Figure 2.1 Sorting tools in the workshop

The correct identification and selection of hand tools and equipment for vehicle maintenance

There are lots of different types of tools used to carry out vehicle **maintenance**. Sometimes you may only need to use **hand tools**.

It is really important that you pick out and use the correct tools for the job. This will save you a lot of time and effort. For example, if you choose the correct spanner size and shape before starting a task, you won't have to keep going to your toolbox to find the right one.

Hand tools and their uses

Spanners

A spanner is a hand tool that is used to hold or twist a nut or bolt. There are many different sizes and shapes of spanners.

Table 2.1 Types of spanners and their uses

Type of spanner	Use	Example
Combination	To tighten nuts and bolts of the same size which might be hard to reach	On a battery terminal
Open ended	When you cannot get to the nut or bolt with a ring or combination spanner	Replacing an engine oil pressure switch
Ring	When you need to tighten nuts and bolts of different sizes	On a vehicle exhaust joint
Ratchet (or Speed)	When you can access the nut and bolt easily and need to do the job quickly	Removing nuts and bolts quickly
Pipe	When you need to tighten brake or fuel pipes	Changing a brake pipe
Adjustable	In an emergency when you do not have the correct spanner size	On a vehicle breakdown

Screwdrivers

A screwdriver is a tool for tightening screws. Screwdrivers come in different shaft lengths and blades to suit the type of work you are doing in or on the vehicle.

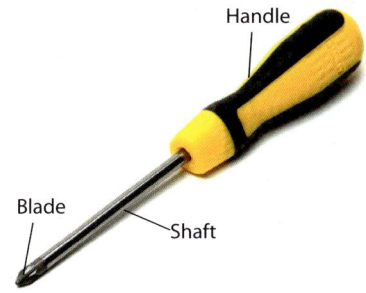
Figure 2.2 A Phillips screwdriver

Table 2.2 Different types of screwdrivers and their uses

Type of screwdriver	Use	Example
Flat blade	When you need to **slacken** or tighten flat-head screws	Pipe clips
Phillips	When you need to slacken or tighten cross-head screws	Brake disc locating screw
Pozidriv	When you need to slacken or tighten flat, cross-head screws	Plastic car interior trim clips
Torx	When you need to slacken or tighten torx-head screws	Fuel filter clamp
Ratchet	For speed. This tool also has different blades that can be attached	• Pipe clips • Brake disc locating screw • Plastic car interior trim clips
Electrical	When you need to reduce the risk of electrical short circuits	Tightening **electrical terminals**

Hammers

A hammer is a tool used to hit an object in a controlled way. Hammers are used to apply **force** when removing and replacing components. Care must be taken when choosing the type of hammer to prevent damaging the component being struck.

Working life

Rachel has been asked to remove and replace the rubber door seals on a car brought into the workshop. She removes the old door seals with a pair of pliers. She then fits the new rubber seals and hammers them in place using a lump hammer. She finishes the paperwork and hands the keys back to the owner.

The next day the owner returns complaining of rainwater coming into the car.

1   Rachel cannot understand why there is a problem. Can you?

Safe working

Never strike two metal hammer heads together as they can shatter and cause injury to the people close by.

Table 2.3 Different hammer types and their uses

Type of hammer	Use	Example
Ball pein	For general engineering operations	Used to strike a chisel
Lump	For heavy engineering operations	Removing seized shackle springs from the eyes of leaf springs by striking the pin with a metal drift
Copper/hide	To protect components to be hit	Removing and replacing wheel bearings
Rubber	To protect components to be hit	Removing and replacing rubber seals
Neoprene	To protect components to be hit	Removing and replacing body trim

Did you know?

Engineers mainly use 'cold' chisels. They are called this because they cut cold steel used for removing welded panels.

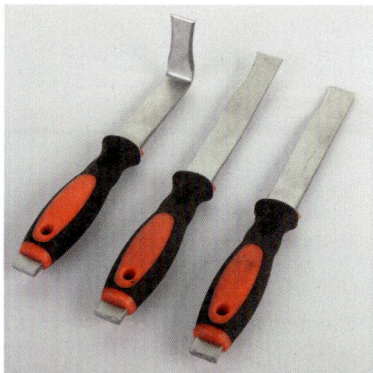

Chisels

A chisel is a tool with a specially shaped cutting edge. Chisels are used to remove or cut small components. They can also be used to remove waste or damaged materials. For example, a chisel could be used to remove a seized nut from a **stud**.

To do this, hold the sharp end at an angle of at around 45 degrees, and hit the other end with a ball pein hammer.

Figure 2.3 'Cold' chisels

Right

Wrong

Figure 2.4 One of these chisels has a 'mushroom' head because it has been used too much

Saws

You will come across two different types of saw – a hacksaw and a junior hacksaw.

Hacksaws are used to cut metal and non-metal materials. The blade is inserted into the frame with the pointed part of each tooth facing forwards.

When using a saw, it is important that you keep the material to be cut as tight to the vice as possible. This will prevent vibration, noise and damage to the hacksaw blade.

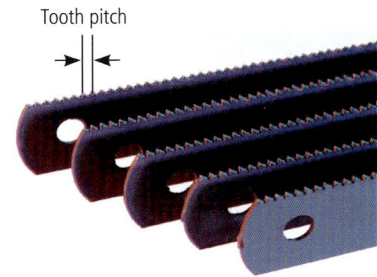

Tooth pitch

Figure 2.5 Hacksaw blades

Table 2.4 Examples of hacksaws

Type of saw	Use	Example
Hacksaw	For fabricating and cutting metal	Removing old exhaust pipes
Junior hacksaw	For small cutting jobs or where access to cutting is difficult	Trimming pipes and removing seized bolts

The number of teeth on the blade affects what type of material it can cut. As the number of teeth per blade length increases, the harder the material it can be used for. A general **tooth pitch** for a hacksaw is 24 teeth per 25 mm.

Key terms

Force – direct pressure on an object.

Stud – a threaded rod of steel.

Tooth pitch – the distance between the point of one tooth and the point of the next.

Working life

Steven is sawing a piece of metal in a vice. This is making a lot of noise and he keeps breaking hacksaw blades.

1   What could he be doing wrong?

Allen keys

Allen keys are used to turn Allen screws. These types of screws are **hexagon** shaped and are used in lots of different ways on a vehicle.

Allen keys come in a variety of sizes as shown in Figure 2.6.

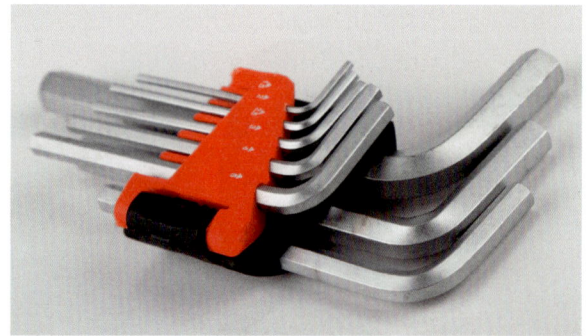

Figure 2.6 Allen keys

Socket sets

A socket set is a box set of lots of different sockets which can be clipped onto a **ratchet wrench**. This is then used to remove and replace nuts and bolts. The tools in the set join together in different ways to help you gain access to nuts and bolts in all areas of a vehicle.

Table 2.5 Examples of socket set tools

Type of socket set	Use	Example
Sockets	By placing tightly over nuts and bolts of varying sizes to connect with wrench	Removing and replacing wheel nuts
Extensions	With a socket to gain better access to a nut or bolt	Removing and replacing spark plugs
Universal joint	With the extension, the universal joint allows the extension to turn at an angle when slackening and tightening nuts and bolts	Removing and replacing exhaust manifolds
Breaker bar	To slacken very tight nuts and bolts when attached to a socket; this can also be called a 'cracker' bar	Removing wheel nuts
Ratchet	To turn sockets quickly to slacken and tighten nuts and bolts	Removing spark plugs

Pliers and grips

Gripping, **turning**, **crimping** or cutting, are all operations that can be done with pliers. Examples for each type are shown in Table 2.6.

Table 2.6 Examples of pliers and grips

Types of pliers/grips	Use	Example
Long nose 	When you need to grip something in a confined space	Removing blade fuses
Engineers/bull nose 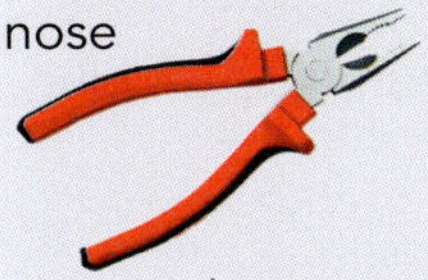	When you need to temporarily hold together or squeeze components together	Removing and replacing split pins
Side cutters/snips 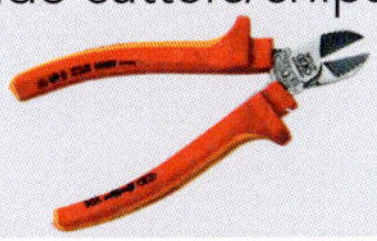	When you need to cut or trim thin metal or plastic components	Cutting electrical wire
Pipe grips 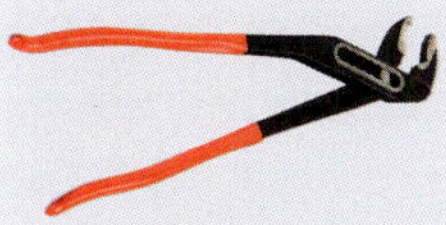	When you need to slacken and tighten pipe nuts or squeeze pipes together	Tightening water pump nuts
Vice grips 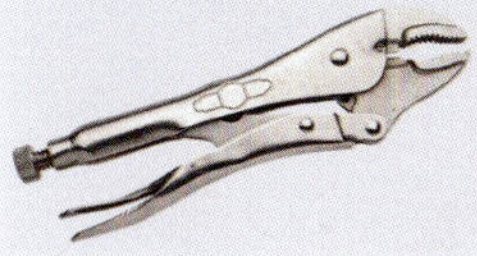	When you need to hold one or more components semi-permanently	Removing broken studs

As shown in the table, each pair of grips has a specific use so it is important to choose the correct ones for the job.

Torque wrench

The torque wrench is the most important tool when tightening nuts and bolts.

A torque wrench limits the amount of turning force used to tighten bolts and nuts. This prevents damage to the threads or loosening of the joined components.

Always reduce the torque setting to zero after use to ensure that the torque wrench stays **calibrated**.

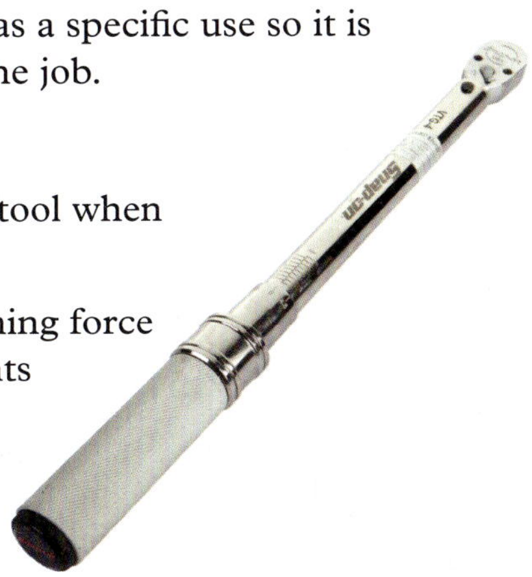

Figure 2.7 A torque wrench

Key terms

Gripping – holding something tightly/securely.

Turning – changing the direction of something.

Crimping – to make into ridges by pinching together.

Workshop manual – gives you all the information you need to repair a car.

Calibrated – a comparison between measurements.

Safe working

If you use the wrong grips for removing broken studs that are tight, the stud will not come out and will need to be taken out with a drill.

Did you know?

Before you use a torque wrench you must set it. To do this twist the bottom part of the handle until it reaches the correct setting. You can find the setting values listed in **workshop manuals**. During use the torque wrench will release with a 'click' on reaching its set value.

Find out

Use a torque wrench to tighten the nuts on a wheel correctly.

Did you know?

Torque is turning force, for example, the amount of force applied by the technician turning a wrench.

Measuring equipment hand tools

You will need to use measuring equipment for various vehicle maintenance operations.

Table 2.7 Examples of measuring equipment

Type of measuring equipment	Use	Example
Tape measure	For measuring long lengths up to 10 m	Measuring a vehicle's length and width
Feeler gauges	For measuring clearances	Measuring spark plug gaps
Steel rule	For measuring short lengths up to 300 mm	Measuring brake pad thickness
Tyre tread depth gauge	For measuring depths	Measuring tyre tread depths

Selecting the correct measuring equipment is very important. The accuracy required will determine which tool to use.

Why is accuracy in measuring so important?

- Tyre tread depth – if you measure a tyre tread depth incorrectly the tyre may be illegal.
- Measuring brake pad thickness – if you measure brake pad thickness incorrectly there may not be enough lining to last until the next service or replacement.

The **accuracy** of a tape measure and a steel rule is 0.5 mm because this is the smallest measurement written on the measuring rule.

Key term

Accuracy – how exact a measurement can be.

Choose a vehicle in the workshop and measure between its wheel centres on the same side (see Figure 2.8). This is called the wheelbase. Then check in the vehicle handbook to confirm this size.

The tyre tread depth must measure a minimum of 1.6 mm across three-quarters of the tread around the whole tyre circumference. Check the tyres on the same vehicle using the tread depth gauge.

Tyre	Passenger side front	Passenger side rear	Driver side front	Driver side rear
Measurement	mm	mm	mm	mm

Figure 2.8 The importance of measuring accurately

Equipment

While you will need your own hand tools, you should be supplied with workshop equipment to carry out your work. You will be able to find the following equipment in most workshops.

Lifting equipment

Vehicles and heavy components will need to be raised off the ground from time to time when they are being worked on. The following equipment is supplied to do this safely.

Table 2.8 Equipment used to raise vehicles and heavy components

Type of lifting equipment	Use	Example
Jacks 	For lifting vehicles or transmission parts	Lifting a vehicle to change a wheel
Ramps/hoists 	For lifting the vehicle fully off the ground	Lifting a vehicle to change the oil
Cranes 	To lift heavy vehicle units	Replacing an engine
Axle stands 	To prevent injury if a hydraulic jack were to give way	Holding the vehicle off the ground
Chocks 	To prevent the vehicle from moving	Keep the vehicle stationary while jacking up

Did you know?

It is your responsibility to look after your personal protective equipment (PPE) and report any damaged PPE to your employer or tutor.

Whenever you remove a car wheel using a jack, you must:

- wear correct PPE
- put the vehicle handbrake on
- fit chocks to the opposite wheel you are replacing
- jack the vehicle in a safe place
- put the axle stand under the vehicle in a safe place.

Air tools

Air tools are also called pneumatic tools. They are run by compressed air from a gas compressor. To reduce the risk of electric shock in the workshop, air tools are often used.

The following air tools should be available within your workshop.

Table 2.9 Air tools and their uses

Type of air tool	Use	Example
Air lines	To transmit air from a compressor to an air tool	Inflating tyres outside a workshop
Tyre inflator/gauge	To inflate and check the pressure of tyres	Inflating tyres during a service
Air wrench	To slacken nuts and bolts quickly	Removing wheel nuts
Air ratchet	To slacken and tighten nuts and bolts quickly	Removing a cylinder head
Blow gun	To clean off parts and clean inside pipes	Removing dirt from windscreen washer pipes

Bench tools

For safety, some tools have to be fixed to a bench or the workshop floor. These are called bench tools. Using these tools will give you a free hand to hold the parts steady when carrying out tasks.

> **Safe working**
>
> Whenever air tools are used you have to remember the air is under great pressure. Make sure you use safety goggles when working with air tools.

Table 2.10 Types of bench tools and their uses

Type of bench tool	Use	Example
Bench drill 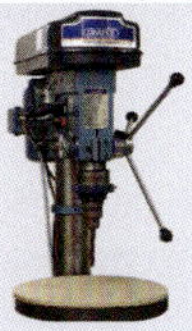	To prevent you having to hold the drill in your hands (fixed to the bench)	Drilling holes to remove broken studs off the vehicle
Bench grinder 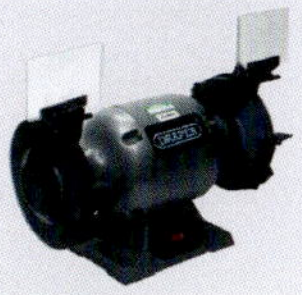	To sharpen tools or remove metal waste from components	Sharpening a chisel

Working life

Emile is about to drill a hole in a piece of metal. He notices that the bench drill he is about to use is not bolted down to the bench.

1   What could happen if he uses the drill?

Find out

Look around your workshop and list five items of equipment which are portable and five which are not.

Portable electric tools

Sometimes air tools may not be available in a workshop. This means electric tools will have to be used instead.

Table 2.11 Types of portable electrical tools and their uses

Type of portable electric tool	Use	Example
Hand drill 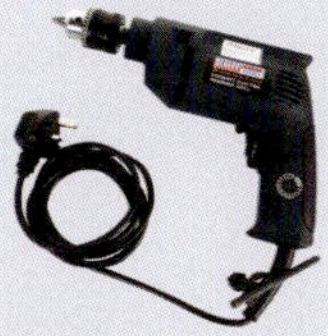	To drill holes where components are not removed from the vehicle	Removing broken studs from the engine
Extension lead 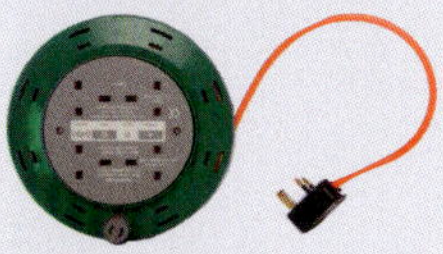	To transmit electrical voltage across the workshop	Using a portable electric drill
Component cleaner/cleaning station	To remove grease and oil from parts removed from the vehicle	Clean wheel bearings
Hand lamp 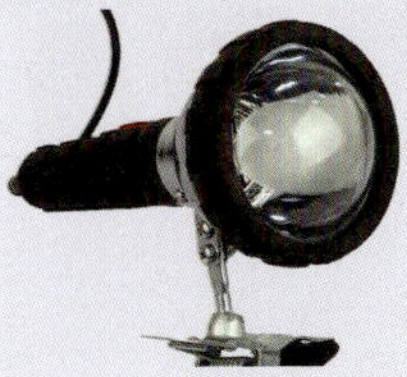	To provide a safe light source	Inspecting under the vehicle body for corrosion

Emergency

If someone receives an electric shock, do not touch them. Turn off the power supply to the tool they are holding. Then try to remove it from them using a wooden handle. See page 24 for further information on first aid for electric shock.

Specialist tools

Some tools are needed to carry out a specific operation. These are called specialist tools. The tools below have one specific use but they may be used to carry out different tasks.

Table 2.12 Specialist tools and their uses

Type of specialist tool	Use	Example
Headlamp aligner	To align headlamp aim	Align headlights following an accident
Wheel alignment equipment	To set the steering alignment	Align the steering following the vehicle hitting a curb
Filter removal strap	To remove filters	Remove an oil filter
Waste oil drainer	To drain waste oils	Drain gearbox oil and store safely before collection
Exhaust extraction	To remove poisonous gases from the workshop	Extract exhaust gases when vehicle is running in the workshop

CHECK YOUR PROGRESS

1 What is a torque wrench used for?
2 Name three types of measuring equipment and their use.
3 Name four tools you would find in a socket set.

The correct preparation, use and maintenance of hand tools and equipment for vehicle maintenance

You need to prepare hand tools and equipment before use.

This might include:

- getting tools out of your toolbox
- preparing more complex specialist tools
- preparing a vehicle for maintenance.

You also need to use tools correctly. This will make sure you do each job safely. Finally, you need to make sure your tools and equipment are kept in good condition.

Preparation, maintenance and use of hand tools

Before you start any vehicle maintenance, you must prepare yourself and the tools for use.

This will include:

- making sure you wear the correct PPE for the work to be done
- making sure you use the correct VPE
- checking the tool is not damaged before using it
- reporting any damage to tools before or after use.

Tables 2.13, 2.14 and 2.15 show the correct preparation, use and maintenance for the hand tools you will use.

Find out

Look in your toolbox. Line up your spanners starting with the smallest to the largest. Check to see if you have any missing.

Table 2.13 Preparation, use and maintenance of spanners, screwdrivers and hammers

Hand tool	Preparation	Correct use	Maintenance
Spanner	Get the correct size and shaped spanner for the nuts and bolts to be turned.	Pull the spanner towards you when slackening and tightening nuts and bolts.	Check the spanner for jaws spreading or wear inside the ring from incorrect use.
Screwdriver	Always use the correct point for the screw head being turned.	Do not use for any other purpose than to turn screws.	Check the blades are not worn and the handles are not damaged from misuse.
Hammer	Ensure you select the correct hammer for the correct task.	Hold the hammer at the end of the handle to apply more pressure. Always use safety goggles when hammering.	Check for loose hammer heads and damaged handles.

Table 2.14 Preparation, use and maintenance of chisels, saws and Allen keys

Hand tool	Preparation	Correct use	Maintenance
Chisels	Remove any 'mushroom' heads before use and ensure cutting edge is sharp. Ensure you have a good pair of safety goggles.	Point the chisel at an angle of about 45 degrees and keep your eye on the chisel head when striking with a hammer. Always wear goggles.	Remove any 'mushroomed' heads after use and ensure cutting edge is sharp.
Saws	Make sure the blade is tight and the teeth are facing forwards before use.	Cut on the forward stroke and keep the material being cut as close to the vice as possible.	Replace broken blades or blades with broken teeth.
Allen keys	Get the correct size Allen key for the screws to be turned.	Always make sure the key is secure in the recess during the slackening and tightening of Allen screws.	Check for rounding of the hexagon-shaped keys at their outer edge.

Find out

Check your hacksaw to make sure the teeth are facing away from the handle and that they are not damaged.

Table 2.15 Preparation, use and maintenance of socket sets, pliers and grips, torque wrenches and measuring equipment

Hand tool	Preparation	Correct use	Maintenance
Socket sets	Get the correct size socket for the nuts and bolts to be turned and the correct extension for easy access.	Always pull the socket turning device towards you when slackening and tightening nuts and bolts.	Check for worn or cracked sockets and replace damaged ratchets.
Pliers and grips	Use the correct pliers and grips for the type of work carried out.	Use a firm grip to hold the pliers and squeeze together as far down the handles as possible for greater force.	Check cutting and gripping edges for sharpness.
Torque wrench	Set the torque wrench to the correct setting.	Pull the wrench towards you until the wrench 'clicks' and release pressure. Tighten nuts and bolts in the correct sequence.	Ensure the torque wrench is reset to zero after use.
Measuring equipment	Use the correct measuring tool for the accuracy required.	Make sure the measuring tool is on the correct starting point of the item to be measured and view the measuring scale at a right angle.	Store carefully where the measuring equipment cannot get damaged.

Find out

Check the torque wrench in your toolbox to make sure it is set at zero.

Tightening sequences

Tightening and slackening **sequences** are used to prevent damage when removing and replacing vehicle parts. During removal, any warm parts may warp or bend. If this happens when you replace them they may not provide a **fluid type seal**. Also during replacement, if the parts are not tightened evenly, they may not be **aligned** correctly. This could cause slackening of the fixing device holding them in place.

Removing nuts and bolts

When removing a number of nuts or bolts holding a component together, always slacken in a spiral order moving inwards (towards the centre).

Figure 2.9 The order in which nuts and bolts should be removed from a cylinder head (unless otherwise stated by the manufacturer)

Replacing nuts and bolts

When replacing a number of nuts or bolts holding a component together, always tighten in a spiral order moving outwards (from the centre out).

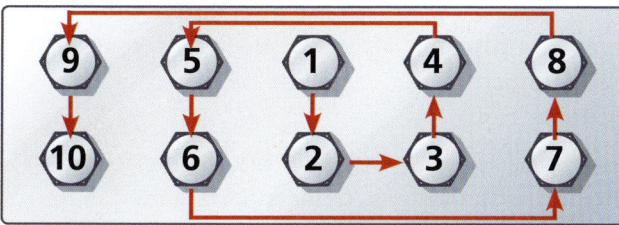

Figure 2.10 The order in which nuts and bolts should be replaced on a cylinder head (unless otherwise stated by the manufacturer)

If the nuts or bolts are an equal distance from the centre, for example a wheel held on to a hub, slacken in 'an equal and opposite' order (see Figure 2.11).

Key terms

Sequences – this is the order for removing and replacing things.

Fluid type seal – a seal which stops water and oil leaking.

Aligned – if two holes are lined up so a bolt will go in easily they are said to be aligned.

Did you know?

A fixing device can refer to a nut, bolt, Allen screw or anything that holds two or more components together.

Figure 2.11 A wheel held on a hub – the nuts and bolts are an equal distance from centre

Preparation, maintenance and use of equipment

It is your employer's responsibility to provide you with equipment that is safe to use. However, you must prepare, use and maintain the equipment in a safe way to prevent injury to others and yourself. Table 2.16 outlines how you should correctly use the equipment provided.

Table 2.16 How to use equipment correctly

Equipment	Preparation	Correct use	Maintenance
Lifting equipment	• Make sure component/vehicle will not move during lifting. • Use correct lifting point. • Make sure correct PPE and VPE is used.	• Lift slowly and keep checking device is secure. • Make sure the component/vehicle is held securely by mechanical means when it is lifted.	• Check for any fluid leaks in jacks, cranes or hoists. • Store lifting and holding devices tidily to prevent trips and falls.
Air tools	Ensure air lines are not obstructing pathways, to prevent trips and falls.	Wear the correct PPE. Goggles must be used with all air tools as compressed air may have tiny dust particles within it that could blind you.	• Check for any air leaks. • Lubricate air tools regularly as water can get into the air supply and damage the air tool.
Bench tools	• If you use a bench drill, use a secure hand vice to hold the material being drilled. • If you use a bench grinder, make sure the work rest is no more than 3 mm from the grinding wheel.	• Ensure all guards are in place and all PPE is worn. • Stand to the side of the machine when started as the equipment has to speed up quickly which could cause it to break.	• Check for any damage to electrical leads. • Check for damaged jaws on the drilling machine or damaged wheel on the grinder. • Make sure all guards are in good order.
Portable electric tools	Make sure any extension leads used will not cause injury through colleagues tripping over them.	Make sure the correct PPE is worn. Goggles will be needed for drills. Apron and gloves are worn when cleaning off vehicle parts with a parts cleaner.	• Check for any damage to electrical leads. • If the equipment does not work, check for fuses which may be blown.
Specialist tools	Ensure the equipment is calibrated and set correctly before use. Headlamp and wheel alignment equipment can only measure accurately if they are set up correctly.	• Take care not to damage the equipment during use. • Confirm the readings after recording them to make sure you are measuring accurately.	• Check equipment for defects or damage. • Take care to store tools correctly after use to prevent damage.

Report any damaged equipment to your line manager/tutor.

Safe working

Never point air lines at people as air can go through the skin and cause severe injury.

Reporting of damage or defects

If you do not report damaged equipment, the next time you or someone else comes to use the equipment it will still be damaged. This will make the job to be carried out take longer or be unsafe. Remember: it is your responsibility to report damaged equipment under the Health and Safety at Work Act.

CHECK YOUR PROGRESS

1 Name three hand tools. Describe how they should be maintained.
2 Name three types of equipment. Describe how each type of equipment should be maintained.
3 Why is it important to tighten nuts and bolts in the correct order?

The identification, selection and use of materials and consumables for vehicle maintenance

Ferrous metals

Ferrous metals are metals which contain iron. The different types of ferrous metals (iron and steel) are defined by the amount of **carbon** they contain. The more carbon a metal contains, the harder and more brittle it will be. This is because carbon itself is brittle. Table 2.17 lists some different types of ferrous metal and the percentage of carbon in each.

Table 2.17 The percentage of carbon in different ferrous metals

% carbon	Ferrous metal
3.00	Cast iron
1.20	High carbon steel
0.50	Medium carbon steel
0.25	Mild steel
0.01	Wrought iron

Facts about ferrous metals:

* Ferrous metals are magnetic.
* If ferrous metals are exposed to the **atmosphere** and are unprotected for long periods, oxidisation (rust) will occur.

Key terms

Consumables – things that can be used up.

Carbon – a non-metallic element.

Atmosphere – the air surrounding the Earth.

Did you know?

Ferrous metals are ones which contain iron and are magnetic.

Find out

Write the headings 'Ferrous' and 'Non-ferrous' on a sheet of paper. Use a magnet to help you find ten ferrous and ten non-ferrous vehicle components in your workshop. Make a note of them under the correct headings.

Uses of ferrous metals

Table 2.18 Common uses of ferrous metals in car components

Typical motor vehicle component	Typical alloy steel material
Gudgeon pins	Low carbon, nickel steel
Valves	Nickel-chromium steel
Crankshaft	Nickel-chromium-molybdenum steel
Road springs	Silicon manganese steel

Non-ferrous metals

These tend to be soft and do not hold up to large **stresses**. Table 2.19 gives some examples of non-ferrous metals and their properties.

Table 2.19 Non-ferrous metals and their properties

Metal	Appearance	Properties
Aluminium	White	Soft, **ductile**, good conductor of electricity and heat, very light
Copper	Reddish brown	Soft, ductile, good conductor of electricity and heat
Tin	Silver white	**Malleable** and ductile
Lead	Silver white	Soft, malleable, unaffected by acid and has hardly any elasticity

Aluminium and its alloys

Pure aluminium is not usually used in vehicle manufacture because it is too soft.

If unprotected and exposed to the atmosphere for long periods of time, a powdery oxide forms on the surface of the alloy. This must not be wiped off as the alloy will continue to corrode (rust).

Table 2.20 lists two types of aluminium alloy and their use in vehicle manufacture.

Table 2.20 Aluminium alloys and their uses

Aluminium alloy	Property	Use
Cast alloys	Easy to cast	Pistons and cylinder heads
Wrought alloys	Can be cold worked and **extruded**	Frames and body panels, wheels and body mouldings

---

Key terms

Stresses – forces which put pressure on the shape of an object.

Ductile – capable of being easily shaped or bent.

Malleable – easily formed by pressure.

Retain – to hold onto.

Extruded – when hot, the metal is forced under pressure through a die to provide a specific shape.

Did you know?

An aluminium alloy is a mixture of aluminium and other materials that are strong, hard, **retain** strength at high temperatures and are resistant to corrosion.

Glass

There are two main types of glass used in motor vehicle construction:

- **Laminated glass** is where a thin sheet of plastic is sandwiched between two layers of glass. This is used for windscreens. If a stone hits the windscreen, the glass will crack but not distort the vision for the driver. The driver can continue to drive until a replacement screen is fitted.

- **Toughened glass** is used for side windows and tailgates. If a stone hits the glass, the pane will shatter into small fragments and should stay in position and not cover the occupants in glass.

Figure 2.12 Laminated glass

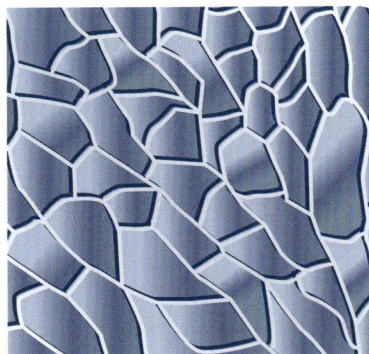

Glass

Polyvinyl butyral

Figure 2.13 Toughened (safety) glass

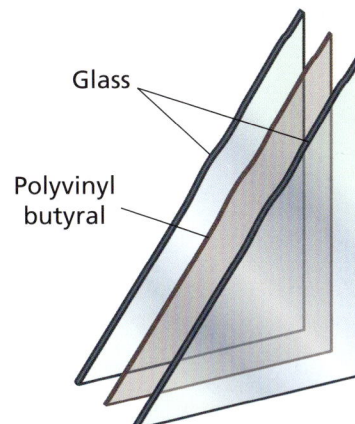

Figure 2.14 Broken toughened (safety) glass

Plastics

Plastics are man-made materials. They can be formed into a variety of shapes with the use of heat and pressure.

The two types of plastics are:

- thermosetting – when this type of plastic has set it cannot be reshaped after reheating
- thermoplastic – if reheated, this type of plastic can be reshaped.

Three common plastics used in vehicle manufacture are nylon, **PTA** and **PVC**. Table 2.21 lists their properties and uses in vehicles.

Did you know?

Toughened glass is sometimes known as safety glass.

Table 2.21 The properties and uses of nylon, PTA and PVC

Materials	Properties	Vehicle application
Nylon	• Strong heat resistance • Low **coefficient of friction**	• Bushes and bearings • Speedometer drive gear • Steering and suspension components • Electrical switches
PTA	• Self-lubricating	• Light duty bearings in small electric motor drive gears
PVC	• Good insulator • Can be coloured • Chemically resistant • Reasonably heat resistant	• Electrical cable insulation • Upholstery • Fuel pipelines

Key terms

PTA – purified terephthalic acid.

PVC – polyvinyl chloride.

Coefficient of friction – a value for the resistance an object encounters when moving over or against another object.

Glass fibre

Glass fibre is a popular material used in vehicle bodies and panels. This is because it is light and can be moulded to many different shapes. Sports cars that have a metal chassis normally have a glass fibre body. Some vehicles use glass fibre for bonnets and tailgates to reduce weight on body panels.

The disadvantage of glass fibre is that it is brittle (breaks easily). This means it cannot be used for the vehicle chassis itself.

Rubber

Rubber is used in many places on the vehicle. It absorbs shock and is resilient. Suspension joints, engine and exhaust mountings are just some examples of where rubber is used. The main use of rubber on a vehicle is in the construction of tyres.

Fluids

Many types of fluids are used to maintain vehicles. Table 22.2 shows the different types available along with their uses and how you can identify them.

Did you know?

Resilient means something will bounce back.

Figure 2.15 Vehicle maintenance chemicals in a storage rack

Table 2.22 Fluids and their use

Fluid	Use	Identification
Lubricants	Engine, gearbox, final drive and power steering systems	Thin light-brown oil, getting thicker for gearbox and final drive. Automatic gearboxes and power assisted steering systems have a pink-coloured thin oil
Hydraulic oils	Hydraulic jacks and lifting hoists	Thin light-brown oil
Antifreeze	Engine coolant	This can be blue/green or orange/pink in colour depending on the type. It is mixed with water before putting it in the engine cooling system
Brake fluid	Braking system and hydraulic clutch	Thin yellow/brown oil
Washer fluid	Windscreen washers	Blue fluid that smells of alcohol

Table 2.23 shows typical examples of the quantities used when replacing fluids and lubricants. You must look at the vehicle manufacturer information found in a driver handbook. This will tell you the correct specifications for each vehicle you work on.

Table 2.23 Fluids and lubricants – replacement quantities

Fluid	Safety	Specifications	Quantity	Mixture	Replacement
Lubricants	Do not get old oils on the skin	10w40 – refers to the oil thickness when cold and hot	Small engines approximately 5 litres	n/a	Every 12 months or 6000 miles
Automatic gearbox and PAS fluid	Do not get old oils on the skin	Dexron III – mineral-based oil	Approximately 3 litres	n/a	Every 5 years
Antifreeze	Do not spill on paintwork	Silicate-free organic acid	Approximately 6 litres of coolant	33% antifreeze 67% water	Every 2 years
Brake fluid	Do not spill on paintwork	DOT 4 – this refers to the boiling point of the fluid	Approximately 2 litres	n/a	Every 2 years
Washer fluid	Do not drink	Alcohol-based with anionic **surfactant**	Approximately 4 litres of mixture	25% fluid 75% water (in summer)	Before washer bottle is empty

Adhesives

An adhesive, or glue, is a mixture in a semi-liquid state that bonds items together. Table 2.24 shows the types of adhesives available and how to apply them.

Key term

Surfactant – a substance which allows liquids to be easily mixed.

Table 2.24 Adhesives – their uses and applications

Type	Use	Application
Contact	Trim material to door cards	Apply to both surfaces and leave to become 'tacky'. Bring together and apply pressure
Pressure-sensitive adhesive	Carpets and other interior trim	Apply to both surfaces and use pressure to hold the two contact areas together
Synthetic adhesives	Rear view mirror to windscreen	Mix two parts together and apply to surfaces – leave to dry
Instant	Nut and bolt threads	Apply directly to the threads and leave to dry
Hot adhesives	Metal to metal bonding such as sports car chassis	Applied in molten form from a heating device

⚠ **Safe working**

Take care that instant adhesives such as 'super glue' do not touch the skin. They can stick your fingers together.

Safe working

Make sure you keep skin contact to a minimum when using sealants.

Sealant

Sealant is used between mating parts and is applied to gaskets. Sealants are mainly flexible rubber-based products. They may need to be applied with a sealant gun.

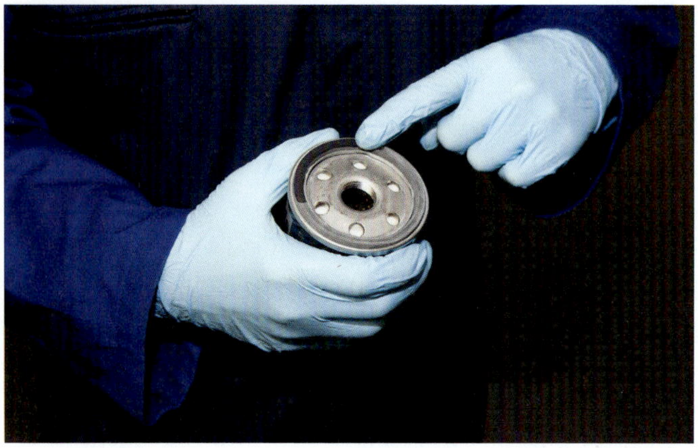

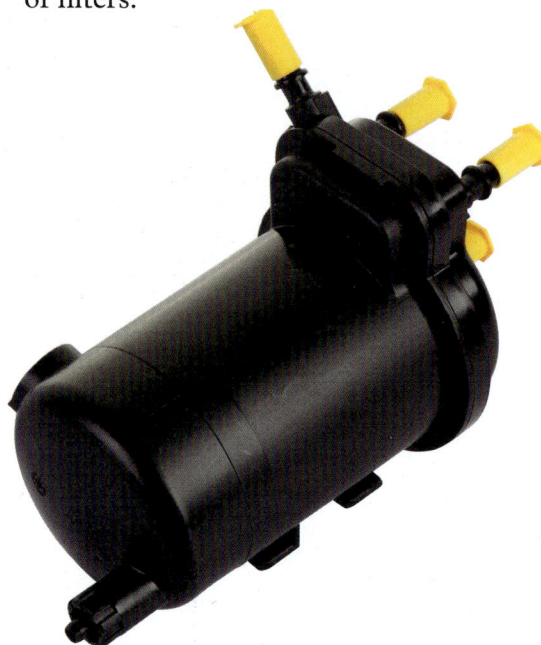

Figure 2.16 An oil filter

Filters

A filter is used to prevent unwanted dirt particles entering the parts of the vehicle that could cause damage. Table 2.25 shows examples of where you might find the different types of filters.

Figure 2.17 A fuel filter

Table 2.25 Filters

Filter	Use	Damage prevention
Oil	To clean dirty oil within the engine	Dirty oil may block oil-ways, which could cause the engine to seize up
Air	To stop dust particles entering the engine	Dust will score the inside of the engine, causing it to wear out more quickly
Fuel	To stop dirt particles and water entering the fuel system	Fuel injection parts have small holes for the fuel to flow through. These must not be blocked with dirt
Pollen	To stop unwanted dust particles entering the vehicle cabin	Pollen and dust in the driver cabin may cause the driver and passenger health problems

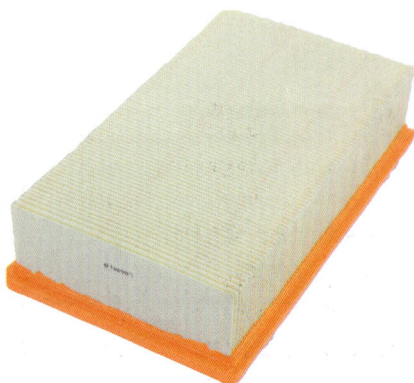

Figure 2.18 An air intake filter

Aerosol sprays

Aerosol sprays are mainly used during the maintenance of vehicles. The list below shows the types of spray available:

- Brake cleaner – used to remove brake dust and dirt from the brake linings
- Penetrating oil – to free off seized nuts and bolts and to lubricate locks and hinges
- Spray paint – used to touch up the vehicle body following minor damage
- Spray adhesive – contact adhesive for vehicle trim components
- De-icer – to remove ice from the windscreen in freezing conditions
- Upholstery cleaner – for cleaning vehicle interior and carpets.

Figure 2.19 A mechanic spraying brake cleaner on assembly to remove dust

Gaskets

A gasket or seal is used whenever two components that hold fluids come together. This is because it is not possible to make each part exactly to prevent leakage. A gasket and/or sealing compound will be used for this purpose.

For general sealing, gaskets are made of **resin-impregnated paper**. This is cut into the same shape as the **mating faces** of the joint. Other materials are used including rubber, metal, cork, felt and plastic materials.

> **Key terms**
>
> **Resin-impregnated paper** – paper which has had glue added during manufacture.
>
> **Mating faces** – the areas which need to be joined together.

Seals

Where there is one fixed and one rotating part which need to be **fluid tight**, the most common seal is a lip seal.

CHECK YOUR PROGRESS

1 Name two types of glass used in the manufacture of a motor vehicle.
2 Name two types of fluids used in vehicles and describe their uses.
3 What is the quickest way to check if a metal is ferrous?

FINAL CHECK

1 Which one of the following is not PPE?

  a dust mask

  b overalls

  c wing cover

  d latex gloves

2 The pitch of a saw tooth is measured:

  a from one end of the blade to the other

  b from the tip of one tooth to the tip of the next

  c from the bottom of the blade to the top

  d from the saw handle to the frame

3 Side cutting pliers are used to:

  a tighten exhaust pipe nuts

  b remove blade fuses

  c cut electrical wire

  d remove broken studs

4 A steel rule is used to:

  a measure brake pad lining thickness

  b measure tread depth

  c measure the width of a vehicle

  d measure spark plug gaps

5 Which one of the following tools is used for cleaning vehicle parts?

  a tyre inflator

  b air ratchet

  c air wrench

  d blow gun

6 Which one of the following should not be carried out as maintenance on an electrical tool?

  a check for damaged wires

  b check for cracked casing on the appliance

  c ensure appliance is clean by washing it in water

  d check for fuses which may be blown

7 A ferrous metal with 3% carbon is known more commonly as:

  a high carbon steel

  b cast iron

  c wrought iron

  d mild steel

8 A ferrous metal is:

  a one which does not rust

  b one which is gold in appearance

  c one which contains iron

  d one which is not magnetic

9 An alloy is:

  a pure aluminium

  b a mixture of a metal and another substance

  c rubber compound

  d an adhesive

10 A thermosetting plastic is one which:

  a remains permanently set after heating

  b remains temporarily set after heating

  c is flexible and can bend without breaking

  d is used as a substitute for rubber

GETTING READY FOR ASSESSMENT

The information contained in this chapter, as well as continued practical assignments in your centre or workplace, will help you to prepare for both the end-of-unit tests and diploma multiple-choice tests. This chapter will also help to prepare you for choosing and using tools, equipment and materials correctly and safely.

You will need to be familiar with:

- Safe working with tools, equipment and materials
- The correct identification and selection of hand tools and equipment for vehicle maintenance
- The correct preparation, use and maintenance of hand tools and equipment for vehicle maintenance
- The identification, selection and use of materials and consumables for vehicle maintenance

You now need to apply the knowledge you have gained in this chapter in your day-to-day working activities. For example, you are required to have knowledge and understanding of the preparation, identification and selection of hand tools and equipment for vehicle maintenance. The chapter has given examples of how to do this by outlining the importance of preparation, correct use and maintenance. You have also been made aware of the importance of wearing PPE, VPE, checking the tool for the task is not damaged before and after use, and the need to report any damages to tools before or after use.

You have also gained an awareness of the different types of hand tools, their uses and the importance of using the correct tool for the job.

You should now use this knowledge in your workshop by working safely, preparing for the task by selecting the correct tools and materials for the job, and reporting any damages to tools and equipment immediately.

Before trying a theory end-of-unit test or diploma multiple-choice test, make sure you have reviewed and revised any key terms that relate to the topics in that unit. Be sure to read all questions fully and take time to digest the information so that you are confident about what the question is asking you. With multiple-choice tests, it is very important that you read all of the answers carefully, as it is common for the answers to be very similar and this may lead to confusion.

For practical assessments, it is important that you have had sufficient practice and that you feel that you are capable of passing. It is best to have a plan of action and work method that will help you. Make sure that you have prepared for the task and have the appropriate tools and equipment to hand. It is also wise to check your work at regular intervals. This will ensure that you are working correctly and help to avoid problems developing as you work.

When you are doing any practical assessment, always work safely throughout the test. Make sure that you observe all health and safety requirements and that you use the recommended personal protective equipment (PPE) and vehicle protection equipment (VPE) at all times. When using tools, make sure you are using them correctly and safely.

Good luck!

3 The automotive industry

The automotive industry is an exciting industry to work in. The skills and knowledge that you gain can be used all over the world.

This chapter will outline various types of job roles and the organisations which make up the maintenance and repair sector of the automotive industry. It will also highlight the various types of vehicle to be repaired.

This chapter covers:

- Safe working within the automotive industry
- The types of organisations that make up the maintenance and repair sector of the automotive industry
- The different types of vehicle within the automotive industry
- The technical and non-technical job roles available within the automotive industry

WORKING PRACTICE

You need to be safe when you work within the automotive industry. You must always use personal protective equipment (PPE) and you also need to think about the possibility of crush or bump injury. You will also come into contact with chemicals such as lubrication oils, grease, coolant and cleaners containing solvents. Make sure that your selection of PPE will protect you from these hazards.

Personal Protective Equipment (PPE)

Safety goggles/glasses reduce the risk of small objects or chemicals coming into contact with the eyes.

Overalls provide protection from coming into contact with oils and chemicals.

Safety gloves provide protection from oils and chemicals. They also protect the hands when handling objects with sharp edges.

Barrier cream protects the skin from old engine oil, which can cause dermatitis and may be carcinogenic (a substance that can cause cancer).

Safety boots protect the feet from a crush injury and often have oil- and chemical-resistant soles. Safety boots should have a steel toe-cap and steel mid-sole.

Safety helmet protects the head from bump injuries when working under cars.

The types of organisations that make up the maintenance and repair sector of the automotive industry

Organisations in the maintenance and repair sector of the automotive industry include:

- retail operations
- vehicle sales (new and used)
- franchised dealer service, repair and parts
- independent repairers
- fast fit operations
- body repair/refinishing
- parts supply.

Retail operations

All organisations which sell anything within the automotive industry are known as retail operations.

Examples of what they might sell include:

- vehicles, including cars, trucks and vans
- accessories, including spoilers and big bore exhausts
- parts for the vehicles that are being sold
- car servicing.

Vehicle sales (new and used)

Normally, new vehicles are purchased from **franchised dealers**. However, because of their high cost, more and more new vehicles are being bought straight from the **manufacturer**'s website.

An alternative to a franchised dealership is a used vehicle supermarket. This is where hundreds of vehicles are sold at a reduced profit compared to a trade salesman selling fewer vehicles for a higher **profit**.

> **Key terms**
>
> **Franchised dealers** – a franchised dealership sells new and used vehicles for one manufacturer, for example BMW, Volkswagen, Ford, Toyota, etc. The dealership is normally listed by the franchised owner's name or location followed by the manufacturer's name, for example 'Chambers Ford'.
>
> **Manufacturer** – a person or business that makes things for use or sale.
>
> **Profit (£)** – the selling price of vehicle (£) minus the price you bought the vehicle for (£).

Figure 3.1 Vehicle sales (new and used)

> **Working life**
>
> James bought a car for £500. He sold it quickly for £550 because he needed the money.
>
> 1   How much profit did he make?

Table 3.1 Vehicle sales

Company	Profit	Dealership selling price	Dealership purchase price
Mason Motors	**£450**	**£9000**	**£8550**
Whittaker's of Warwickshire		£7000	£6200
Lord Motors		£22,000	£20,500
Sykes Motors		£13,000	£12,350
Feeney Motors		£22,000	£20,500

Figure 3.2 A non-franchised dealership can sell new and used vehicles for several different manufacturers

Franchised dealer service, repair and parts

This type of organisation specialises in looking after one make of vehicle. It may:

- sell this type of vehicle only
- sell and service this type of vehicle only
- sell, service and repair this type of vehicle only
- supply part sales for this type of vehicle only
- sell, service, repair and supply part sales for this type of vehicle only.

The advantage of repairing only one make of vehicle is that all the special tools and parts needed for repair will be available. This normally means that work on newer vehicles will be carried out at a franchised dealership.

Vehicle warranty repair is a main source of income for a franchised dealer. Every new vehicle comes with a vehicle warranty. This is the manufacturer's guarantee to replace any faulty main components for an agreed length of time. Manufacturers often insist that their warranty will only be valid if the vehicle is serviced or repaired at an approved franchise dealer.

Working life

It has been five years since Mr Simpson purchased an Audi TT from his local Audi dealership. As his car was new he didn't feel it needed to go into an approved dealership each year for a service. The warranty on the Audi TT was for the first five years from new. This year (the fifth year) Mr Simpson found out that there was a fault with the window wiper motor. He contacted Audi's warranty department and they arranged for the motor to be repaired. On this occasion his wiper motor was repaired.

1 Why would Audi have been in a position to refuse to repair the motor?

Independent repairers

As vehicles get older, the warranty starts to run out. It is usually cheaper to get the vehicle repaired and serviced at an **independent repairer**. This is because the parts do not have to be approved by the manufacturer. The high costs of keeping specialised equipment in good repair and the rental of large premises are not passed on to the customer.

An independent repairer may:

- repair any type of vehicle
- carry out MOT inspection
- replace parts.

Figure 3.3 The clutch fluid gearbox and final drive levels are checked as part of a service

Fast fit operations

Vehicles are much more reliable than they used to be. For this reason, there are now many successful fast fit organisations that replace vehicle parts that wear out.

Fast fit operations may:

- replace exhausts
- replace/repair tyres
- replace suspension dampers
- replace batteries
- replace clutches
- carry out service and MOT.

Did you know?

When fast fit organisations were first set up, they dealt with the replacement of tyres and exhausts only.

Working life

Jay works as a vehicle fitter at a fast fit operation. His main responsibilities include:

- spotting worn and faulty components in line with the **company standards**
- advising customers of work required to make sure the vehicle is in a **roadworthy** condition
- removing and refitting components and carrying out tyre-related adjustments in line with the **work instructions**
- repairing tyres in line with the technical standards and work instructions
- contributing to **good housekeeping routines**
- placing goods and materials into storage and assisting in stock routines as directed
- understanding and following the **company's policies and procedures**.

1   While Jay is changing a wheel that has a worn tyre, he notices damage to the vehicle suspension. What should he do?

2   If Jay does not put new stock in its correct place following delivery, what could be the consequences?

Key terms

Independent repairer – vehicle repair business which is owned by an individual and not by a large manufacturer.

Company standards, policies and procedures – rules written down by an employer for all employees to follow.

Roadworthy – a vehicle which is safe to use on the road.

Work instructions – instructions on how to complete a task.

Good housekeeping routines – keeping your workshop tidy and putting away tools after use.

Figure 3.4 Paint refinishing work

Body repair/refinishing

If a car is damaged it may need **cosmetic** and/or **structural repairs**. Cosmetic and structural repairs are completed in a body shop.

Types of repair include:

- vehicle insurance claims
- warranty body repairs
- paint refinishing work.

These kinds of repairs are expensive. This is because of the high cost of the equipment needed to repair the vehicle body and the paint booth costs.

A paint booth is a dust-free enclosed space that the vehicle or parts of the vehicle can be painted in. It has a special fume extraction unit to remove dangerous paint fumes and an oven to dry the paint quickly.

Key terms

Cosmetic repairs – for example, a scratch to the paintwork. These do not cost as much as structural repairs.

Structural repairs – a main repair which is likely to cost a lot of money, such as repairing a dented door.

Car insurance – provides protection/security against any costs from a road traffic incident, for example injury to the people involved or damage to the vehicle. It is compulsory to have car insurance within the UK. It is also referred to as motor insurance.

Table 3.2 Examples of equipment found in a body shop

Equipment		Uses
Fume extraction equipment		Removes dangerous fumes from the workshop
Lifts		Lifts vehicles in the air
Tracking equipment		Lines up wheels after suspension damage
Paint booth		Keeps paint fumes contained and speeds up the drying process of wet paint; reduces dust settling on wet paint
Body jig		The alignment measurement of a chassis after a collision

If the vehicle has been in an accident, the repair may be covered by the vehicle's **car insurance**. Then the cost of the repair will be paid to the body shop by the car insurance company.

Parts supply

Your workshop, dealership or body shop may not have the part needed for a repair. If this happens then your manager may contact a parts supplier and ask for the part to be delivered. This might cause a delay in the repair job. Your manager will contact the customer to let them know about the delay.

A vehicle owner may be able to repair their car themselves. Or they might want to add an accessory to it. They can buy these parts straight from the part supplier.

Potential parts supply sales include:

- customers whose vehicles are in the workshop (including warranty repairs)
- additional accessories sold alongside new and used car sales
- private retail operations/customers via the retail counter
- other businesses (trade customers).

Figure 3.5 Spare parts should be stored neatly so you can easily find out whether the parts you need are available

Figure 3.6 Car door speakers

Table 3.3 Types of customer

Type of customer	Example
Private retail	You are carrying out a small repair on your own vehicle and do not work in the motor trade.
Trade customer	You work for a garage carrying out a repair on vehicles.
In house	Your vehicle workshop has a parts department attached to it.

Find out

You work for an independent repairer and you are carrying out a service on a vehicle. Which type of parts customer would you be?

CHECK YOUR PROGRESS

1 Name three types of organisation that make up the maintenance and repair sector of the automotive industry.
2 What is the difference between a franchise dealer and an independent repairer?
3 What is a trade customer?

The different types of vehicle within the automotive industry

In the automotive industry, you will be asked to work on one or more types of the following vehicles:

- cars
- vans (car-derived and specialist)
- rigid trucks
- tractor unit and trailer heavy goods vehicles
- motorcycles and mopeds.

Cars

Cars come in a variety of different sizes and shapes. They are covered in more detail in Chapter 4. One of the most popular shaped cars is the family saloon shown in Figure 3.7.

Figure 3.7 A saloon car

Vans

There are two main types of van – a specialist van, shown in Figure 3.8 and a car-derived van. The specialist van has a larger carrying capacity and is designed and built as a van with a dedicated chassis.

The car-derived van is part car (front) and part van (rear). An example is shown in Figure 3.9. Examples of use include:

- parts delivery
- parts collection
- mobile repairs.

Figure 3.8 A van

Rigid trucks

A rigid truck is the next largest vehicle for carrying loads after the van. Figure 3.10 shows a rigid breakdown recovery vehicle. Other rigid vehicles may have a container fixed to the back which will protect the load being transported.

Figure 3.9 A car-derived van

Did you know?

'Derived' means developed from something else.

Figure 3.10 A rigid breakdown recovery vehicle

Tractor unit and trailer heavy goods vehicles

Heavier vehicles use a tractor unit to pull a trailer. Examples of use include:

* a car transporter (see Figure 3.11)
* parts delivered from the vehicle manufacturer to large franchise dealerships.

With these heavier vehicles the tractor will be joined to the trailer by a fifth wheel coupling. This helps to improve the steering for the long load.

Figure 3.11 A car transporter

Motorcycles and mopeds

For journeys where access is difficult and where loads to be carried are not large, the motorcycle and moped are the best way of moving around. Examples of use include:

* minor roadside vehicle breakdowns where large parts and recovery are not required for repair
* delivery of parts
* mobile repairs.

Did you know?

A fifth wheel coupling is a pin and socket that attaches the tractor unit to the trailer on a heavy vehicle.

Find out

See how many different types of vehicle breakdown services there are and list them all in a table like the one below. State what colours are showing on the vehicles and the type of vehicles they use.

Organisation	Colour of vehicle	Type of vehicle

Figure 3.12 A motorcycle

Did you know?

Mopeds are less powerful than motorbikes. They have slower maximum speeds because of their smaller engines, which are no larger than 50cc.

CHECK YOUR PROGRESS

1 What does 'derived' mean?
2 What is the difference between a rigid truck and a tractor unit and trailer heavy goods vehicle?
3 What is a fifth wheel coupling?

The technical and non-technical job roles available within the automotive industry

Tables 3.4 and 3.5 list the technical and non-technical job roles within the automotive industry. They also give an example of the type of work you may be required to carry out in each job role.

Table 3.4 Technical job roles at Swan Motors, Newtown

Job role	Technical	Type of work carried out
Tony Wan Apprentice technician	Technical	Constantly learns from the skilled technicians how to become a competent technician
Philip Heskey Vehicle technician	Technical	Maintains and repairs vehicles in the workshop
Emma York Vehicle technician	Technical	Maintains and repairs vehicles in the workshop
Ewan Philips Vehicle technician	Technical	Maintains and repairs vehicles in the workshop
Kani Singari Vehicle examiner/MOT tester	Technical	Carries out inspections on vehicles to make sure they are safe
Dwight Greenep Body repair technician	Technical	Repairs and replaces body panels in the body shop
Susan Mullen Vehicle refinisher technician	Technical	Prepares and paints vehicle body panels in the body shop
George Colclough Parts adviser	Technical	Supplies parts to the public and the motor trade
Ben Rosen Valetor	Technical	Prepares and cleans vehicles for sales and body shop customers
Martin Daniels Autoglazing technician	Technical	Repairs and replaces vehicle glass panels
Bill Peters Apprentice	Technical	Constantly learns from the skilled technicians to become a competent technician
Philipe Cussons Senior technician	Technical	Communicates with vehicle manufacturers to solve difficult problems on vehicles

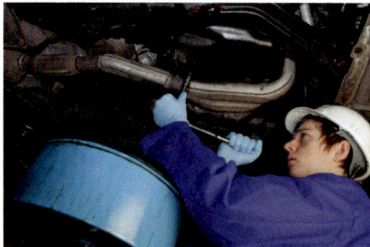

Figure 3.13 A vehicle technician at work

Figure 3.14 A parts adviser

Table 3.5 Non-technical job roles at Swan Motors, Newtown

Job role	Non-technical	Type of work carried out
Roberta Arnolds Vehicle sales person	Non-technical	Sells new and used vehicles to the public and the motor trade
Catherine Corson Workshop/body shop controller	Non-technical	Supervises technicians in the various technical departments
David Lord Parts and sales manager	Non-technical	Supervises parts and sales staff in the various departments
Aaron Samson Vehicle damage assessor	Non-technical	Checks vehicles following an accident to see how much it will cost to repair them
Jayne Davies Customer service adviser	Non-technical	Liaises with customers and workshop staff
Mia Klukowski Receptionist	Non-technical	Meets and greets customers in large dealerships and points them to the department they need
Richard Westwood Accounts clerk	Non-technical	Responsible for wages and the money going in and out of the business
Irine Welland After sales manager	Non-technical	Manages the whole dealership and its staff

Find out

Read through the job roles listed in Tables 3.4 and 3.5. Make a note of the roles you find interesting and find out more about them. Then make a note of the qualifications you need to do these jobs.

Working life

Below is a typical half-day for Tony, an apprentice technician:

8.30 a.m.	Arrives at Bodgitt & Scarper	10.00 a.m.	Removes faulty components and refits new ones
8.35 a.m.	Puts on PPE and makes a brew	11.10 a.m.	Asks his mentor to check his work
8.45 a.m.	Collects job card from supervisor	11.15 a.m.	Mentor drives the car out of the workshop using VPE
8.55 a.m.	Prepares tools and work area for jobs to be completed	11.20 a.m.	Returns the keys to reception along with the completed job card
9.05 a.m.	Collects vehicle's keys from reception		
09.10 a.m.	Asks his mentor to drive the vehicle into the workshop	11.30 a.m.	Breaks for 15 minutes
09.20 a.m.	Inspects the vehicle and confirms the repairs with his mentor	12.30 a.m.	Tidies workshop area and returns the tools to their correct place and disposes of any waste correctly (housekeeping)
09.40 a.m.	Returns to reception and advises the customer of required repairs	1.00 p.m.	Returns any parts not required to the stores
9.45 a.m.	On agreement from customer collects parts from stores	1.20 p.m.	Lunch break

Make a timetable for a typical day in your workshop.

A typical organisation structure within the automotive industry

In an organisation all job roles are important and give the business structure. Knowing where you belong in an organisation's structure helps you to see:

- who is responsible for what department
- who you, your manager and their manager should report to
- possible job role progression within the organisation.

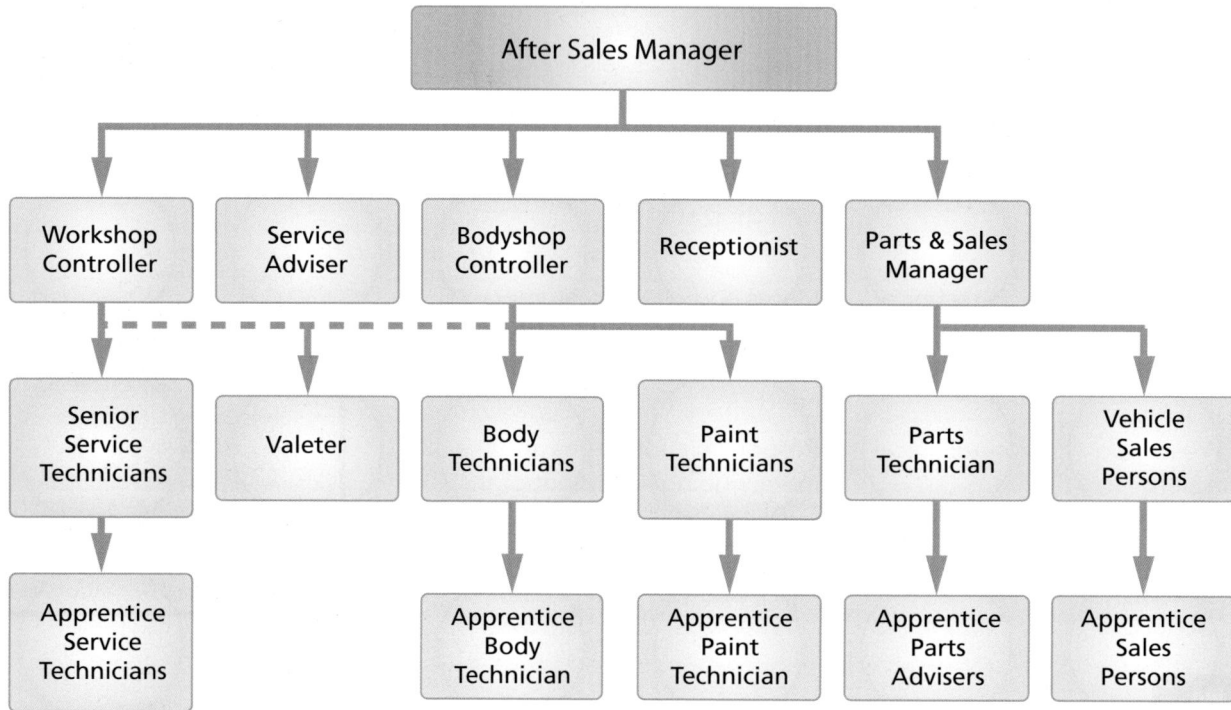

Figure 3.15 The organisational structure for Swan Motors, Newtown

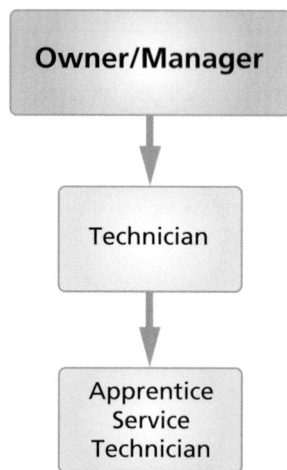

Figure 3.16 A typical organisational structure for an independent repairer

The opportunities for career progression and development within the automotive industry

There are many opportunities for an apprentice and technician to progress within the automotive industry. They include:

- senior technician
- supervisor and manager
- vehicle damage assessor
- technical trainer and assessor
- warranty personnel.

Senior technician

Once you have gained years of experience working as a technician, you may want to progress to a higher level. However, you may still want to continue to work on vehicles. You could progress into the role of a senior technician.

This role is suited to someone who has a high technical ability and who can solve problems. In this role you will also be required to:

- talk to vehicle manufacturers to gain the most up-to-date information
- talk to customers to let them know what is wrong with their vehicle in terms they can understand.

Figure 3.17 A senior technician using a scan tool

After senior technician, you could progress to the role of workshop controller (see below) and after that, to the role of after sales manager.

Vehicle damage assessor

Body shops have to correctly **estimate** the cost to repair accident damage. This is to make sure the insurance company is not overcharged. It also ensures that the body shop makes a profit following the repair of an accident.

After gaining years of experience as a body shop or paint technician, you may want to progress to the role of a vehicle damage assessor. This person is **responsible** for estimating collision damage correctly and giving this information to the insurance companies.

A vehicle damage assessor also ensures the safe repair of the vehicle and provides a correct repair method that aligns to the researched method.

After the post of vehicle damage assessor, you could progress to the role of workshop controller and after that, to the role of after sales manager.

Key terms

Estimate – a rough calculation or guess. For example, a vehicle may need a wing, a door and a bonnet replacement plus painting. It is estimated that this will cost £600.

Responsible – something could be your responsibility, for example 'you are responsible for keeping your work area tidy'.

Supervise – to oversee someone doing their job.

Working life

Johnny is an apprentice body technician.

1   What is his next job role before applying to be a vehicle damage assessor?

Supervisor and manager

If you are good at communicating and like to be given responsibility you might want to progress beyond senior technician to be a supervisor or workshop controller. In this role you may have your own team of technicians to **supervise**. This could be in any of the departments within a franchise dealership or large independent repairer.

If you are talented in this role you may want to progress even further. You could become the manager of a vehicle maintenance and repair organisation.

Figure 3.18 A technical trainer helping an apprentice

Technical trainer and assessor

If you don't want to work towards a management job, you may want to pass on your knowledge to other people by teaching them. As a technical trainer you will need to create and run training courses so that other technicians can repair vehicles or units correctly.

Part of your role will be to confirm that the technician has gained enough knowledge to repair the vehicle or unit. You will do this by checking their work to see if they are doing it correctly.

Warranty personnel

Vehicles can be booked in for work that is related to warranty. As an experienced technician, you may want to monitor warranty claims. This will involve:

- making sure that the parts for replacement are available
- making sure any work done is correctly recorded
- getting payment for all costs.

CHECK YOUR PROGRESS

1 What is the role of a vehicle examiner?
2 Name three opportunities for progression within the automotive industry.
3 Who eventually pays for the work done on a vehicle under warranty?

FINAL CHECK

1 Which of the following are not fast fit operations?

  a replace exhausts

  b replace/repair tyres

  c replace suspension dampers

  d diagnose faults on an engine system

2 The profit on a vehicle sold is equal to the:

  a selling price of vehicle (£) + price you bought vehicle for (£)

  b selling price of parts (£) + price you bought vehicle for (£)

  c selling price of vehicle (£) – price you bought vehicle parts for (£)

  d selling price of vehicle (£) – price you bought vehicle for (£)

3 Which of the following statements is not true?

  a a paint booth has an oven which can dry painted panels faster

  b a paint booth is used for welding body panels

  c a paint booth has fume extraction to remove dangerous paint fumes

  d a paint booth extracts dust from the atmosphere to ensure clean paint surfaces

4 Which one of the following provides protection/security against any costs from a road traffic incident?

  a insurance

  b assurance

  c warranty

  d guarantee

5 You are carrying out a small repair on your own vehicle and do not work in the motor trade. What type of customer are you to a parts supplier?

  a trade customer

  b in house customer

  c private retail customer

  d specialist retail customer

6 A vehicle that looks like a car at the front and a van at the back is called a:

  a walk-through van

  b specialist van

  c car-derived van

  d rigid van

7 The job role that involves preparing and cleaning vehicles for sales and body shop customers in a large dealership is a:

  a valetor

  b paint technician

  c body technician

  d service technician

8 Which of the organisational structures below shows that of a dealership?

  a manager > apprentice > workshop controller > technician

  b manager > technician > apprentice > workshop controller

  c manager > workshop controller > technician > apprentice

  d manager > apprentice > technician > workshop controller

9 What is an estimate?

  a a rough calculation or guess of a price

  b something that could be your responsibility

  c an exact calculation of the cost of a job

  d a quotation

10 Which one of the following lists the job titles in the correct order?

  a senior technician, apprentice, technician, manager

  b apprentice, technician, senior technician, manager

  c manager, apprentice, technician, senior technician

  d apprentice, senior technician, technician, manager

GETTING READY FOR ASSESSMENT

The information contained in this chapter, as well as continued practical assignments in your centre or workplace, will help you to prepare for both the end-of-unit tests and diploma multiple-choice tests.

You will need to be familiar with:

- Safe working within the automotive industry
- The types of organisations that make up the maintenance and repair sector of the automotive industry
- The different types of vehicle within the automotive industry
- The technical and non-technical job roles available within the automotive industry

You now need to apply the knowledge you have gained in this chapter in your day-to-day working activities. For example, you are required to know and understand the types of organisation that make up the maintenance and repair sector of the automotive industry. The chapter has given you this information through examples of currently operating organisations, explaining the differences in the make up of each type of organisation and the service they provide.

You have also been made aware of the types of repairs/services supplied, sales and profit, and the customers expected to be seen in the differing organisations.

The knowledge you have gained has given you an understanding of the types of organisations and career prospects in the automotive industry. You should use this knowledge when planning your future career.

This chapter has given you an overview of job roles and the organisations which make up the maintenance and repair sector of the automotive industry and the types of vehicle to be repaired. It has provided you with the basic knowledge that will help you with both theory and practical assessments.

Before trying a theory end-of-unit test or diploma multiple-choice test, make sure you have reviewed and revised any key terms that relate to the topics in that unit. Be sure to read all questions fully and take time to digest the information so that you are confident about what the question is asking you. With multiple-choice tests, it is very important that you read all of the answers carefully, as it is common for the answers to be very similar and this may lead to confusion.

Although for this chapter there is no practical assessment, assignments will have to be completed to make sure you have understood the information given. Ensure you plan the assignment before you begin and check your work at regular intervals. This will help you to be sure that you are working correctly and help to avoid problems developing as you work.

Good luck!

4 Light vehicle construction & maintenance

There are many different kinds of light vehicle layouts, body types and methods of construction used in the manufacture of light vehicles.

In this chapter you will find out about them, how to carry out routine maintenance, which tools to select for the job and the procedures and methods to follow when carrying out effective maintenance.

This chapter covers:

- Safe working on light vehicle construction and maintenance

- The body types for a range of vehicles

- The engine and driveline configurations for a range of vehicles

- The main body parts found on light vehicles

- Inspecting systems and components during basic routine vehicle maintenance

- Appropriate sources of information required to carry out basic routine vehicle maintenance

- The main components and systems that require routine maintenance on a modern vehicle

- The correct procedures when inspecting, replacing fluids and service items, and lubricating and adjusting systems and components

WORKING PRACTICE

You need to be safe when you work within the automotive industry. You must always use personal protective equipment (PPE) and you also need to think about the possibility of crush or bump injury. You will also come into contact with chemicals such as lubrication oils, grease, coolant and cleaners containing solvents. Make sure that your selection of PPE will protect you from these hazards.

Personal Protective Equipment (PPE)

Safety helmet protects the head from bump injuries when working under cars.

Overalls provide protection from coming into contact with oils and chemicals.

Safety mask protects against dust inhalation

Safety gloves provide protection from oils and chemicals. They also protect the hands when handling objects with sharp edges.

Barrier cream protects the skin from old engine oil, which can cause dermatitis and may be carcinogenic (a substance that can cause cancer).

Safety goggles reduce the risk of small objects or chemicals coming into contact with the eyes.

Safety boots protect the feet from a crush injury and often have oil- and chemical-resistant soles. Safety boots should have a steel toe-cap and steel mid-sole.

To reduce the possibility of damage to the car, always use the appropriate vehicle protection equipment (VPE):

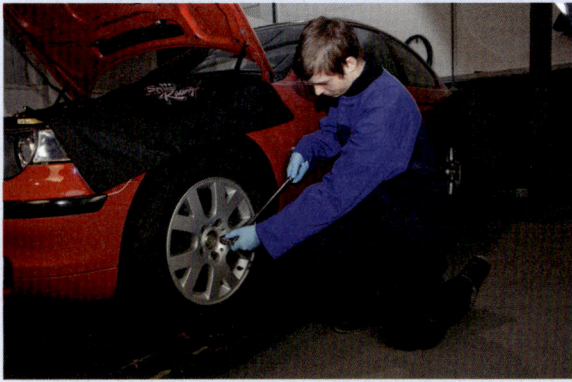

Wing covers

Seat covers

Steering wheel covers

Floor mats

If appropriate, safely remove and store the owner's property before you work on the vehicle. Before returning the vehicle to the customer, reinstate the vehicle owner's property. Always check the interior and exterior to make sure that it hasn't become dirty or damaged during the repair operations. This will help promote good customer relations and maintain a professional company image.

Vehicle Protective Equipment (VPE)

Safe Environment

During the repair or maintenance of light vehicle electrical systems you may need to dispose of certain waste materials such as used engine oil. Under the Environmental Protection Act 1990 (EPA), you must dispose of them in the correct manner. They should be safely stored in a clearly marked container until they are collected by a licensed recycling company. This company should provide you with a waste transfer note as the receipt of collection.

To further reduce the risks involved with hazards, always use safe working practices including:

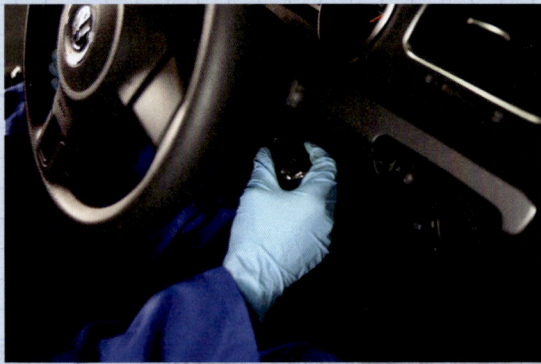

1. Immobilise vehicle by removing the ignition key. If possible, allow the engine to cool before starting work.

2. Prevent the vehicle moving during maintenance by applying the handbrake or chocking the wheels.

3. Follow a logical sequence when working. This reduces the possibility of missing things out and of accidents occurring. Work safely at all times.

4. Always use the correct tools and equipment. (Damage to components, tools or personal injury could occur if the wrong tool is used or a tool is misused.) Check tools and equipment before each use.

5. Following the replacement of any vehicle components, thoroughly road test the vehicle to ensure safe and correct operation. Make sure that all work is correctly recorded on the job card and vehicle's service history, to ensure that any maintenance work can be tracked.

6. If components need replacing, always check that the parts are the correct quality and type for the vehicle if it is still under guarantee. Inferior parts or deliberate modification might make the warranty invalid. Also, if parts of an inferior quality are fitted, this might affect vehicle performance and safety.

Preparing the car

Dial gauge

Engine oil and filter replacement tools

Torque wrench

Vehicle hoist

Tyre pressure gauge

Tread depth gauge

Safe Working

- Always clean up any fluid spills immediately to avoid slips, trips and falls.
- Always disconnect the ignition when you check the tension of drive belts. You should also remove the ignition key when working under the bonnet of a vehicle. This will prevent anyone jumping in and starting it while your hands are near parts that move when the engine turns.
- Always use correct manual handling techniques when removing and refitting heavy components.
- Always remove all sources of ignition (for example, smoking) from the area. Have a suitable fire extinguisher to hand when working on petrol or diesel fuel systems.
- Always work in a well-ventilated area.
- Always use exhaust extraction when running engines in the workshop.
- Always wear safety goggles and a particle mask when working on components such as brake linings, fuel and exhaust systems.

The body types for a range of vehicles

The vehicle's body is there to accommodate the driver and passengers. It must look good and allow enough space for luggage. The vehicle body must provide fixing points and access for the **power train**. It must also protect the driver and passengers from all weather conditions and accidental damage.

There are many different shapes and sizes of light vehicles, as shown in Table 4.1.

Table 4.1 Light vehicle body types

Type	Design
Family saloon	This has room for four or more passengers. It has a large area for luggage at the rear.
Estate	The roof of this vehicle extends to the back of the car. This increases the space inside the vehicle. The folding rear seats provide room for more luggage. A large rear door called the tailgate allows large items to be easily loaded into the vehicle.
Hatchback	A hatchback has a tailgate and comes in either three door or five door versions. The three door hatchback is also referred to as a 'hot hatch' and is normally thought of as small and quick.
Coupé	This two door vehicle normally seats only two people – the driver and the passenger. Some manufacturers make 2 + 2 coupés which have rear seats. Coupés are often referred to as two door sports or sports cars.
Convertible	These have a removable metal or cloth roof. The vehicle can be converted into an open air car by folding the roof into the front of the boot. You may also hear convertibles called cabriolets or rag tops.
MPV – multi-purpose vehicle	This is more regularly known as a people carrier. It is much taller than a standard car and it provides more passenger and luggage room. Many multi-purpose vehicles have seven seats.
4 x 4 – all terrain vehicle	This type of vehicle is built to be driven on or off road. The body is raised to give a better ground clearance when driving over rough ground. It also has four-wheel drive.

The engine and driveline configurations for a range of vehicles

The engine and transmission positions can be different depending on the type of vehicle. Where it is located depends on what it is used for and the vehicle body design.

Other factors which affect the way the car is made include:

- cost
- access for maintenance
- how much load the vehicle has to carry
- weight balance
- vehicle control.

The engine can be put in the front, the rear or the middle of the vehicle. Examples of engine position are shown in Table 4.2.

Key term

Transverse – the engine is positioned sideways when looking from the front.

Table 4.2 Transverse engine layout

Type of transverse engine layout	Engine position
Front engine – front-wheel drive – transverse	Key: Engine, Clutch, Gearbox, Final drive, Drive shafts
Rear engine – rear-wheel drive – transverse	Key: Engine, Clutch, Gearbox, Final drive, Drive shafts
Mid engine – rear-wheel drive – transverse	Key: Engine, Clutch, Gearbox, Final drive, Drive shafts

A vehicle layout can be divided into two categories – front- or rear-wheel drive. Four-wheel drive vehicles distribute power to all wheels.

Key terms

Longitudinal – the engine is positioned in-line with the greater length of the vehicle body.

Transmission (system) – made up of the clutch, gearbox, **drive shafts** and final drive.

Drive shaft –the drive shaft is used when passing the drive from the final drive to the hub.

Table 4.3 Longitudinal engine position

Type of longitudinal engine layout	Drive configuration
Front engine – front-wheel drive – longitudinal	Key: Engine, Clutch, Gearbox, Final drive, Drive shafts
Front engine – rear-wheel drive – longitudinal	Key: Engine, Clutch, Gearbox, Final drive, Drive shafts
Front engine – four-wheel drive – longitudinal	Key: Engine, Clutch, Gearbox, Final drive, Drive shafts

Find out

Walk around your workshop, look at the vehicles there and find out whether they are front-, rear- or four-wheel drive. Alternatively, make a list of vehicles to find out about.

You will come across many terms used to describe a power train or driveline layout. For example, transverse is used when the engine and **transmission** run the width of the vehicle. Longitudinal refers to the engine and transmission running the full length of the vehicle.

The main body parts found on light vehicles

Basic non-structural body panels

The following panels are referred to as **non-structural**:

- bonnet
- wing
- boot lid
- tailgate
- door.

These panels are positioned on the vehicle as shown in Figure 4.1.

Key terms

Non-structural – this panel does not provide strength as part of the **chassis**.

Chassis – the main structure of the vehicle which has to withstand all the forces if a collision occurs.

Find out

Identify the non-structural panels listed above on a vehicle in your workshop.

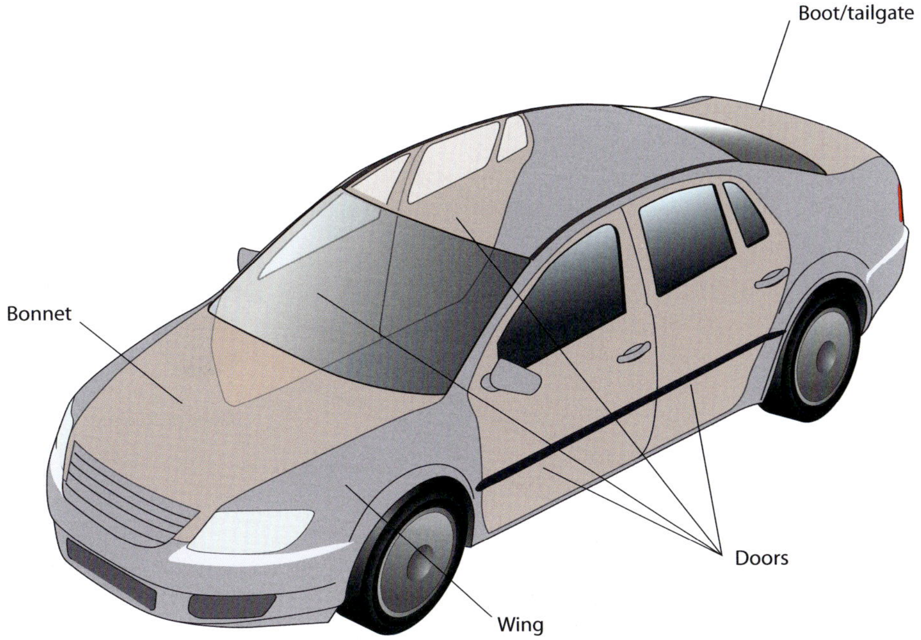

Figure 4.1 Non-structural body panels

Vehicle bodies are made up from lots of different materials. Steel is the most popular material used. On some vehicles a combination of materials are used.

Table 4.4 Vehicle body materials

Material	Use
Steel	Cost-effective chassis construction
Aluminium	Doors – sports cars
Reinforced plastic	Bonnets – sports cars

Main trim components

On light vehicles the chassis provides the vehicle strength. The non-structural parts protect pedestrians from falling into the engine compartment and passengers and load from falling out of the vehicle. Other trim components give a luxury finish and some are necessary to fulfil legal requirements.

The following trim components can be found on the interior and exterior of the vehicle:

- bumper
- headlamp units
- rear light units
- front windscreen
- door drop glass
- dashboard
- parcel shelf
- seat belts
- door mouldings
- head lining.

Did you know?

A boot lid on a saloon car is the same as the tailgate on a hatchback or estate, except the boot lid does not have a rear window.

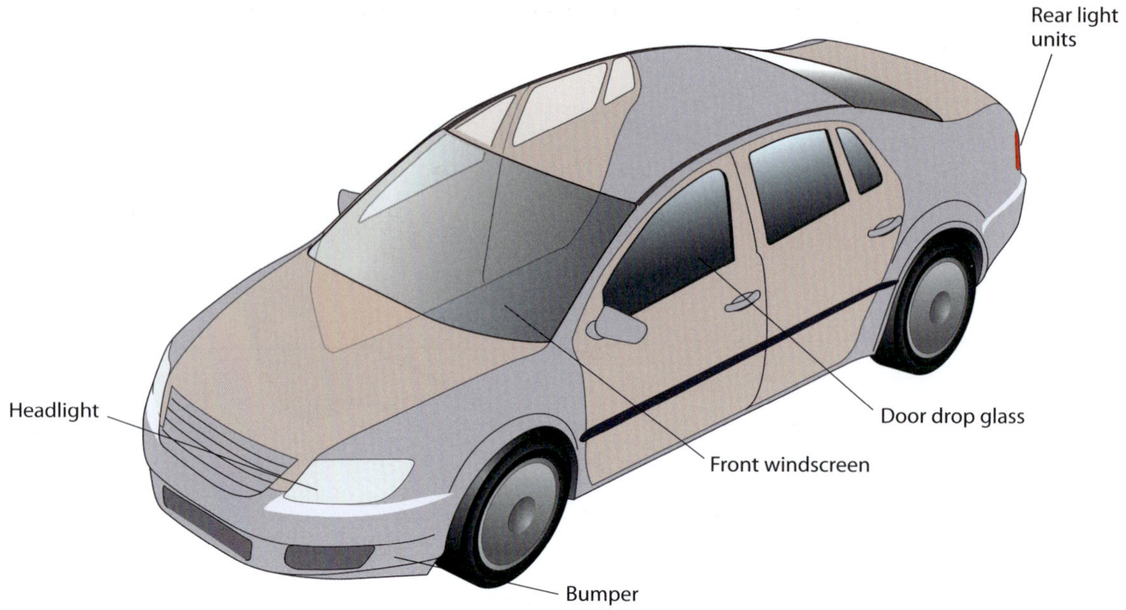

Rear light units

Headlight

Door drop glass

Front windscreen

Bumper

Figure 4.2 Outside trim components

Dashboard

Seat belts

Headlining

Door mouldings

Figure 4.3 Inside trim components

Table 4.5 describes a feature of each trim component and states the legal requirement.

Table 4.5 Trim components and their legal requirements

Trim component	Feature on a typical vehicle	Legal requirement
Bumpers	Attached to the front and the rear of the vehicle. These absorb impact in the event of a slight collision	Must be secure with no sharp edges
Headlamp units	Contain the headlamp, main beam and sidelights in one unit	All lights work Lenses must not be cracked
Rear light units	Contain the indicators, brake lights, fog lights, reverse lights and rear sidelights in one unit	All lights work Lenses must not be cracked
Front windscreen	To allow the driver to see through clearly	Must not be cracked in wiped area or chipped
Door drop glass	To allow the driver and passengers to see through clearly and provide a means of ventilation for the vehicle	Must not be shattered Front windows cannot be made of mirror glass
Dashboard	To hold all the instruments that the driver needs	Must clearly light up the speedometer
Parcel shelf	Provide a means of storing light goods under the rear tailgate	n/a
Seat belts	To prevent injury to the driver and passengers if a collision occurs	Must not be frayed and lock under severe braking
Door mouldings	To provide a good appearance to the inside of the vehicle doors	n/a
Headlining	To provide a good appearance to the inside of the vehicle roof	n/a

CHECK YOUR PROGRESS

1　State three purposes of the vehicle body.
2　Name five different body styles and their uses.
3　What vehicle units go together to make the power train?
4　Name four internal trim components.
5　Name four external trim components.
6　Name three non-structural vehicle panels.

Inspecting systems and components during basic routine vehicle maintenance

You will need to use your senses to establish the condition of the vehicle systems when you carry out routine maintenance. Some examples are shown in Table 4.6.

Table 4.6 Using your senses to assess the condition of the vehicle

Sense	Example of routine inspection check carried out
Visual (see)	• Look for lights not operating correctly • Look for oil leaks under the engine
Aural (hear)	• Listen for the horn working correctly • Listen for loose alternator drive belt squealing
Smell (sniff)	• Smell for fuel leaks • Smell oil leaking on to the exhaust manifold
Touch (feel)	• Feel for wear on a disc brake surface • Feel for movement in steering joints

Functional assessment

Functional assessment is where you carry out a **simulation** of system operation. You can carry out a simulation by connecting a test instrument to the vehicle system to be checked.

Examples of functional testing include:

*   brake roller testing
*   coolant pressure testing
*   wheel balancing on a machine.

Appropriate sources of information required to carry out basic routine vehicle maintenance

When carrying out routine maintenance, you need to:

*   use various sources of information to help you do the job
*   record information on the condition of the vehicle and the servicing activities that you carry out.

To do this, you will use the documents listed below.

Job card

The job card gives you all the information you need to complete routine maintenance. It will be given to you by your supervisor. It will tell you what type of maintenance is required.

Key term

Simulation – where you recreate an inspection activity to mirror the real thing.

The job card can contain:

- job number
- date
- customer details
- vehicle details
- customer's instructions
- description of work carried out
- parts and consumables used and costs
- labour – hours and costs
- total cost
- customer authorisation (signature)
- technician's name and signature.

In-vehicle service record

Every new vehicle is supplied with a service book. It is kept by the owner. Every time you service a vehicle, you should stamp the service book with the vehicle's mileage, the date and the type of service carried out. This will form part of the vehicle's service history information.

Vehicle handbook

The vehicle handbook is an instruction manual for the driver. It is produced by the vehicle's manufacturer. It also gives information needed for inspection and service operations including:

- oil and fluid capacities
- tyre pressures
- bulb sizes
- vehicle dimensions.

It is essential to refer to the vehicle handbook when you are servicing a vehicle.

Manufacturer's workshop manuals

Every manufacturer produces a detailed instruction manual on how to repair, inspect and service their vehicles. These are now more commonly produced in electronic format. They include:

- step-by-step guides for the diagnosis and repair of each vehicle system
- all **specifications** relating to the vehicle for inspection and service.

Generic manufacturer data

Automotive data information for vehicles is available from various companies in paper form, on CD or online. This is used by smaller garages who do not work on just one make or model of vehicle.

Key term

Specification – the technical information provided by the manufacturer. It gives details of how the vehicle works to its optimum (best) performance.

Did you know?

Information produced in paper form is often referred to as a 'hard copy'.

Vehicle identification number (VIN)

The vehicle identification number (VIN) is made up of 17 characters. It is unique to the vehicle. The VIN tells you:

- when the vehicle was built
- how it is constructed
- the correct spare parts required during inspection and servicing.

The VIN is printed on the chassis plate and can be found in any of the following places:

- under the bonnet near to the bonnet catch
- on the bulkhead
- on the inner wing
- stamped into the floor next to one of the front seats.

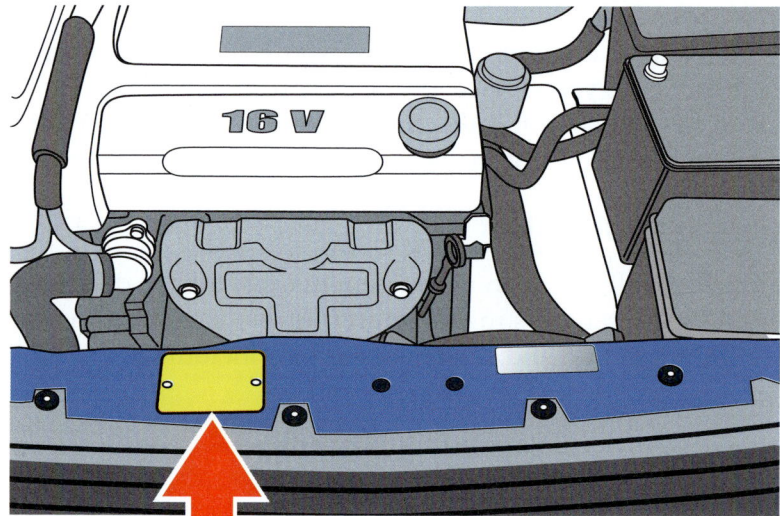

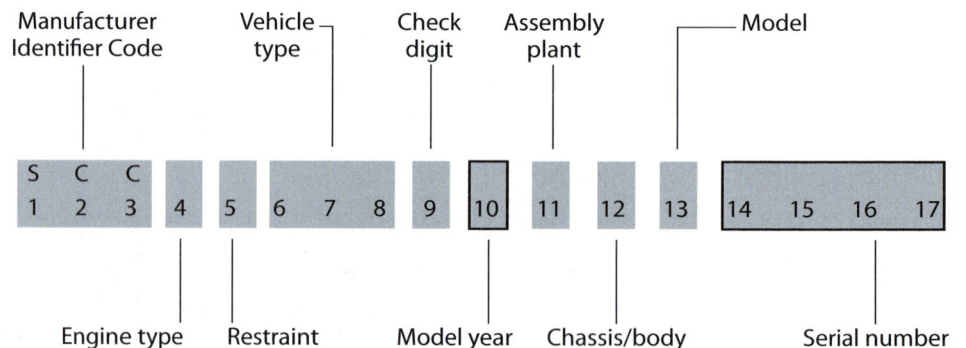

Figure 4.4 Vehicle identification number (VIN)

In newer vehicles the chassis plate may also be at the foot of the windscreen. You need to check the vehicle handbook or manufacturer's repair specifications to confirm the location.

Routine maintenance inspection sheets

A routine maintenance inspection sheet lists the checks that should be made. It also gives the order in which they should be completed. They are put together by the manufacturer in a certain order to make sure that inspection processes are carried out correctly.

CHECK YOUR PROGRESS

1 What does an aural inspection mean?
2 The job card sets out all the information required for the inspection or service. Write down five pieces of information that you will find on a job card.
3 What does VIN mean?

The main components and systems that require routine maintenance on a modern vehicle

Routine maintenance involves measuring, testing and looking for general wear and tear. The tests are carried out on the ground, when the vehicle is in the air and then with the vehicle at chest height. When you carry out routine maintenance you will recognise whether the components fit the specified requirements. The two main types of routine maintenance are:

• the interim service
• the full service.

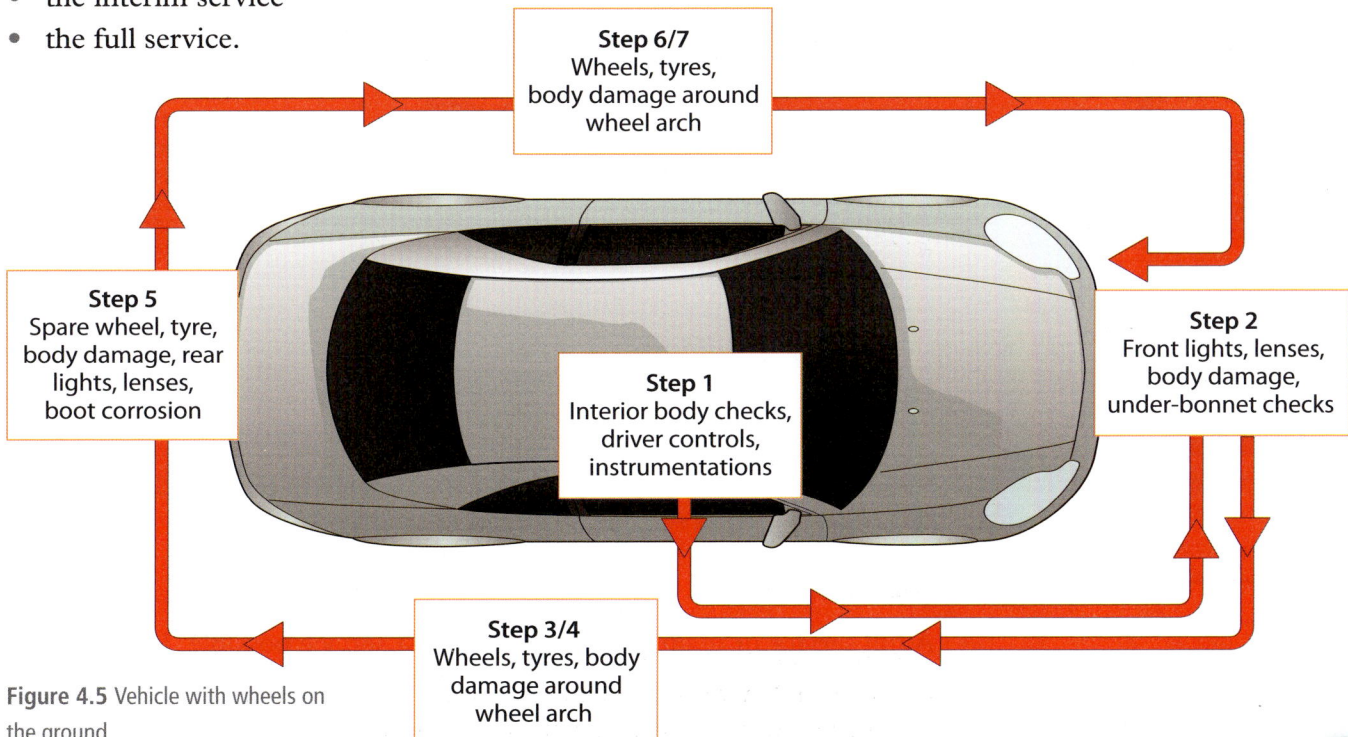

Step 6/7
Wheels, tyres, body damage around wheel arch

Step 5
Spare wheel, tyre, body damage, rear lights, lenses, boot corrosion

Step 1
Interior body checks, driver controls, instrumentations

Step 2
Front lights, lenses, body damage, under-bonnet checks

Step 3/4
Wheels, tyres, body damage around wheel arch

Figure 4.5 Vehicle with wheels on the ground

Step 8/9
Rear suspension wear and security, chassis damage and corrosion, brake pipe corrosion and security, fuel lines

Step 6/7
Boot corrosion, rear suspension wear and security, exhaust security

Step 3
Drive shafts, oil leaks from drive train, exhaust condition and security

Step 1/2
Suspension wear, steering joints, tyre wear, oil leaks, corrosion and security of engine

Step 4/5
Exhaust condition and security, chassis damage and corrosion, rear suspension wear and security, brake pipe corrosion and security

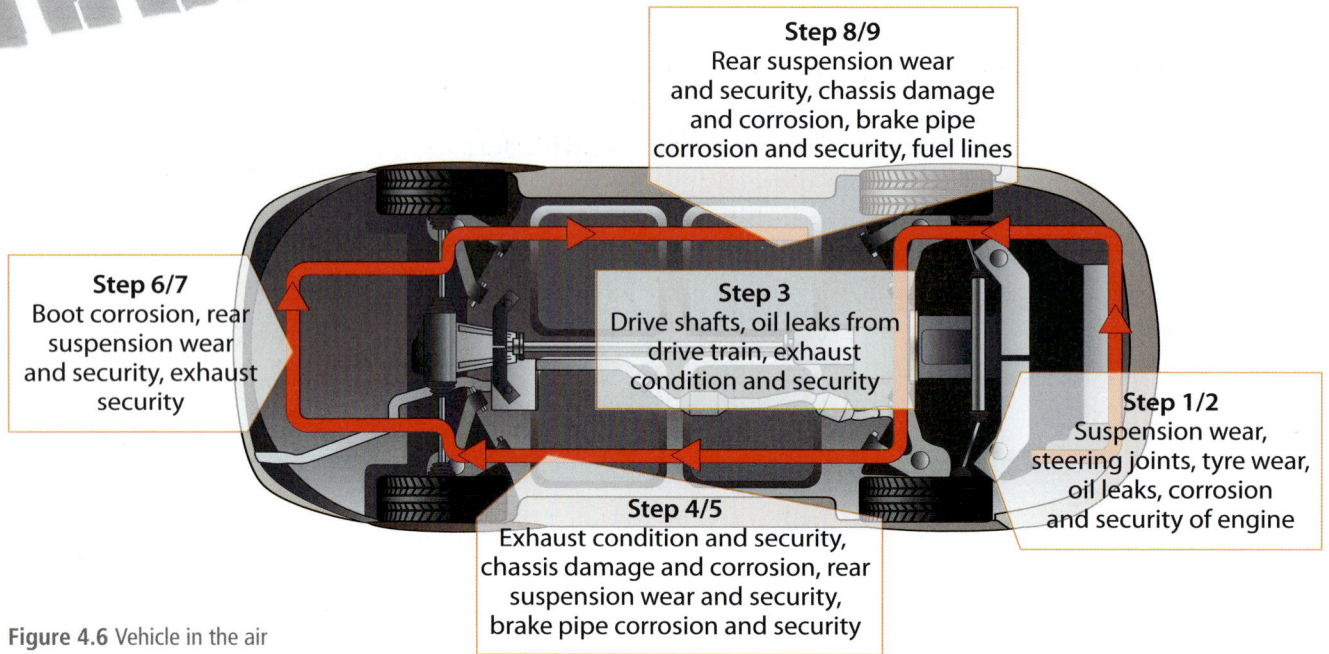

Figure 4.6 Vehicle in the air

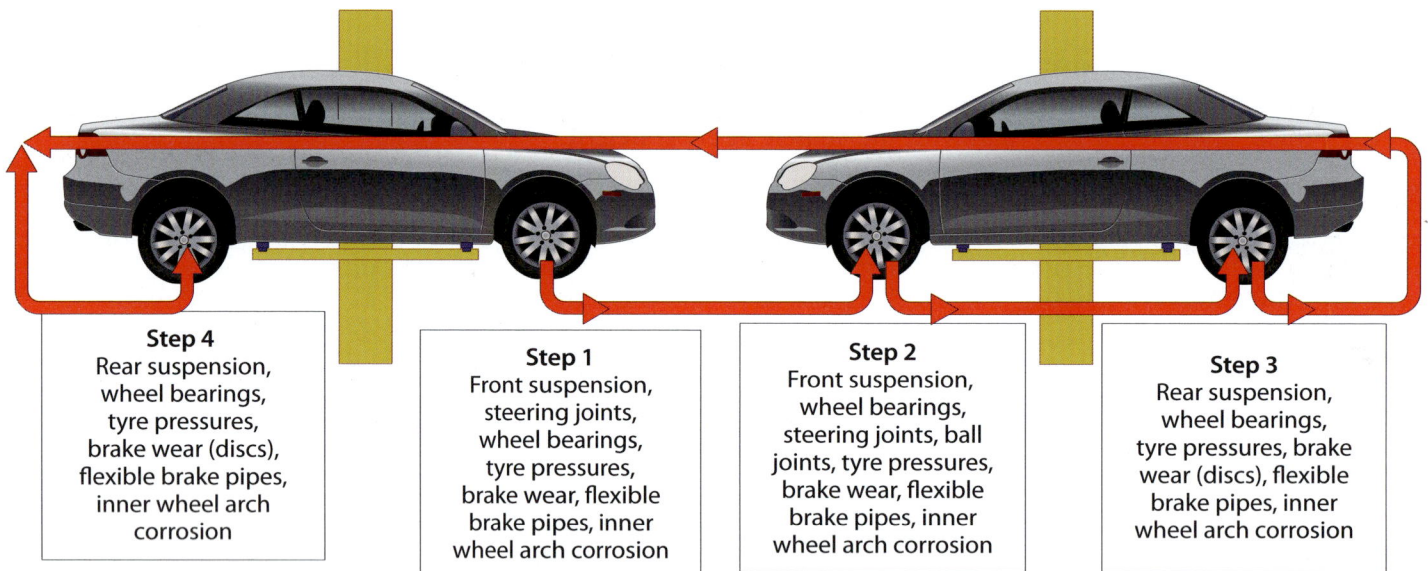

Step 4
Rear suspension, wheel bearings, tyre pressures, brake wear (discs), flexible brake pipes, inner wheel arch corrosion

Step 1
Front suspension, steering joints, wheel bearings, tyre pressures, brake wear, flexible brake pipes, inner wheel arch corrosion

Step 2
Front suspension, wheel bearings, steering joints, ball joints, tyre pressures, brake wear, flexible brake pipes, inner wheel arch corrosion

Step 3
Rear suspension, wheel bearings, tyre pressures, brake wear (discs), flexible brake pipes, inner wheel arch corrosion

Figure 4.7 Vehicle at chest height

Key term

Replenishments – new vehicle parts and fluids used during routine maintenance of vehicles.

Interim service

An interim service is a service between two full services. It requires fewer **replenishments** and less in-depth inspection of vehicle systems. For example, brakes would not need to be inspected.

Full service

A full service or main service is the largest service possible. It requires the replacement of all filters and fluids and an in-depth inspection of brakes.

Table 4.7 shows a summary of the components that require routine maintenance inspection.

Table 4.7 Components and their routine maintenance inspection

Component	Routine maintenance inspection required
Tyres	Wear, condition and pressure
Wheels	Damage, buckling and rim size
Brakes	Wear, fluid level, fluid leaks, corrosion of pipes, condition of hoses, the parking brake and correct adjustment
Suspension and steering	Security of components, wear of joints, suspension damper operation, power steering leaks and level
Electrical	Battery, alternator and drive belts, horn, front and rear wipers
Lighting	Function of side and rear lamps, number plate lamp, headlamps, dip and main beam control, boot lamp (on and off), interior lamps, indicators, hazard lamps, front and rear fog lamps and warning lamps
Engine compartment	Coolant leaks and level, oil leaks and level, washer fluid level, brake fluid level, PAS fluid level, drive belt and battery
Transmission	Clutch operation and adjustment, drive shafts, joints, rubber boots, fluid leaks
Vehicle exterior	Bodywork, paintwork, trim, doors and door locks, wing mirror condition, bonnet release
Vehicle interior	Seats (condition and adjustment), seat belts, driver controls, warning lamps, wing mirror operation

The correct procedures when inspecting, replacing fluids and service items, lubricating and adjusting systems and components

Tyres

Tyres must be able to:

- support the vehicle load
- provide a smooth ride for the occupants
- transmit drive and braking forces to the road surface in all weather conditions.

Inspecting tyres for wear and condition

Checklist				
PPE	VPE	Tools and equipment	Consumables	Source information
• Steel toe-capped boots • Overalls • Latex gloves	• Wing covers • Steering wheel cover • Seat covers • Floor mat covers	• Tread depth gauge • Tyre pressure gauge • Compressor	• Air from compressor	• Tyre pressures

1. Tyre walls and carcass – inspect for bulges, cracks or tears in the walls or distortion of the tyre carcass.

2. Tyre tread cleanliness – make sure no objects are embedded in the tread.

3. Tyre wear – use a tread depth gauge. Tread depth must be a minimum of 1.6 mm across three-quarters of tyre width around the whole circumference.

4. Tyre direction of rotation – if the tyres are rotational, check the arrows of rotation point towards forward rotation of the vehicle.

5. Tyre valves – check to make sure they are not damaged and that dust caps are fitted.

6. Tyre pressures – make sure these are in line with the manufacturer's specifications.

7. Spare tyre – don't forget to check it.

Checking tyre pressures

Check tyre pressures using a pressure gauge on cold tyres.

The pressure will vary depending on the vehicle's load and speed. A vehicle driven at speed with a full load needs a greater tyre pressure than a vehicle that is driven at low speed with a light load. You will find the required pressure values for the inflated tyres in the vehicle handbook.

The emergency spare wheel should be set to the highest recommended pressure for the vehicle. If it needs to be used, air can be let out by the person replacing the wheel, to suit the vehicle use.

Some emergency spare wheels have the pressure setting marked on the wheel itself.

> **Did you know?**
>
> As tyre temperature increases through use, the tyre pressure increases.

> **Did you know?**
>
> A chemical-filled aerosol is used on vehicles without an emergency spare wheel. If a tyre gets a puncture, this chemical forms a seal around the puncture area. The driver can then re-inflate the tyre and get to the nearest tyre specialist for a lasting repair.
>
> The emergency aerosol kit has an expiry date and you will need to check this date during an inspection.

Wheels

A wheel should:

- bear the vehicle load
- provide a sealed mounting for the tyre
- make the vehicle look attractive.

Inspecting wheels

Table 4.8 Wheel inspection

Check	Required inspection result
Rim	Look for damage and any signs of cracks on the inside and outside of the wheel.
Buckled wheel	Spin the wheels by hand and check for the wheel rotating normally.
Size	Make sure that the wheels are the same size on the same axle.

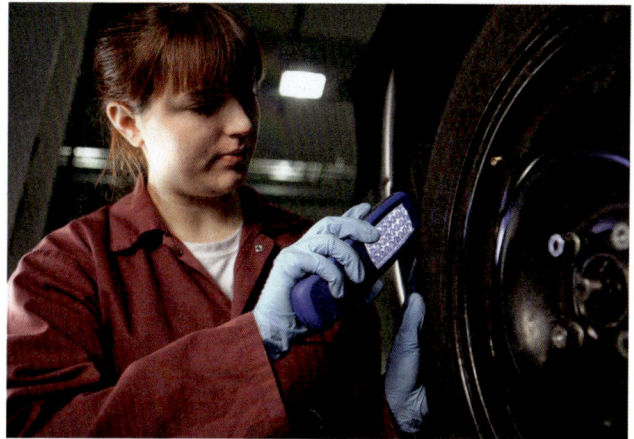

Figure 4.8 Checking a wheel for damage

CHECK YOUR PROGRESS

1 What is the legal requirement for tyre tread depth?
2 What tool is used to measure tread depth?
3 What happens to the tyre pressure as the temperature increases?

Brakes

Brakes should:

- slow the vehicle down
- stop the vehicle
- keep it stationary.

Routine maintenance of the system involves checking for wear, fluid leaks, fluid level, corrosion of pipes, condition of hoses and correct adjustment.

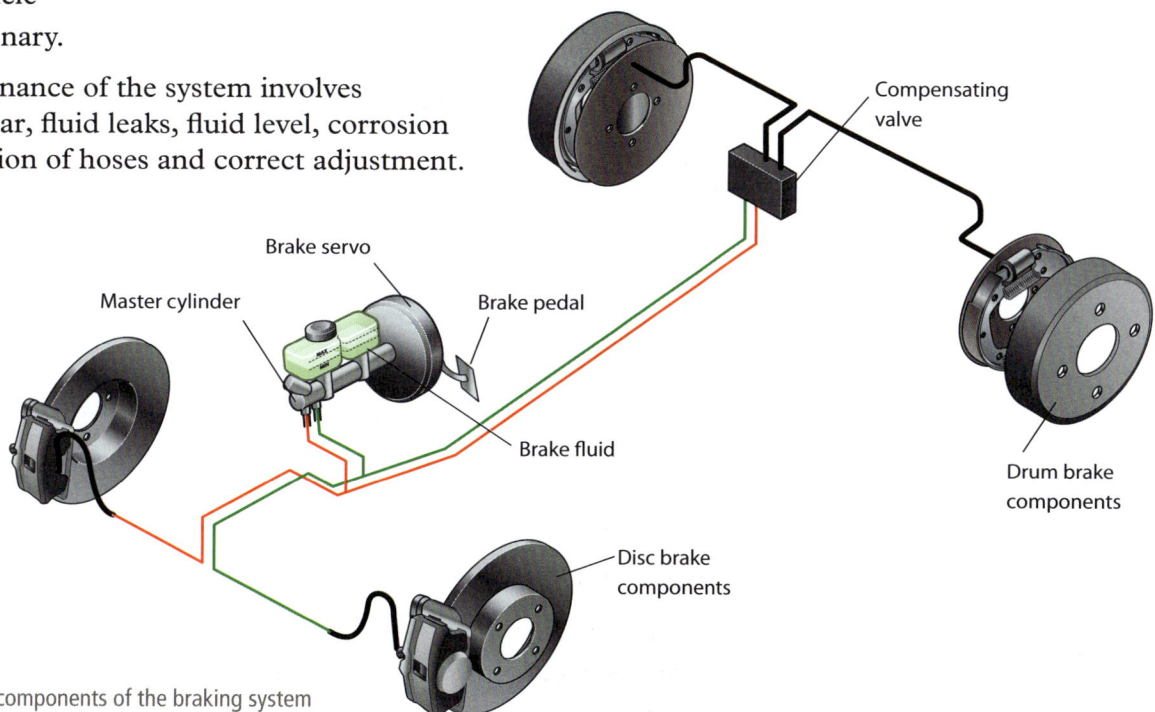

Compensating valve

Brake servo

Master cylinder

Brake pedal

Brake fluid

Drum brake components

Disc brake components

Figure 4.9 The main components of the braking system

Checking for disc brake wear by removing the pads

Checklist				
PPE	**VPE**	**Tools and equipment**	**Consumables**	**Source information**
• Steel toe-capped boots • Overalls • Latex gloves • Dust mask • Safety goggles	• Wing covers • Steering wheel cover • Seat covers • Floor mat covers	• Hoist • Torque wrench • Lever • Bleed tube and bottle • Piston retraction tool	• Brake pads • Brake fluid • Copper grease • Brake cleaner • Squeal-reduction shims (if fitted)	• Minimum brake pad thickness • Brake fluid type

1. Remove the brake fluid reservoir cap.

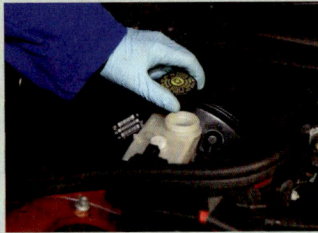

2. Raise the vehicle. Remove the wheel and position the steering to allow access to the disc assembly.

3. Spray brake cleaner on the assembly to remove any dust.

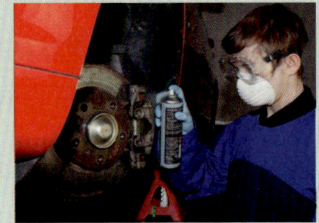

4. Remove the fixings holding the pads in place, then remove and inspect the pads.

5. Fix the bleed pipe and bottle to the bleed nipple on the caliper and slacken the bleed nipple.

6. Use the piston retraction tool to push back the piston and tighten the bleed nut.

7. Smear a small amount of copper grease on the back of the new pads.

8. Inspect the brake linings for wear. Visually check that the brake pads are not greasy and measure the thickness of the brake linings to make sure there will be enough material to last through to the next service.

9. Fit the new pads with any squeal-reduction shims, taking care not to get grease on the linings.

10. Replace the fixings holding the pads in place.

11. Remove the bleed pipe and refit the wheel.

12. Pump the brake pedal until it goes firm.

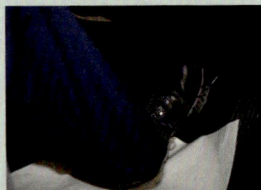

13. Check the brake fluid level and top up if necessary to the maximum level.

14. Carry out the same procedure on the other side of the vehicle on the same axle.

15. Refit the brake fluid reservoir cap.

⚠ Safe working

If you find a brake pad fault on one side of a vehicle, you must replace all the pads on the same axle.

Checking for drum brake wear

Over time, the drum and brake shoes become worn. Inspecting the drum brake involves checking the condition and wear of the brake shoes and drums.

Brake fluid level check

The brake fluid level must always be between the maximum and minimum marks on the reservoir.

An instrument panel warning light (see Figure 4.12) will light up if the level is below the minimum mark. This is activated by a switch in the brake fluid reservoir.

Figure 4.10 Checking the brake shoes

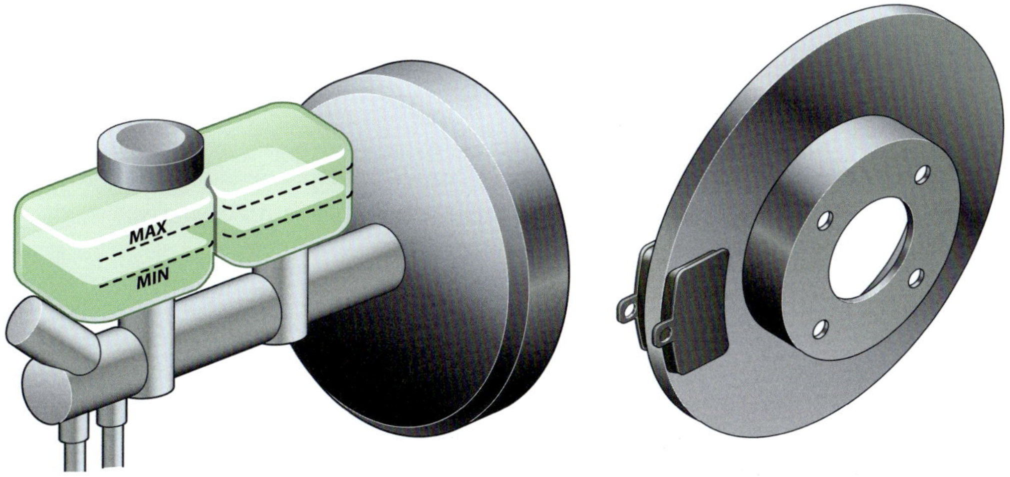

Figure 4.11 The brake fluid level must be between the 'MAX' and 'MIN' indicators

Figure 4.12 Brake fluid level warning indicator light

You must check the brake fluid level during every inspection and the brake fluid must be replaced every two years. This is because most brake fluids are **hygroscopic**.

Brake fluid must be changed at the intervals recommended by the manufacturer. If not, the driver will only become aware that there is a problem when the brakes feel 'spongy' and the car fails to stop.

Key term

Hygroscopic – will absorb water from the air over a period of time which reduces the boiling point and may cause brake failure. For this reason, manufacturers state that brake fluid must be changed on a regular basis (every two years).

Did you know?

If the water content in the brake fluid reaches the 3% mark, the boiling point of the fluid will be reduced and air bubbles will be produced.

Safe working

Avoid spilling or splashing brake fluid on to painted surfaces. This is because it can work like paint stripper and may spoil the surface. If there is a spill, wash it off straight away with water.

Pre-checks

Before you check and adjust the brake fluid level:

- refer to the vehicle manual to check the type of brake fluid used
- make sure you follow all specifications given
- use brake fluid from an unopened container.

Brake fluid safety warning symbols and their meanings are shown in Figure 4.13.

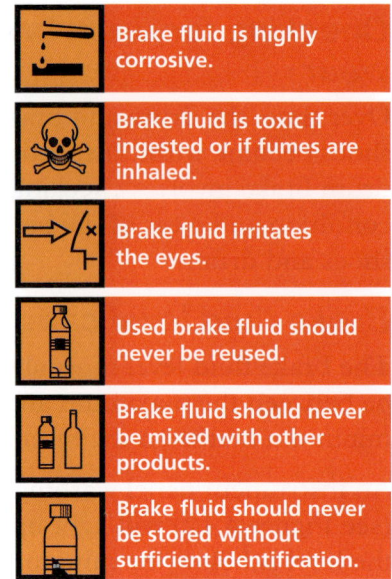

Brake fluid is highly corrosive.

Brake fluid is toxic if ingested or if fumes are inhaled.

Brake fluid irritates the eyes.

Used brake fluid should never be reused.

Brake fluid should never be mixed with other products.

Brake fluid should never be stored without sufficient identification.

Figure 4.13 Brake fluid safety warning symbols

Checking and adjusting brake fluid level

Checklist				
PPE	VPE	Tools and equipment	Consumables	Source information
• Steel toe-capped boots • Overalls • Latex gloves	• Wing covers • Steering wheel cover • Seat covers • Floor mat covers	• Suitable cloth	• Brake fluid	• Brake fluid type

Figure 4.14 Different types of brake fluid

Did you know?

Most light vehicles use DOT 3 or 4 brake fluid. DOT stands for Department of Transport.

1. Find the master cylinder.

2. Check the level of the brake fluid. (You should be able to see this through the reservoir (see Figure 4.11 on p. 91).) The brake fluid should be at the 'full' level.

3. If the brake fluid is below this level, unscrew the cap and add the specified brake fluid until it reaches the full mark.

4. Replace the brake fluid cap and check that it is secure.

Checking the hydraulic circuit for leaks and corrosion

Table 4.9 Hydraulic circuit inspection

Check	Required inspection result
Metal pipes	• No leaks • No corrosion • Securely mounted • Not kinked
Flexible pipes	• Fitted correctly • Not twisted • Not swelled • Not worn or been rubbing against rotating parts

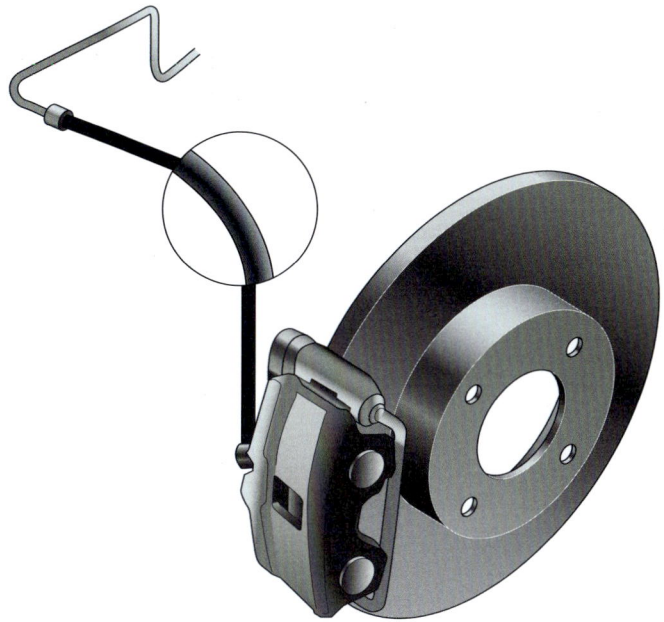

Inspecting the parking brake

Table 4.10 Inspecting the parking brake (handbrake)

Check	Required inspection result
Cables	• The cable slides in its sheath smoothly • No damage to either the cable or the sheath
Travel/wear	• Handbrake between four and six clicks from off to full operation • Handbrake completely releases

Figure 4.15 The correct alignment of the brake hose

Excessive **handbrake travel** may be due to worn linings and/or the automatic adjuster on a drum brake system not working. If the handbrake does not release fully, mechanical linkages may have seized, due to corrosion, or there may be a fault with a handbrake cable.

Key term

Handbrake travel – the amount the lever can be lifted by the driver. The normal rule is that it should not go above four 'clicks' when the handbrake button is not pressed in during lifting.

Figure 4.16 Checking the handbrake cable

Did you know?

An automatic wear take-up mechanism pushes the shoes closer to the drum as the brake linings wear. You should check the operation of the automatic wear take-up mechanism and the wheel cylinders at the same time as the brake drum.

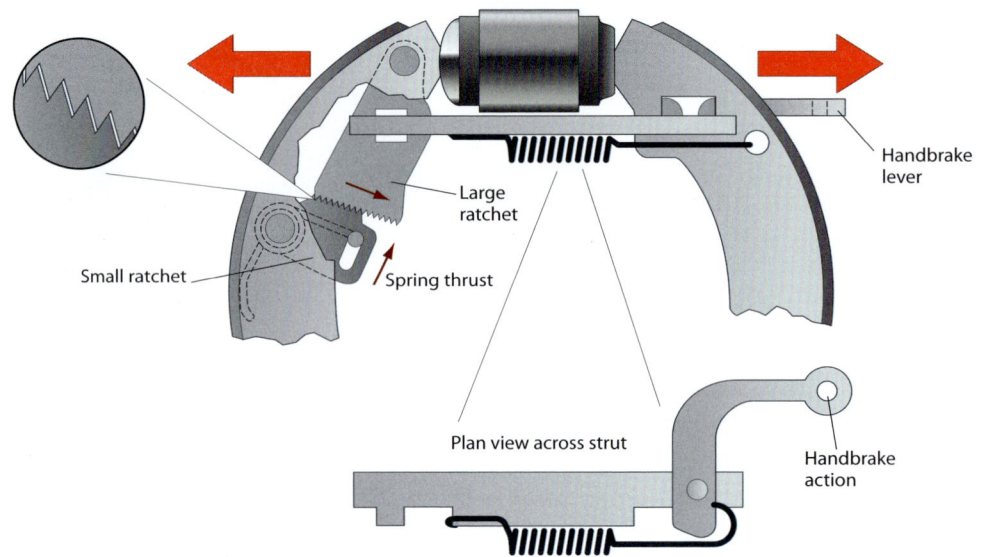

Figure 4.17 Checking the automatic wear take-up system

When inspecting the automatic wear take-up system, you need to check that:

- it is positioned correctly
- it is clean
- when the brakes are applied, the ratchet mechanism holds the shoes in an outward position.

You carry out this check by spinning the drum and watching the drum's speed. If the drum does not slow down, this shows that there is no resistance from the brake shoes. This tells you that the brakes are not correctly adjusted.

CHECK YOUR PROGRESS

1 What is meant by hygroscopic?
2 Name two checks on the braking system.
3 If you find a brake pad fault on one side of a vehicle, what must you do?

Suspension and steering system

The suspension system provides a smooth ride for the driver, passengers and the load. The steering system of a vehicle allows the driver to control the direction of the vehicle. Both systems tend to have components that become insecure and suffer joint wear.

The main areas that you need to check for wear and damage are:

- swivelling and pivoting joints
- springs
- dampers
- linkages.

Checking the suspension system

Table 4.11 Checking the suspension system

Check	Required inspection result	
Vehicle ride (trim) height – place a tape measure from the hub centre to the base of the wheel arch	The measurements need to be the same on each axle and within the manufacturer's recommendations	
Dampers – vehicle on the ground	The vehicle should return to its natural position after pushing down on the suspension. The vehicle should bounce one and a half times	
Dampers – underneath the vehicle with an inspection light	No leaks or corrosion	
Suspension springs	Not broken and are located correctly in their housings	
Suspension bushes – using a lever or pry bar	No wear, damage or splits	
Suspension joints – using a lever or pry bar	No excessive wear	
Wheel bearings	No excessive movement or rumbling	

Checking the steering system

Table 4.12 Checking the steering system

Check	Required inspection result	
Steering column – push the steering wheel at right angles to the column	No wear when force applied upwards and downwards Steering secure	
Steering column – push the steering wheel in line with the column	No wear when force applied forwards and backwards Steering secure	
Steering noises – rotate the steering wheel from lock to lock with the engine running	No abnormal noises and smooth operation of the power assisted steering (PAS) system	
Steering arms – vehicle at chest height	No excessive movement in the steering arms	
Steering rack	Securely mounted to the body or subframe	

Check	Required inspection result	
Steering rack gaiters – squeeze them to check if they have any power steering fluid in them	No splits or leaks	
Locking devices and fixings	Present and secure	
Power assisted steering (PAS) system visual check	No leaks	
PAS hoses	Hoses are secure, not damaged or chafing and are positioned correctly	
PAS fluid level	Oil level reading maximum on dipstick when engine is cool	

Checking and adjusting the PAS fluid levels

Before you check and adjust the PAS fluid level, make sure:

- the engine is cool
- the handbrake is on
- the vehicle is on level ground
- the gears/transmission are/is in neutral.

Checking and adjusting the PAS fluid levels

Checklist				
PPE	VPE	Tools and equipment	Consumables	Source information
• Steel toe-capped boots • Overalls • Latex gloves	• Wing covers • Steering wheel cover • Seat covers • Floor mat covers	• Suitable cloth	• PAS fluid	• None required

1. Unscrew and remove the cap to the power steering reservoir. The cap normally has a dipstick attached.

2. Wipe off the dipstick and replace the cap.

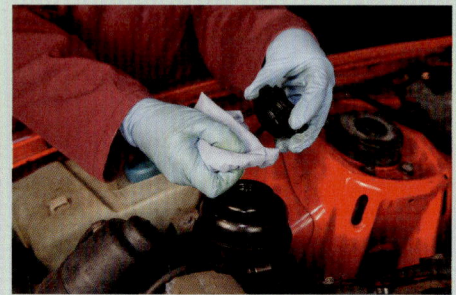

3. Remove the cap and inspect the level of the fluid on the dipstick.

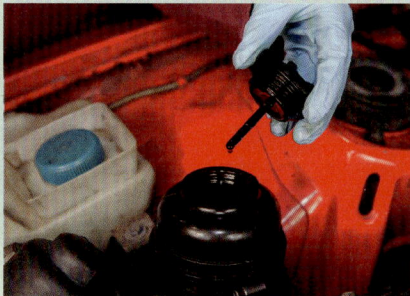

4. Make sure you read the correct marking on the dipstick as specified in the vehicle handbook or manual.

5. Top up the system if the fluid has dropped below the marking on the dipstick.

⚠ **Safe working**

Overfilling will result in fluid spraying out of the top of the reservoir and onto the engine and other components.

CHECK YOUR PROGRESS

1 Name two checks on suspension dampers.
2 Name two checks before checking the power steering system level.
3 Name two tests on the power steering system.

Electrical system

The electrical systems on a vehicle include:

- the battery
- the alternator
- the warning lamps
- the horn
- the front and rear wipers.

Battery

The battery is the main source of power for all the vehicle's electrical and electronic systems. See p.286 for more information about batteries. There are two main types of battery:

- maintenance-free (MF) batteries
- batteries which require routine maintenance.

You must check the label on the top of the battery before any routine maintenance. It will state 'maintenance free' if it does not require maintenance.

Figure 4.18 Battery charge warning light

Checking the battery

Table 4.13 Checking the battery

Check	Required inspection result
Battery casing (bottom of battery) and the cover (top of battery)	No cracks or breakages
Battery cover (top of battery)	Clean and has no water or dirt on it. Wipe clean any deposits from the battery cover.
Battery hold-down mounting	Must be secure – use sockets to tighten any loose mountings
Battery terminals	Well connected and secure, no corrosion
Battery cables	Not frayed – fully insulated

Figure 4.19 shows all the things you should be looking for when checking a battery.

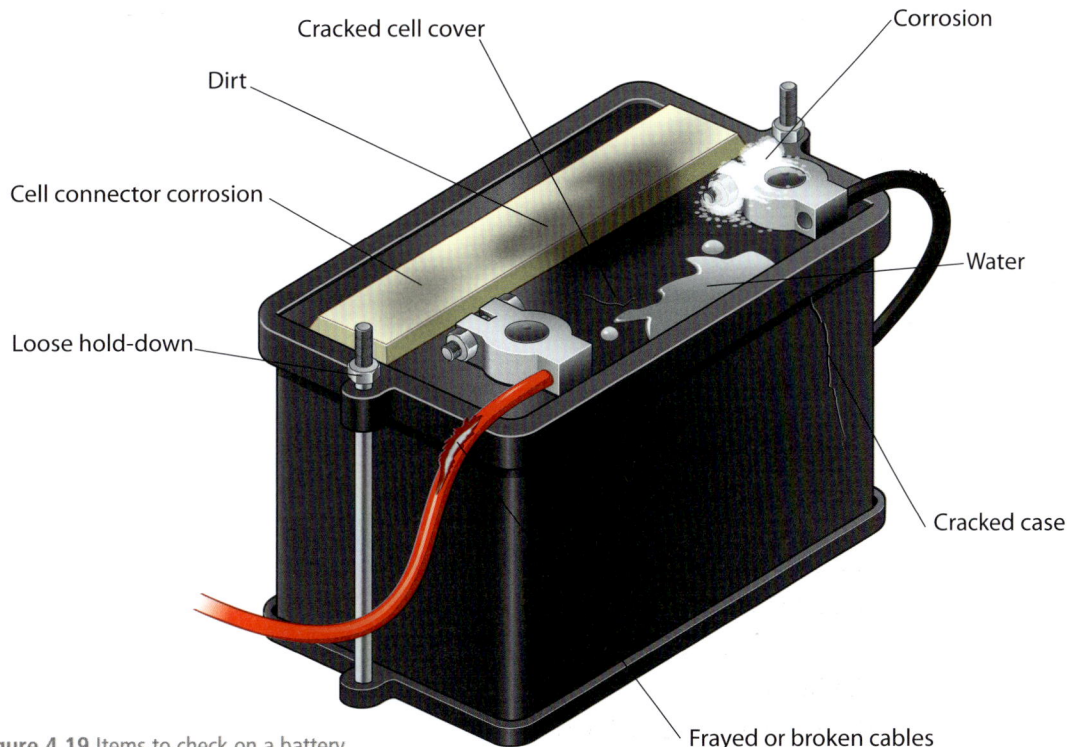

Cracked cell cover
Dirt
Corrosion
Cell connector corrosion
Water
Loose hold-down
Cracked case
Frayed or broken cables

Figure 4.19 Items to check on a battery

Checking the alternator and drive belt

Checks to carry out on the alternator and drive belt include:

- **visually** inspecting:
  - the belt for signs of ageing
  - and for cracks in the rubber
  - checking the alternator – the warning light will go out when the vehicle is running
- **listening**:
  - to confirm if the belt is slipping (often slipping belts will make a loud squealing noise which is more notable on initial start up when cold)
- **functionally inspecting**:
  - the tightness of the belt (if the belt is too loose it will slip and not drive the fan or auxiliary equipment correctly, which could lead to the engine overheating)
  - the security of the electrical connections into the alternator.

Adjusting the alternator drive belt

You can usually adjust the belt manually by following the manufacturer's information systems.

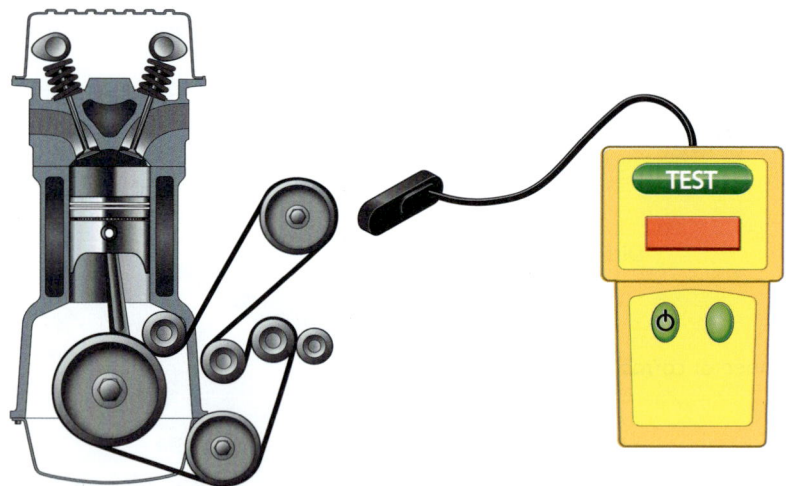

Figure 4.20 Measuring tension

Key term

Deflection – the amount of distance that the belt moves when pressure is put on it.

The tension must be checked on the longest section of the belt where the **deflection** is at its greatest. The effects of incorrect tension are shown in Table 4.14.

Table 4.14 The effects of incorrect tension in the fan belt

Belt too slack	Belt too tight
• Excessive squealing noise during acceleration • Reduced output of the alternator due to slippage of the belt	• Extra tension, causing premature failure of alternator bearings

The horn

The horn warns other road users and pedestrians a vehicle is approaching. Pressing the horn will check its operation. The 'aural' check will be to ensure there is an unbroken loud tone.

The wiper system

The wiper system uses a rubber blade to move dirt particles and water that land on the windscreen away from the swept area. See Chapter 8 (page 296) for more information. There are two types of blade, as shown in Figure 4.21.

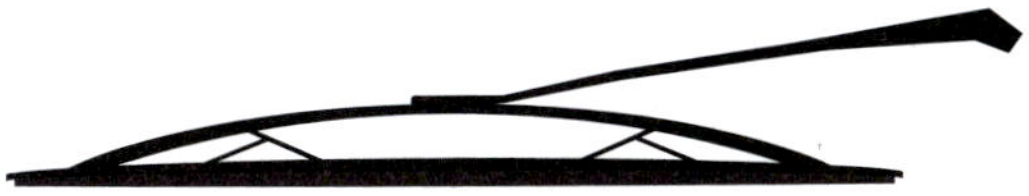

See Chapter 8 (page 296)

Checking the wiper system

During the service, you are required to complete the wiper checks shown in Table 4.15.

> **Safe working**
>
> Make sure there is no one under the bonnet when checking the operation of the horn. The sound may startle the person working in the engine compartment.

Conventional blade – incorporates a metal frame

Flexible blade – incorporates a stiffener which adapts to the different types of windscreen

Figure 4.21 Types of wiper blades

> **Did you know?**
>
> The flexible type of blade has a longer service life than the usual blade. The different types of wiper blade are not compatible and cannot be mixed.

Table 4.15 Wiper checks

Check	Required inspection result
Wiper blades	Must not be split or cracked
Wiper operation	Need to operate at all speeds
Wiper travel	Must not travel past the screen or touch each other
Wiper condition	Must clear the screen completely and not leave a thin film of water on the screen
Washer jets	Must not be blocked and the washer fluid must reach the whole screen

Washer fluid

You need to top up the washer fluid when carrying out routine maintenance. You will do this by adding the fluid to the washer bottle until it reaches the top of the filler neck.

> **Did you know?**
>
> Washer fluid is a mixture of an alcohol-based solvent and water. This prevents freezing in cold weather.

CHECK YOUR PROGRESS

1 Name the two types of motor vehicle battery.
2 What could fail if an alternator drive belt was too tight?
3 Name three checks on the wiper system.

The lighting system

The lighting system allows the driver to see and the vehicle to be seen. See Chapter 8 (pages 308–314) for more information.

Checking the lighting system

You should check that the front, rear and interior lights work correctly. Figure 4.22 shows the front lights on a light vehicle and Figure 4.23 shows the rear lights.

When checking lights you should also think about:

* All forward-facing continuous use lights must show white light only.
* All rear-facing continuous use lights must show red light only.

The reverse light is white because it indicates that the vehicle is moving towards you. The flashing indicator lights and hazard warning lights are amber so they are different from the front and rear lights.

Did you know?

If the brake or tail lamp bulb has failed, take care to fit the new bulb correctly. If you don't, the brake light will come on during tail lamp use and the tail lamp will come on during brake light use.

Figure 4.22 Front lights

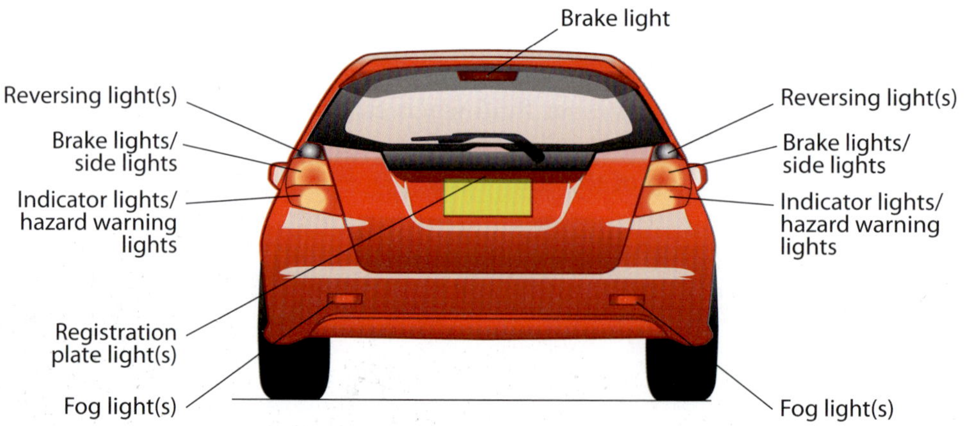

Figure 4.23 Rear lights

Figure 4.24 shows the interior lights in a light vehicle.

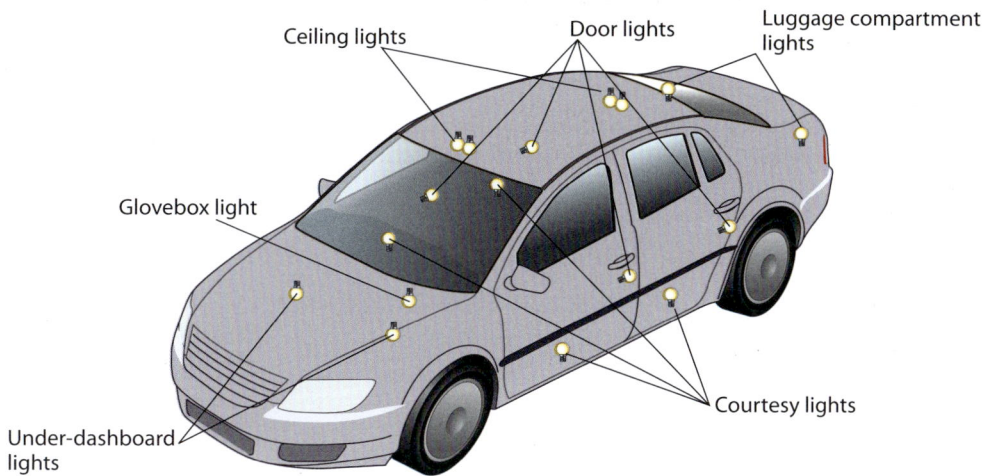

Figure 4.24 Interior lights

While you are conducting the vehicle light checks, also make sure that you check that the warning lights on the interior instrument panel are working correctly (see Figure 4.25). You need to check the:

- battery charge warning light
- oil pressure warning light
- engine temperature warning light
- STOP warning light
- brake circuit fault warning light
- seat belt reminder light
- hazard warning light indicators, also with buzzer function
- parking brake warning light
- dipped headlight indicator
- front and rear fog lamp light
- main beam warning light.

Did you know?

Most vehicles have a self-test procedure on the instrument panel. This allows you to check all the warning and indicator lights.

Figure 4.25 Warning lights on the interior instrument panel

Other checks that you need to carry out on lights are shown in Table 4.16.

Table 4.16 More checks on lights

Check	Required inspection result
Security	The headlight and rear light units are secure
Water ingress	Water has not entered the system
Reflectors	In good condition
Plastic or glass lenses	Not broken, cracked or obscured
Reflector lenses	Not cracked or deteriorated
Bulb contacts	Not corroded

Safe working

Always follow all safety instructions. Be careful when handling bulbs as they are very hot when in use and before they cool down.

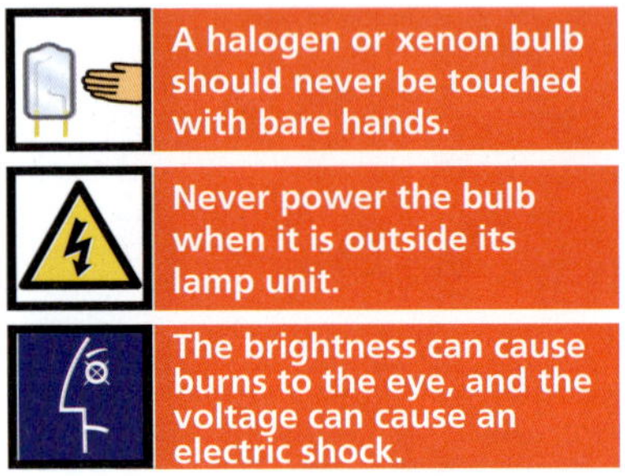
A halogen or xenon bulb should never be touched with bare hands.
Never power the bulb when it is outside its lamp unit.
The brightness can cause burns to the eye, and the voltage can cause an electric shock.

Figure 4.26 Safety symbols for the lighting system

For the safety symbols you will see when servicing lights, see Figure 4.26.

Light bulbs contain a gas around the metal filament inside the sealed glass bowl. This is to prevent the filament burning out. Most light bulbs contain a halogen gas, but xenon is another type of gas that can be used. A xenon light bulb gives a bluer, brighter light. See Chapter 8, pages 309–310, for more information on bulbs.)

CHECK YOUR PROGRESS

1. State three front lights and three rear lights that require checking.
2. Why do you need to take extra care fitting a brake light bulb?
3. Name five warning lights.

Did you know?

Water expands as it freezes. If the coolant freezes in the spaces within the engine it will expand and the engine block may crack. Leaks will happen and it will be very expensive to repair.

Safe working

Always allow an engine to cool down before adjusting coolant levels. This is because at higher temperatures the coolant is under pressure. If you open the pressure cap, the engine coolant will expand and force boiling liquid over you.

Engine compartment

The engine provides the source of power for the vehicle. Servicing the engine will ensure that it lasts longer and provides the best performance.

Coolant level check

The coolant level should be checked regularly because it helps the engine to maintain a constant temperature.

- During hot weather, engine coolant removes heat from a vehicle. This prevents the engine from overheating.
- When the weather is very cold, coolant prevents the engine from freezing.

Before you check and adjust the coolant level:

- Check the specifications given in the vehicle manual and service manual. This information will tell you at what mileage to change the coolant and if it has been changed recently.
- Park the vehicle on a level surface.
- Make sure the vehicle has been at rest for a considerable time and is cool.

- Antifreeze is toxic and must be disposed of properly at a fluid recycling centre.
- Avoid skin contact.
- Avoid spilling it on painted finishes as it attacks paintwork.

Figure 4.27 A coolant pressure cap

Checking the coolant level

Checklist				
PPE	VPE	Tools and equipment	Consumables	Source information
• Steel toe-capped boots • Overalls • Latex gloves	• Wing covers • Steering wheel cover • Seat covers • Floor mat covers	• Suitable cloth • Measuring jug	• Coolant (water and antifreeze mix)	• Coolant capacity • Ratio of water to antifreeze mix

1. Open the bonnet.

2. Find the coolant reservoir.

3. Make a note of the coolant 'cold' or 'cool' level.

4. If the level is below cold/cool, you will need to add coolant.

5. First use a suitable cloth to wipe around the filler cap to prevent dirt entering the system.

6. Remove the coolant pressure cap.

7. Add the coolant and watch to see if the level increases. If it does not there may be a leak.

8. If there is a leak, repair it before continuing.

9. Replace the cap securely.

The life span of antifreeze is usually about two years. This means that every two years the coolant should be drained and replaced with a new mixture of equal parts of water to antifreeze (50:50). Over a period of time, antifreeze loses its strength and the freezing point of the coolant will be at a higher temperature.

Figure 4.28 Tools and equipment for engine oil and filter replacement

Engine oil and filter replacement

Engine oil cools and lubricates the components of a vehicle's engine. Over a period of time, the oil will gather particles of dirt and debris from engine wear. These particles can get between moving parts and cause more wear and even engine failure. At the same time, the oil becomes diluted with unburnt fuel.

Therefore, the engine oil and oil filter must be changed regularly following the manufacturer's specifications. This is to ensure the engine lasts for as long as possible.

Replacing the engine oil and filter

Checklist				
PPE	VPE	Tools and equipment	Consumables	Source information
• Steel toe-capped boots • Overalls • Latex gloves	• Wing covers • Steering wheel cover • Seat covers • Floor mat covers	• Hoist • Oil filter • Wrench • Oil drainer • Exhaust extraction	• Oil • Oil filter • Sump washer	• Oil capacity

1. Open the bonnet. Remove the oil filler cap.

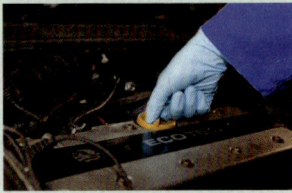

2. Lift the vehicle safely. Position the oil drainer under the sump plug and filter.

3. Remove the sump plug using the correct socket.

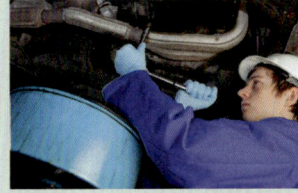

4. Remove the filter using the oil filter wrench. Allow the oil to drain fully.

5. Wipe clean the sump bung and mating surface. Replace the bung with a new washer.

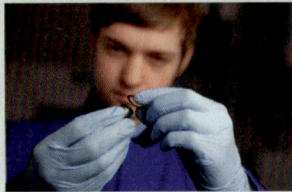

6. Wipe clean the filter mating surface which has the rubber seal attached. Apply a thin layer of new engine oil to the engine oil filter seal before installing the engine oil filter.

7. Tighten the oil filter by hand until it touches the mating surface. Then tighten the filter by two-thirds of a turn.

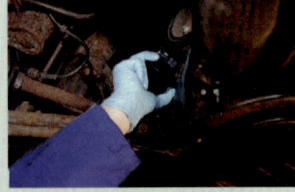

8. Lower the vehicle safely ensuring the drainer is clear from underneath the vehicle.

9. Fill the engine with new oil, using only the amount stated by the manufacturer.

10. Replace the oil filler cap and dispose of any oil-soaked gloves used for PPE.

11. Run the engine for five minutes using the exhaust extraction equipment. (Do not rev the engine as the oil needs to circulate in the system.) Check for any oil leaks from under the vehicle.

12. Check the oil level using the dipstick.

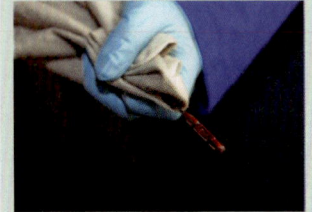

```
Open bonnet ◄──────────────┐
     │                      │
     ▼                      │
Remove dipstick        Run engine and leave
     │                   5 minutes
     ▼                      ▲
Wipe dipstick               │
     │                 Replace oil cap
     ▼                      ▲
Insert dipstick             │
     │                 Top up oil level with
     ▼                 recommended lubricant
Remove dipstick             ▲
     │                      │
     ▼                 Remove oil cap
Oil between MAX and MIN     ▲
and as near to MAX as   NO  │
possible? ──────────────────┘
     │ YES
     ▼
Insert dipstick
     │
     ▼
Close bonnet
```

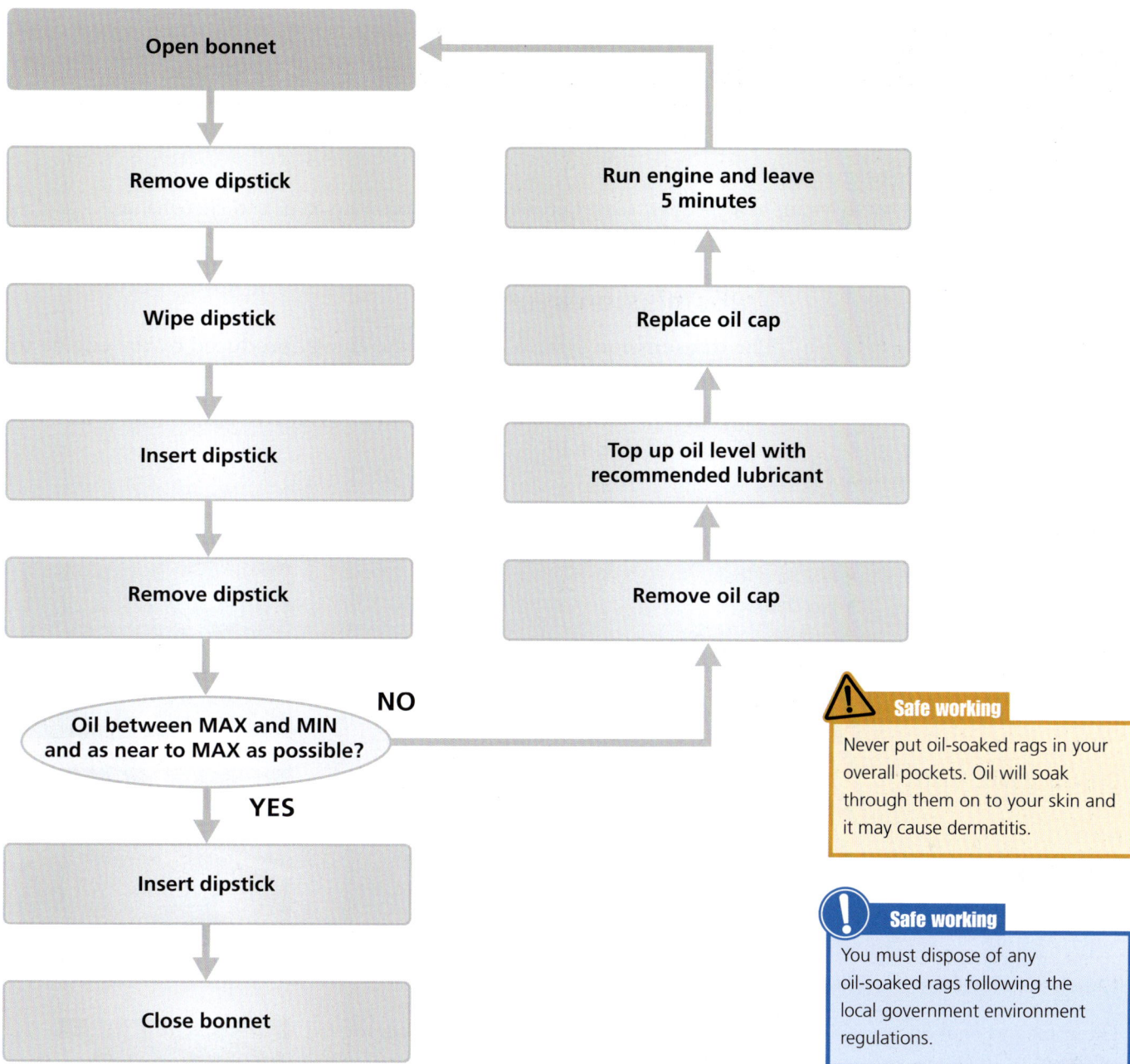

Figure 4.29 Checking engine oil level

Safe working
Never put oil-soaked rags in your overall pockets. Oil will soak through them on to your skin and it may cause dermatitis.

Safe working
You must dispose of any oil-soaked rags following the local government environment regulations.

It is both your and your employer's responsibility to follow the local government environmental regulations when you dispose of old engine oil filters.

Other engine compartment checks

There are other checks which have been described earlier in this chapter that are carried out at the same time as the coolant level check and oil filter change. The brake fluid level, power assisted steering fluid level and the washer bottle checks are all carried out in the engine compartment. These checks need to be carried out to make sure the maintenance inspection is completed as efficiently as possible.

1 What happens to water as it freezes?
2 Why does oil need to be changed?
3 Why should you not rev an engine straight after filling it up with new oil following an oil and filter change?

Transmission system

The transmission system transfers the energy produced by the engine to the vehicle's wheels. See Chapter 7 for more information.

Inspection of the transmission system involves five main checks, as outlined in Table 4.17.

Table 4.17 Transmission system checks

Check	Required inspection result	
Transmission and axle casing	• No fluid leaks • No damage • Secure	
Drive and prop shafts	• No excessive play • No gaiters split	
Clutch fluid level	Fluid must be on maximum level	
Gearbox and final drive oil levels	Oil must trickle from the level plug when it is removed	
Wheel bearings	• No play • No noises	
Clutch operation	Smooth take up of drive and gear operation	

Clutch fluid

Clutch fluid is used to:

- lubricate the internal parts of the hydraulic clutch operation assembly
- reduce friction in the hydraulic system
- assist in smoother gear changing by providing a progressive fluid pressure in the hydraulic operation assembly.

It must be changed every two years as this is essentially brake fluid, which is hygroscopic (see page 91).

Clutch fluid level check

Checklist				
PPE	VPE	Tools and equipment	Consumables	Source information
• Steel toe-capped boots • Overalls • Latex gloves	• Wing covers • Steering wheel cover • Seat covers • Floor mat covers	• Suitable rag	• Clutch fluid (brake fluid)	• Clutch fluid (brake fluid) type

1. Turn the engine off.	2. Locate the clutch fluid reservoir and remove the cap.	3. Check the fluid level. If the fluid is not to the top, add more fluid.	4. Replace the clutch fluid cap and check that it is secure.

Vehicle body exterior and interior inspection

An inspection of the vehicle body exterior and interior should be completed at regular intervals in line with the manufacturer's service recommendations. This is because the exterior body is subject to corrosion if there is any damage to paintwork. The interior is subject to wear and tear.

Body exterior inspection

A vehicle body exterior inspection involves checking the:

- registration plate
- vehicle identification number (VIN)
- doors
- locks
- hinges
- bonnet
- body panels
- windows.

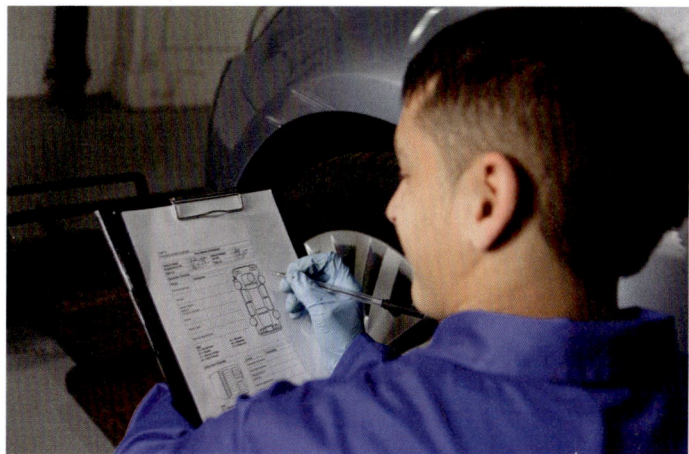
Figure 4.30 The vehicle body should be inspected regularly

Table 4.18 Body exterior inspection

Check	Required inspection result
Registration plate	Matches the job card
Vehicle identification number (VIN)	Matches the job card
Doors	Open and close securely from inside and outside
Locks	Lock and unlock using key or remote control
Door hinges	No excessive movement
Bonnet release	Operates easily and smoothly
Body panels	No scratches, marks or dents
Windows	No cracks or stone chips

Maintenance of the exterior body

- Lubricate the bonnet release mechanism.
- Lubricate the door locks.
- Lubricate the door hinges.

Chassis inspection

When carrying out an under body chassis inspection, look for signs that the vehicle has been lifted or supported incorrectly. This can be the cause of many faults.

Table 4.19 Chassis inspection

Check	Required inspection result	
Under-seal	Not damaged or cracked	
Chassis rails	Straight and smooth	
Floor pan	Not damaged and corroded	

Vehicle interior inspection

Table 4.20 Vehicle interior inspection

Check	Required inspection result	
Seat belts – particularly around anchorages, buckles and loops	No webbing cuts or obvious signs of fraying	
Seat belt operation	• Smooth when pulled • Lock when pulled (do not pull too sharply as this may damage the locking mechanism and the belt may not work properly in the event of an accident. This is a directive from VOSA)	
Mirrors	• Secure • Not cracked	
Seats	• Secure • Hinges lock in place	

Items on the vehicle to be checked following routine maintenance

Following the routine maintenance of a vehicle, you should check that you have:

- carried out the customer's instructions in full
- removed all grease and oil marks from the interior and exterior of the vehicle
- made sure that the clock is set and radio stations are restored if the battery has been disconnected
- washed the vehicle and removed all VPE
- checked the vehicle for any damage that may have been a result of poor workmanship
- completed all paperwork.

The paperwork that you need to complete as part of your inspection and service procedures includes:

- routine maintenance inspection sheet
- in-vehicle service book
- job card.

Customer satisfaction

If there is additional work to be carried out, you need to let the customer know. Discuss a convenient time that suits the customer to return the vehicle for the work to be completed. It is good practice to:

- provide routine maintenance date stickers letting the customer know when the inspection is due
- ask the customer if they would like to pre-book inspection
- ask the customer to complete a customer satisfaction survey. This will help you to gain feedback and improve on your level of service.

CHECK YOUR PROGRESS

1 How is the gearbox oil level checked?
2 What items require lubricating on the exterior body during routine maintenance?
3 What could be a cause of chassis damage?

1 Which one of the following is not a non-structural panel?

  a bonnet
  b boot
  c chassis
  d door

2 VIN is short for:

  a very important number
  b vehicle identification number
  c very important notice
  d vehicle identification notice

3 Which one of the following assessment methods would you use to test a horn?

  a visual
  b taste
  c smell
  d aural

4 Which one of the following statements is true?

  a tyre pressure increases as the tyres get warmer
  b tyre pressures remain constant as the temperature rises
  c tyre pressures reduce as the tyres get warmer
  d tyre pressures increase as the tyres get colder

5 Hygroscopic means:

  a absorbs air
  b absorbs brake fluid
  c absorbs water
  d absorbs heat

6 Which of the following checks are associated with suspension routine maintenance?

  a checking dampers
  b checking springs
  c checking swivelling or pivoting joints
  d all of the above

7 Which of the following checks are not completed on external lights?

  a headlamp operation
  b brake warning lamp operation
  c indicator operation
  d fog light operation

8 Which of the following needs to be observed when carrying out an oil change?

  a engine is cool
  b all tools at hand
  c vehicle is on a flat surface
  d all of the above

9 Which one of the following would be incorrect routine maintenance on a vehicle?

  a fill the clutch fluid level to the maximum in the reservoir
  b check wheel bearings for wear
  c fill the gearbox to the top of the casing
  d fill the differential until the oil just trickles out of the level plug

10 Which one of the following checks would be incorrect routine maintenance on the vehicle body exterior?

  a check the operation of door locks
  b lubricate the bonnet catch
  c check the hinges for stone chips
  d check the VIN plate matches the job card

GETTING READY FOR ASSESSMENT

The information contained in this chapter, as well as continued practical assignments in your centre or workplace, will help you to prepare for both the end-of-unit tests and diploma multiple-choice tests.

You will need to be familiar with:

- Safe working on light vehicle construction and maintenance
- The body types for a range of vehicles
- The engine and driveline configurations for a range of vehicles
- The main body parts found on light vehicles
- Inspecting systems and components during basic routine vehicle maintenance
- Appropriate sources of information required to carry out basic routine vehicle maintenance
- The main components and systems that require routine maintenance on a modern vehicle
- The correct procedures when inspecting, replacing fluids and service items, and lubricating and adjusting systems and components

You now need to apply the knowledge you have gained in this chapter in your day-to-day working activities. For example, you are required to know and understand the main components and systems that require routine maintenance on a modern vehicle. This chapter has given examples of how to do this by outlining what is involved when carrying out routine maintenance, which includes measuring and testing. You have also been made aware that tests can be carried out on the ground, in the air and at chest height, and that the two main types of routine maintenance are the interim service and full service. You have also gained an awareness of the main components and their routine maintenance inspection.

You should now use this knowledge in your workshop by working safely, using a checklist to ensure you have the correct PPE, VPE, tools and equipment, consumables and sources of information for the task.

This chapter has given you an introduction and overview to light vehicle construction and maintenance, providing you with the basic knowledge that will help you with both theory and practical assessments. Before you try a theory end-of-unit test or multiple-choice test, make sure you have reviewed and revised any key terms and read all the questions carefully. Take time to digest the information so that you are confident about what the question is asking you. With multiple-choice tests, it is very important that you read all of the answers carefully, as it is common for two of the answers to be very similar, which may lead to confusion.

For practical assessment, it is important that you have had enough practice and that you feel that you are capable of passing. Before you begin a task make sure you have the correct PPE, VPE, tools and equipment to hand and that you have a plan to follow, along with the equipment and information you need to complete the task. You could, for example, be asked to assist in replacing the engine oil and filter on a modern light vehicle. For this you would need to wear steel toe-capped boots, overalls and latex gloves. You would need to protect the vehicle using wing covers, seat, steering wheel and floor mat covers. To carry out the task you will need a hoist, oil filter, wrench, oil drainer and exhaust extraction. Consumables include oil, an oil filter and a sump washer. Before commencing the task you should refer to source material for oil capacity levels.

Read the chapter again to confirm you have completed and understood any tasks. This will help you to be sure that you are working correctly and will help you to avoid problems developing as you work.

When you are doing any practical assessment, always work safely throughout the task. Make sure that you observe all health and safety requirements and that you use the recommended personal protective equipment (PPE) and vehicle protection equipment (VPE). When using tools, make sure you are using them correctly and safely.

Good luck!

5 Light vehicle engine systems

This chapter will give you an introduction to light vehicle engine systems, components and operation. It provides the basic knowledge that will help you with both theory and practical assessments.

It will help you to identify the main components used in both spark ignition (petrol) and compression ignition (diesel) engines. It will also introduce the basic operating principles of light vehicle engine mechanical, lubrication, cooling, ignition, fuel and exhaust systems. Finally, it will help you plan a systematic approach to light vehicle engine maintenance and inspection.

This chapter covers:

- Safe working on light vehicle engine systems
- Petrol and diesel engine systems
- Spark ignition and compression ignition engine operation
- Liquid cooling and lubrication systems
- Spark ignition systems
- Fuel systems
- Exhaust systems
- Routine engine maintenance

The engine is the heart of the vehicle and provides the energy to run all other systems. You will come across many hazards when you work with engine systems. You must take care during maintenance and repair to avoid dangers to yourself and others which could result in injury, or even death. You must always use personal protective equipment (PPE) and you also need to think about the possibility of crush or bump injury. You will also come into contact with chemicals such as lubrication oils, grease, coolant and cleaners containing solvents. Make sure that your selection of PPE will protect you from these hazards.

Personal Protective Equipment (PPE)

Safety goggles reduce the risk of small objects or chemicals coming into contact with the eyes.

Overalls provide protection from coming into contact with oils and chemicals.

Safety gloves provide protection from oils and chemicals. They also protect the hands when handling objects with sharp edges.

Barrier cream protects the skin from old engine oil, which can cause dermatitis and may be carcinogenic (a substance that can cause cancer).

Safety boots protect the feet from a crush injury and often have oil- and chemical-resistant soles. Safety boots should have a steel toe-cap and steel mid-sole.

Safety helmet protects the head from bump injuries when working under cars.

To reduce the possibility of damage to the car, always use the appropriate vehicle protection equipment (VPE):

Wing covers

Seat covers

Steering wheel covers

Floor mats

If appropriate, safely remove and store the owner's property before you work on the vehicle. Before you return the vehicle to the customer, reinstate the vehicle owner's property. Always check the interior and exterior to make sure that it hasn't become dirty or damaged during the repair operations. This will help promote good customer relations and maintain a professional company image.

Vehicle Protective Equipment (VPE)

Safe Environment

During the repair or maintenance of light vehicle engine systems you may need to dispose of certain waste materials such as engine oil, oil filters and coolant (antifreeze). Under the Environmental Protection Act 1990 (EPA), you must dispose of them in the correct manner. They should be safely stored in a clearly marked container until they are collected by a licensed recycling company. This company should provide you with a waste transfer note as the receipt of collection.

To further reduce the risks involved with hazards, always use safe working practices including:

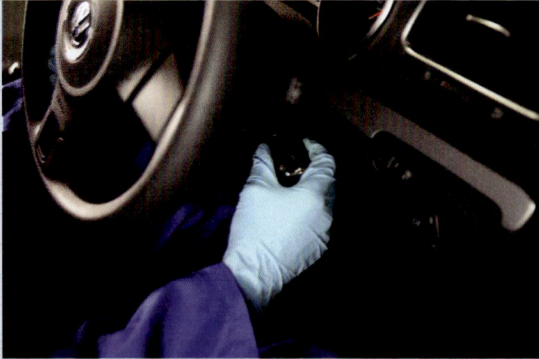

1. Immobilise vehicle (by removing the ignition key). If possible, allow the engine to cool before starting work.

2. Prevent the vehicle moving during maintenance by applying the handbrake or chocking the wheels.

3. Follow a logical sequence when working. This reduces the possibility of missing things out and of accidents occurring. Work safely at all times.

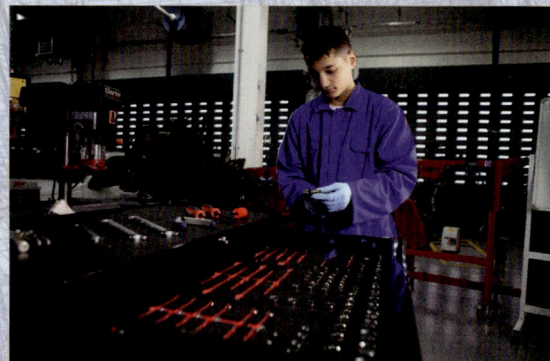

4. Always use the correct tools and equipment. (Damage to components, tools or personal injury could occur if the wrong tool is used or a tool is misused.) Check tools and equipment before each use.

5. Following the replacement of any vehicle components, thoroughly road test the vehicle to ensure safe and correct operation. Make sure that all work is correctly recorded on the job card and vehicle's service history, to ensure that any maintenance work can be tracked.

6. If components need replacing, always check that the parts are the correct quality and type for the vehicle if it is still under guarantee. Inferior parts or deliberate modification might make the warranty invalid. Also, if parts of an inferior quality are fitted, this might affect vehicle performance and safety.

Preparing the car

Tools

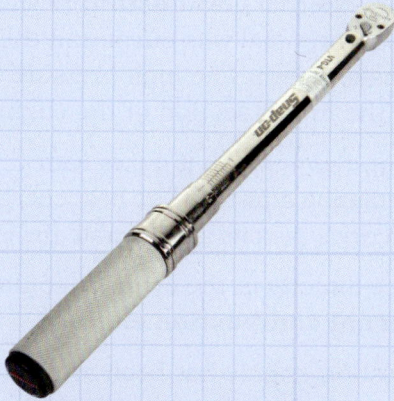
Torque wrench

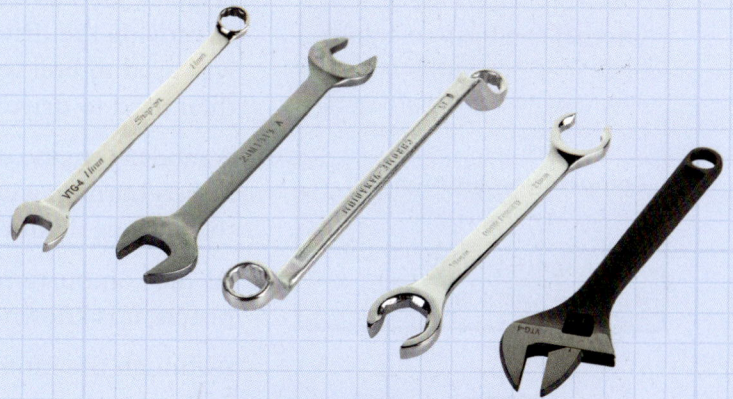
Spanners

Ratchet and sockets

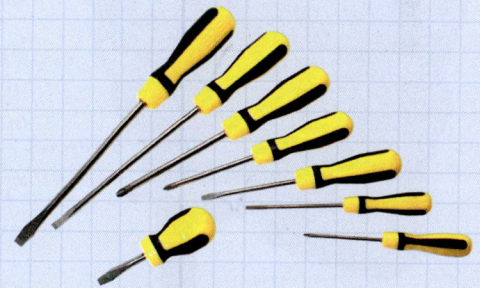
Screwdrivers

Safe Working

- If you are using a ramp or vehicle hoist, always check that the car is evenly positioned, secure and its weight does not exceed any safe working loads (SWL).
- Always clean up any fluid spills (especially engine oil) immediately to avoid slips, trips and falls.
- Always allow the cooling system to cool down before you work on it.
- Always isolate (disconnect) the electrics before you work on a car ignition system.
- Always remove all sources of ignition (e.g. smoking) from the area and have a suitable fire extinguisher to hand when you are working on petrol or diesel fuel systems.
- Always use exhaust extraction when you are running engines in the workshop.
- Always wear heatproof gloves when you work on exhaust system components.
- Always wear safety goggles when you remove and fit exhaust system components.

Petrol and diesel engine systems

The engine of a car uses heat and electricity to make it work. Early vehicles such as trains used an external **combustion** (a steam engine) to provide the drive for movement. Coal was burnt and the heat created was used to boil water and make steam. The high pressure steam was then used to drive pistons in the motor section of the engine.

A modern engine uses petrol or diesel, which it mixes with air. This mixture is then burnt in a cylinder containing a piston. The heat energy given off by the burning mixture is used to drive pistons and operate the engine.

As the combustion takes place directly in the cylinder containing the piston, it is known as an **internal combustion engine**.

Although the internal combustion engine has been around for over 150 years, and many developments have been made, the overall operation has changed very little.

Even though petrol and diesel engines use different fuel types and ignition methods, many of the same components exist and perform the same function.

In the following explanations, if a difference exists between the function or operation of engine mechanical components, it will be stated if it applies to **compression ignition (CI)** or **spark ignition (SI)**.

Spark ignition and compression ignition engine systems – components and purpose

Cylinder block

The cylinder block forms the main body of the engine. Cylinder blocks are usually made from cast iron or aluminium. It contains machined cylinders which house the piston and connecting rod assemblies.

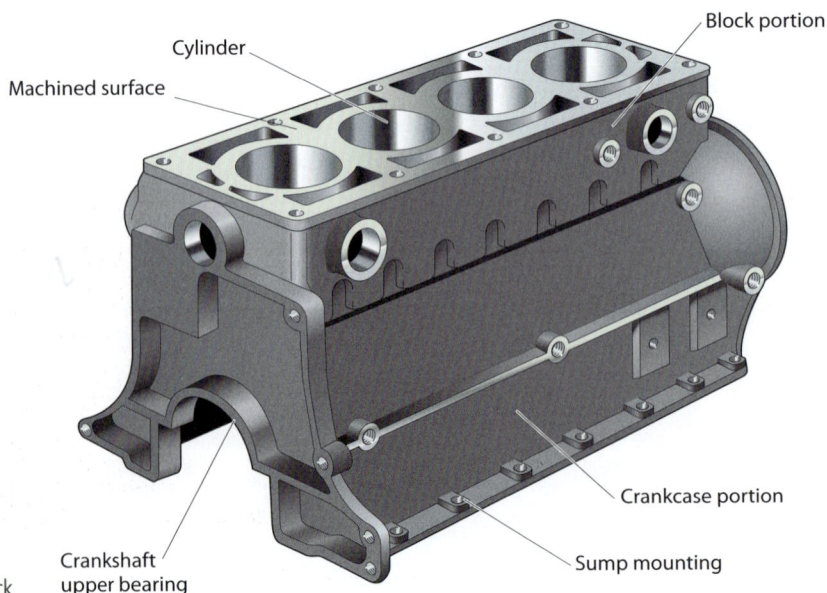

Figure 5.1 A modern light vehicle engine

Labels: Block portion · Cylinder · Machined surface · Crankcase portion · Sump mounting · Crankshaft upper bearing

Figure 5.2 A cylinder block

Cylinder head

The cylinder head is the top part of the engine. It contains the **inlet and exhaust valves**. It may also house the valve drive mechanism. The components of the valve drive mechanism can include:

- camshaft followers
- rocker arms
- valve springs.

The valve drive mechanism is often called the **valve train**.

Many cylinder heads have machined recesses called **combustion chambers**. This is where the burning of the air/fuel mixture takes place. They will also house the valves.

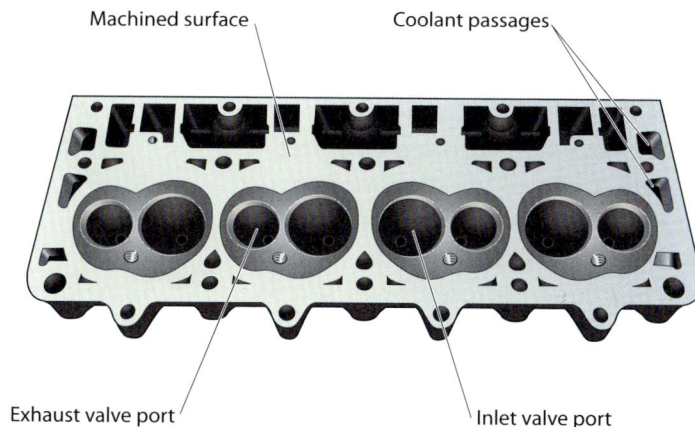

Machined surface   Coolant passages

Exhaust valve port   Inlet valve port

Figure 5.3 A cylinder head

The amount of air and fuel that can be squashed into the combustion chamber when the piston reaches the top of its stroke is known as the **compression ratio**.

Engine sump

The sump is the lower part of the engine. It is a container designed to store engine oil. The sump also forms a cover to house the lower engine components such as the crankshaft, big ends and oil pump.

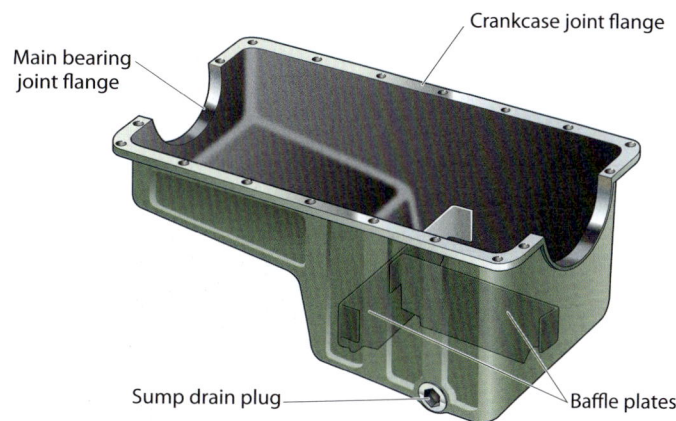

Crankcase joint flange

Main bearing joint flange

Sump drain plug   Baffle plates

Figure 5.4 An engine sump

Key terms

Capacity – the volume of a cylinder when its piston is at the bottom of its stroke.

Stroke – the movement of the piston as it travels up and down.

Inlet and exhaust valves – components that open and close as the engine operates, allowing air and fuel into the engine and exhaust gas out.

Valve train – a series of mechanisms used to open and close the inlet and exhaust valves.

Combustion chamber – an area above the piston where the combustion takes place.

Compression ratio – the difference in volume above the piston when it is at the bottom of its stroke and when it is at the top of its stroke.

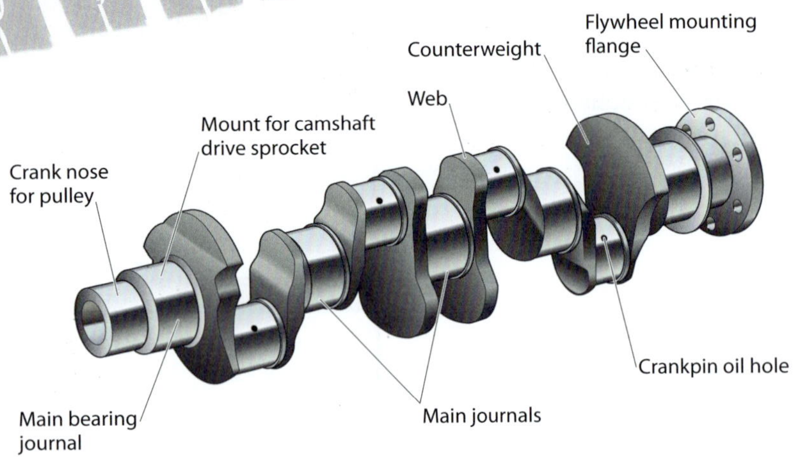

Figure 5.5 A crankshaft

Crankshaft

The crankshaft is the engine's main **driving shaft**. It is usually mounted in the lower part of the cylinder block and has **offset drive pins** that connect to the bottom of the connecting rod (sometimes called 'big ends'). The purpose of the crankshaft is to turn the up and down motion of the piston and connecting rod into a round and round movement. This rotational movement is used to provide drive from the engine.

Key terms

Driving shaft – a rotating rod that powers movement.

Offset drive pins – rods connected to the main crankshaft by arms set at right angles.

Bearing surface – a rotating support placed between moving parts to allow them to move easily.

Small end – a bearing surface found at the top end of the connecting rod.

Big end – a bearing surface found at the bottom end of the connecting rod.

Kinetic energy – movement energy.

Ring gear – a large toothed gear around the edge of the flywheel.

Connecting rods

The connecting rod or 'con rod' is a strong metal shaft designed to join the piston to the crankshaft. It has a **bearing surface** at either end to allow movement. At the top where it joins the piston it is often called the '**small end**' and at the bottom where it joins the crankshaft it is often called the '**big end**'.

Figure 5.6 A connecting rod

Pistons and rings

The piston is a strong, lightweight metal component. It is designed to fit securely inside the cylinder block. The piston is forced downwards because of the high pressure created in the combustion chamber. As the piston is forced downwards the energy from the combustion chamber is transferred to the connecting rod which then turns the crankshaft.

To avoid pressure escaping from the piston and cylinder walls, seals called rings are fitted into gaps in the sides of the piston. It is common for two rings called 'compression rings' to form this seal. A third, lower ring called an 'oil control ring' is used to help keep lubrication oil below the piston.

Figure 5.7 A piston

Camshaft

The camshaft is the component that is designed to open the valves. It is a metal shaft with **lobes** (lumps) machined into the surface. The camshaft is turned at half the speed of the crankshaft. As it rotates, the lobes push on mechanisms that force the inlet and exhaust valves open. As the lobes release the valves, they go back to the closed position with the aid of return springs.

Figure 5.8 A camshaft

Valves

Valves are mounted in the cylinder head and are opened by the camshaft mechanism. This will allow air and fuel to enter and exhaust gas to leave. The valves are kept shut by springs when not needed, so that the combustion chamber can be sealed.

Figure 5.9 Valves

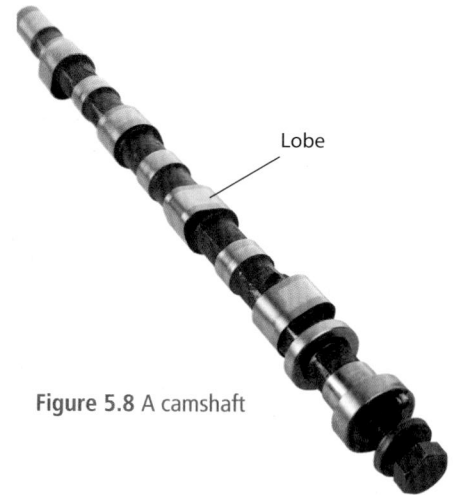

Inlet and exhaust manifolds

A series of ducts and pipe work called inlet manifolds are used to transfer air (and sometimes fuel) from outside the car into the engine. Exhaust manifolds transfer exhaust gases away. Both inlet and exhaust manifolds are designed to have smooth and clear passageways. This allows the movement of air and exhaust gases to take the easiest route in or out of the engine. This will give good engine performance.

Figure 5.10 Exhaust manifold

Flywheel

The flywheel is a large, heavy, metal disc that is bolted onto one end of the crankshaft. As the crankshaft turns, the spinning **kinetic energy** is stored in the flywheel. This energy keeps the engine rotating on the non-power strokes. Because of its location, the flywheel is often a good place to mount the clutch. A large toothed gear (known as the ring gear) is fitted around the edge of the flywheel. When the ignition key is turned, the starter motor connects to the **ring gear** and rotates the crankshaft to start the engine.

Figure 5.11 A flywheel

Front drive pulley

The front drive pulley is mounted on the opposite end of the crankshaft to the flywheel. This pulley can be used to provide drive for various pieces of **auxiliary** engine equipment, such as alternators, cooling fans and air conditioning compressors, when connected to a rubber belt.

Key term

Auxiliary – in addition to the main components, providing a supporting role.

Gaskets and seals

No matter how well engine components are made, imperfections in contact surfaces can allow leakage (both in and out) of fluids or gases. To help seal surfaces between the various engine components, thin sheets of material called gaskets are used. Common examples of some gaskets used on engines include:

- head gaskets
- sump gaskets
- rocker cover gaskets
- inlet and exhaust manifold gaskets.

Rotating shafts, such as camshafts and crankshafts, need to be fitted with 'lip' seals to prevent the leakage of lubricating oil.

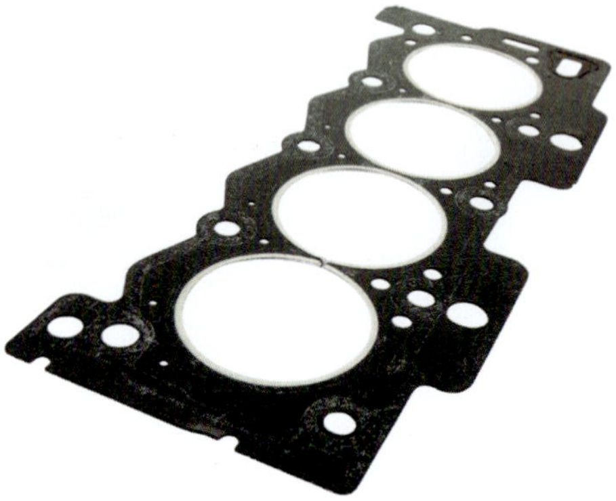

Figure 5.12 A cylinder head gasket

Turbo charger

To improve the performance of engines, air is sometimes forced into the cylinder above the piston. The more oxygen in the cylinder, the more fuel that can be burned, and therefore the more energy that can be released. A turbo charger is an exhaust-gas-driven compressor. It takes the incoming air and raises its pressure, forcing it through the inlet valve and into the combustion chamber. The fitting of turbo chargers in spark ignition engines tends to be restricted to high performance petrol engines. However, they are a more common design feature of standard diesel (compression ignition) engines.

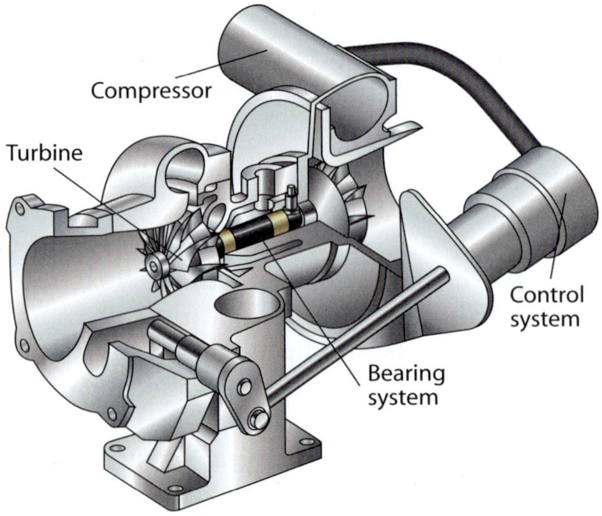
Compressor
Turbine
Control system
Bearing system

Figure 5.13 A turbo charger

Table 5.1 Key differences between petrol and diesel engine components

Engine component	Petrol engine	Diesel engine
Cylinder head	Contains a combustion chamber	Sometimes flat with no combustion chamber
Cylinder block	Lightweight and strong	Heavier and stronger construction
Sump	Similar to diesel	Similar to petrol
Crankshaft	Lighter construction	Stronger construction
Connecting rods	Lighter construction	Stronger construction
Pistons and rings	Flat or domed piston crown	Combustion chamber may be designed into piston crown
Camshaft	Similar to diesel	Similar to petrol
Valves	Similar to diesel	Similar to petrol
Flywheel	Similar to diesel	Similar to petrol

Starting and charging systems

Starting system

To start the combustion process inside the engine, the crankshaft must first be turned.

This used to be done by turning a starting handle at the front of the engine connected to the crankshaft. This required effort and skill and not everybody was able to start their car using this method.

Modern systems use an electric motor which is operated by the driver turning the key in the ignition. This operates as a switch. When the motor turns it connects with a large gear on the side of the engine flywheel called the ring gear. The electric motor is able to turn the crankshaft at high speed, starting the engine. Very little effort is required by the driver.

Charging system

The energy needed to run the electrical components on a car is stored in a battery. The amount of energy the battery stores is not limitless and when used it will eventually go flat.

To keep the battery topped up or 'charged', a continuous supply of electricity needs to be available to the car's electrical systems. An engine-driven electricity generator called an **alternator** is used for this purpose on modern cars. (On early cars, a dynamo was used.)

> **Key term**
>
> **Alternator** – a modern type of electrical generator used to charge the car battery.

CHECK YOUR PROGRESS

1   Name five engine mechanical components.
2   Why are gaskets and seals used between engine component surfaces?
3   What is the purpose of the crankshaft?

Spark ignition and compression ignition engine operation

The processes going on inside the internal combustion engine work on a cycle called the Otto Cycle. It is named after Nikolaus August Otto, who developed it.

The Otto Cycle is made up of four main processes. Each has an official name and a simplified name (given below in brackets).

1. induction (suck)
2. compression (squeeze)
3. power (bang)
4. exhaust (blow).

No matter what design of internal combustion engine is used in a car, these four main processes must occur in turn for the engine to operate.

The Otto Cycle can be broken down into two further categories:

• two-stroke (rarely used on cars)
• four-stroke.

A **stroke** is the movement of the piston as it travels from top-dead-centre (**TDC**) to bottom-dead-centre (**BDC**) or from bottom-dead-centre to top-dead-centre.

Key terms

Stroke – the movement of the piston as it travels up and down.

TDC – top-dead-centre (where the piston has reached the highest point of its stroke).

BDC – bottom-dead-centre (where the piston has reached the lowest point of its stroke).

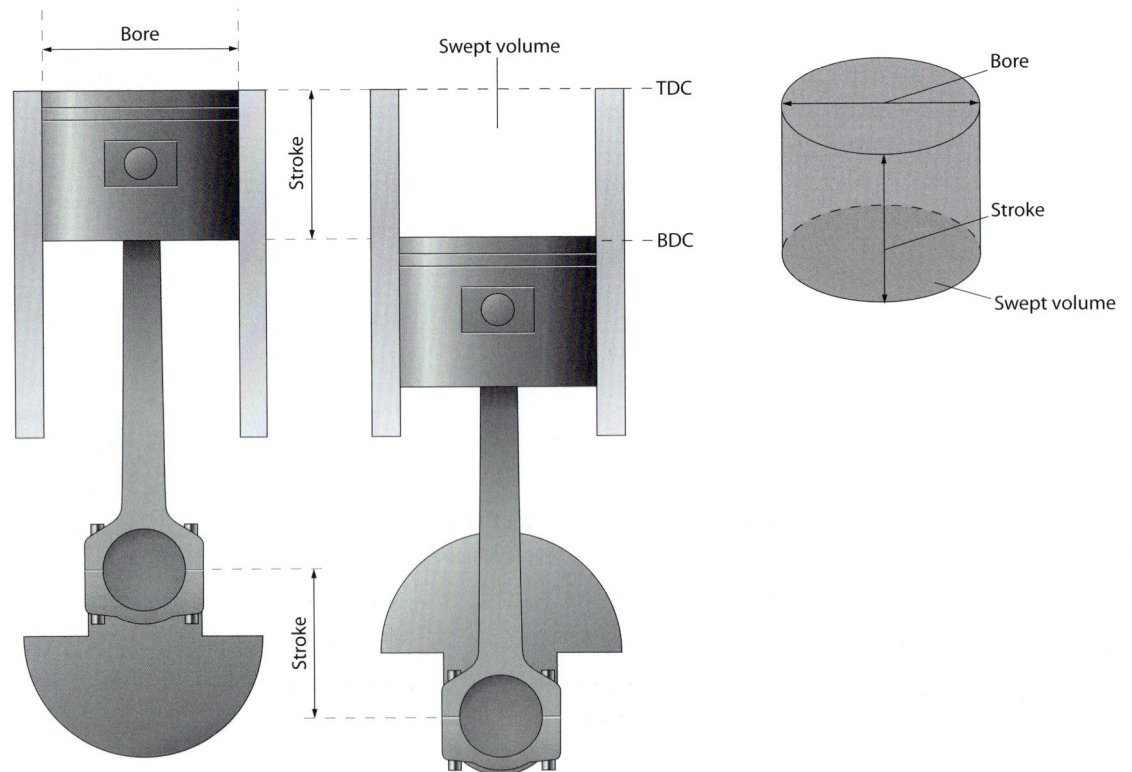

Figure 5.14 Piston and connecting rod movement

The operating cycles for four- and two-stroke engines

The four-stroke operating cycle of a spark ignition (SI) petrol engine

1. Induction (suck)

As the piston moves downwards, air and fuel are **sucked** into the cylinder through the open inlet valve.

2. Compression (squeeze)

As the piston reaches the bottom of its stroke, the inlet valve closes and the piston begins to move up the bore. Air and fuel are **squeezed** into the combustion chamber. Just before top-dead-centre the spark plug ignites the mixture.

3. Power (bang)

The burning air/fuel mixture expands and creates a very high pressure, pushing the piston downwards (**bang**). This downwards movement turns the crankshaft.

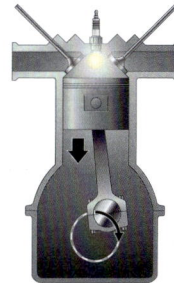

4. Exhaust (blow)

When the piston reaches the bottom of its power stroke, the exhaust valve opens and the piston moves upwards, **blowing** exhaust gas out of the cylinder.

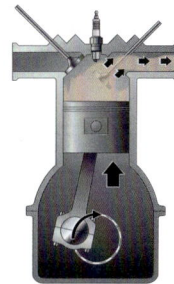

To complete one full cycle of operations (induction, compression, power, exhaust), the crankshaft of a petrol engine must make two full revolutions and the piston will move up and down twice (four strokes).

The power stroke is the only active stroke in the four-stroke cycle. To keep the engine turning on the other strokes, the flywheel stores kinetic (movement) energy to help turn the crankshaft.

Figure 5.15 The four-stroke operating cycle of a spark ignition (SI) petrol engine

Did you know?

When a gas is squashed, the temperature of the gas goes up. For example, when you use a bicycle pump to inflate a tyre the pump and hose get very hot as the pump is operated.

The four-stroke operating cycle of a compression ignition (CI) diesel engine

The four main processes of the Otto Cycle are also used in a compression ignition (CI) engine:

1. induction (suck)
2. compression (squeeze)
3. power (bang)
4. exhaust (blow).

However, in a compression ignition engine, the way that the air and the fuel are taken into the engine and ignited is different from in a petrol engine.

Stronger and heavier mechanical engine components are needed in a compression ignition (CI) engine because of the higher pressures and stresses involved. However, it is a more efficient way of generating energy from fuel. This means a compression ignition (CI) engine will run longer on diesel than a spark ignition (SI) engine will run on the same amount of petrol.

Induction (suck)

As the piston moves downwards, only air is drawn into the cylinder through the open inlet valve.

Compression (squeeze)

As the piston reaches the bottom of its stroke the inlet valve closes and the piston begins to move up the bore. The air is squashed into a very small combustion chamber where it becomes very hot. Just before top-dead-centre the injector sprays diesel into the superheated air.

Power (bang)

The burning air/fuel mixture expands and creates a very high pressure, pushing the piston downwards. This downwards movement turns the crankshaft.

Exhaust (blow)

When the piston reaches the bottom of its power stroke, the exhaust valve opens and the piston moves upwards, pushing exhaust gas out of the cylinder.

To complete one full cycle of operations (induction, compression, power, exhaust), the crankshaft of a diesel engine must make two full revolutions and the piston will move up and down twice (four strokes).

As with a petrol engine, a flywheel is needed to keep the engine turning over on the non-power strokes.

Did you know?

Rudolph Christian Karl Diesel is credited with the design of the compression ignition (CI) engine.

Find out

In groups, create a poster which shows the differences between spark ignition (SI) and compression ignition (CI) engines.

The two-stroke operating cycle of a spark ignition (SI) petrol engine

Two-stroke engines are rarely used on cars because they are inefficient and highly polluting.

A two-stroke engine completes a full cycle of operation in just two strokes (the piston moves up and down once). With a four-stroke engine the entire process takes place above the piston, but a two-stroke engine also makes use of the area below the piston (the crankcase) to speed up the operation.

Instead of valves, most two-stroke engines use **ports** (holes in the cylinder walls) that are opened and closed (covered and uncovered) by the piston as it moves up and down.

It is important that you imagine what is going on above and below the piston at the same time when you are asked to describe the operation of a two-stroke engine. There are six processes involved in the operating cycle of a two-stroke engine.

> **Key terms**
>
> **Ports** – holes machined in the cylinder walls of a two-stroke engine.
>
> **Scavenging** – the process in the combustion chamber that helps remove exhaust gases with the aid of the incoming air charge.

Induction (suck)

Induction takes place below the piston. As the piston moves upwards, air, fuel and lubricating oil are drawn into the crankcase through the open inlet port.

Pre-compression

As the piston moves downwards, it covers up the inlet port, squashing the air, fuel and lubricating oil in the crankcase and raising its pressure slightly.

Transfer

As the piston continues to move downwards, a transfer port is opened which allows the air, fuel and lubrication oil to be forced upwards into the area above the piston. (The new charge of air, fuel and lubrication oil helps push exhaust gases out through the open exhaust port. This is known as **scavenging**.)

Compression (squeeze)

As the piston continues to move upwards, the transfer port and the exhaust port are closed. Air, fuel and lubricating oil are squashed in the combustion chamber. Just before top-dead-centre the spark plug ignites the mixture.

Power (bang)

The burning air/fuel mixture expands and creates a very high pressure, pushing the piston downwards. This downwards movement turns the crankshaft.

Exhaust (blow)

As the piston continues to move downwards, the exhaust port is uncovered. Exhaust gases begin to escape and, as the transfer port is uncovered, the new incoming charge of air, fuel and lubricating oil is able to help push out the exhaust gas (scavenging).

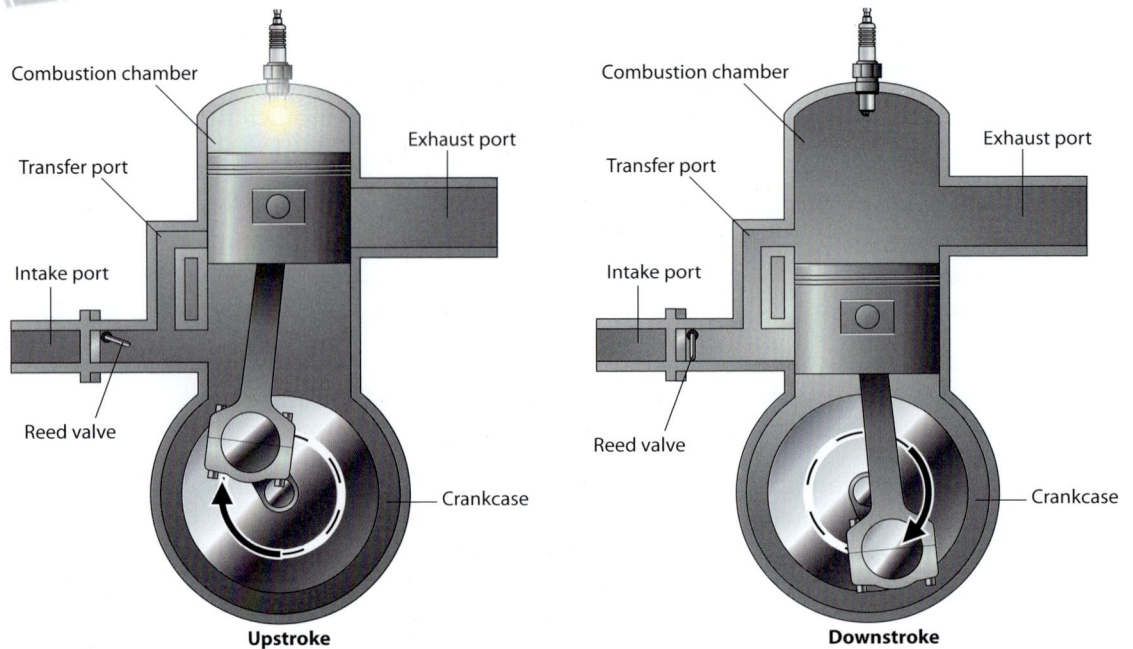

Figure 5.16 A two-stroke spark ignition (SI) petrol engine

Find out

Investigate vehicle types that use, or have used, two-stroke operating system technology.

To complete one full cycle of operations (induction, compression, power, exhaust), the crankshaft of a standard two-stroke engine only makes one full revolution and the piston will move up and down once (two strokes).

Firing order

In a **multi-cylinder** engine, the **firing order** of each cylinder (when the fuel is ignited) is staged at different points along the crankshaft to make the overall operation smoother.

A typical crankshaft layout and firing order for a four-cylinder engine is shown in Figure 5.17.

Key terms

Multi-cylinder – an engine with more than one piston.

Firing order – the order in which the cylinders receive the sparks as the crankshaft rotates.

Bore – the diameter of an engine cylinder.

Did you know?

If an engine is running at 6000 rpm, the piston will move down the **bore** and back up again 100 times a second.

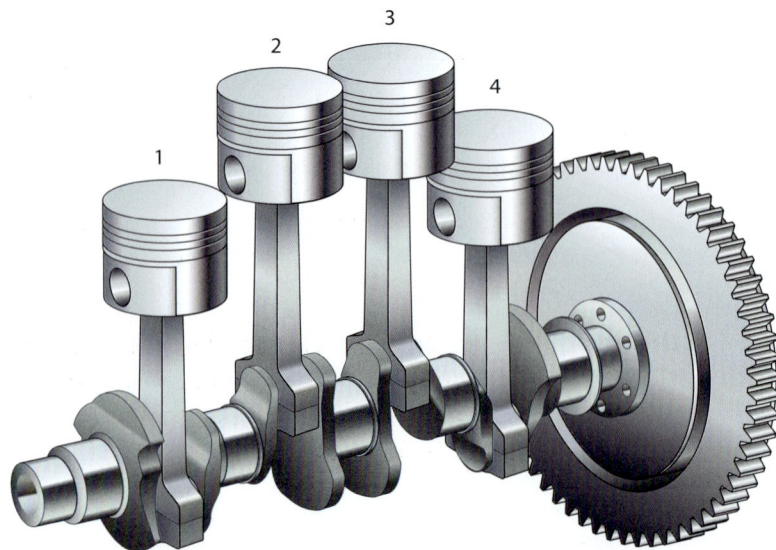

Figure 5.17 A crankshaft with pistons – the typical firing order is 1, 3, 4, 2

Valve timing

An engine's cycle of operations happens very fast when the engine is running. This means that air and fuel must get into the engine, be ignited and burn, and exhaust gases be removed from the engine very quickly. Because of this, certain operations take place early and finish late during the four-stroke cycle. This helps to give extra time for the processes to take place.

The opening and closing of the inlet and exhaust valves are timed to coincide with piston movements.

- To give as much time as possible for air and fuel to enter the cylinder, the induction valve is opened slightly early (while the piston is just coming to the end of its exhaust stroke) and is closed late (just as the piston is starting its compression stroke).
- To give as much time as possible for exhaust gases to leave the cylinder, the exhaust valve is opened slightly early (while the piston is just coming to the end of its power stroke) and is closed late (just as the piston is starting its induction stroke).
- When a valve opens early, it is called **valve lead**.
- When a valve closes late, it is called **valve lag**.

Due to this valve timing, there is a period when both the inlet and exhaust valves are open at the same time. This is called **valve overlap**. One advantage of a valve overlap period is that the incoming air and fuel help to push out the exiting exhaust gases, and the exiting exhaust gases help to draw in fresh air and fuel.

The valve movements can be shown on a timing diagram as shown in Figure 5.18.

Ignition timing

As with valve timing, the spark from the plug also has to happen early. It needs to do this so that the fuel is given time to burn. As the flame spreads the full pressure of the expanding gases pushes on the piston crown at exactly the right time.

As the engine runs faster and faster, the spark has to happen earlier in the cycle to allow for the fuel burn time. This is called **ignition advance**.

A similar thing happens with a diesel engine. As a diesel engine has no spark plug to ignite the fuel, the injection of the fuel must happen early. As the engine runs faster and faster, the injection has to happen earlier in the cycle each time. This is called **injection advance**.

> **Did you know?**
>
> Diesel fuel can be injected into the combustion chamber or into a small area of the cylinder head called a swirl chamber. When injected into the combustion chamber, this is known as a direct injection. When injected into a swirl cylinder, this is known as indirect injection.

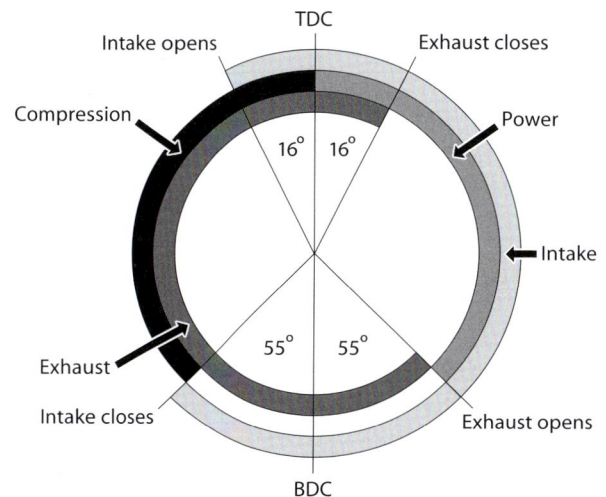

Figure 5.18 An engine valve timing diagram

> **Key terms**
>
> **Valve lead** – where a valve (inlet or exhaust) opens early.
>
> **Valve lag** – where a valve (inlet or exhaust) closes late.
>
> **Valve overlap** – the period when both inlet and exhaust valves are open at the same time.
>
> **Ignition advance** – the automatic bringing forward of the ignition spark in relation to the engine speed.
>
> **Injection advance** – the automatic bringing forward of the fuel injection in relation to the engine speed.

CHECK YOUR PROGRESS

1 What is valve lead?
2 What is valve lag?
3 What is valve overlap?

Liquid cooling and lubrication systems

The need for a cooling system

An internal combustion engine produces energy by releasing heat. During normal operation it is common for combustion temperatures to reach 2500°C. A large amount of this heat leaves the engine by the exhaust, but enough is left over to melt engine components. A cooling system is therefore needed to help remove unwanted heat, so that engine components are not melted.

The process of moving heat is called transfer. Heat is transferred in three ways:

1. Conduction is the transfer of heat energy through solid objects.
2. Convection is the transfer of heat energy through liquids or gases.
3. Radiation is the transfer of heat energy through a vacuum or space.

Engine cooling can be done in one of two ways:

- **Air cooling** works by directing air over cooling fins on the outside of the engine. This allows the heat energy to be transferred into the air and away from the engine. For air cooling to work, the surface area of the outside of the engine has to be increased and cooling fins included in the design of the outside of the engine. As airflow over the engine is restricted by the car's bodywork, devices are used to force air through the cooling fins and help with cooling. These devices are covers, ducts and fans.
- **Liquid cooling** works by transferring the heat energy to a liquid surrounding the engine. A liquid coolant is pumped around the outside of the engine and the heat is transferred away from the engine. The heat energy goes through a radiator before being transferred into the surrounding air.

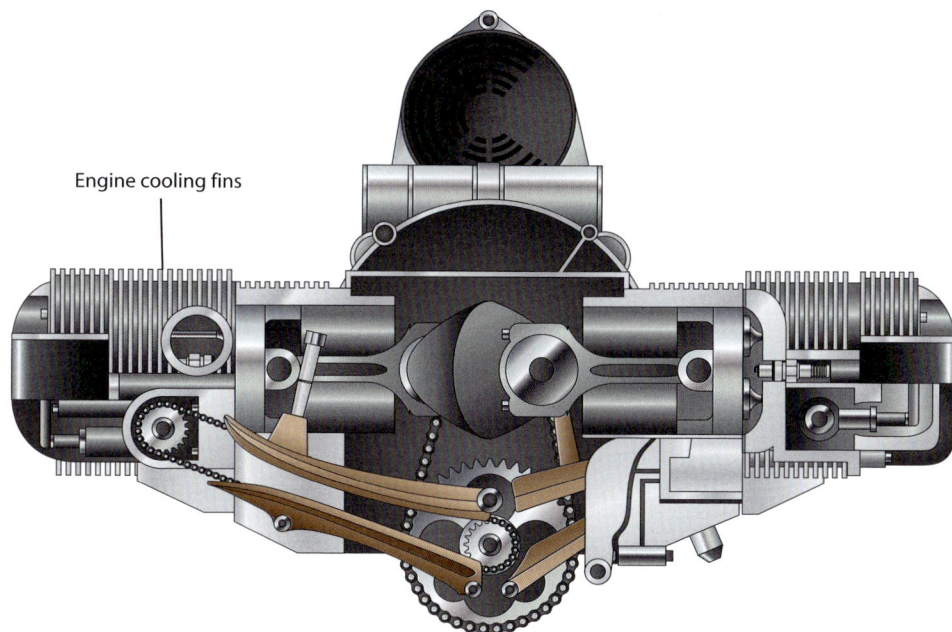

Engine cooling fins

Figure 5.19 An air cooled engine

Liquid cooling system components

Coolant – water and antifreeze mixture

Water is one of the best liquids at absorbing heat. This is why older engine cooling systems used water. However, there are problems with using water (on its own) to cool an engine system.

- Water boils at 100°C and freezes at 0°C (under normal conditions). If water is allowed to boil, bubbles begin to form in the liquid. These bubbles do not conduct heat as well as water.

- If water is allowed to freeze, ice crystals form which take up more space than water. Water expands (becomes larger) when it freezes because ice particles become fixed into a pattern. The size of all the ice crystals is so large that, if the water is trapped inside the engine when it freezes, enough force can be created to crack the metal of the engine block.

- Water mixes with oxygen in the air and this mixture comes into contact with metal. The water/oxygen mixture acts on the metal and starts to break it down. This is a chemical process known as corrosion. Corrosion eats away at metal engine components and creates a 'sludge' that can block the pipes and waterways of an engine cooling system.

To help reduce the problems caused by temperature and corrosion, antifreeze and other chemicals are mixed with the water, to provide protection against freezing, boiling and corrosion. Liquid mixtures of water and chemicals such as antifreeze are known as coolants.

Find out

Make a list showing the advantages and disadvantages of air and liquid cooling.

Did you know?

Because early engine systems used water as a **coolant**, some of the components still use this word (for example, water pump).

Key term

Coolant – water and antifreeze mixture.

Many modern engine coolants come pre-mixed from the manufacturer. However, if you need to mix antifreeze with water yourself, Table 5.2 shows the amounts to be used. The table also shows how much protection against freezing the coolant will provide.

Table 5.2 Percentage of protection against freezing provided by coolant

Percentage of antifreeze added to water (%)	Percentage of water (%)	Protection from freezing down to X°C
25	75	-10
33	67	-15
40	60	-20
50	50	-30

Radiator and radiator cap

The heat absorbed by the liquid coolant is transferred into the surrounding air using a radiator.

A radiator is a series of metal tubes surrounded by **cooling fins**. It is mounted away from the engine in a position that allows airflow to pass over it and heat to **dissipate** into the surrounding air.

The pipe work in the centre of the radiator zigzags back and forth so that the coolant passing through it is exposed to the air for as long as possible. This makes the radiator efficient at getting rid of its heat. If the pipe work is mounted vertically (top to bottom) the radiator **core** is known as 'upright'. If the pipe work is mounted horizontally (side to side) the radiator core is known as 'crossflow'.

Figure 5.20 A cooling system radiator (upright)

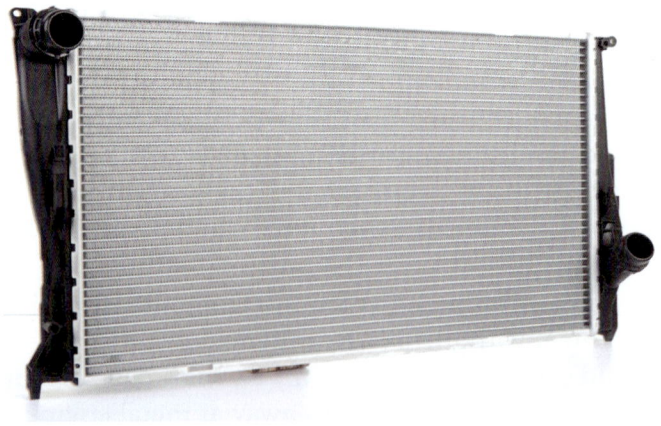

Figure 5.21 A cooling system radiator (crossflow)

Under normal conditions, even the addition of antifreeze would not raise the boiling point of the coolant enough to stop it from boiling at some point in the engine system.

To raise the boiling point of the coolant more, the cooling system is pressurised. Pressure has a direct effect on the boiling point of water:

- If the pressure is lowered the boiling point is lowered.
- If the pressure is raised the boiling point is raised.

A radiator cap is used to pressurise the cooling system. The radiator cap seals the cooling system creating pressure.

When the engine is running the coolant warms up and expands. Water expands (becomes larger) when it heats up because the particles move further away from each other. The water, which now takes up more room, has nowhere to go (because the radiator cap is sealing the system) so pressure increases. This pressure increase will also raise the boiling point of the liquid coolant in the system.

To make sure that the pressure does not continue to rise past safe limits, the radiator cap contains a spring-loaded valve. When a pre-set pressure is reached, the valve will open allowing some of the coolant to escape to an expansion tank (lowering the pressure).

When the engine cools down, another valve in the radiator cap opens to allow any expelled coolant to be drawn back in from the expansion tank. This keeps the system topped up.

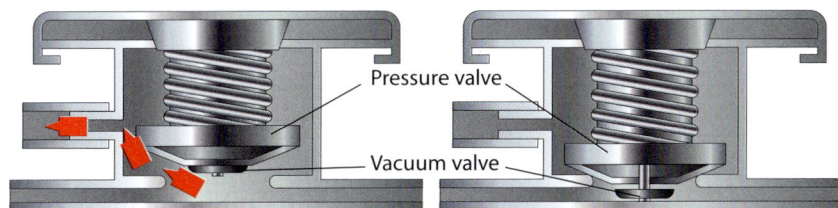

Figure 5.22 Cooling system pressure cap, with relief valve open (left) and pressure valve open (right)

Pressure valve
Vacuum valve

> **Did you know?**
>
> If you were to make a cup of tea at the top of Mount Everest, the water would boil at 69°C because the atmospheric pressure is so low.

> **Did you know?**
>
> By raising the cooling system pressure by 1 bar (approximately 15psi) the coolant boiling point will be around 121°C.

> **Working life**
>
> Chris has been asked by his boss to check the antifreeze on a customer's car. The engine is still hot.
>
> 1 What should Chris do?
>
> 2 a) What might happen if he undoes the radiator cap?
>
>   b) Why would this happen?

Expansion tank

The expansion tank is a plastic container designed to act as a **reservoir** for excess coolant. It is connected to the cooling system by a pipe at the radiator cap valve.

As the cooling system warms up, any coolant that is allowed to escape past the radiator valve is transferred to the expansion tank. As the cooling system temperature falls, any losses from the system are allowed to be drawn back in from the expansion tank. This keeps the system topped up. As the cooling system is sealed, if the coolant needs topping up it is done using the expansion tank.

> **Key terms**
>
> **Cooling fins** – thin pieces of corrugated (ridged) metal designed to increase surface area.
>
> **Dissipate** – spread out in many directions. For example, heat dissipates through air.
>
> **Core** – the central pipe work and cooling fins of a radiator.
>
> **Reservoir** – a supply or source of something.

The expansion tank will normally have a maximum and a minimum reading indicated on the side to show that the correct quantity of coolant is present. This level should be checked when the engine is cold.

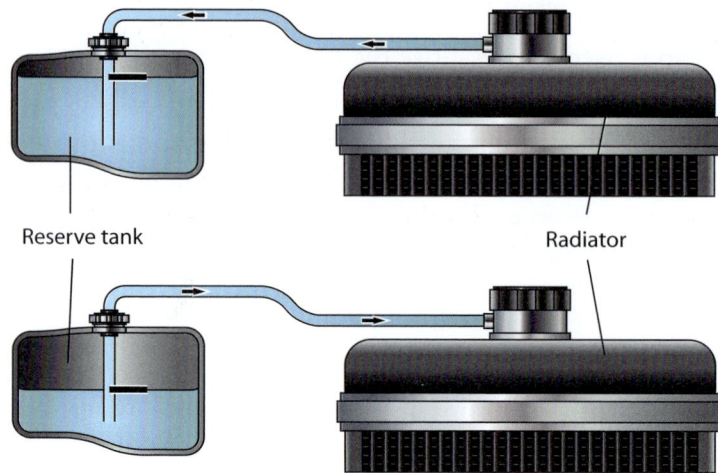

Reserve tank   Radiator

Figure 5.23 An expansion tank

Thermostat

An engine will run at its best when the overall temperature is around 100°C (the boiling point of water).

When an engine starts from cold it is not running at its best (it is not as efficient as it could be). This means the engine will consume higher levels of fuel, produce more emissions and more wear will be done to the engine.

Therefore it is important for an engine to warm up quickly so that the engine runs at its best as soon as possible. A temperature controlled valve called a thermostat is fitted to help with the rapid warm up of engines. The thermostat restricts the flow of water until a certain temperature has been reached.

A thermostat works on the expansion of gases or liquids (usually wax). The thermostat is positioned in the cooling system so that when the thermostatic valve is closed it will stop the flow of coolant into the radiator.

When the engine is cold, the thermostatic valve is closed, but as the engine warms up the wax expands (becomes larger). This opens the valve and allows the coolant to circulate freely through the system and radiator.

If the temperature in the cooling system falls, the wax in the thermostat contracts (becomes smaller). This closes the valve and once again reduces the flow of coolant through the radiator.

Most thermostats operate at approximately 90°C.

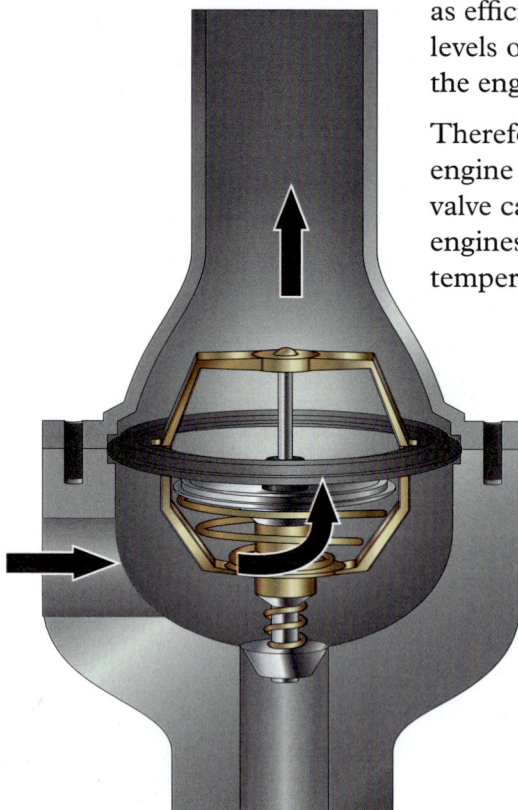

Figure 5.24 A wax type thermostat

Pipes and hoses

A series of pipes and hoses is used to move coolant from the engine to the radiator. Rubber hoses are used to connect the engine to the radiator rather than rigid metal pipes. This is because the radiator is usually mounted solidly to the vehicle body but the engine is normally allowed to move slightly on rubber mountings. If rigid metal pipes were used between the engine and radiator, shaking could lead to fractures and leakage from the pipes. Therefore rubber hoses are used to connect the fixed radiator and movable engine, so that any shake or bend is cushioned.

Figure 5.25 A cooling system radiator hose

Gaskets and sealing rings

Gaskets and seals are used in engine cooling systems so that leaks do not occur between surfaces that are in contact.

Water pump and drive belt

In modern systems an engine-driven water pump is used to circulate coolant around the engine and radiator and maintain a relatively even temperature. The water pump is normally mounted in the engine block and, when turned, a small **impeller** is used to drive coolant around the system. The impeller is turned by a pulley which is connected to a rubber drive belt driven by the engine crankshaft. This rubber drive belt (sometimes called the fan belt) is normally operated by the front drive pulley of the crankshaft. However, some manufacturers are now using the camshaft timing belt to turn the water pump.

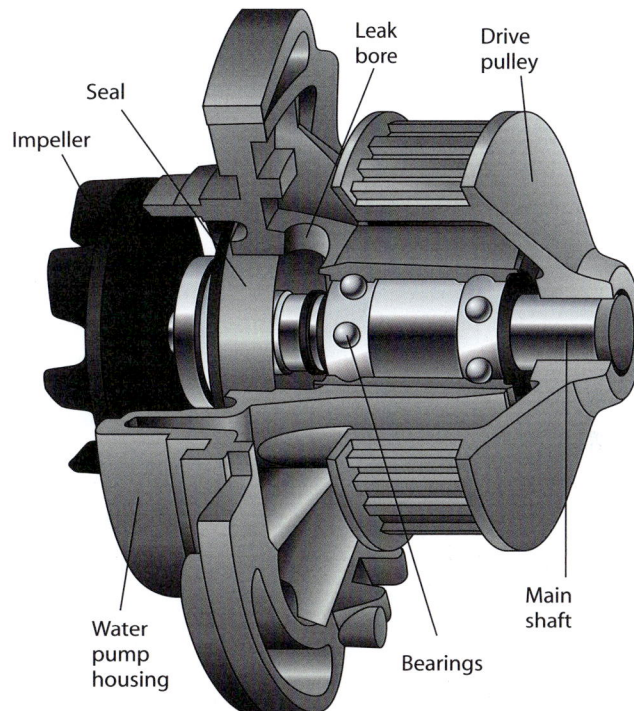

Figure 5.26 A cooling system water pump

Cooling fan – mechanical and electric

There are some situations when a car is in use where there is not enough airflow over the radiator (for example, if a car is sitting in traffic). The engine can overheat if this happens for a long period. Cooling fans are included in car designs to blow or draw air through the radiator when airflow is slow.

Early systems had a fan mounted on the water pump pulley, so that as the pump was turned by the 'fan belt' it also operated the cooling fan. By running the cooling fan all the time, **drag** occurred and overcooling occurred.

Key terms

Impeller – a part of a pump which rotates to move liquid.

Drag – retarding (holding back) force exerted by air or other fluid surrounding a moving object.

A drive system known as **viscous coupling** was developed. This used a liquid or wax to provide varying amounts of slippage between the pulley and the fan, depending on the engine temperature. This way, if the engine was cold, a lot of slip occurred and the fan hardly turned. As the engine temperature increased, the viscous coupling created more drag, allowing the fan to spin faster.

Many modern cars now have a cooling fan driven by an electric motor. Because of this the fan can be bolted directly to the radiator unit. A temperature sensitive switch is mounted in the cooling system, and when the engine temperature reaches a preset value, the switch will close, operating the electric motor and driving the cooling fan. As the engine cools down the switch will open, turning off the electric motor. Because the cooling fan now only works when needed, this reduces load on the system, in this way improving its overall efficiency.

Temperature sending unit

Cooling fan

Thermostat switch

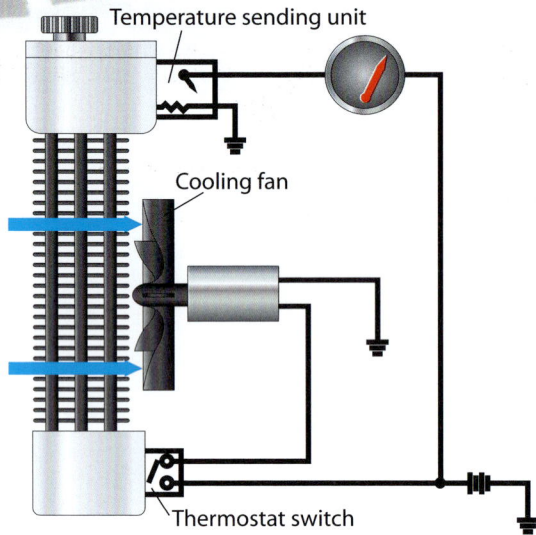

Figure 5.27 Electric cooling fan circuit

Key term

Viscous coupling – a drive coupling that uses a fluid to transmit drive.

Vehicle heater

A heater can be made by adding a second radiator inside the car and connecting it to the engine's cooling system. Hot coolant from the engine is directed by a series of valves or controls to pass through this radiator (or heater matrix). An electric fan can then be used to circulate air through it and this warms the passenger compartment.

Table 5.3 The purpose of cooling system components

Cooling system component	Purpose
Radiator	Transfers heat to the surrounding air
Radiator cap	Seals the cooling system to raise the pressure
Expansion tank	An overflow container for the coolant
Thermostat	Regulates the flow of coolant to control temperature
Pipes and hoses	Connect the engine to the radiator
Water pump	Helps circulate the coolant around the engine and through the radiator
Cooling fan	Draws air through the radiator to help with cooling
Heater matrix	A small radiator used to warm the passenger compartment

CHECK YOUR PROGRESS

1 List three functions of antifreeze.
2 What regulates the temperature of the cooling system?
3 List the three methods of heat transfer.

Lubrication systems

The moving metal parts inside the engine rub against each other at considerable speed during normal operation. (Remember, a piston moving inside the cylinder can travel up and down many hundreds of times a second.) The rubbing of the surfaces creates friction, and friction creates large amounts of heat. The heat can be great enough to melt the surfaces of metal components so that they become fused together and no longer work. When this happens the engine is said to have 'seized'.

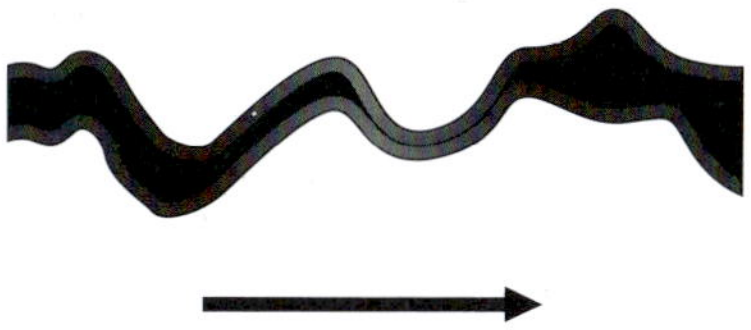

Figure 5.28 Boundary lubrication

Oil is used within a lubrication system to overcome this problem. It is pumped around the engine so that the metal surfaces rubbing against each other run on a thin film of oil. This keeps them apart and reduces friction.

There are two main types of lubrication:

- **Boundary lubrication** – a thin film of oil will coat the surfaces of the mechanical components. This coating is very hard to penetrate. Because of this, when the two surfaces rub against each other, they slide on this thin boundary film and hardly make contact.
- **Hydrodynamic lubrication** – the movement of engine parts forces the component to ride up on a wedge of oil, as shown in Figure 5.29.

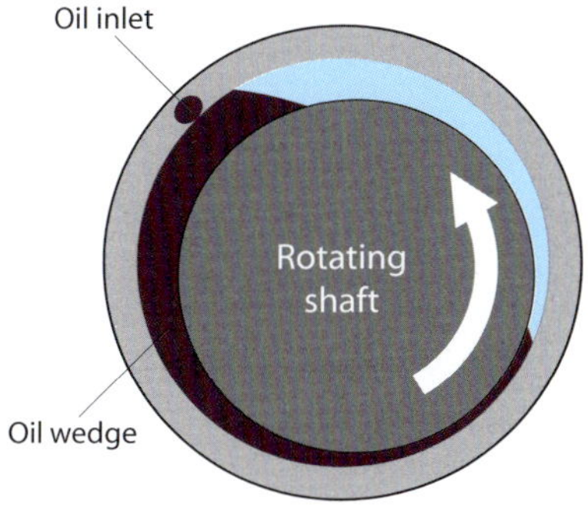

Figure 5.29 Hydrodynamic lubrication

Lubrication oil not only helps to reduce friction, it will also cool engine parts and trap dirt, taking it away for removal by the oil filter. Lubrication oil can also provide a coating which helps reduce corrosion of metal components.

The main components in engine lubrication systems

Lubricants

There are two main kinds of lubricant:

- **Natural** – crude oil is a naturally occurring substance found underground. It is formed from ancient plant and animal life that has decayed over millions of years under extreme heat and pressure. Crude oil is pumped from underground and then processed in a refinery. Impurities are taken out during the refining process and chemicals are added to make it suitable for use as engine oil.
- **Synthetic** – synthetic oil is man-made from a series of chemicals in laboratories in factories. A fully synthetic oil can be manufactured to have different properties which improve its lubricating ability. The processing of synthetic oils can be very expensive and this is reflected in the price.

Figure 5.30 Engine oils

How oils are classified

- **SAE** – the Society of Automotive Engineers. This organisation classifies the measurement of **viscosity**. The SAE number shows how thick or thin an oil is. The lower the number, the runnier it is and the higher the number, the greater its resistance to flow.

- **API** – the American Petroleum Institute. This organisation classifies the quality of oil. The classification is given as a two letter code. The first letter (S or C) indicates whether it is suitable for a spark ignition engine or a compression ignition engine. The second letter (drawn from a chart in alphabetical order) shows the quality of the oil – the further through the alphabet the letter is, the higher the quality.

- **Multigrade** – a blend of oils, often used for engine lubrication. They are given an SAE rating showing their viscosity when they are cold and hot. For example, an SAE 10W/40 shows the oil has a viscosity of 10 when cold (often followed by the letter W and tested down to -18°C) and a viscosity of 40 when hot (tested up to a temperature of 99°C).

Engine sump

The sump is normally bolted to the lower part of the engine. It is used as a reservoir to hold the lubrication oil. From here oil is picked up, pumped around the engine and returned when it has done its job. Some sumps contain 'baffles'. These are metal plates that help prevent the oil from sloshing around when the car is moving. The sump is normally made from thin steel or aluminium. It is exposed to airflow under the engine to help cool the oil.

Oil pump and strainer

Oil has to be pumped upwards and around the engine under pressure. This is because the sump sits at the bottom of the engine.

An **oil strainer** is fitted in the lowest part of the sump. It is attached to a tube which joins it to the oil pump. The oil pump, driven by the engine, draws up oil through the strainer and pickup pipe, and forces it around the engine. Four different types of oil pump construction are shown in Figures 5.31 to 5.34.

The oil returns to the sump through passageways under the force of gravity when it has completed its function.

Key terms

Viscosity – a liquid's resistance to flow (sometimes referred to as thick and thin).

Oil strainer – a basic filter made from wire mesh.

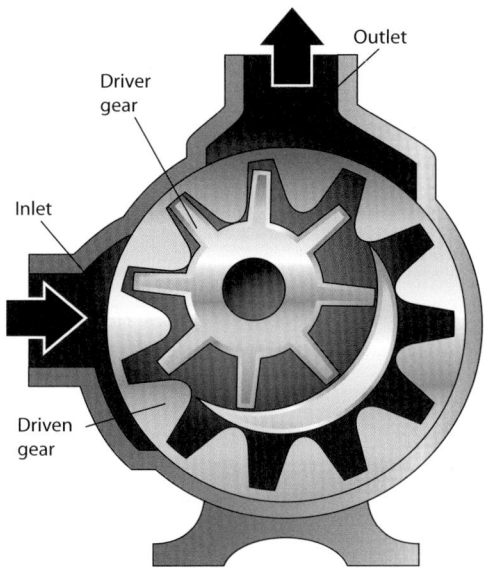

Figure 5.31 Crescent type engine oil pump

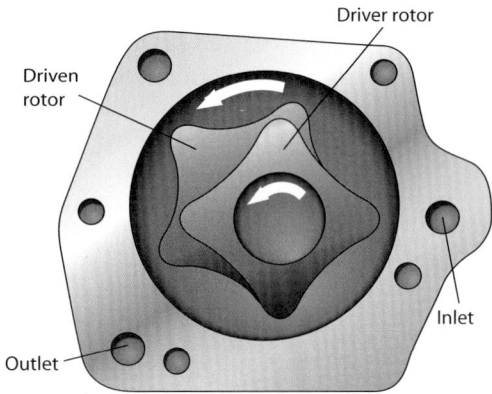

Figure 5.32 Eccentric gear type engine oil pump

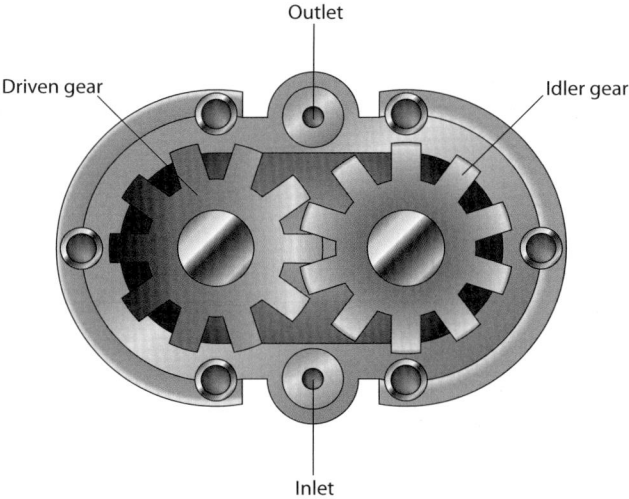

Figure 5.33 Gear type engine oil pump

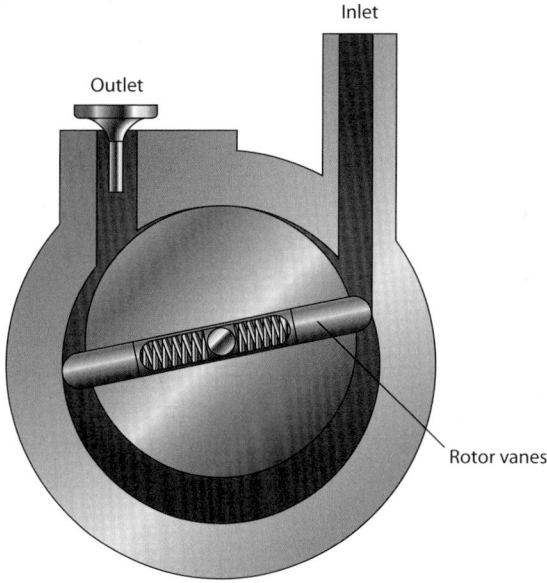

Figure 5.34 Vane type engine oil pump

Pressure relief valve

An oil pump must be capable of providing enough oil pressure, even when the engine is running slowly. As the engine speeds up and the pump turns faster, oil flow and pressure rises. An oil pressure relief valve is fitted to prevent pressure becoming too much.

The oil pressure relief valve consists of a spring-loaded plunger which is normally fitted on the outlet of the oil pump.

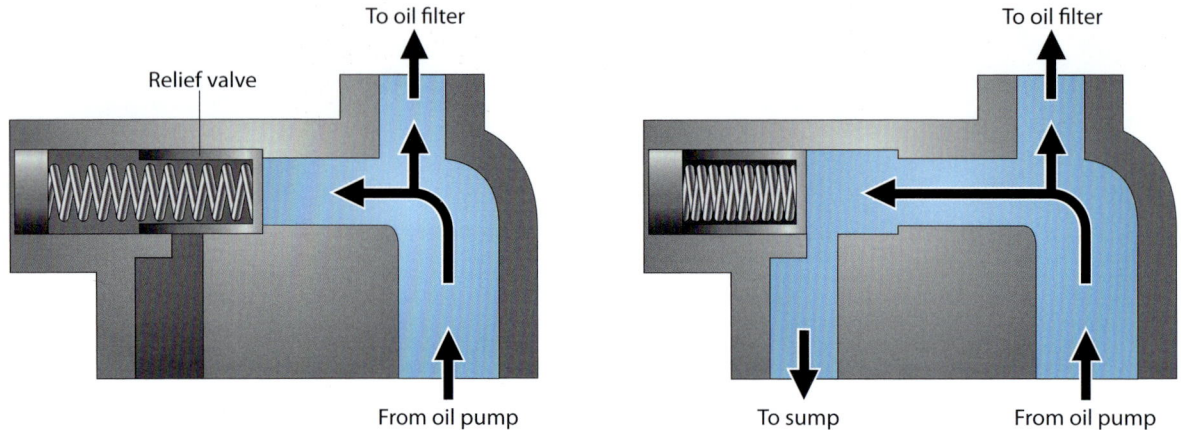

Figure 5.35 Pressure relief valve

As the engine runs and the oil pump operates, spring pressure holds the plunger closed. When engine speed increases and oil pressure rises, the plunger is pushed off its seal. Any excess oil or pressure is then returned to the sump. In this way a constant oil pressure is maintained under all engine operating conditions.

Oil filter

One of the functions of engine oil is to remove dirt. A filter is included in the lubrication system. This will try to maintain a clean flow of oil to the engine's mechanical parts. A replaceable pleated paper element is fitted after the oil pump. This is designed to strain small particles of dirt and debris from the lubrication oil. The paper element is usually contained in a housing which is easy to access. This makes it easier to replace during routine maintenance.

Two main types of oil filter system are used:

- **Full flow lubrication system** – oil is pumped from the sump through a filter element, and then on to the rest of the engine. In this way, all of the oil travelling to the engine has been filtered.
- **Bypass lubrication system** – a much finer oil filter is used that can remove more dirt than a full flow system. It is connected to a passage next to the one which takes the lubricating oil to the rest of the engine. When oil is pumped from the sump, only some of the oil goes to the engine, while the rest is forced through the filter element and back to the sump. Because of this the oil in the sump is always relatively clean.

Did you know?

The filter can become blocked if it is not changed for a long time. However, a small valve is located inside the full flow oil filter. This will allow oil through even if the filter element gets blocked. This way the engine will still receive oil – dirty oil is better than no oil.

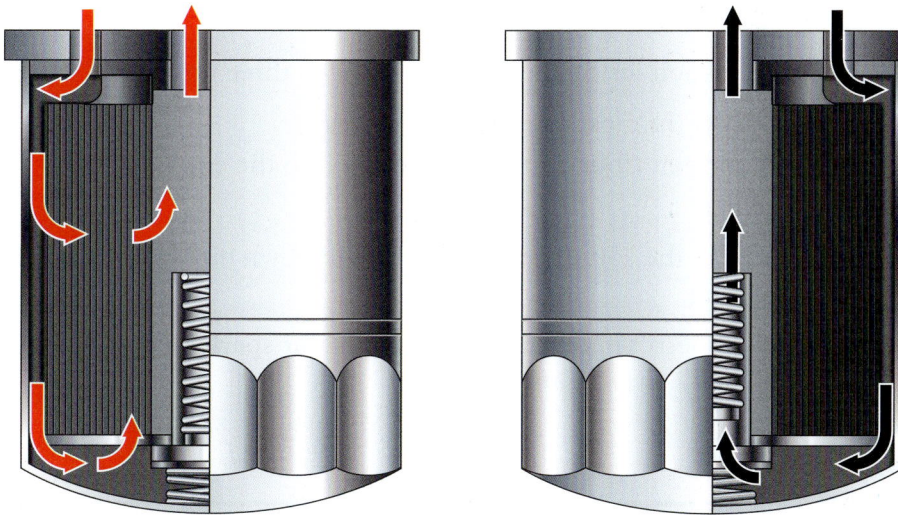

Figure 5.36 Full flow oil filter

Oil galleries

Oil galleries are passageways that are cast in the cylinder block and head during manufacture. They direct oil to vital engine components and return it to the sump when finished.

Table 5.4 The purpose of cooling system components

Lubrication system component	Purpose
Oil strainer	Helps remove large particles of dirt in the sump
Oil pickup tube	Transfers oil to the oil pump
Oil pump	A mechanical pump turned by the engine to create oil flow and pressure
Pressure relief valve	A pressure limiting valve which returns excess oil to the sump
Oil filter	A paper element designed to remove dirt particles from the oil
Galleries	Passageways cast into the cylinder head and block
Sump	The oil container at the bottom of the engine

CHECK YOUR PROGRESS

1 What does API stand for?
2 What does SAE stand for?
3 What is oil used for?

Figure 5.37 The triangle of fire

Did you know?

Glow plugs are found in a diesel engine and operate for a few minutes on start-up. They help warm the air in the combustion chamber to assist with cold starting.

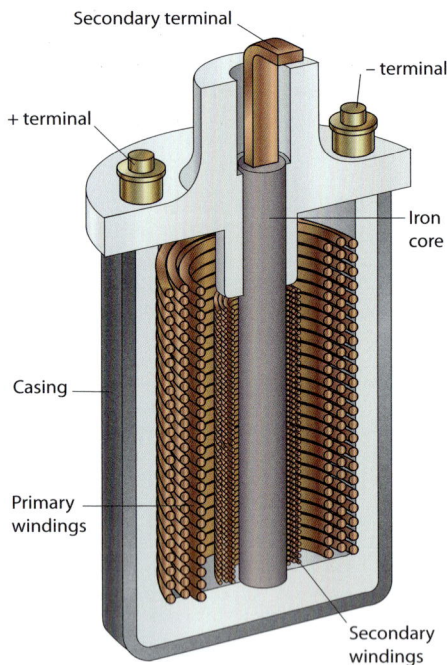
Figure 5.38 Ignition coils

Key terms

Copper windings – copper wire that has been wound into a coil.

Primary – the large loosely wound coil in the low tension circuit.

Secondary – the thin tightly wound coil in the high tension circuit.

Spark ignition systems

To create combustion inside the engine, you need three important things:

- a source of fuel (in most cases petrol or diesel)
- a source of oxygen (found in the air drawn into the engine)
- a source of heat.

In a petrol engine, the source of heat is provided by a high voltage spark created at the plug.

A diesel engine doesn't need an ignition system or spark plugs because it uses superheated air to ignite the fuel.

Working life

Jean-Luc is working on a running engine when he accidentally touches a plug lead.

1 What might happen?

2 Why would this happen?

3 What might the result be?

Components and purpose

Ignition coil

The ignition coil is a step-up transformer. It is designed to take the 12V found in the car battery and increase it to many thousands of volts. This high voltage lets the spark jump the gap of the plug and ignite the air and fuel.

Voltage is electrical pressure. If the voltage from a large coil of copper wire is squeezed into a thin coil of copper wire, pressure is increased (the voltage goes up).

Magnetic fields are used to create this electric voltage.

- An invisible magnetic field is produced if an electric current is passed through a copper wire.
- An electric current is created in a coil of copper wire if a magnetic field is moved through it.

In an ignition coil two **copper windings** are used, one inside the other. If electric current is passed through the first large coil (**primary**), a strong magnetic field is produced. This also surrounds the second much thinner coil (**secondary**). If the feed is switched off to the primary coil, electric current stops flowing and the magnetic field collapses. The collapsing magnetic field moves through the secondary thin coil and high voltage (pressure) is squeezed into this circuit. The high voltage can now be directed to the correct plug and create a spark.

HT leads

High tension (HT) leads are the wires which carry the high voltage electricity from the ignition coil to the spark plugs. They are highly **insulated** to reduce the possibility of electrical leakage (short circuit). They can be separated into plug leads and the main coil lead (sometimes called the **king lead**).

LT leads

The **low tension (LT)** circuits and leads are those that carry the 12V electricity from the battery. They are normally a standard size copper wire. They have thinner insulation as the voltage is much lower than that found in a high tension circuit.

Spark plugs

Spark plugs in petrol engines ignite the fuel and start the rapid burn at precisely the correct time. The gap at the tip of the spark plug creates an **open circuit** and no electric current can flow until a high enough voltage is produced to overcome the **resistance** of the air gap. (Voltage or pressure is increased until it reaches a point that a spark can be forced to jump across the plug gap.)

For this purpose two **electrodes** are used:

- a centre electrode, normally fed from the high tension ignition lead
- an earth electrode, usually made as part of the spark plug shell.

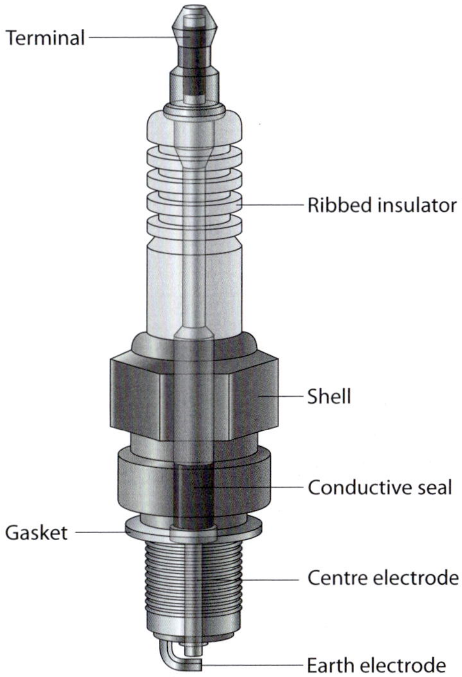

The tips of these electrodes are machined to precise shapes.

Terminal
Ribbed insulator
Shell
Conductive seal
Gasket
Centre electrode
Earth electrode

Figure 5.39 Spark plug

Spark plug dimensions

Many spark plugs are made so that they can be used across a range of cars. When they design engines, vehicle manufacturers take into account the size and shape of spark plugs and also how hot the spark plug will run. This is known as heat range.

Most new spark plugs do not come 'ready gapped' from the plug manufacturer. They need setting before they are fitted to a particular engine. You should follow the manufacturer's recommendations to set the spark plug gaps. Use feeler gauges to accurately measure the gap and adjust if necessary.

Key terms

High tension (HT) – the high voltage electricity created by the ignition coil secondary circuit.

Insulated – coated in a material to help prevent the leakage of electric current.

King lead – the main high tension lead leaving the ignition coil.

Low tension (LT) – the low voltage electricity from the battery that flows through the primary circuit of the ignition coil.

Open circuit – an electrical circuit with a break or a gap in it.

Resistance – something which slows electricity down.

Electrodes – the metal tips that are found at the gap of the spark plug.

Did you know?

The size indicated on a spark plug socket refers to the diameter of the plug thread. It is not a measurement across the flats of the nut surface (as found on standard sockets which are used for nuts and bolts).

Table 5.5 Spark plug specifications

Specification	Meaning
Thread diameter	Spark plugs are made with a number of different thread sizes, most commonly 10, 12, 14 and 18 mm.
Spark plug reach	The reach of a spark plug is the length of the threaded section that is screwed into the cylinder head. These are normally described as long reach and short reach.
Spark plug gap	The air gap found between the electrodes at the tip of the spark plugs has been accurately calculated by the manufacturer during the engine design process.
Heat range	The ideal operating temperature of the spark plug is normally between 500°C and 850°C.

Heat range

If the temperature of a spark plug is too low, it can become dirty (known as **fouling**). This makes it **misfire**.

If the temperature is too high, electrode and/or piston damage can occur (components can melt).

The heat range of a spark plug is determined by its shape and design.

The length of the ceramic **insulator** will help to make the spark plug run either 'hot' or 'cold':

- **Hot** – a long insulator around the electrode means that heat has to travel much further before it can be transferred to the spark plug walls and into the cylinder head. Because of this the spark plug will run hot.

- **Cold** – a short insulator around the electrode means that heat does not have to travel so far before it can be transferred through the spark plug walls and into the cylinder head. Because of this it will run cold.

You must always use spark plugs with the correct part number otherwise engine damage could occur.

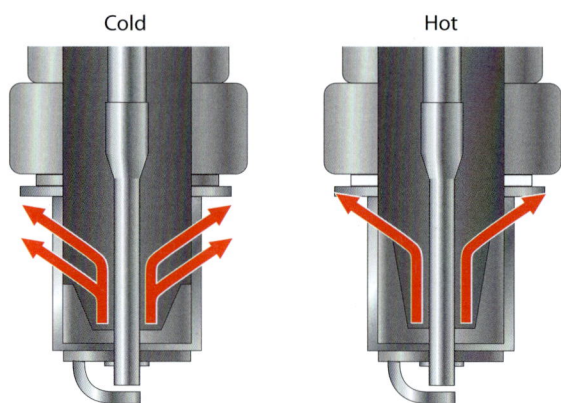

Figure 5.40 Spark plug heat range

Key terms

Fouling – a build-up of combustion products around the electrodes of the spark plug.

Misfire – when a spark plug does not get a spark, or the spark happens at the wrong time.

Insulator – a ceramic coating around the centre electrode of the spark plug.

Find out

Choose a vehicle in your workshop and investigate the manufacturer's recommended spark plug:

- brand (who makes the original spark plugs for the manufacturer)
- type (part number)
- electrode gap.

Conventional ignition systems

Early ignition systems used a mechanical switch known as a **contact breaker** (sometimes called **points**). Contact breakers were used to turn the primary circuit of the ignition coil on and off. They were normally opened and closed from their own camshaft, which turned at half the speed of the crankshaft. This contact breaker camshaft would have a lobe for each cylinder requiring a spark.

When the engine was cranked or running, the contact breakers would open and close.

When closed the coil would charge up and electric current flowed through the primary winding, making a magnetic field. As soon as the contact breakers opened, current flow stopped and the magnetic field in the primary circuit collapsed. As the magnetic field collapsed a high voltage was created in the secondary winding and the spark was produced.

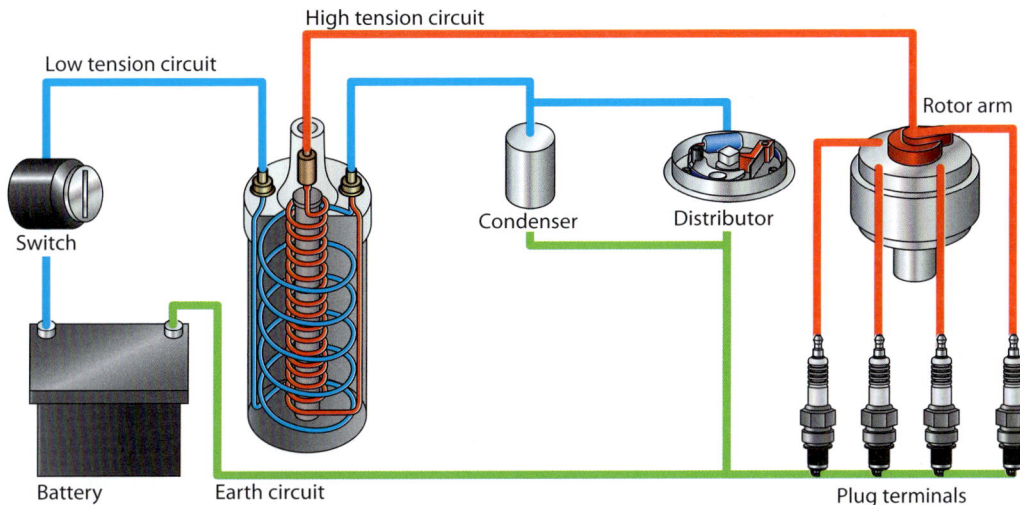

Figure 5.41 A conventional ignition circuit

Condensers

As the contact breakers open and close, sparks can occur, causing damage to the points. A component called a condenser is fitted to help prevent this. It acts as a temporary storage for the electric current, preventing **arcing** and increasing the lifespan of the contact breakers. When the contact breaker gap is so wide that the spark can no longer jump between the faces of the points, the condenser discharges back through the primary winding. This will help speed up the collapse of the magnetic field and improve the performance of the ignition coil.

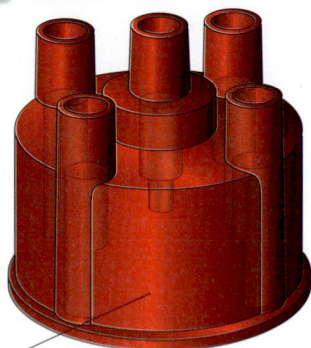

Distributor cap

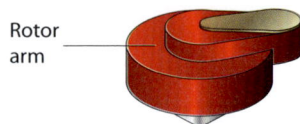

Rotor arm

Figure 5.42 Distributor cap and rotor arm

Distributor cap and rotor arm

Once the high voltage has been created in the coil, it moves through the king lead to the distributor cap. The high-voltage electricity enters the distributor cap through a centre contact, which touches a rotating arm inside the distributor. As this rotor arm turns, it will point to the correct spark plug lead in the engine's firing order.

Electronic ignition systems

In an electronic ignition system, many of the same parts used in a conventional ignition system exist. The main difference is that contact breaker points are no longer used.

Electronic control units (ECU)

Modern ignition systems now use electronic control to switch the primary circuit off and on. The mechanical contact breakers have been replaced by an electronic switch with no moving parts (in most cases a **transistor**) that can be contained inside a computer known as an electronic control unit (ECU).

The advantages of using an electronic system to switch the primary circuit of the ignition coil are:

• no mechanical wear is produced
• reaction time is far quicker
• greater accuracy
• better ignition timing and reliability.

Figure 5.43 An engine management electronic control unit

Table 5.6 Ignition system components and their purposes

Ignition system component	Purpose
Spark plug	Creates the spark in the cylinder to ignite the air/fuel mixture
Ignition coil	Transforms battery voltage (12V) into many thousands of volts to create the spark at the spark plug
Plug leads	Highly insulated wires used to transfer high voltage electricity to the spark plugs
Contact breakers	A mechanical switch used to turn the ignition coil primary circuit on and off
Condenser	A temporary storage of electricity used to help prevent arcing at the contact breaker points
Distributor cap and rotor arm	Used to direct high voltage electricity to the correct spark plug lead
Electronic control unit (ECU)	An electronic switching device used to replace the contact breaker points in a modern ignition system

CHECK YOUR PROGRESS

1 What does HT mean?
2 What tool is needed to check spark plug gaps?
3 What are diesel glow plugs used for?

Fuel systems

An internal combustion engine needs a source of oxygen (which it gets from air) and fuel (petrol or diesel). The air and the fuel need to be mixed and then ignited by a heat source inside the combustion chamber. This creates the high pressures needed to drive the pistons and provide power.

- **Spark ignition (SI)** – with a spark ignition (petrol) engine, air and fuel are usually mixed in the air intake or induction manifold before they are taken into the combustion chamber.
- **Compression ignition (CI)** – with a compression ignition (diesel) driven engine, only air is drawn through the induction system. Fuel is introduced once it has entered the combustion chamber.

Air filtering

Air entering the engine must be clean and free from dirt and dust particles. The engine's mechanical components could wear very quickly if dirt and dust are allowed to enter. As air enters the intake system, it passes through an air filter. The air filter can be made of a specially designed paper which is folded backwards and forwards (pleated) to increase its surface area. Alternatively, it could be made from foam-like material which is sometimes soaked with oil so that dust and dirt sticks to it.

Air filter housing (air box)

Air filter

Fresh air induction pipe

Figure 5.44 Air filter system

The fuel system

The fuel system for both petrol and diesel engines starts with a storage container called the tank. A method is needed to transfer fuel from the tank to the engine through pipes (often called 'fuel lines'). This can be done using a fuel pump, which can be electrical or mechanically driven. A fuel filter is then used to remove any small particles of dirt and moisture before the fuel enters the engine.

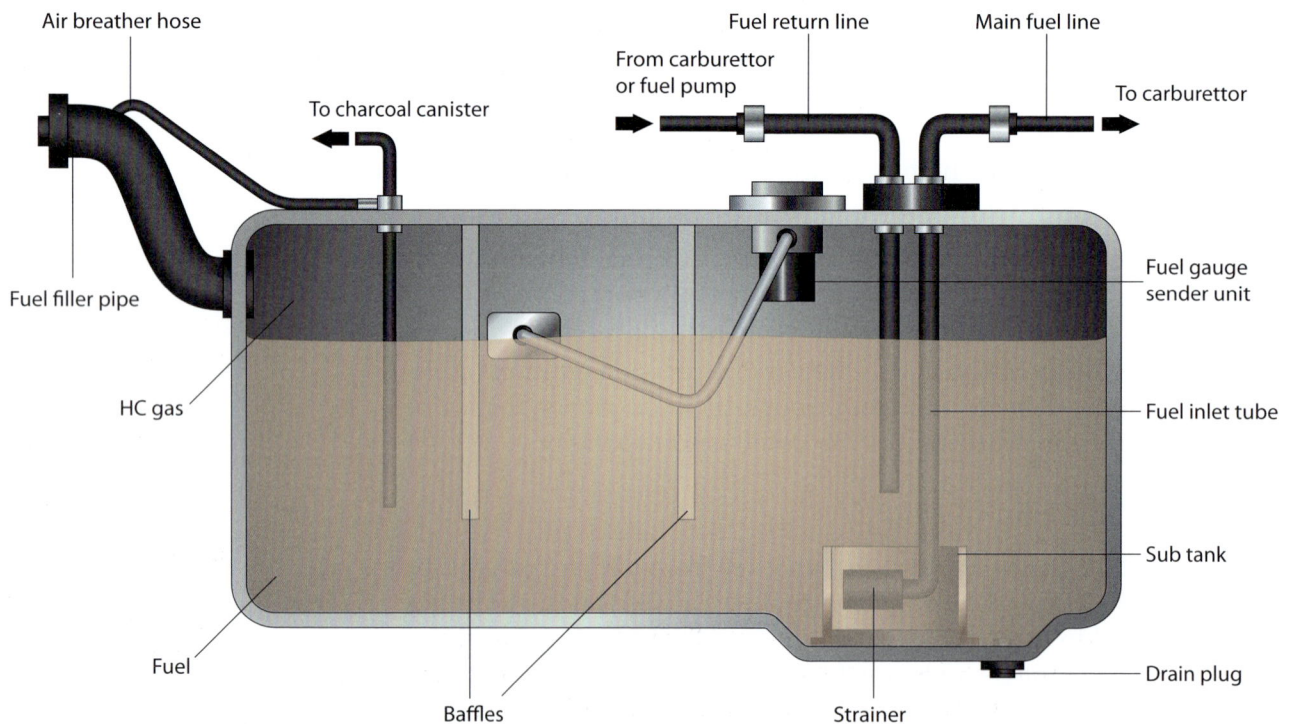

Air breather hose

Fuel return line

Main fuel line

From carburettor or fuel pump

To charcoal canister

To carburettor

Fuel gauge sender unit

Fuel filler pipe

HC gas

Fuel inlet tube

Sub tank

Fuel

Drain plug

Baffles

Strainer

Figure 5.45 A fuel tank

A mechanism is needed to mix the air and fuel so that it can be burnt in the combustion chamber. This can be done by a number of different methods including:

* carburettor
* single point fuel injection
* multi-point fuel injection.

On older petrol driven cars, a carburettor is used.

A carburettor uses the movement of air and different air pressures to mix air and fuel in the correct quantities.

The main components of a carburettor are:

* **The venturi** – designed to lower air pressure in the inlet tube. The venturi is a restriction which causes air to speed up. As air speeds up, its pressure will fall.

The amount of air passing through the venturi at any one time is controlled by a mechanical flap. This is known as the throttle **butterfly**.

Engine speed on a petrol engine is controlled by the amount of air entering the combustion chamber. The more air that goes in the engine, the faster the engine will run. The 'butterfly' is connected to the accelerator pedal (either with a cable or electronically). This enables the driver to control the amount of air and therefore the speed of the engine.

* **The float chamber** – a small reservoir of petrol to one side of the venturi. It holds an accurate amount of fuel ready to be mixed with the air in the carburettor. The fuel is kept at a constant level by a float and needle valve. (It works in a similar way to a ballcock system found in a toilet cistern.)

An accurately sized tube known as the main jet is directed upwards out of the float chamber. It sticks out into the narrowest section of the venturi.

When the engine is running, a low pressure (depression) is created by the downward movement of the piston drawing air in through the venturi of the carburettor. As it passes through the restriction it speeds up and air pressure drops, creating a pressure lower than atmosphere. Petrol is drawn up the main jet and out into the air stream, just like sucking on a straw. The fast moving air breaks the petrol up into small droplets – a process called **atomisation**.

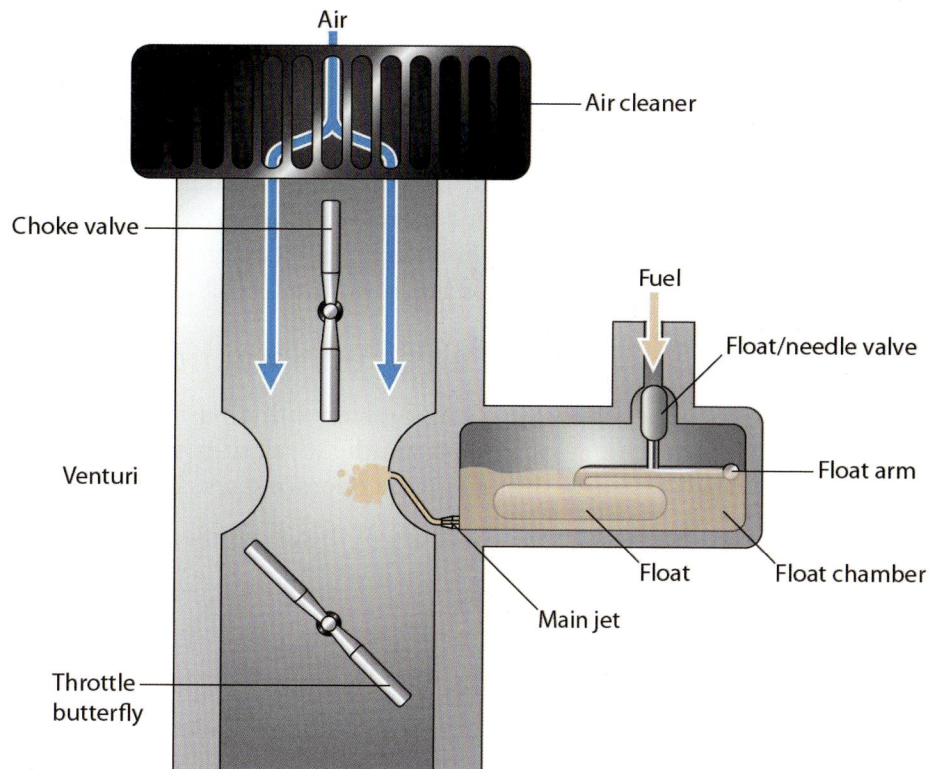

Figure 5.46 A carburettor

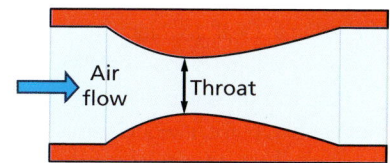

Figure 5.47 The Venturi section of a carburettor

Key terms

Venturi – a narrowing of the air intake (usually in a carburettor) designed to speed up airflow and lower air pressure.

Butterfly – a mechanical flap that controls the amount of air entering a petrol engine.

Atomisation – the process of braking petrol up into tiny droplets or a spray.

Find out

Find and copy a labelled drawing of a basic fixed choke (venturi) carburettor. It should include: venturi; butterfly valve; main jet; float chamber; float and needle valve.

Key term

Stoichiometric – a balanced chemical reaction used to represent the correct air/fuel ratio.

Did you know?

Because petrol is liquid it can have considerable weight (1L of petrol will weigh approximately 1 kg), whereas air weighs relatively little in comparison. If you were to burn 1 kg of petrol, 14.7 kg of air would also have to be used for correct stoichiometric operation. It is sometimes easier to think of this as volume. Ideal stoichiometric value by volume is approximately 11000 to 1 (meaning that if 1L of petrol is burnt, it would use approximately 11000L of air).

An ideal mixture of air and fuel is needed so that an engine can run properly. This way, when it is burnt in the cylinder, a chemical reaction takes place that uses all of the combustible elements and little waste is left over. This is normally referred to as the **stoichiometric** value or air/fuel ratio. The ideal air/fuel ratio for a standard petrol engine is 14.7 to 1 by mass (or weight). This means one part petrol is mixed with the equivalent mass (weight) of 14.7 parts air. To help represent the ideal air/fuel ratio the Greek letter lambda is used (λ).

Petrol fuel injection

The problem with a basic carburettor is that it is very difficult to maintain the ideal air fuel ratio of 14.7 to 1 by mass over a large range of engine operating speeds and conditions.

Because of this, fuel injection has been developed.

Single point (or throttle body) fuel injection

Early developments in electronic fuel injection included single point or **throttle body** fuel injection.

A throttle body of similar size, shape and position as a carburettor is mounted on the inlet manifold. A single large fuel injector is positioned above the throttle butterfly. Intake air is regulated by the throttle

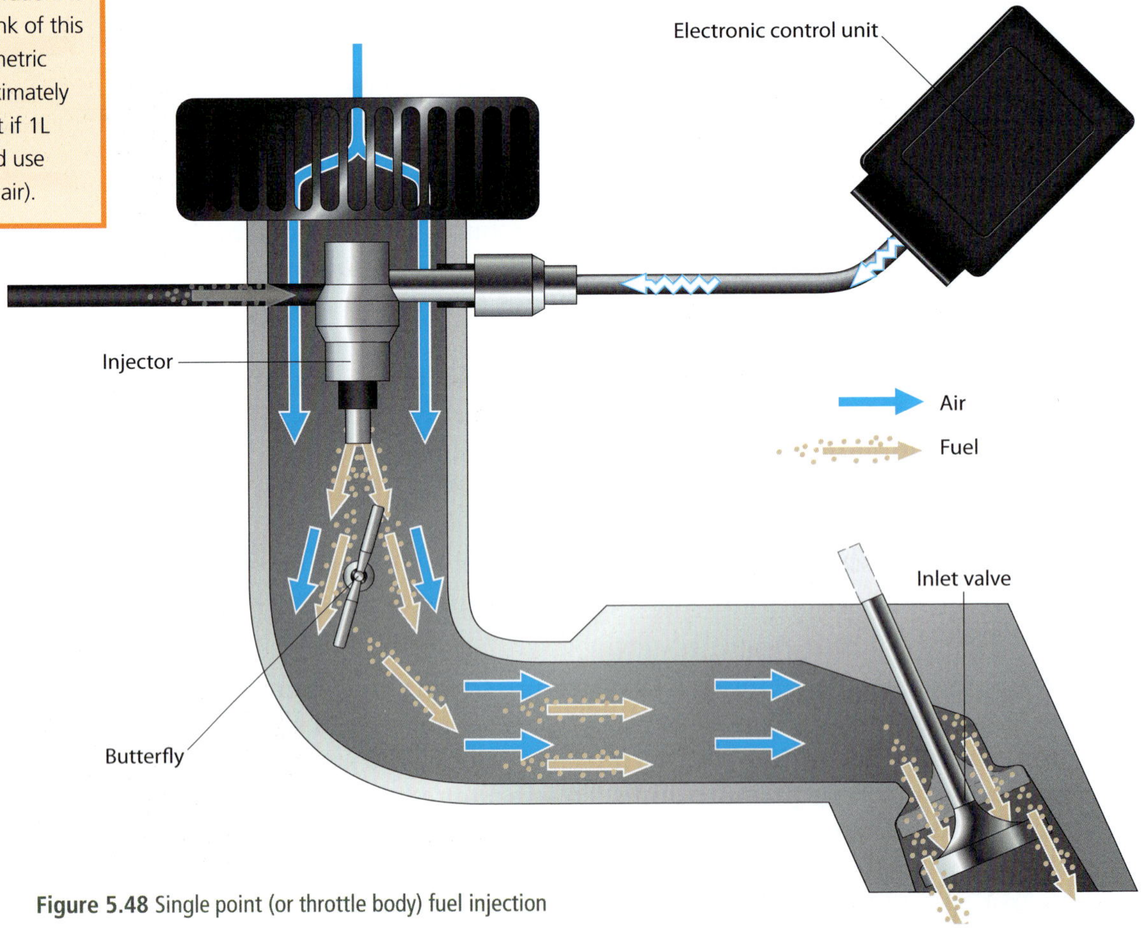

Figure 5.48 Single point (or throttle body) fuel injection

152

butterfly and an electronically controlled fuel injector is used to atomise varying quantities of fuel into the air stream. This mixture of air and fuel travels along the intake manifold and into the cylinder.

Multi-point (or electronic) fuel injection

In multi-point fuel injection each cylinder has its own fuel injector mounted in the inlet manifold, just before the intake valve. The air intake tract now only draws air, and fuel is injected at the last moment before it enters the cylinder. In this way the quantity of fuel and its timing can be accurately controlled.

Multi-point fuel injection uses an electric pump to transfer fuel from the tank to the engine. After passing through a fuel filter to remove dirt particles, it arrives at a pipe connected to the fuel injectors. This is called the **fuel rail** – the fuel is held here under pressure.

Air measurement sensors mounted in the inlet manifold work out how much air is being drawn into the engine. They then send this information to an electronic control unit (ECU – see page 148). The ECU can now calculate the correct quantity of fuel to be injected to get the ideal air/fuel ratio.

The injectors are then opened for a set time period. This allows more or less fuel to be injected depending on engine requirements.

Key terms

Throttle body – an intake component containing the butterfly flap.

Fuel rail – a single fuel pipe that all of the fuel injectors are connected to.

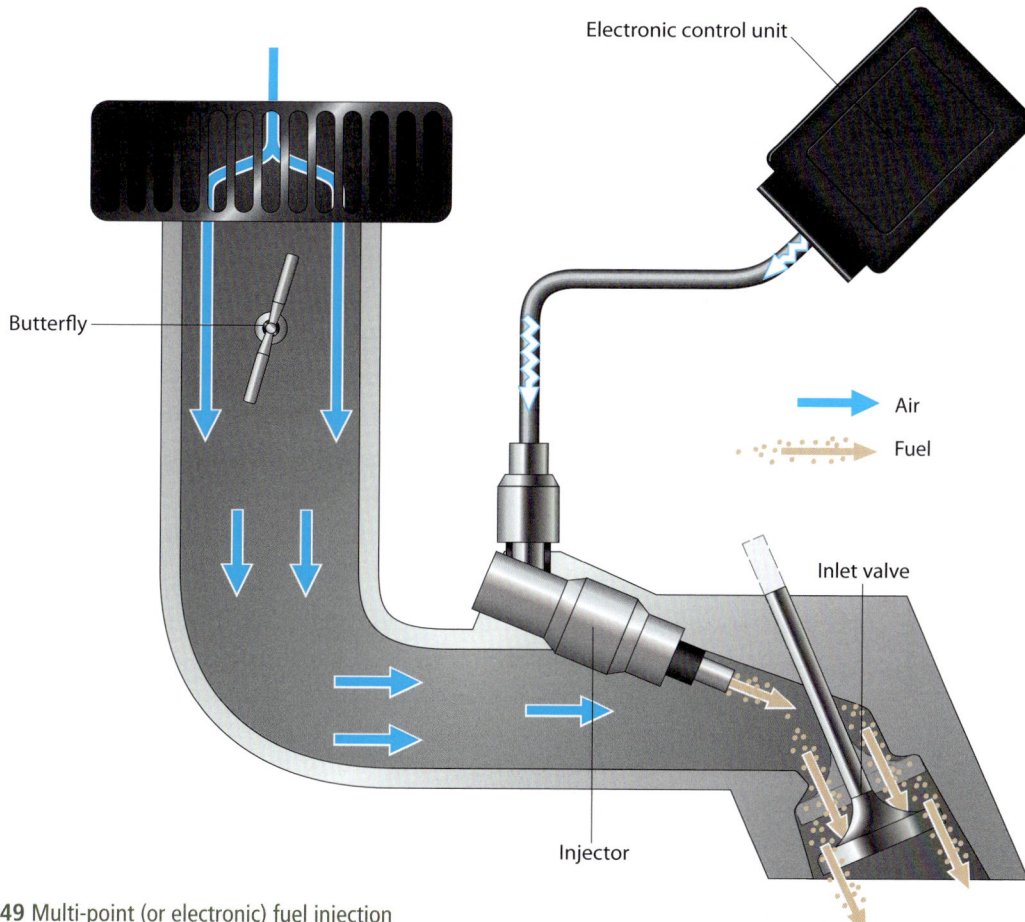

Figure 5.49 Multi-point (or electronic) fuel injection
(Note: This diagram shows only one cylinder. However, multi-point injection involves several cylinders, each with its own injector.)

Diesel injection

With a diesel engine, the fuel taken from the tank is put under extreme pressure by a pump.

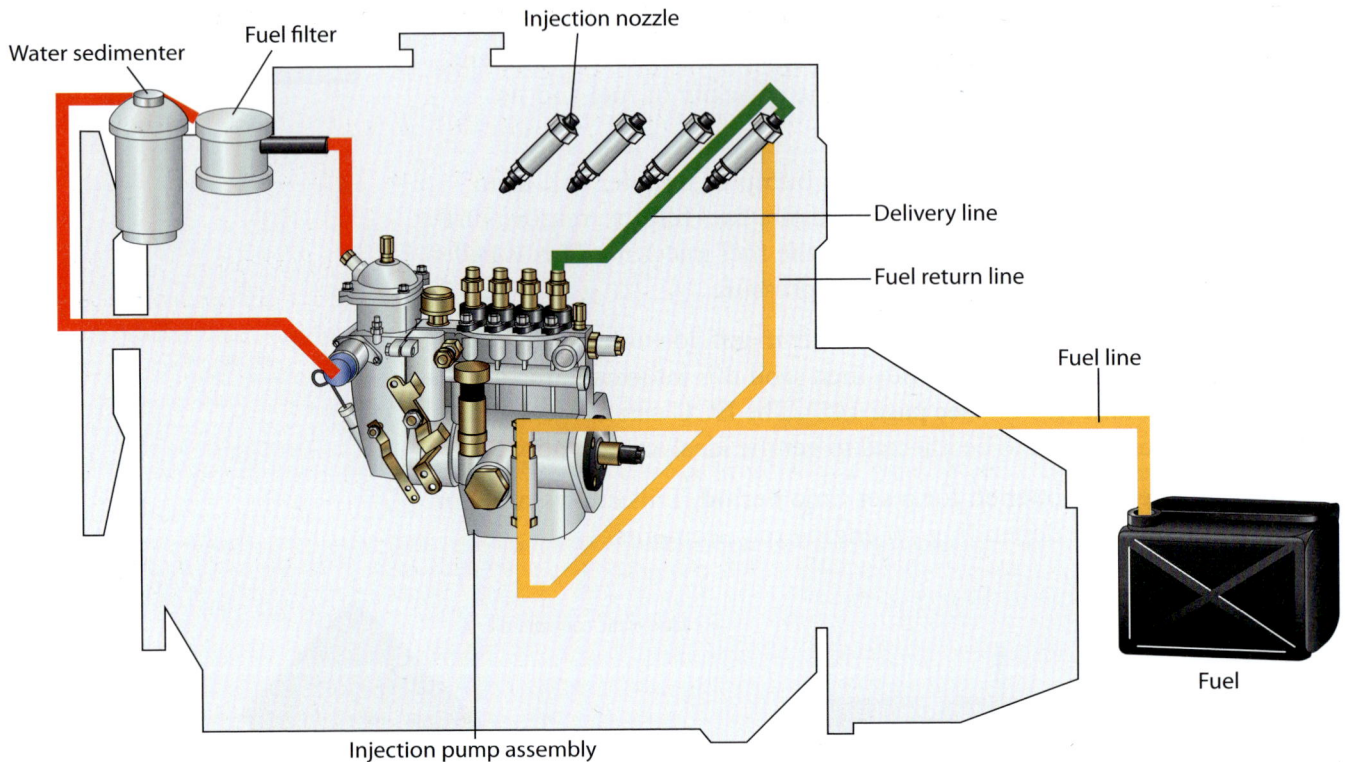

Figure 5.50 Diesel fuel injection

On mechanical systems, the injectors open and spray fuel into the combustion chamber when the fuel pressure reaches a preset value. The high pressure injection pump is timed and **sequenced** so that the diesel is injected into the correct cylinder at exactly the right time.

In a modern **common rail** diesel engine, high pressure fuel is stored in a container (a rail or **accumulator**). Electronically controlled fuel injectors open to spray this high pressure fuel into the combustion chamber when switched by computer (ECU).

Engine speed on a diesel engine is controlled by the amount of fuel entering the combustion chamber. The more fuel that enters the engine, the faster the engine will run. The accelerator pedal is connected to the fuel injection system (either with a cable or electronically). In this way, the driver is able to regulate the amount of fuel and therefore the speed of the engine. As the air is not regulated in a diesel engine, only the fuel, ideal air/fuel ratios are difficult to maintain. When the engine is running slowly it will be weak (too much air), and when running quickly it can be rich (too little air).

Table 5.7 Fuel system components and their purposes

Fuel system component	Purpose
Fuel tank	Provides a storage area for the fuel (both petrol and diesel)
Fuel filter	Helps remove small dirt particles from the fuel before it travels to the engine
Fuel pump	Transfers fuel under pressure from the tank to the engine
Carburettor	A component used to mix petrol and air in the intake system of a spark ignition engine
Single point fuel injection	A form of fuel injection which uses a single injector to mix petrol and air in the intake system of a spark ignition engine
Multi-point fuel injection	A form of fuel injection which uses an injector for each cylinder of the engine. The injectors atomise fuel at the intake valve just before it enters the cylinder of a spark ignition engine
Electronic control unit (ECU)	A small computer used to control electronic fuel injection systems
Diesel injection pump	A high pressure mechanical pump which supplies diesel to the injectors at the correct time and in the correct order
Common rail injection	A modern form of diesel injection which operates on a similar principle to the multi-point fuel injection of a petrol engine

Fire precautions

Petrol is volatile. This means it is unstable and turns into a vapour easily at room temperature. The fumes given off by petrol are highly flammable.

Diesel fuel is much more stable than petrol. The temperature has to be increased before it will start to vaporise. However, care still needs be taken when working on diesel fuel systems and you must avoid all sources of ignition.

Handling and disposing of materials

You must consider environmental issues when you dispose of used materials while working on fuel systems (both petrol and diesel). You should not throw away fuel soaked rags where there is a possibility that they could be ignited. If fuel is spilt, make people around you aware of the hazard and clean up immediately.

Fuel should never be tipped down the drain. Petrol and diesel will float on water and if they enter the drain system they will travel considerable distances. This could pose a fire risk for some miles around.

See page 117 for more information on the safe handling and disposal of materials.

Safe working

When working on fuel systems you must avoid all sources of ignition (i.e. do not smoke, only work on a cold engine).

Emergency

Keep an appropriate fire extinguisher handy, ready to use in case of emergency.

When changing a fuel filter on either petrol or diesel engines, the systems should be de-pressurised. The easiest way to de-pressurise a fuel system is to isolate the fuel pump (if it is an electric pump, this can be done by removing the fuse) and start the engine. The engine should run until the fuel pressure is used up, and then it will stall.

Maintenance precautions

The fuel systems on both petrol and diesel vehicles should be kept very clean. If you need to dismantle the fuel system, it is important that no contamination (dirt, moisture or foreign matter) is allowed to enter.

Change fuel and air filters

Air and fuel filters should be replaced at manufacturers' specified intervals.

Figure 5.51 Fuel filter

Changing a diesel fuel filter

Checklist			
PPE	**VPE**	**Tools and equipment**	**Source information**
• Steel toe-capped boots • Overalls • Latex gloves	• Wing covers • Steering wheel covers • Seat covers • Foot mat covers	• Spanners • Screwdrivers • **Lint-free** cloth • Container to catch spilt fuel • Fire extinguisher	• Manufacturer's technical data • Filter manufacturer instructions

1. Place a container under the filter to catch any spilt fuel.

2. Remove the filter carefully to prevent contamination from dirt.

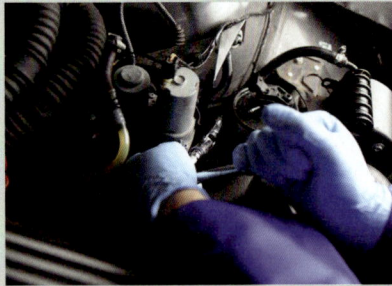

3. Clean the area around the filter with a lint-free cloth.

4. Wipe a small amount of diesel around the new seal to help the sealing process. Fit and tighten the filter.

5. Bleed air from the diesel system.

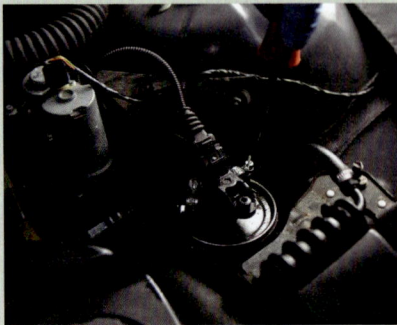

6. Undo the bleed screw found on the filter housing and operate the hand **priming** pump until air-free fuel flows from the bleed screw.

7. Tighten the bleed screw and clean up any spilt fuel.

8. Run the engine and check correct function and operation.

Safe working

You must always make sure it is safe before you start or run a car to check for correct function and operation.

- Make sure that the car is not in gear.
- Make sure that nothing could be trapped in the operating engine.
- Connect exhaust extraction equipment.
- Make sure that others are aware of your actions so they are not harmed.

Did you know?

Some manufacturers' design systems are known as **self-purging**. This means that as you crank the engine, the fuel return hose to the tank takes away any air bubbles.

Key terms

Lint-free – a non-fluffy cloth.

Priming – the initial filling of the system.

Self-purging – self-bleeding (removing the air from the system).

CHECK YOUR PROGRESS

1   List three ways of mixing petrol and air.
2   What controls the speed of a diesel engine?
3   What controls the speed of a petrol engine?

Exhaust systems

After the air and fuel has been burnt inside the combustion chamber, the waste products (exhaust gases) must be removed and transferred to the atmosphere. The chemical process that takes place during combustion converts the air and fuel into pollutants. These could be harmful if breathed in. A system of steel pipe work is designed to duct (send) these exhaust gases past the passenger compartment and out of the back of the car.

Some exhaust systems contain a catalytic converter. This is a component which uses a chemical reaction to make some of these pollutants less harmful.

The exhaust system will also contain a number of silencers. These help to reduce the amount of noise made by the engine during the combustion process.

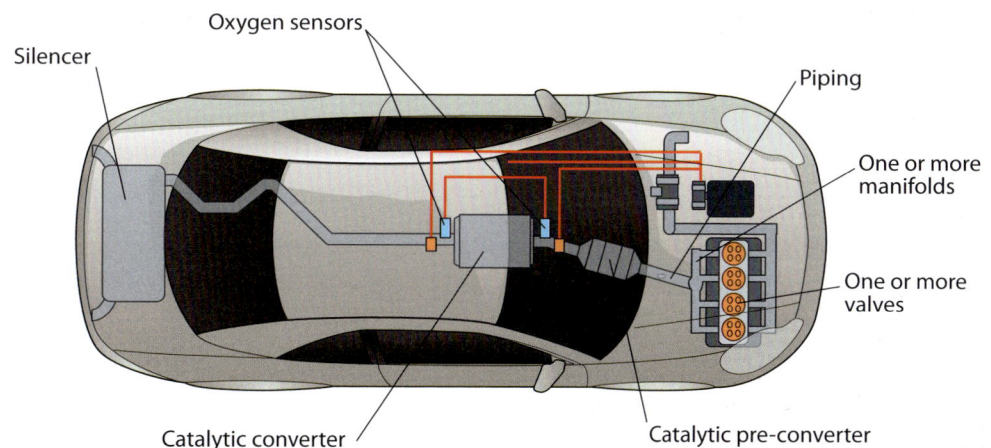

Figure 5.52 Exhaust system

Exhaust system components

Front pipe

The front pipe is the first part of the exhaust system after the manifold. It is carefully shaped so it does not restrict the flow of exhaust gas leaving the engine.

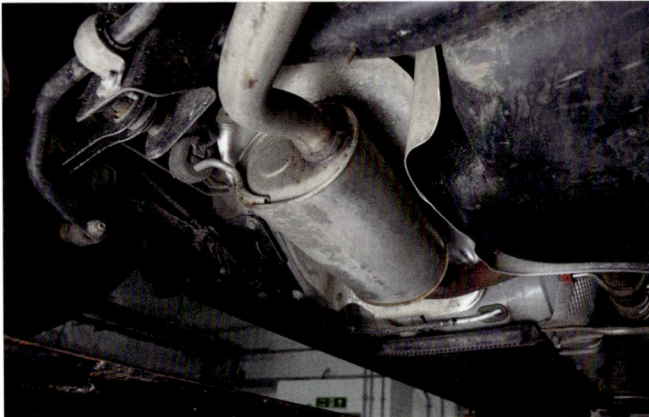

Figure 5.53 Exhaust silencer

Silencers

The noise of the engine's combustion process is also conducted away by the exhaust system. To help reduce this noise, the exhaust pipe is fitted with silencers. Two main types of silencer exist:

- **Absorption** – usually contains a glass fibre material. As the exhaust gases pass through it, it helps to remove high-pitched noises.
- **Expansion** – usually a larger box containing baffles (metal plates and deflectors). These slow the exhaust gases down, allowing them to expand and remove low-pitched noises.

Catalytic converter

> **Key term**
>
> **Catalytic** – a chemical reaction that will bring about a change in a substance.

The **catalytic** converter takes harmful exhaust gas pollutants and turns them into less harmful emissions. It is positioned in the exhaust system as close as possible to the engine. It may look like a silencer from the outside. The inside of a catalytic converter is usually made of a ceramic honeycomb. It can be coated in precious metals such as platinum, palladium and rhodium. When these metals come into contact with the exhaust gases, they start and continue a catalytic reaction. This converts the pollutants at a temperature of between 600°C and 800°C.

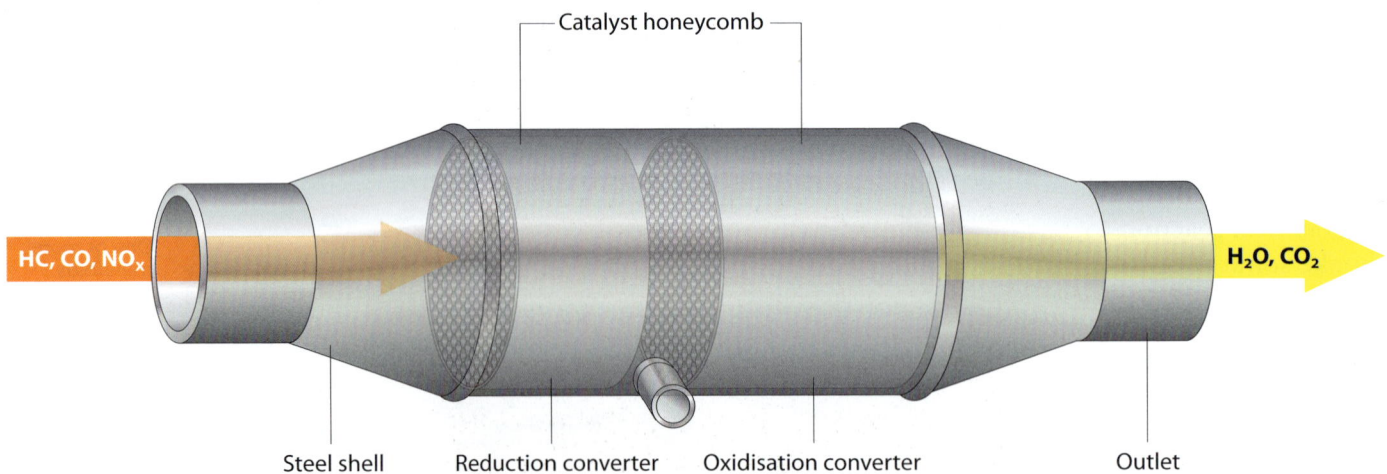

Figure 5.54 A catalytic converter

> ⚠️ **Safe working**
>
> You must take care when working around the catalytic converter as flammable items can be easily ignited, and you could burn yourself.

Lambda sensor

A catalytic converter needs the engine to be running with the correct air/fuel ratio so that it can work correctly. An oxygen sensor, known as a Lambda sensor, is fitted in the exhaust system before the catalytic converter. It makes sure that the fuel injection system is operating within fuelling limits.

The Lambda sensor measures the amount of oxygen in the exhaust gases as they leave the engine. It 'tells' the engine management whether it is running too rich or too weak, so that adjustments can be made. (If there is too much oxygen in the exhaust gases the engine is running weak. If there is too little oxygen in the exhaust gases the engine is running too rich.)

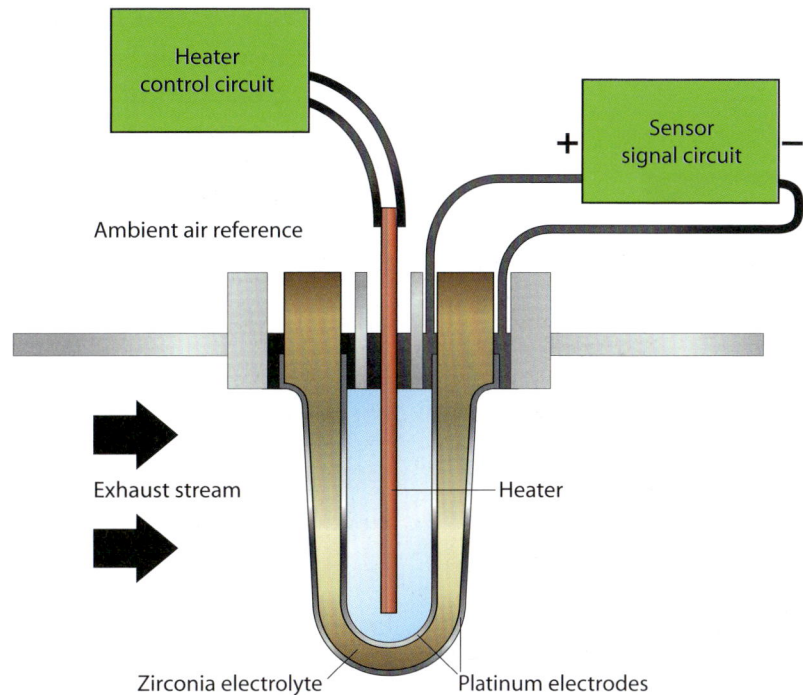

Figure 5.55 Lambda sensor

Tailpipe

The tailpipe is the short metal tube that sticks out from the end of the rear silencer box. It directs exhaust gases away from the vehicle body.

Exhaust system brackets and mountings

An exhaust system must be supported by brackets that are mounted to the vehicle's body because of its length. The brackets are normally connected by rubber mountings. This reduces exhaust vibrations being transferred to the vehicle body.

Figure 5.56 A car exhaust tailpipe

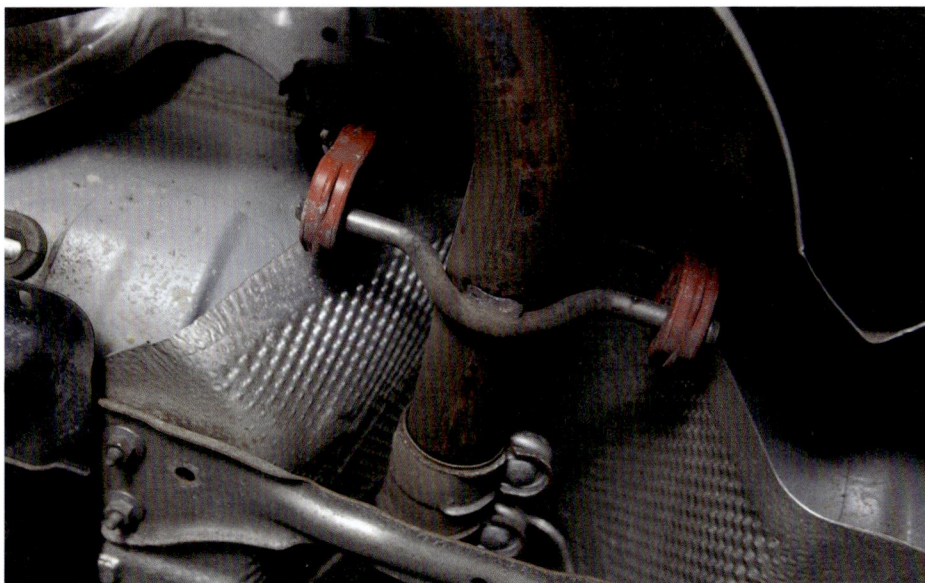

Figure 5.57 An exhaust in its mounting

Exhaust system joints and gaskets

Most exhaust systems are made in sections. This allows easy fitting. In addition, if one part of the exhaust fails due to damage or corrosion, individual sections can be replaced instead of the entire system. Each individual section joint must have a good seal to prevent the escape of exhaust gases. Exhaust joints can be tubular, with one sitting inside the other and a clamp to hold them in place, or a flat metal face (**flange**) with a gasket in between.

Table 5.8 Exhaust system components and their purposes

Exhaust system component	Purpose
Lambda sensor	Measures the amount of oxygen in the exhaust gases
Front pipe	Directs exhaust gases away from the engine
Absorption silencer	Helps remove high-pitched exhaust noise
Expansion silencer	Helps remove low-pitched exhaust noise
Catalytic converter	Converts harmful exhaust pollutants into less harmful pollutants
Tailpipe	Directs exhaust gases away from the back of the car
Brackets/ mountings	Supports/attaches the exhaust system to the vehicle body
Gaskets	Make gas-tight seals at joints to prevent the leak of exhaust gases

Exhaust emissions

During the combustion process, the air/fuel mixture is burnt within the cylinder. This releases heat energy and changes its chemical make-up.

Air is made up of approximately:

- 78% nitrogen
- 21% oxygen
- 1% other gases.

It is the oxygen from air that is used during the combustion process. Nitrogen is an **inert** gas which does not support combustion. Therefore, in theory, it should pass through the system unchanged.

Fuel (both petrol and diesel) are hydrocarbons (HC). Depending on how they are **refined**, they are made up of approximately:

- 84% carbon
- 14% hydrogen
- 2% other chemical elements.

It is the hydrogen from the petrol or diesel that is flammable and provides the source of fuel. The carbon is not used during the combustion process and is a by-product that can often be seen as black soot.

If perfect combustion could be achieved, the exhaust emissions would contain carbon dioxide (CO_2), water (H_2O) and nitrogen (N), as shown in Figure 5.58. These elements are made up from the chemicals shown above. Unfortunately, perfect combustion is rarely possible and the elements combine to produce pollutants that are dangerous to both the environment and human health.

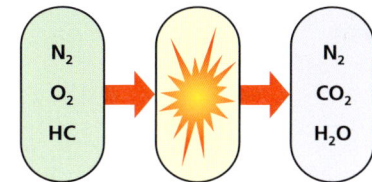

Figure 5.58 Perfect combustion – a stoichiometric reaction

Table 5.9 Exhaust emissions

Emission	Cause and effect
O_2 (oxygen)	Oxygen which has passed through the combustion process unchanged. It is not considered a harmful emission.
HC (hydrocarbons)	Hydrocarbons are unburnt fuel that has passed through the combustion process unchanged. This is an environmental pollutant and its release to the atmosphere should be restricted.
CO (carbon monoxide)	Carbon monoxide is partially burnt fuel (petrol or diesel that has started to burn and then gone out). It is usually caused by a lack of oxygen or rapid cooling. Carbon monoxide has no odour or taste but is harmful to health. If breathed in, it will replace the oxygen carried around the body by blood cells. This then starves the body's organs of oxygen and can lead to death.
CO_2 (carbon dioxide)	Carbon dioxide is actually a product of good combustion. However, it is a 'greenhouse gas' and therefore damaging to the environment, even though it is not considered directly harmful to health.
NO_X (oxides of nitrogen)	Oxides of nitrogen are produced under the extreme heat of combustion when temperatures are higher than 1800°C. This is a harmful pollutant to both the environment and health. It can lead to breathing problems if it is inhaled.
Particulate matter	This is a further by-product of the combustion process and is particularly connected to diesel engines. It is often called soot.

Did you know?

Carbon monoxide can be likened to a burning piece of wood. If it burns away to ash then combustion is complete. However, if it goes out halfway through and you are left with charcoal, then it has only partially burnt.

Emergency

Exhaust extraction must be used, or adequate ventilation provided, when an engine is running in a workshop. This is because of the exhaust gas pollutants that are considered dangerous to health. If exhaust gases are allowed to build up in a confined area, this could make you extremely unwell or even cause death.

Figure 5.59 An extraction pipe attached to a car exhaust

CHECK YOUR PROGRESS

1 Name four exhaust pollutants.
2 What does the word stoichiometric mean?

Routine engine maintenance

You must have a good source of technical data when carrying out routine maintenance on light vehicle engine systems. This will provide you with correct measurements and specifications. Sources of technical data can include:

* paper-based workshop manuals
* technical data books
* manufacturers' technical bulletins
* electronic-based technical data
* Internet-sourced information.

Routine maintenance of engine liquid cooling and lubrication systems

Manufacturers make recommendations about when cooling and lubrication systems should be serviced. These recommendations can be based on a time or mileage covered by the car.

Normally it is necessary to start and run the engine after you have checked and replaced engine components. Always follow recommended health and safety procedures.

Engine oil and filter change

Checklist			
PPE	VPE	Tools and equipment	Source information
• Steel toe-capped boots • Overalls • Latex gloves	• Wing covers • Steering wheel covers • Seat covers • Foot mat covers	• Spanners • Screwdrivers • Oil filter remover • Measuring jug • Lint-free cloth • Oil drainer • Torque wrench	• Technical data • Filter and oil manufacturer instructions • Service schedule • Customer's service history

1. Open the bonnet. Remove the oil filler cap.

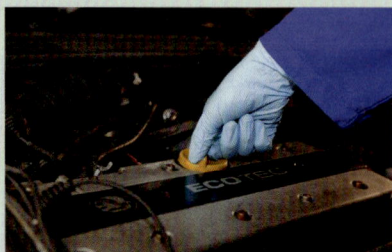

2. Lift the vehicle safely. Position the oil drainer under the sump plug and filter.

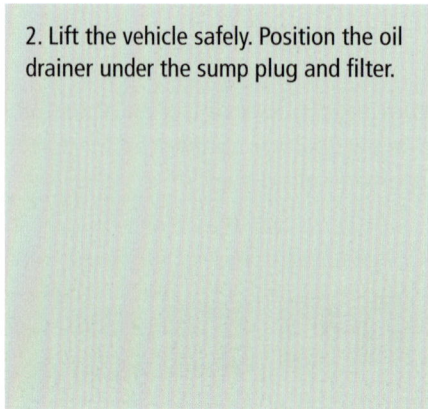

3. Remove the sump plug using the correct socket.

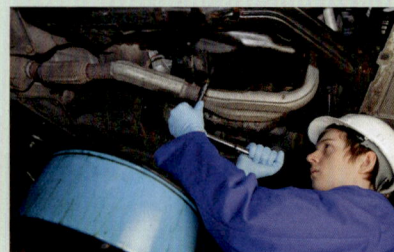

4. Remove the filter using the oil filter wrench. Allow the oil to drain fully.

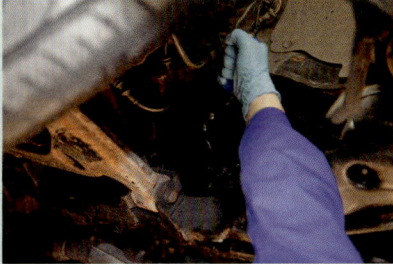

5. Wipe clean the sump bung and mating surface. Replace the bung with a new washer.

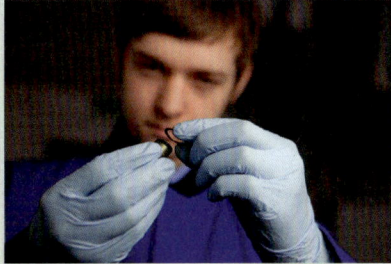

6. Wipe clean the filter mating surface which has the rubber seal attached. Apply a thin layer of new engine oil to the engine oil filter seal before installing the engine oil filter.

7. Tighten the oil filter by hand until it touches the mating surface. Then tighten the filter by two-thirds of a turn.

8. Lower the vehicle safely ensuring the drainer is clear from underneath the vehicle.

9. Fill the engine with new oil, using only the amount stated by the manufacturer.

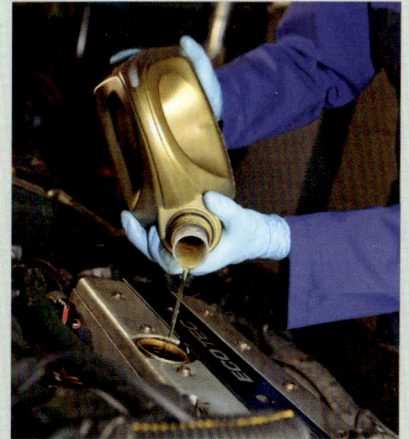

10. Replace the oil filler cap and dispose of any oil-soaked gloves used for PPE.

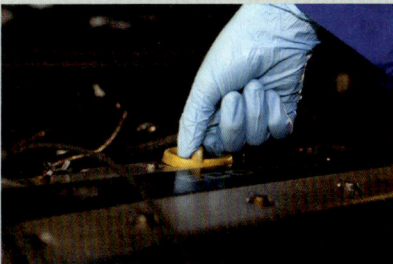

11. Run the engine for five minutes using the exhaust extraction equipment. (Do not rev the engine as the oil needs to circulate in the system.)

12. Check the oil level using the procedure described on page 107.

Coolant checks

Checklist			
PPE	**VPE**	**Tools and equipment**	**Source information**
• Steel toe-capped boots • Overalls • Latex gloves • Goggles	• Wing covers • Steering wheel covers • Seat covers • Foot mat covers	• Hydrometer • Measuring jug	• Technical data • Antifreeze manufacturer instructions

1. Check coolant level, strength and condition during each scheduled service or routine maintenance.

2. Coolant level should be checked when the engine is cold and the vehicle is standing on a level surface.

3. If the coolant level is low you should top it up with the correct type and quantity recommended by the manufacturer.

4. Coolant strength can be checked with a tool called a **hydrometer**. Take a sample of coolant using a hydrometer and compare the **specific gravity** reading with a freezing point temperature chart.

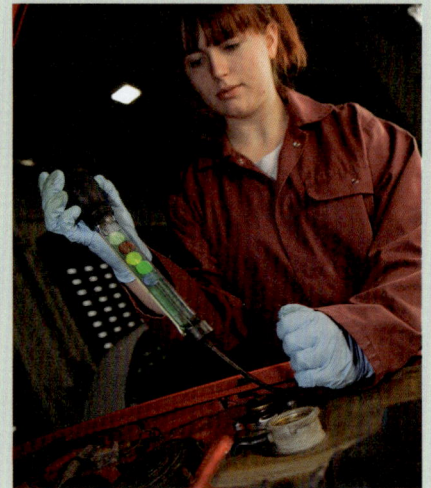

5. If the freezing point is not low enough, antifreeze should be added to increase coolant strength.

6. At this time the coolant's condition can also be visually inspected.

7. If the coolant's condition is poor or antifreeze strength is low, then you should recommend that it is drained, flushed and replaced.

8. Replace the radiator cap, run the engine and check for leaks.

Key terms

Hydrometer – a tool that measures the density of a liquid. It can be used for checking antifreeze strength.

Specific gravity – a measurement of a liquid's density when compared with water.

Safe working

Never undo a sealed cooling system cap when an engine is hot. The sudden pressure drop will make the coolant boil and you could be seriously injured.

Replacing radiators and thermostats

Checklist			
PPE	**VPE**	**Tools and equipment**	**Source information**
• Steel toe-capped boots • Overalls • Latex gloves • Goggles	• Wing covers • Steering wheel covers • Seat covers • Foot mat covers	• Spanners • Socket set • Screwdrivers • Drain tray • Measuring jug	• Technical data • Antifreeze manufacturer instructions • Component part numbers

1. Ensure that you are wearing correct PPE and that VPE is used to protect the vehicle (particularly wing covers).

2. The engine should be cool so that you can safely remove the expansion cap.

3. Drain the coolant into a suitable container and store it until it can be taken away by a licensed recycling company.

6. Unbolt/unclip the radiator and remove.

4. Remove covers and casings to gain access to the radiator and thermostat.

5. Loosen hose clips and remove hoses.

7. When the radiator is refitted, take care that no damage is done to the radiator core.

8. Refit hoses and clips and retighten.

9. Remove the thermostat housing and replace the thermostat. Many thermostats are stamped with an arrow to show the direction of coolant flow. This must be checked to make sure that you have fitted it correctly.

11. When all components have been refitted, refill the system with a coolant consisting of the correct quantities of water and antifreeze.

10. A new gasket should be used when refitting the thermostat housing.

12. The cooling system may require bleeding to remove any air locks. You can often do this by running the engine for a short period of time with the system pressure cap removed.

13. Keep an eye on the level and top up with coolant.

14. When this is complete, refit the pressure cap and run the engine to check for correct operation and any signs of leaks.

! Safe working

You must always follow the manufacturer's procedures when replacing the radiator or thermostat in the cooling system.

Checking cooling systems for leaks

Checklist			
PPE	VPE	Tools and equipment	Source information
• Steel toe-capped boots • Overalls • Latex gloves • Goggles	• Wing covers • Steering wheel covers • Seat covers • Foot mat covers	• Cooling system pressure tester	• Technical data • Cooling system operating pressure (may be found on the radiator cap)

You could use a visual inspection to check for leaks. Alternatively, you could put the system under load using a cooling system pressure tester.

A cooling system pressure tester consists of a pump and gauge that you attach instead of the radiator cap.

1. Ensure that the engine is cold and remove the radiator cap.

2. Attach the pressure tester instead of the radiator cap.

3. Pump the system pressure up until it reaches the recommended operating pressure.

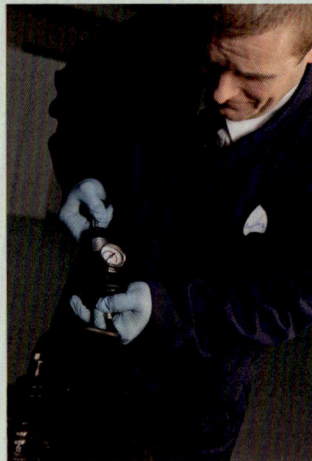

4. Visually inspect the system for leaks.

5. Release the pressure from the pump, check coolant level and refit radiator cap.

Checking and adjusting auxiliary drive belts

Checklist			
PPE	VPE	Tools and equipment	Source information
• Steel toe-capped boots • Overalls • Latex gloves	• Wing covers • Steering wheel covers • Seat covers • Foot mat covers	• Spanners • Screwdrivers • Socket set • Lever bar	• Technical data

Auxiliary drive belts (sometimes called fan belts) can become worn, damaged or loose and may not work properly. This can lead to the loss of drive of an alternator and/or water pump.

1. You should visually inspect the drive belts for damage and/or signs of old age and make sure that they are correctly running in their pulley.	2. You should examine the belts for correct tension. This can be done in a number of ways. (See 'Did you know' below.)	3. If the belt is loose you can often adjust it by moving the alternator on a sliding bracket. Always follow the manufacturer's instructions.	4. If the belt has worn to such an extent that it is running in the bottom of a drive pulley, it should be replaced.	5. The engine should be started, to check for correct operation, if you have replaced or adjusted an auxiliary drive belt.

Did you know?

A skilled technician may be able to feel whether auxiliary drive belts are correctly tensioned. The belt is normally held between a thumb and forefinger and moved up and down so that the tension can be felt. You will need to practise this skill on a number of different cars so that you get to know what it should feel like.

There are a number of tools that can help you tension the belt correctly. A tool called a deflection gauge can be fitted to the belt and the tension checked on readout.

Replace vehicle ignition components

Replacing the ignition coil

Checklist			
PPE	**VPE**	**Tools and equipment**	**Source information**
• Steel toe-capped boots • Overalls • Latex gloves	• Wing covers • Steering wheel covers • Seat covers • Foot mat covers	• Spanners • Screwdrivers • Socket set	• Technical data • Engine firing order

Safe working

There is a high risk of electric shock when you are working with ignition system components. This is due to the voltages involved. You should isolate or disconnect ignition electrics before you start any work.

1. Carefully remove all electrical terminals (low tension and high tension), making careful note of their location and connection type. The ignition coil can be bolted to the vehicle body/engine by a bracket. Alternatively, it can be surface mounted (which is where no bracket is used).	2. It is important that you clean the mounting surfaces if the ignition coil is a surface mounted type. A surface mounted ignition coil may use the metal area to which it is bolted to dissipate heat, and dirt can reduce function. (Dirt is also able to create a short circuit, which is where the high voltage electricity will take a shortcut through the dirt instead of passing through the ignition leads and spark plugs, giving a misfire.)	3. Reconnect the low tension and high tension leads in the correct order, run the engine and check for correct operation.

Remove and refit HT leads

Checklist			
PPE	VPE	Tools and equipment	Source information
• Steel toe-capped boots • Overalls • Latex gloves	• Wing covers • Steering wheel covers • Seat covers • Foot mat covers	• Cleaning materials/cloth	• Technical data • Engine firing order

The high tension leads of a multi-cylinder engine will be assembled in the firing order of each individual cylinder.

1. It is important to make a note of which lead has come from where, its overall length, and its routing around the engine, clips, etc. Make sure you do this before you remove the plug leads.

2. When the leads are removed, lay them out in order. In this way, when you come to refit them they will return to their original position.

3. It is important that you clean any dust or dirt off the spark plug leads once you have refitted them. This reduces the possibility of a short circuit and misfire.

4. Start the engine and check for correct operation.

Did you know?

Spark plug leads are very well insulated, to reduce the possibility of the high voltage electricity shorting out to metal engine components and causing a misfire. If the plug leads become damp or wet, this will make it easier for the high voltage to short out to earth. Water repellent, such as WD40, is sometimes sprayed over the plug leads to help start a damp engine. This is a 'short-term fix' as dirt can now stick to the plug leads, which will also increase the possibility of a short circuit. The best method to use when plug leads are damp is to dry them fully with a clean cloth.

Spark plug change

Checklist			
PPE	**VPE**	**Tools and equipment**	**Source information**
• Steel toe-capped boots • Overalls • Latex gloves	• Wing covers • Steering wheel covers • Seat covers • Foot mat covers	• Socket set • Spark plug socket • Feeler gauges	• Technical data • Spark plug specifications

Periodically spark plugs require replacement and this will be recommended in the manufacturer's service schedule.

1. Isolate the ignition system and clean the area around the spark plugs. This helps to prevent dirt entering the cylinder when the plug is removed.

2. Remove the spark plug leads and lay them out in order. In this way they can be connected correctly when refitted.

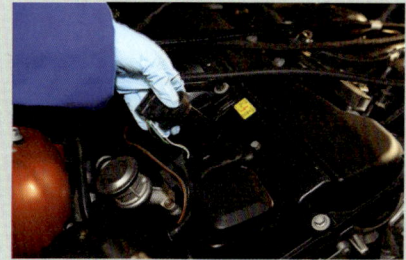

3. Now use the correct size spark plug socket to remove the old spark plugs.

4. Examine the removed spark plugs for damage and condition. This can point to further engine faults.

5. Use the manufacturer's technical data to choose the correct grade and type of spark plug.

6. The manufacturer will also supply a recommended gap for the electrodes. You should check and adjust this using feeler gauges.

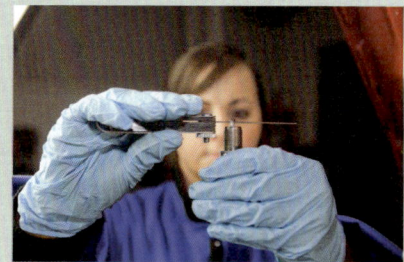

7. The spark plugs should be tightened with a torque wrench and the spark plug leads refitted in the correct order.

8. You can now start the engine and check for correct operation.

Remove and refit exhaust systems

Checklist			
PPE	VPE	Tools and equipment	Source information
• Steel toe-capped boots • Overalls • Latex gloves • Heatproof gloves • Goggles • Ear defenders	• Wing covers • Steering wheel covers • Seat covers • Foot mat covers	• Socket set • Spanners • Screwdrivers • Lever bars • Hammers • Cutting equipment (not oxy/acetylene)	• Technical data

Exhaust systems are supplied in sections. In this way, if one part fails because of damage or corrosion, it can be replaced individually.

Did you know?

Most exhaust systems are made of steel, which has a tendency to corrode. Rust build-up can make it difficult to separate sections or components, so you must take care not to damage the system while you are working. This is particularly the case when working around catalytic converters, which are fragile and easily damaged.

1. Remove the exhaust section.

2. Loosely assemble all components (including any replacement parts) and line them up so that they are in the correct position. Keep them away from bodywork, brake pipes, fuel lines and electric wires.

3. Tighten all joints and connections. Then run the engine to check the exhaust system for leaks.

4. Road test the car to make sure there are no noises or vibrations and that everything works correctly.

Procedures for replacing a gasket

Checklist			
PPE	VPE	Tools and equipment	Source information
• Steel toe-capped boots • Overalls • Latex gloves	• Wing covers • Steering wheel covers • Seat covers • Foot mat covers	• Socket set • Spanners • Screwdrivers • Gasket scrapers/cleaners • Torque wrench	• Technical data

The joints between many engine components are sealed with gaskets. If these gaskets have been disturbed by the removal of the component, they should be replaced.

1. You can remove any old gasket material with a specialist gasket scraping tool.

2. Finish the cleaning of the surfaces with the use of emery paper or sandpaper. Be careful not to cause any damage.

3. Select the correct gasket and attach to surfaces using appropriate sealant.

4. Refit the components and tighten down evenly following the manufacturer's recommendations.

5. Check the engine for correct function and operation.

Cylinder heads removal/refitting

Checklist			
PPE	**VPE**	**Tools and equipment**	**Source information**
• Steel toe-capped boots • Overalls • Latex gloves	• Wing covers • Steering wheel covers • Seat covers • Foot mat covers	• Socket set • Spanners • Screwdrivers • Gasket scrapers/cleaners • Torque wrench	• Technical data • Timing belt replacement instructions • Head bolt tightening torque/sequence

So that you can access component parts, you may need to remove the cylinder head of either a petrol or diesel engine. Different cars and engine types will vary in the way that they are put together. This means that you will have to follow manufacturer's instructions. A generalised procedure is shown below.

1. Make sure that you are wearing appropriate PPE and that VPE is used to protect the vehicle (particularly wing covers).

2. Disable the engine by disconnecting the battery or removing the key to help prevent it being accidentally turned over while being worked on.

3. Follow a logical sequence using the correct tools and information data.

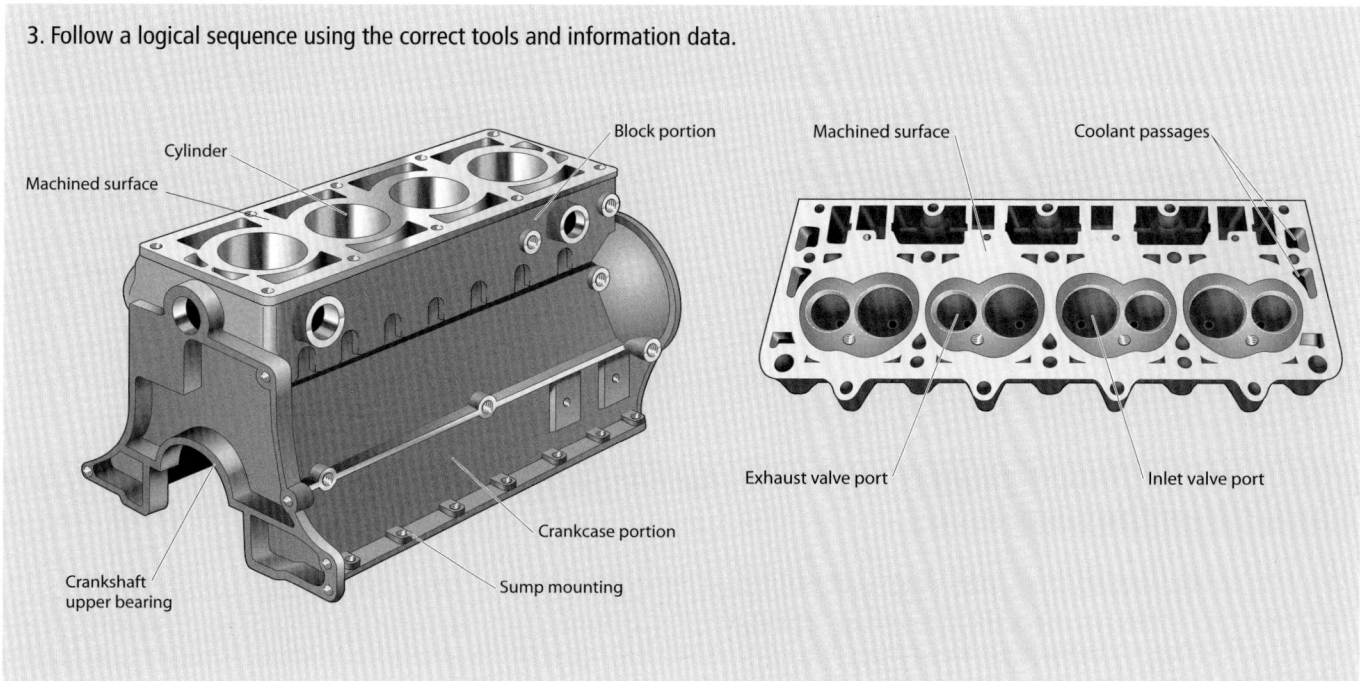

4. Remove any engine covers and gain access to the cylinder head.

5. Any camshaft timing should be set. You can now remove the camshaft timing belt.

6. Make sure that you store any removed components safely so that they are not damaged and can be replaced in the correct order.

7. Undo cylinder head bolts and remove the cylinder head.

8. Any engine repair work can now be done.

9. When refitting the cylinder head, make sure the surface of the cylinder block and cylinder head are completely flat and free of dirt and damage.

10. A new cylinder head gasket must be used and the cylinder head carefully lifted back onto the block.

11. Cylinder head bolts must be tightened in accordance with the manufacturer's instructions (some manufacturers specify that head bolts must be replaced). This is both the torque setting and the order in which they are tightened.

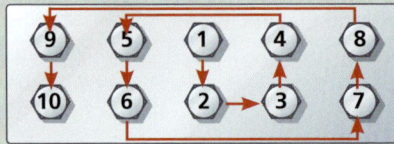

12. Camshaft timing should be checked and cam belts or cam chains refitted.

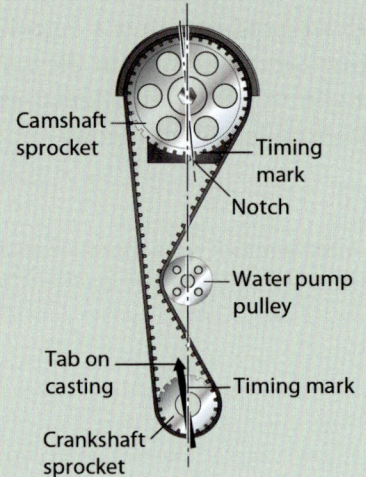

13. You should turn the engine over by hand, with a spanner on the crankshaft. Check that all components operate correctly and that camshaft timing is correct.

14. Correctly refit any covers.

15. You can now start the engine and check for correct operation. Make sure there are no leaks.

16. Vehicle components and bodywork should be cleaned so that grease and fingerprints are removed.

CHECK YOUR PROGRESS

1 Why do you need to mark spark plug leads before you remove them?
2 What tool is used to check the strength of antifreeze?
3 Can you reuse an old cylinder head gasket?
4 Why shouldn't you use hammers to remove catalytic converters?

FINAL CHECK

1 What does the process of combustion require?

  a a source of fuel
  b a source of oxygen
  c a source of heat
  d all of the above answers

2 Which of the following processes is not included in the Otto Cycle?

  a induction
  b convection
  c power
  d exhaust

3 Which of the following means the piston has reached the top of its stroke?

  a TCD
  b BCD
  c TDC
  d BDC

4 A diesel engine can also be referred to as a:

  a external combustion engine
  b spark ignition engine
  c conduction ignition engine
  d compression ignition engine

5 Engine oil is used to:

  a reduce friction
  b help collect dirt particles
  c cool the engine
  d all of the above answers

6 Which of the following is not a function of antifreeze?

  a lowering the boiling point
  b raising the boiling point
  c lowering the freezing point
  d reducing corrosion

7 What is the ideal air/fuel ratio for a petrol driven car?

  a 14.7 : 1
  b 11 : 1
  c 18.7 : 1
  d 4.7 : 1

8 The component that 'steps up' the voltage in a spark ignition (SI) engine system is:

  a spark plug
  b distributor
  c contact breakers
  d coil

9 Which of the following is not considered a harmful exhaust pollutant?

  a oxygen
  b oxides of nitrogen
  c carbon monoxide
  d hydrocarbons

10 Cylinder head bolts should be tightened with a:

  a spanner
  b air gun
  c box wrench
  d torque wrench

GETTING READY FOR ASSESSMENT

The information contained in this chapter as well as continued practical assignments in your centre or workplace will help you to prepare for both the end-of-unit tests and diploma multiple-choice tests. This chapter will also help to prepare you for working on light engine systems safely.

You will need to be familiar with:

- The mechanical components of engines, their purpose and operation (both spark ignition (SI) and compression ignition (CI) engines)
- The four-stroke engine operating cycle and the differences in operation of spark ignition (SI) and compression ignition (CI) engine systems
- The two-stroke engine operating cycle
- Technical terms associated with the size, volume and component location/position of internal combustion engines
- The need for and operation of an engine cooling and lubrication system
- The properties of engine oil
- The properties of antifreeze
- Methods for replacing coolant and lubrication oil
- The components and function of a spark ignition (SI) engine system
- Safety precautions when working on spark ignition (SI) engine systems
- The composition of air/fuel mixtures including air/fuel ratios
- The basic operation of carburettors
- The basic operation of fuel injection systems (petrol and diesel)
- The function of exhaust system components
- Chemical constituents of exhaust emissions, including their impact on the environment and health

This chapter has given you an introduction to and overview of light vehicle engine systems. It has provided you with the basic knowledge that will help you with both theory and practical assessments.

Before trying a theory end-of-unit test or diploma multiple-choice test, make sure you have reviewed and revised any key terms that relate to the topics in that unit. Be sure to read all questions fully and take time to digest the information so that you are confident about what the question is asking you. With multiple-choice tests, it is very important that you read all of the answers carefully, as it is common for the answers to be very similar and this may lead to confusion.

For practical assessments, it is important that you have had sufficient practice and that you feel that you are capable of passing. It is best to have a plan of action and work method that will help you. Make sure that you have sufficient technical information, in the way of vehicle data, and appropriate tools and equipment. It is also wise to check your work at regular intervals. This will help you to be sure that you are working correctly and to avoid problems developing as you work.

When undertaking any practical assessment, always make sure that you are working safely throughout the test. Ensure that all health and safety requirements are observed and that you use the recommended personal protective equipment (PPE) and vehicle protection equipment (VPE) at all times. When using tools, make sure you are using them correctly and safely.

Good luck!

6 Light vehicle chassis systems

This chapter will give you an introduction to light vehicle chassis systems, components and operation. It provides the basic knowledge that will help you with both theory and practical assessments.

You will learn how to identify the main components used in light vehicle wheels and tyres, braking, steering and suspension. The chapter will also introduce the basic operating principles of light vehicle chassis system components.

By working through this chapter, you will be able to plan a systematic approach to the inspection and maintenance of wheels, tyres, brakes, steering and suspension.

This chapter covers:

- Safe working on light vehicle chassis systems
- Tyres and wheels – construction and maintenance
- Routine maintenance and the replacement of road wheels and tyres
- Braking systems – components and maintenance
- Routine maintenance on vehicle braking systems
- Steering and suspension systems – components and maintenance
- Routine maintenance on steering and suspension systems

WORKING PRACTICE

Chassis systems include the operation of:

- steering
- suspension
- braking
- wheels and tyres.

Each of these systems is essential to the correct and safe operation of the car. You must take care during maintenance and repair to these systems to avoid danger to yourself, drivers, passengers and other road users. If you don't take care, it could result in injury or even death.

You must always use personal protective equipment (PPE) and you also need to think about the possibility of crush or bump injury. You will also come into contact with chemicals such as lubrication oils, grease, brake dust, brake fluids and cleaners containing solvents. Make sure that your selection of PPE will protect you from these hazards.

Personal Protective Equipment (PPE)

Safety goggles reduce the risk of small objects or chemicals coming into contact with the eyes.

Overalls provide protection from coming into contact with oils and chemicals.

Safety gloves provide protection from oils and chemicals. They also protect the hands when handling objects with sharp edges.

Barrier cream protects the skin from lubrication grease, which can cause dermatitis and may be carcinogenic (a substance that can cause cancer).

Safety boots protect the feet from a crush injury and often have oil- and chemical-resistant soles. Safety boots should have a steel toe-cap and steel mid-sole.

Safety helmet protects the head from bump injuries when working under cars.

To reduce the possibility of damage to the car, always use the appropriate vehicle protection equipment (VPE):

Wing covers

Seat covers

Steering wheel covers

Floor mats

If appropriate, safely remove and store the owner's property before you work on the vehicle. Before you return the vehicle to the customer, reinstate the vehicle owner's property. Always check the interior and exterior to make sure that it hasn't become dirty or damaged during the repair operations. This will help promote good customer relations and maintain a professional company image.

Vehicle Protective Equipment (VPE)

Safe Environment

During the repair or maintenance of light vehicle chassis systems you may need to dispose of certain waste materials such as brake fluid. Under the Environmental Protection Act 1990 (EPA), you must dispose of them in the correct manner. They should be safely stored in a clearly marked container until they are collected by a licensed recycling company. This company should provide you with a waste transfer note as the receipt of collection.

To further reduce the risks involved with hazards, always use safe working practices including:

1. Immobilise the vehicle by removing the ignition key. Where possible, allow the engine to cool before starting work.

2. Prevent the vehicle moving during maintenance by applying the handbrake or chocking the wheels.

3. Follow a logical sequence when working. This reduces the possibility of missing things out and of accidents occurring. Work safely at all times.

4. Always use the correct tools and equipment. (Damage to components, tools or personal injury could occur if the wrong tool is used or a tool is misused.) Check tools and equipment before each use.

5. Following the replacement of any vehicle components, thoroughly road test the vehicle to ensure safe and correct operation. Make sure that all work is correctly recorded on the job card and vehicle's service history, to ensure that any maintenance work can be tracked.

6. If components need replacing, always check that the parts are of the correct quality and type for the vehicle if it is still under guarantee. Inferior parts or deliberate modification might make the warranty invalid. Also, if parts of an inferior quality are fitted, this might affect vehicle performance and safety.)

Preparing the car

Tools

Torque wrench

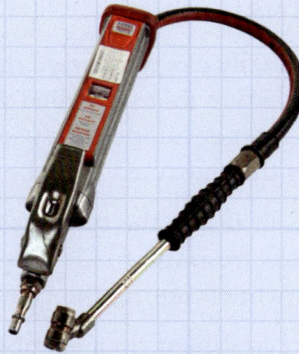

Tyre pressure gauge

Tyre tread depth gauge

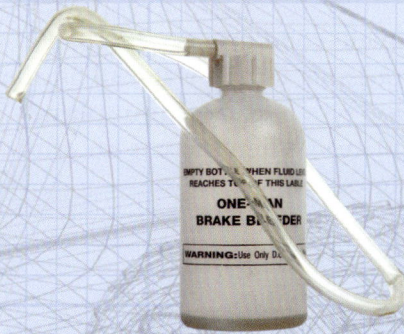

Brake bleeding equipment

Wheel alignment equipment

Tyre removal machine

Safe Working

- If you are using a ramp or vehicle hoist, always check that the car is evenly positioned and secure, and that its weight does not exceed any safe working loads (SWL).

- Always clean up any fluid spills (especially brake fluid) immediately to avoid slips, trips and falls.

- Always take care when you are handling worn tyres because sharp steel wires could be exposed.

- Always use exhaust extraction when you are running engines in the workshop.

- Ensure that the car is correctly loaded and on level ground when you are checking steering alignment settings.

- Always wear a particle mask and do not allow brake dust to become airborne.

- Always cover brake rollers when not in use.

Tyres and wheels – construction and maintenance

Tyres

Originally car tyres were made of solid rubber. The first **pneumatic** tyre (a tyre that was filled with air) was invented by R.W. Thomson. It was then developed for use on bicycles in 1888 by J.B. Dunlop. Soon after this, air filled tyres began to be used on motor vehicles.

The air inside the tyre has a slight springing effect. This helps to cushion shocks from the road. This also gives passengers a more comfortable ride while keeping the tyre in good contact with the road surface.

Car tyres must do a number of different jobs. They must:

- be able to carry the weight of the car, passengers and cargo
- grip the road, under all conditions
- last a long time
- be safe.

Tyres are made of rubber so that they provide good grip on the road surface. However, if the road is wet, it is possible for the water to lift the tyre away from the road. If this happens the tyre is said to **aquaplane**. Control is lost if the tyre on the car aquaplanes. When there is no grip between the tyre and the road, the car cannot:

- steer
- accelerate
- brake.

When the tyre is made, a tread pattern is moulded into the design to help the tyre move water out of the way. This helps it to stay in contact with the road and maintain grip.

The grooves in the tyre tread in Figure 6.2 have been designed to help move water out of the way when it rolls along the road.

Key terms

Pneumatic – using air pressure to work.

Aquaplane – when water lifts the tyre away from the road surface and grip is lost.

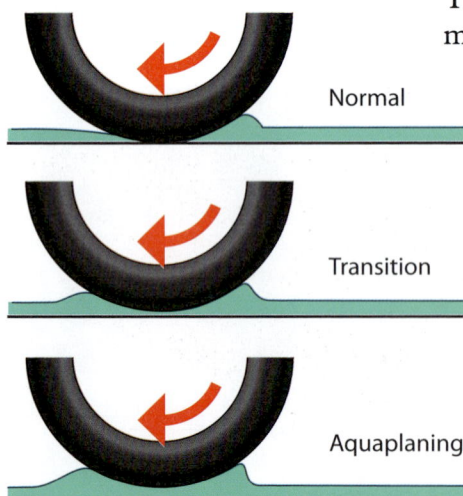

Figure 6.1 Aquaplaning

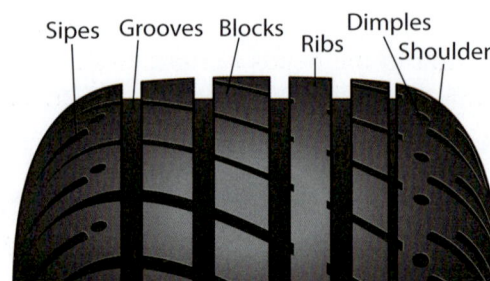

Figure 6.2 Tyre treads

Building the tyre

The two main types of tyre are **cross-ply** and **radial-ply**. They are named because of the way they are built. Radial-ply is the most common.

Both cross-ply and radial-ply tyres can be inflated with air, sealed between the rim and the tyre (**tubeless**), or using an inner tube (**tubed**).

The basic design of these tyres is the same. Two solid rings of steel wire are used to hold the tyres in place on the wheel rim. These rings of steel wire are called **beads**.

Material or fabric known as **plies** is wrapped around these beads. This forms the body of the tyre. Plies are long strips of material commonly made from rayon or nylon but they may also contain aramid fibres.

Figure 6.3 The main parts of a tyre

> **Did you know?**
>
> Aramid is a type of very strong man-made fibre also known as **Kevlar**.

Key terms

Cross-ply – a tyre construction method where the casing plies are wrapped at an angle of approximately 40° around the bead wires.

Radial-ply – a tyre construction method where the casing plies are wrapped at an angle of 90° around the bead wires.

Tubeless construction – a tyre construction method that uses a soft inner rubber liner to seal the air against the wheel rim.

Tubed construction – a tyre construction method that uses an inner tube similar to that used on a bicycle to hold air in the tyre.

Bead – a number of steel wire hoops. The casing plies are wrapped around them.

Plies – flat strips of material that are wrapped around bead wires to form the tyre casing.

Kevlar – a very strong man-made organic fibre.

Deform – to become badly out of shape.

Lifespan – how long something will last.

The way that the plies are wrapped around the beads creates the tyre's construction type (i.e. cross-ply or radial).

Radial tyres

A radial tyre is the most popular type of tyre. It is made by wrapping the plies around the beads at right angles (or 90°). This gives a softer, more flexible construction which allows the tyre to **deform** easily as it moves over the road surface or as the car goes round a bend. This flexibility helps keep the tyre tread in contact with the road surface, giving good grip. It also helps to get rid of heat, reducing wear and improving the **lifespan** of the tyre.

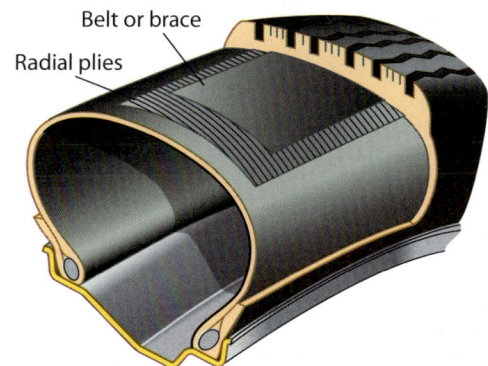

Figure 6.4 A radial tyre

Key term

Belt – a material layer fitted between the tyre casing plies and the tread to help the tyre keep its shape.

Because the tyre is flexible it can be badly deformed under acceleration and braking. To help prevent this, a belting material is placed around the outside edge of the tyre just below the tread. This **belt** material helps support the tyre tread and stops it from changing shape during inflation, acceleration and braking.

Table 6.1 Radial tyre components and their purposes

Radial tyre component	Purpose
Bead wires	Solid loops of steel wire which are designed to hold the tyre in place on the rim
Soft rubber inner liner	Helps to seal air inside the tyre
Plies	Pieces of material wrapped at 90° to the bead wires. They form the main body of the tyre
Sidewall	Soft rubber edges to the tyre. They are designed to provide protection to the plies and allow the tyre to flex
Belt	A material that sits around the outside of the tyre just beneath the tread. It stops it deforming too much as the tyre is inflated or drives along the road
Tread	A pattern moulded into the surface of the tyre, to provide grip in wet conditions

Did you know?

Radial tyres have flexible sidewalls and so the weight of the car can make them bulge outwards. This might make you think the tyre pressures are too low.

Correctly inflated tyres

Tyre pressures should be checked regularly with an accurate pressure gauge. Do this when the tyre is cold. This means that pressures are best checked before the car has been driven any distance.

Working life

The tyres on Rob's car looked soft, so he pumped them up until the sidewalls no longer bulged.

1 What would the car feel like to drive?

2 What are the possible dangers involved?

Did you know?

A radial tyre has less rolling resistance. It moves along the road easier than a cross-ply tyre. This means cars with radial tyres are more economical to run because they use up less fuel.

Cross-ply and bias-belted tyres

Cross-ply and bias-belted tyres are made in a similar way.

A cross-ply tyre is an older construction type. It is less popular for use on common everyday cars. It is made by wrapping the plies around the beads at an angle of approximately 40° so that the plies layer up on top of each other in a criss-cross pattern. This gives a very strong and stiff type of construction.

Find out

Why do you think that cross-ply tyres are less popular for normal road cars than radial tyres?

Cross-ply tyres are good for cars that are going to be used off-road. When used on a normal road, the plies rub against each other. This creates heat which can lead to the tyre wearing out rapidly. When the car tries to go around a corner, the stiff construction means that the sidewalls of the tyre do not flex easily and it has less grip.

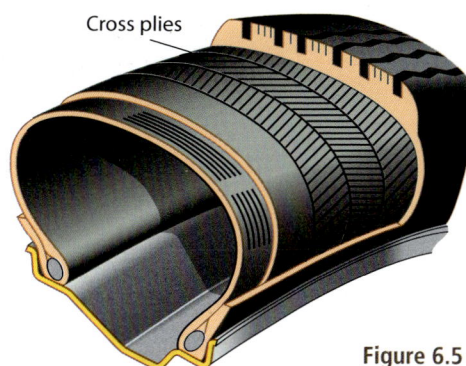

Cross plies

Figure 6.5 A cross-ply tyre

Table 6.2 Cross-ply tyre components and their purposes

Cross-ply tyre component	Purpose
Bead wires	Solid loops of steel wire are designed to hold the tyre in place on the rim
Soft rubber inner liner	Helps to seal air inside the tyre or can also help to protect an inner tube
Plies	Pieces of material wrapped at approximately 40° to the bead wire. They form the main body of the tyre
Sidewall	Soft rubber edges to the tyre. They are designed to provide protection for the plies
Breaker	A material that sits around the outside of the tyre and helps bond the casing plies to the tread
Tread	A pattern moulded into the surface of the tyre, to provide grip in wet conditions

Rubber moulding

Once the main body of the tyre has been made, it is placed in a mould. It is then injected with rubber to form the tread pattern and sidewalls. The tread pattern will be of a specific design depending on how the tyre will be used. The rubber of the sidewall is designed to help protect the plies from damage. The moulded rubber is then put under high pressure and temperature to create the overall tyre. This is known as **vulcanisation**. The tyre will be checked for quality and **balance** when it is complete.

Tyre size, shape, construction method, design and tread pattern will depend on the type of vehicle and how it will be used. For example:

- vehicle types – car, motorcycle, truck, caravan, trailer
- surface types – on-road or off-road.

Key terms

Vulcanisation – a method for hardening rubber by mixing it with chemicals and heating.

Balance – the balance of a tyre ensures that it has no heavy spots that will create vibration as it rolls along the road.

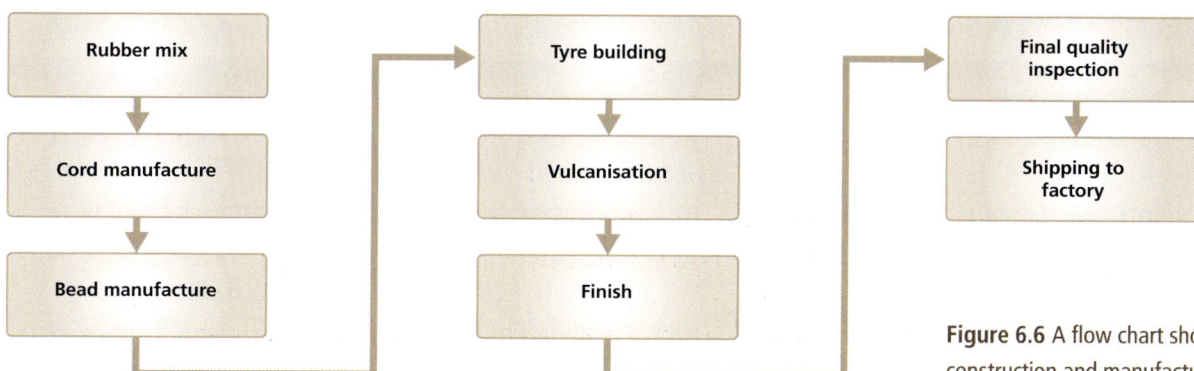

```
Rubber mix ──► Tyre building ──► Final quality inspection
    │               │                    │
    ▼               ▼                    ▼
Cord manufacture  Vulcanisation    Shipping to factory
    │               │
    ▼               ▼
Bead manufacture  Finish
```

Figure 6.6 A flow chart showing tyre construction and manufacturing

1 What is aquaplaning?
2 Name two differences between a radial tyre and a cross-ply tyre.
3 When should you check tyre pressures?

Tube or tubeless?

A tyre must contain air and the air should not escape. Therefore an airtight seal must be formed between the tyre and the wheel rim.

In a tubed tyre, this is done in a similar way to a bicycle. A rubber inner tube is fitted between the tyre and wheel rim. This inner tube must be of the correct size and shape. The tube is then inflated by a **valve** inserted through a hole in the wheel rim.

In tubeless tyres, a soft rubber inner liner forms the main seal to hold the air between the tyre and the wheel rim. (This inner liner is created during the manufacturing process.) The tyre is then inflated by a valve which is sealed into the wheel rim so that no air can escape past it.

Key terms

Tyre valve – an air lock which allows the tyre to be inflated and deflated when connected to specialist equipment.

Deflate – let the air out of something.

Blowout – a sudden or rapid deflation of a tyre, usually caused by a puncture.

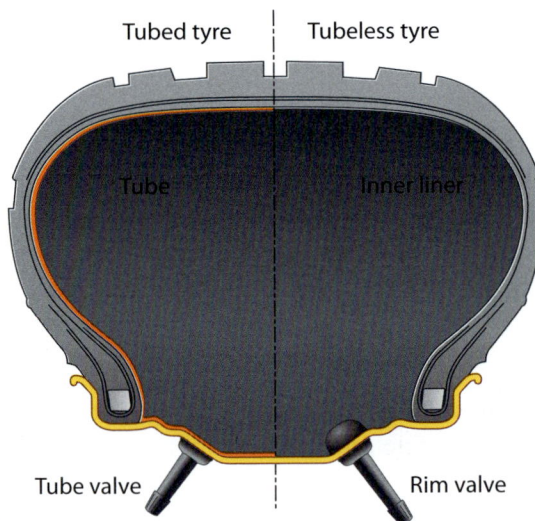

Tubed tyre | Tubeless tyre
Tube | Inner liner
Tube valve | Rim valve

Figure 6.7 Tubed and tubeless tyres

Table 6.3 Advantages of tubeless tyres when compared with tubed tyres

Tubeless tyres	Tubed tyres
Lightweight construction	More components which means this type of tyre will be heavier than a tubeless construction
Fewer components which means there is less to go wrong	More components, requiring a greater amount of maintenance and repair
If punctured, tubeless tyres tend to deflate slowly because the soft rubber inner liner can partially seal against sharp objects	If punctured, tubed tyres tend to **deflate** quickly resulting in a **blowout**

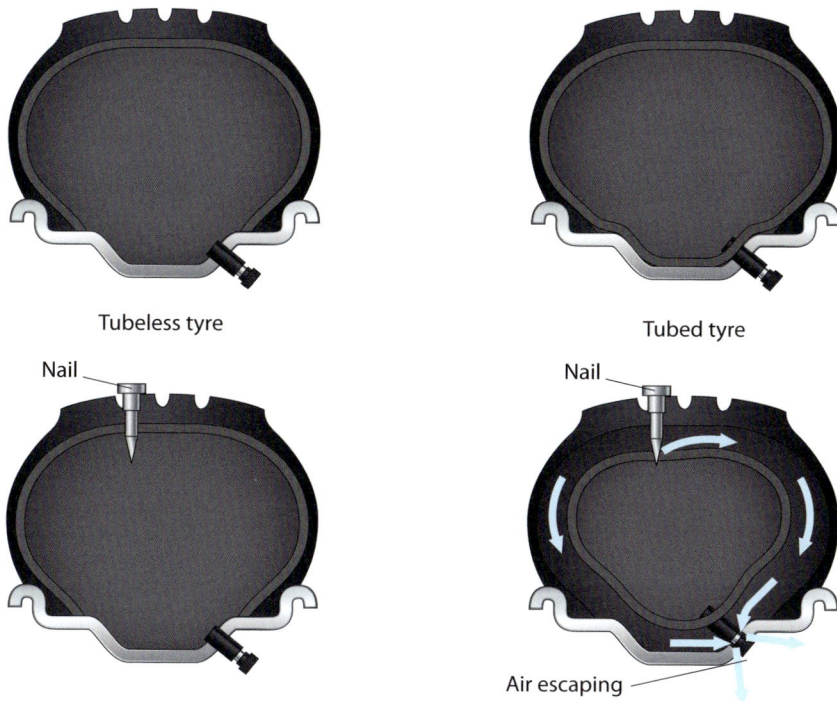

Tubeless tyre

Tubed tyre

Nail

Nail

Punctured tubeless tyre seals around the nail

Air escaping

Inner tube is punctured, causing a blow-out

Figure 6.8 Punctures in tubeless and tubed tyres

Road wheels

There are three types of wheel:

* pressed steel
* light **alloy**
* wire.

Table 6.4 Advantages and disadvantages of wheel construction types

Wheel type	Advantages	Disadvantages
Pressed steel	• Cheap and easy to produce • Can sometimes be repaired if damaged	• Heavy construction adding to vehicle weight • Doesn't look very good (often covered with wheel trims) • Reduced airflow over brakes
Light alloy	• Lighter than steel construction • Looks good • Easy to clean • Good airflow over the brakes	• Expensive to produce • Easily damaged • A target for thieves (often stolen)
Wire	• Very strong construction • Good airflow over the brakes • Looks good	• Needs to be fitted with an inner tube • Difficult to keep in balance • Expensive • Difficult to clean

Safe working

When you carry out tyre repairs, never fit an inner tube to a tubeless tyre.

As the car is driven along, the soft rubber inner liner (which is not designed to come into contact with an inner tube) can rub over the surface of the inner tube. This friction could cause a blowout.

Safe working

If a car tyre is punctured, repairs are possible but the best thing to do is to replace the tyre or tube.

You should not repair a tubeless tyre on or near the sidewall.

Key terms

Alloy – a mixture of metals combined to form a single metallic construction.

Figure 6.9 A space-saver spare tyre

Some common features of road wheels include:

- the hub boss – where plastic hub caps are clipped to the wheel
- the wheel nut boss – where the wheel nuts will bolt the wheel rim to the hub
- a centre hole – sometimes needed so that a hub nut or grease cap can be used
- air vents or openings – so that cooling air can be directed over brake discs and drums
- the wheel rim – the outer edge where the steel bead of the tyre mounts the tyre securely to the wheel rim
- the well section – a dip on the inside of the wheel rim that allows tyres to be fitted.

Some wheels use a removable split rim to make the fitting of tyres easier.

Safe working

A number of people have been killed or seriously injured during the fitting and inflation of split rim road wheels.

When the new tyre is inflated it should be done inside a safety cage.

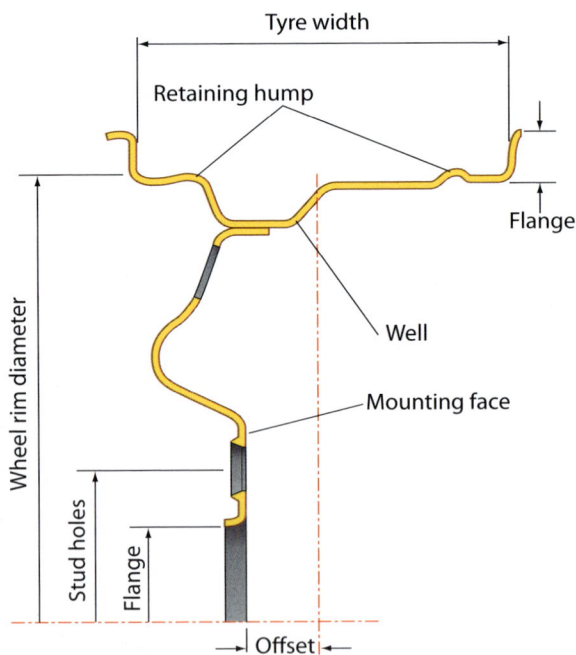

Figure 6.10 Well-based rim

Key terms

Space-saver – an emergency spare wheel which is smaller (to save space) than a standard road wheel.

Well – the dip that is found in the middle of a wheel rim. It allows tyres to be taken off and put back on.

Wheel and tyre terminology

Manufacturers provide markings on the sidewall of the tyre. These mean that the correct tyre size can be chosen for a particular vehicle and operation. These markings are **standardised** and include information such as:

- the size of the rim on which they will fit
- the width of the tyre tread
- the height of the sidewall (this is sometimes referred to as **aspect ratio** or **profile**)
- the type of construction – radial or cross-ply
- a speed rating – the maximum speed that the tyre is designed to operate at, under the correct loading and weight conditions
- the load-carrying capacity of this particular tyre.

Other information can also be included on the sidewall, such as the manufacturer's name, country of origin and direction of rotation.

Key terms

Standardised – the same on everything.

Aspect ratio – the height of the tyre sidewall, measured as a percentage of the tyre tread's width.

Profile – another name for aspect ratio.

ISO – the International Organization for Standardization.

Table 6.5 ISO tyre coding

ISO tyre coding system					
Width of the tyre measured in mm	Aspect ratio measured as %	Construction R for radial	Diameter of the wheel rim measured in inches	Load index showing the maximum weight for that tyre	Speed rating letter code showing the maximum speed capability

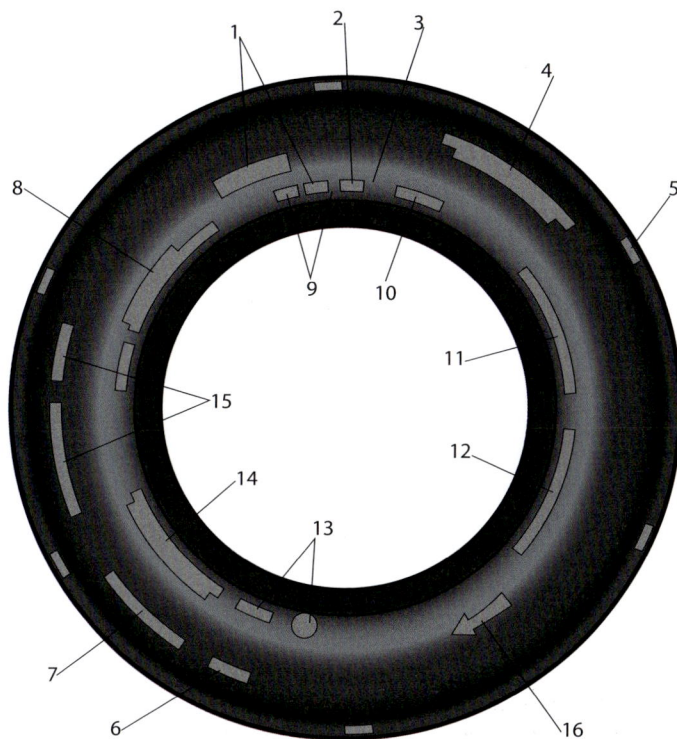

	Information given
1	Type of tyre construction
2	Load index
3	Speed symbol
4	USA quality rating
5	Tread wear indicates into tread pattern
6	Tread pattern
7	Reinforcing mark
8	USA load and pressure specification
9	Tyre size
10	Tyre type
11	Country of manufacture
12	USA compliance symbol
13	European tyre approval
14	Tyre construction
15	Manufacturer's brand name
16	Direction of rotation

Figure 6.11 Tyre side wall markings

Table 6.6 Load index table

Load index	Load kg	Load index	Load kg	Load index	Load kg
65	290	77	412	89	580
66	300	78	425	90	600
67	307	79	437	91	615
68	315	80	450	92	630
69	325	81	462	93	650
70	335	82	475	94	670
71	345	83	487	95	690
72	355	84	500	96	710
73	365	85	515	97	730
74	375	86	530	98	750
75	387	87	545	99	775
76	400	88	560	100	800

Find out

1 List the meanings and sizes of the following tyre sidewall markings:

- 185/65 R 14 86H
- 175/70 R 13 70T
- 305/45 R 18 84ZR

2 Now choose a car in your workshop and record the tyre details. Are all tyres the same size and design? Are the tyres suitable for the car? (Manufacturers' technical data may be required.)

Table 6.7 Speed rating

Speed symbol	MPH	KPH	Speed symbol	MPH	KPH
M	81	130	H	130	210
P	93	150	V	150	240
Q	99	160	W	169	270
R	106	170	Y	187	300
S	112	180	ZR	150+	240+
T	118	190	–	–	–

CHECK YOUR PROGRESS

1 Name three different wheel rim types.

2 Why do some manufacturers use space-savers?

3 Why is the well base of a wheel rim needed?

Routine maintenance and the replacement of road wheels and tyres

Removing and refitting road wheels

Checklist			
PPE	**VPE**	**Tools and equipment**	**Source information**
• Steel toe-capped boots • Overalls • Latex gloves	• Wing covers • Steering wheel covers • Seat covers • Foot mat covers	• Torque wrench • Wheel brace • Socket set • Jack • Axle stands	• Technical data • Wheel nut torque settings • Tyre sidewall information • Direction of rotation

1. Observing all health and safety, including PPE and VPE, apply the handbrake and chock the wheels.

2. Slightly loosen the wheel nuts with the car still on the ground.

3. Examine the jack and axle stands for any signs of damage. Check that the weight of the car doesn't exceed the safe working load (SWL).

4. Jack the car up and place securely on axle stands.

5. Fully undo the wheel nuts and put them somewhere safe.

6. Remove the wheel. Make sure you use correct manual handling techniques.

7. Refit the wheel in the reverse order, making sure that you follow any manufacturer's instructions including direction of rotation.

9. Tighten the nuts to the correct torque in a diagonal pattern.

10. Jack the car up, remove the axle stands and lower the car safely to the ground.

8. Fit wheel nuts with the conical face (angled edge) facing inwards.

Safe working

The wheel nuts should be retightened after the car has been driven for a short distance/time.

Did you know?

If wheel nuts have a tapered surface (an angled edge) this should face the wheel. The tapered conical face of a wheel nut is designed to help centralise the road wheel against the hub. It also increases the surface area where it touches the wheel rim. This helps to improve friction and reduces the possibility of it coming loose.

Inspecting and maintaining tyres

- Check the tyres for correct inflation pressure.
- Examine the tread depth, condition, size and type.
- Check the road wheel for damage and security.

Did you know?

Tyre pressures should only be checked when the tyres are cold. This is because, as a tyre is used, it begins to warm up. The air inside the tyre also warms up and as it does so it expands. Because the air has nowhere to expand to, and it is sealed inside the tyre, the pressure will increase. This means that pressures will always be higher when the tyre is warm.

Tyre pressure

- Always use the manufacturer's recommendations when you are inflating tyres.
- Lots of vehicles have different pressures depending on the load that they are carrying, for example how many people are in the car.
- There should be no difference in tyre pressure between the left-hand and right-hand tyres. This helps the car to handle correctly.
- Some manufacturers have different tyre pressures for the front and the rear tyres. This difference must be maintained to help prevent handling issues such as **oversteer** or **understeer**.

Key terms

Tyre shoulders – the outer edges of the tyre tread.

Understeer – where a vehicle tries to continue in a straight line when it is turned into a corner or bend.

Oversteer – when a car tries to turn further than it should when it is turned into a corner or bend. This usually leads to the back end of the car breaking away and spinning.

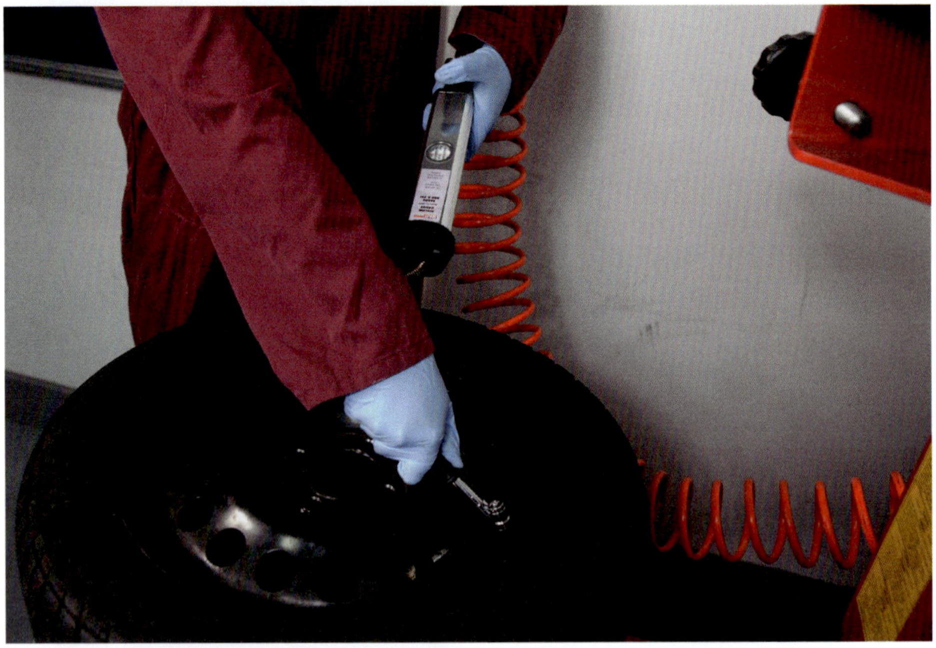

Figure 6.12 Inflating tyres to the correct pressure

Incorrect tyre pressures

Incorrect tyre pressures lead to poor handling. They can also lead to rapid and abnormal tyre wear.

If a tyre is under inflated, the weight of the car will squash the tyre. This lifts the centre of the tread off the road. As the tyre is being driven along the road, it will run mainly on the **tyre shoulders**. There leads to excessive wear around the edges of the tyres.

If the tyre is over inflated, it will expand, lifting the shoulders of the tyre away from the road surface. This leads to excessive wear around the centre of the tyre tread.

Figure 6.13 shows correct tyre inflation, under inflation and over inflation.

Tyre tread depth

Current regulations state that the minimum tyre tread depth for a light vehicle (car) to be used on the public highway in the UK is 1.6 mm of tread depth across the centre three-quarters of the tyre around the entire **circumference**.

This effectively means that an eighth of the tyre tread on either edge can be worn below 1.6 mm. As long as no cord or plies are exposed, it meets the minimum legal requirement.

You can also find the centre three-quarters of the tyre tread using the 'paper fold' method.

- Get a piece of paper and cut or fold it so that it is the same width as the tyre tread.
- When you have a piece of paper that is the same width of the tyre tread, fold it in half.
- Then fold it in half again.
- This now gives you three creases and four sections of paper. Each section represents one quarter of a width of the tyre tread.
- Unfold the paper and then fold in one quarter once more.
- You are now left with three-quarters of the width of the tyre tread represented by the paper, and this can be laid on the tyre tread to show the centre three-quarters.
- You can compare this with the measurement method to check your results.

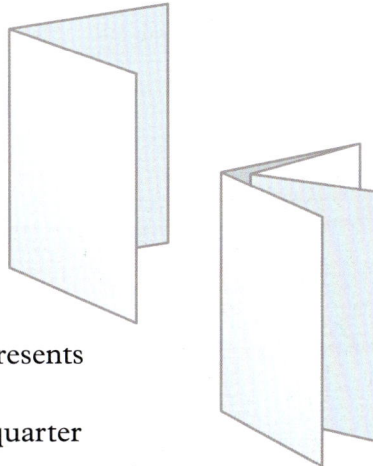

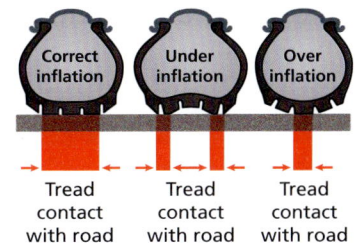

Figure 6.13 Correct tyre inflation, under inflation and over inflation

Key term

Circumference – the outer edge of a circle.

Find out

Find the centre three-quarters of a tyre tread by measurement.

Did you know?

Most tyre tread depth gauges can be calibrated on a flat surface, like a piece of glass. The tyre tread depth gauge is pushed against a piece of glass. It should line up with a calibration line at around 0 mm.

Checking tyre tread depth

Checklist			
PPE	**VPE**	**Tools and equipment**	**Source information**
• Steel toe-capped boots • Overalls • Latex gloves	• Wing covers • Steering wheel covers • Seat covers • Foot mat covers	• Tread depth gauge • Tape measure	• Technical data • Legal specifications

! Safe working

If the tread depth falls below 1.6 mm anywhere in the central three-quarters of the tyre tread, it is below the minimum legal limit and must be replaced.

⚠ Safe working

Do not use the end of the tread depth gauge to hook stones out of the tread or open up splits or cuts. This can damage the measurement tip and make it inaccurate.

⚠ Safe working

Remember that tread wear indicators are only a rough guide and accurate readings should always be taken from a depth gauge.

⚠ Safe working

Car tyres must be the same size on the same axle.

1. Raise and support the car securely. Remember to observe all health and safety, including PPE and VPE.

2. You can measure the amount of tread using a tyre tread depth gauge. Before you use it, the gauge must be calibrated.

3. Calculate the centre three-quarters of the tyre tread.

4. You can now take measurements by pushing the tip of the gauge into the tyre tread.

5. You should use the tyre tread depth gauge at a number of points across the width of the tyre and around the entire circumference.

Did you know?

Any part of the tyre tread pattern that is designed to wear out significantly before the rest of the main body of the tyre tread should not be included in your measurement. This means you do not need to measure tiny markings that will wear out very early.

Tyre wear indicators

Many tyre manufacturers include tread wear indicators in their design. These help to show the level of tyre wear. Small rubber blocks are moulded into the grooves and, when the tread wears down to these, it is approaching the legal limit. Many manufacturers also put markings on the tyre sidewall where the tread wear indicators will be found. This is to stop them being confused with the normal tyre tread. The markings are usually arrows or small letters that say TWI (tread wear indicator).

Cuts

A tyre must be replaced if a cut is found in the tyre carcass that is deep enough to show the ply or cord.

A tyre must also be replaced if a cut is found in the tread that is deep enough to show the plies, longer than 25 mm or 10 per cent of the tread width.

Bulges and lumps

If a bulge or lump is found in the carcass of the tyre, it shows a breakdown in the casing plies. The tyre must be replaced.

Mixing cross-ply and radial tyres on the same car

Wherever possible, tyres of the same construction type should be used on all wheels – so all radial or all cross-ply. If a mixture of cross-ply and radial-ply tyres are used on the same car, cross-ply must be fitted to the front axle and radial-ply must be fitted to the rear. Any other combination is illegal and can severely affect handling or safety. Use the **CBR** method shown in Table 6.8 below to help you remember the legal requirements for mixing tyres.

Table 6.8 CBR – cross-ply, bias-belted, radial

Cover up the letter that represents the tyre type that you don't have and it will show you whether it goes on the front or the rear of the car.

Front ⟵——————————————⟶ Rear		
C	B	R
Cross-ply	Bias-belted	Radial

Fitting a tubeless tyre

Checklist			
PPE	**VPE**	**Tools and equipment**	**Source information**
• Steel toe-capped boots • Overalls • Latex gloves	• Wing covers • Steering wheel covers • Seat covers • Foot mat covers	• Wheel brace • Socket set • Torque wrench • Valve core removal tool • Tyre changer machine • Tyre inflation pump • Jack • Axle stands	• Technical data • Wheel nut torque settings • Tyre sidewall information • Direction of rotation

1. Safely raise the vehicle and support with axle stands, if required. Remember to observe all health and safety, including the correct use of PPE and VPE.

2. Remove the wheel and store the wheel nuts safely.

3. Use a specialist tool to undo the valve core and let the air out of the tyre.

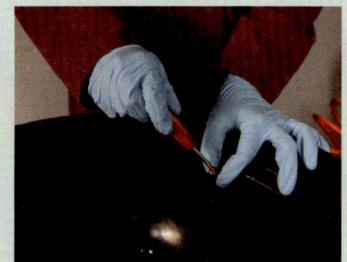

4. You will now use a tyre fitting machine.

5. The bead must then be 'broken'. Use an attachment on the machine to push the tyre bead into the wheel well.

6. Mount the wheel and tyre on the turntable and lock into place.

7. Adjust the removal arm to the wheel rim so that it is held slightly away from the rim, to prevent damage during operation.

8. Push one side of the bead down into the well and lever the other edge up on top of the removal tool.

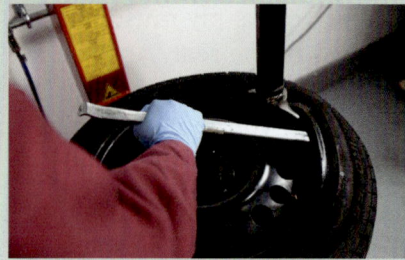

9. Use the foot pedals to rotate the turntable until the first side of the tyre is removed.

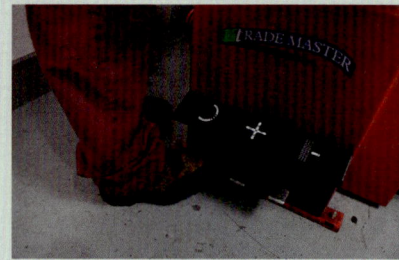

10. Repeat the procedure for the second side.

11. Before you fit the new tyre the wheel rim should be inspected for damage.

12. It is recommended that a new tyre valve assembly is fitted. Cut away the old valve and use a specialist tool to fit the new one securely in the rim.

13. Lubricate the bead and reverse the procedure. Use the removal arm to feed the new tyre into place. (Make sure that you push the bead down into the well to prevent damage.)

14. Re-inflate the tyre to the correct pressure and make sure that the bead has seated correctly on the rim.

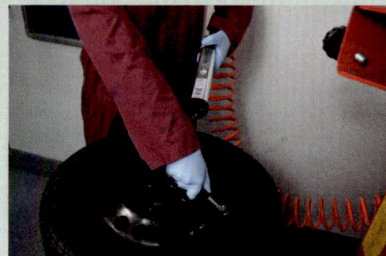

15. Before the tyre is refitted to the car it should be balanced. (See page 196.)

16. Refit the road wheel to the car. Tighten the nuts in the correct order to the manufacturer's specified torque setting.

Safe working

Take care when you re-inflate the tyre. As the bead moves into place against the rim, there is normally a loud bang and fingers could be trapped and debris could fly off.

Key terms

Static imbalance – a heavy spot around the centre line of the wheel and tyre assembly.

Shimmy – an abnormal wobble of a car's wheels

Dynamic imbalance – a heavy spot found towards the edge of the wheel and tyre assembly.

Wheel balance

When a wheel and tyre is made, sometimes heavy spots are created, even though manufacturing processes are extremely good. These heavy spots will have an effect on the movement of the tyre once it is rolling along the road.

A heavy spot in the centre of the tyre is known as **static imbalance**. As the tyre rotates the heavy spot will be thrown outwards by centrifugal force. It will strike the road surface with every rotation, causing a vibration which can also lead to rapid tyre wear.

Sometimes a heavy spot can be found towards the edge or shoulders of the tyre. As the tyre rotates centrifugal force will throw this heavy weight inwards or outwards. This will create a vibration that throws the tyre and steering inward and outward creating a **shimmy**. This shimmy or shake from side to side is known as **dynamic imbalance**.

Figure 6.14 A bicycle valve

Centre line

Add weights here

Figure 6.15 Static wheel imbalance

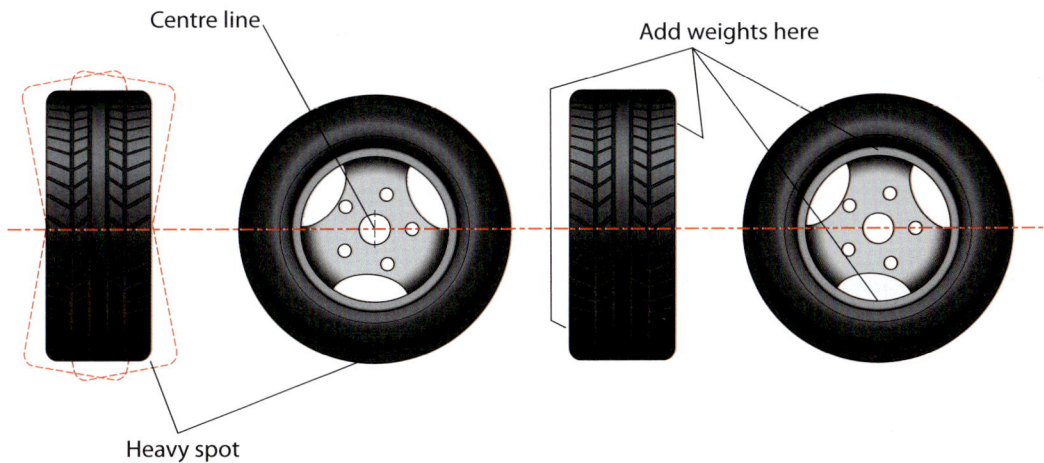

Centre line

Add weights here

Heavy spot

Figure 6.16 Dynamic wheel imbalance

Most imbalance will be felt within a certain speed range. This means that no vibrations are felt when the vehicle is travelling slowly. As the car speeds up a vibration begins. As the vehicle continues to speed up the vibration may well disappear.

To sort out balance problems, a metal weight is fitted to the wheel which will oppose the heavy weight created during manufacture. This means an equal weight is placed opposite to the heavy weight created during manufacture.

The correct position and size of weight needed can be calculated using a balancing machine (see Figure 6.17).

Figure 6.17 A tyre on a balancing machine

Balancing a wheel and tyre assembly

Checklist			
PPE	VPE	Tools and equipment	Source information
• Steel toe-capped boots • Overalls • Latex gloves	• Wing covers • Steering wheel covers • Seat covers • Foot mat covers	• Wheel brace • Socket set • Torque wrench • Wheel balancing machine • Wheel balance weights • Jack • Axle stands	• Technical data • Wheel nut torque settings • Balancing machine readings

1. Safely raise the vehicle and support with axle stands, if required. Remember to observe all health and safety, including the correct use of PPE and VPE.

2. Remove the wheel and store the wheel nuts safely.

3. Visually inspect the road wheel and tyre for any signs of damage. Clean wheel and tyre if required.

4. Before checking for imbalance, remove any old balance weights.

5. Mount the wheel and tyre assembly to the balancing machine, following the manufacturer's instructions.

6. You will now need to input some data about the size and position of the wheel.

7. Wheel rim diameter can be found on the tyre sidewall markings.

8. Wheel rim bead width can be found using a large set of calipers, usually supplied with the balancing machine.

9. The distance of the wheel rim from the machine can be found with a measurement device usually attached to the balancer.

10. When this is done, close the safety cover and motors will spin the wheel and tyre assembly.

11. The machine will then calculate any imbalance and supply data so that you can attach the correct size weights to the wheel opposite any imbalance.

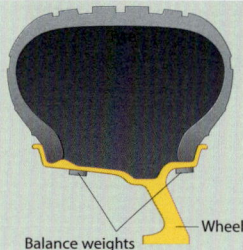

Balance weights — Wheel

12. Rotate the wheel until the correct position has been reached. Fix a weight of the size indicated to the rim. Weights can be clip on, which are hammered into place or glued to the rim.

13. When the weights have been attached to the wheel rim, repeat the whole process until no further weights are needed.

14. Refit the road wheel to the car. Tighten the nuts in the correct order to the manufacturer's specified torque setting.

15. The car should then be road tested for safety and to check that no balancing vibrations exist.

Safe working
To road test a car you must hold a full licence and insurance. The car must be taxed and in a roadworthy condition.

CHECK YOUR PROGRESS

1 Which is the lightest of the road wheel types?
2 Which way round should wheel nuts be fitted?
3 What is the difference between static and dynamic imbalance?

Braking systems – components and maintenance

The engine and transmission system provides the power to make the car move. It is the job of the braking system to slow the car down and stop it. Movement is a form of energy known as **kinetic energy**.

Because the movement energy from the car cannot be destroyed, it must be converted into something else. The brakes do this by changing it into heat.

The parts used to create friction in the braking system, and which turn movement energy into heat, are brake drums and shoes, or brake discs and pads.

A braking system will have one moving part – a rotating brake drum or disc. This is clamped against a stationary part – the brake shoes/backplate or the brake pads/caliper. This creates friction that is turned into heat. The heat created is moved into the surrounding air by a process called **dissipation**.

If the friction materials used in the braking system overheat, no more movement energy can be converted. If this happens the car will no longer slow down, no matter how hard the pedal is pressed. This is called **brake fade**.

> ### Key terms
>
> **Kinetic energy** – the energy of movement.
>
> **Dissipation** – the removal of heat into the air from braking components.
>
> **Brake fade** – a condition that occurs when brake friction material can no longer absorb any more heat and can therefore no longer convert movement energy.

Drum brake systems

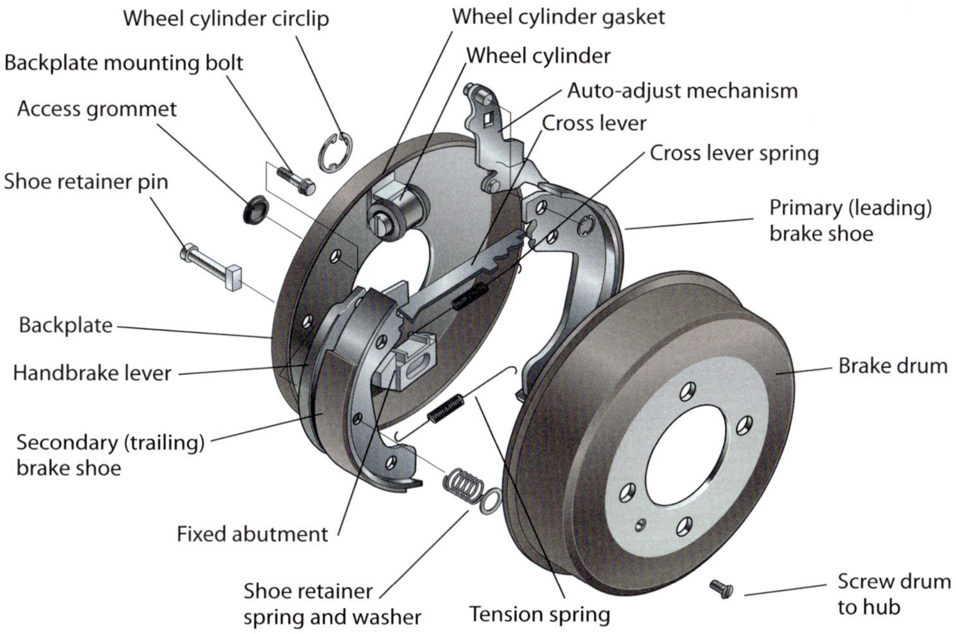

Figure 6.18 Drum brakes

Table 6.9 The main components of a drum brake system

Component	Description
Brake shoes	A pair of curved brake shoes with a friction material bonded (glued) or riveted to the surface
Backplate	Where the brake shoes are attached. The backplate is usually secured to the axle or suspension and does not rotate
Brake drum	A drum made from cast iron which is attached to the wheel hub and rotates with the wheel
Wheel cylinders	One or more wheel cylinders which contain pistons that are operated by hydraulic fluid pressure
Return springs	Springs are mounted across the brake shoes, so that when hydraulic pressure is released from the wheel cylinders, the brake shoes are returned to their rest position

Wheel cylinders

The wheel cylinder sits between the ends of the brake shoes. When it is operated, it forces the brake shoes out into contact with the rotating drum.

There are three main types of wheel cylinder operation:

- The most common has two pistons, one at either end of the cylinder. When they are operated by hydraulic fluid, the brake shoes are forced apart.
- In a twin leading shoe arrangement, two wheel cylinders are used to operate the brake shoes. Each cylinder contains one piston.
- The final wheel cylinder type has only one piston but it is able to slide in a slot created in the backplate. When it is operated, the piston moves out until it is in contact with a shoe. Then the whole cylinder slides until it is in contact with the other shoe.

Adjustment

An adjuster mechanism is included in a drum brake system. As the friction material of the brake shoes wears down, any extra space between the shoe and the drum can be adjusted away. A drum brake adjuster mechanism can be manually operated, set by the technician in the workshop. Alternatively, it can be automatic (self-adjusting).

Handbrake

A handbrake mechanism is easily included in the design of a drum brake. When operated by the driver from inside the car, a cable system pulls on a mechanical lever. This acts against the brake shoes and forces them outwards against the brake drum. This holds the brake drum still and helps to stop the car moving.

Brake shoe layouts

A number of brake shoe layouts are possible depending on design. The main difference between these designs will depend on whether the brake shoes are operated at one end or both ends.

Find out

Investigate a drum brake system on a car in your workshop. Identify the shoes, wheel cylinder and the return springs. Is the adjustment manual or automatic?

Find out

Investigate some handbrake systems on cars in your workshop. Does the handbrake work on the front or rear of the car? How far up does the handbrake lever come (how many clicks on the ratchet mechanism) before the handbrake fully holds?

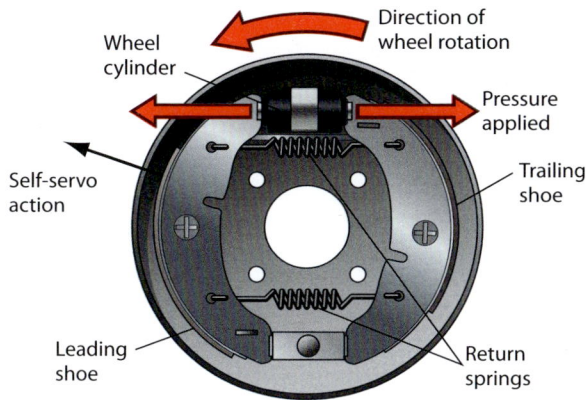

Figure 6.19 Leading and trailing shoe arrangement

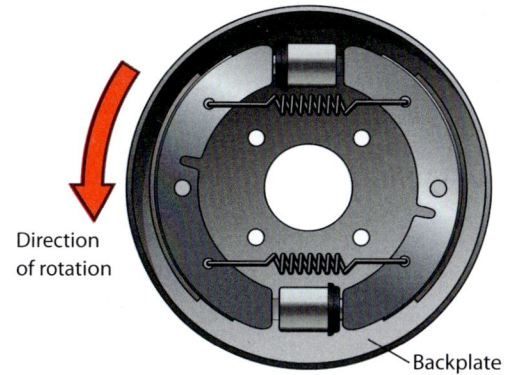

Figure 6.20 Twin leading shoes

Self-servo/self-energising drum brakes

When pushed outwards by the wheel cylinder, one end of the brake shoe will be dragged into contact with the drum by its rotation. This is called **self-servo** action and it helps the operation of the drum brake.

- A brake shoe that is mounted after the wheel cylinder in the direction of rotation of the drum is called the **leading shoe**.
- A brake shoe that is mounted before the wheel cylinder in the direction of rotation of the drum is called the **trailing shoe**.

The leading shoe is dragged into contact with the drum while the trailing shoe is pushed away.

Disc brake systems

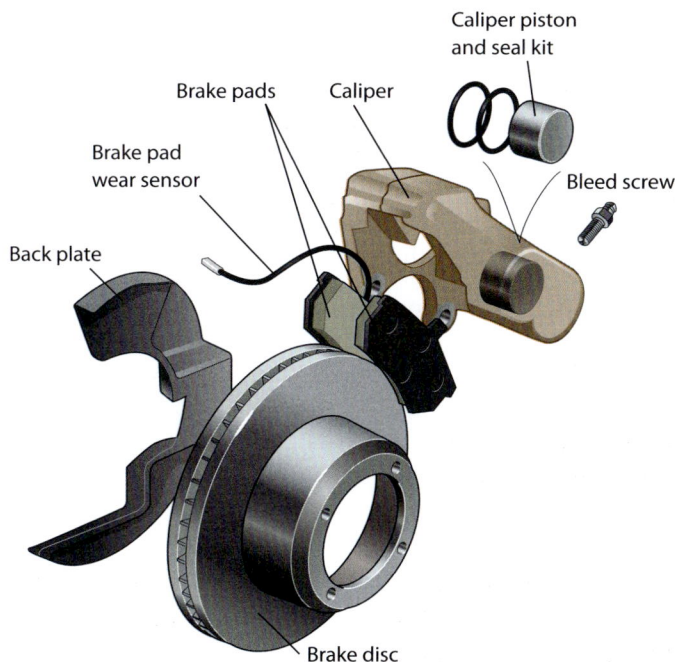

Figure 6.21 Disc brakes

Find out

Draw a single leading shoe arrangement. Label all the components and show the direction of drum rotation.

Key terms

Self-servo – this is a type of braking assistance, created by the rotation of the brake drum.

Leading shoe – the brake shoe mounted after the wheel cylinder in the direction of drum rotation.

Trailing shoe – the brake shoe mounted before the wheel cylinder in the direction of drum rotation.

Table 6.10 The main components of a disc brake system

Component	Description
Brake discs	A steel or cast iron disc which is accurately machined flat and parallel on both sides. The brake disc can be solid or cast with air vents passing through the middle to help improve cooling (these are usually called 'vented discs'). The brake disc is bolted to the wheel hub and rotates with the wheel assembly. (This disc assembly can sometimes be called a rotor.)
Brake pads	Brake pads are a thin metal plate with friction material bonded (glued) to their surface. They are mounted inside a non-rotating brake caliper. When operated by pistons, they clamp against each side of the brake disc to create friction.
Brake calipers	The calipers are components containing pistons which are operated by hydraulic pressure. Fluid pushes the pistons out of the caliper, forcing the brake pads against the discs.

Brake calipers

Brake calipers come in many different designs and shapes. All will be very accurately machined components, with hydraulic fluid passages drilled in them. Inside the caliper is usually a large hole called the caliper bore. A large round piston is fitted inside the caliper bore. The piston slides in and out of the caliper because of hydraulic pressure created by the brake pedal and master cylinder.

Many brake calipers have two pistons, one for each brake pad. However, in some designs there is not enough space behind the wheel and hub to allow for this. In that case, a single piston caliper is used that is able to slide when operated. It clamps the brake pads against the disc.

The small gap between the piston and the bore is filled by a rubber seal. The rubber seal is very tight to make sure the fluid doesn't leak out. However, it is not so tight that it stops the piston from moving. When the brake pedal is released, the elastic nature of the rubber seal is able to slightly **retract** the piston. This frees the pressure on the brake pads and allows the disc to rotate again.

Brake pads are normally held in the caliper by metal pins or clips. They can also have a thin metal plate which sits between the piston and the back of the brake pad. These are called 'anti-squeal shims' and are designed to help reduce any noise created by the braking process.

Did you know?

If no brake drum system is used in the design of a car, the manufacturer has to include a parking brake mechanism within the brake disc caliper. As with the drum brake system, it is a mechanically operated device. When operated by the driver, it pulls on a series of cables and works a lever mechanism inside the brake caliper, to clamp the brake pads firmly against the disc.

Key term

Retract – to take or move back.

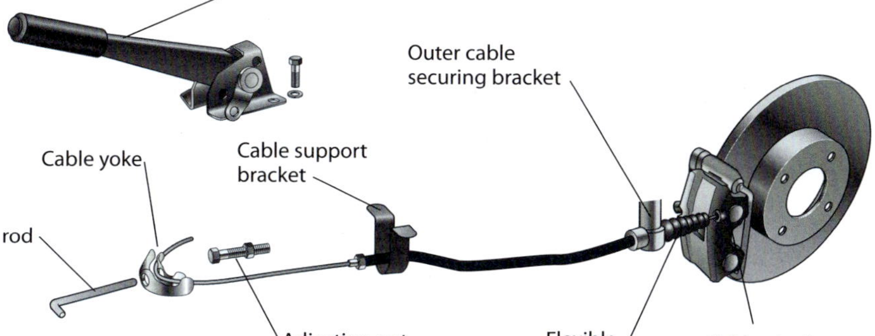

Figure 6.22 Handbrake linkage layouts

Table 6.11 The advantages of drum brakes and disc brakes

Advantages of drum brakes	Advantages of disc brakes
• A drum brake is relatively cheap and easy to produce. • The operation of the leading brake shoe can provide a self-servo effect. This helps the operation of the brake. • A parking brake mechanism is easily included in the design.	• A disc brake has a large surface area exposed to the air. This means that it can dissipate heat easier making it more efficient. The quicker it can get rid of heat, the more movement energy it can convert. • Disc brakes are self-adjusting, so less maintenance is needed. • Most of the disc brake components are more visible. This means they may not need stripping out like a drum brake, so they are easier to inspect for wear. • Brake pedal travel does not increase as the disc heats up.

Hydraulic braking systems

Table 6.12 The main components of a hydraulic operating system

> **Key term**
>
> **Hydraulic** – using liquids to operate.

Component	Description
Brake pedal	The brake pedal is a mechanical lever operated by the driver's foot. It acts upon pistons in the brake master cylinder.
Master cylinder	A brake master cylinder contains one or more pistons. When moved by the brake pedal, the piston/s forces fluid under pressure through pipes and hoses down to the wheel assemblies.
Slave cylinder	A slave cylinder is a general term used to describe the **hydraulic** pistons at the wheel brake assemblies, i.e. wheel cylinders and calipers.
Brake booster/ servo	A vacuum operated mechanism that assists the driver in the application of the brakes. This is a form of power assistance.
Brake pipes and hoses	When a hydraulic fluid leaves the brake master cylinder, it travels to the wheel assemblies using pipes and hoses. Metal brake pipes are rooted around the car body/chassis and are usually clipped securely to prevent vibration and damage. At the wheel assemblies, reinforced rubber hoses are used to direct the fluid to wheel cylinders or brake calipers. Flexible rubber hoses are needed so that they can allow for steering and suspension movement.

Key terms

Single line system – a hydraulic braking system that only uses a single pipe to transfer brake fluid from the master cylinder to the wheel cylinders.

Tandem master cylinder – a brake master cylinder that contains two pistons and is connected to a dual line system.

Dual line system – a hydraulic braking system that uses several outlets and pipes to transfer brake fluid from the master cylinder to the wheel cylinders.

Brake pedal

The brake pedal is anchored by a pivot at one end. When the driver presses it down, the pedal acts as a lever to increase the force. The brake pedal pad must have an anti-slip material attached.

Master cylinder

If a brake master cylinder contains one piston and one outlet, it is called a **single line system**. There is a disadvantage to this type of master cylinder – if a fluid leak occurs, all hydraulic braking effort is lost to the wheels and the car cannot stop.

An improvement on the design of the brake master cylinder is to use two operating pistons, with the hydraulic outlets separated between different wheels. This way, if a hydraulic fluid leak occurs in one part of the system, braking effort is only lost in that particular section. Braking performance will be considerably reduced, but it will still be able to operate. In this way, some form of emergency braking is still available.

If the master cylinder contains two pistons it is usually called a **tandem master cylinder**. The overall design is known as a **dual line system**.

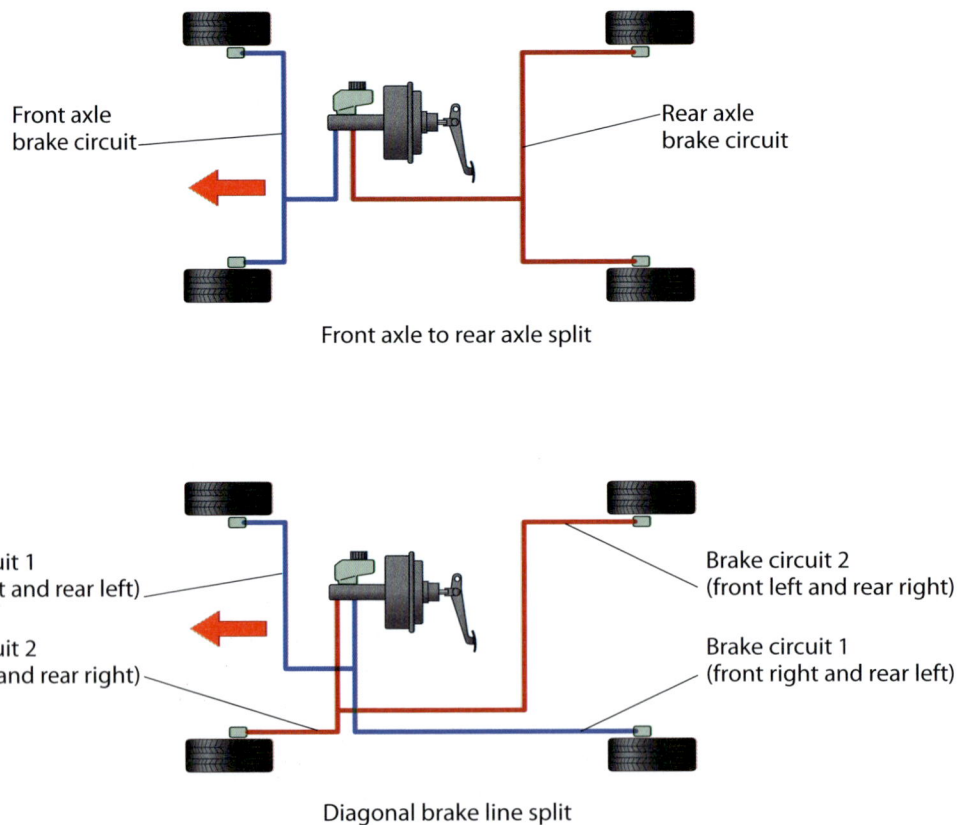

Front axle brake circuit

Rear axle brake circuit

Front axle to rear axle split

Brake circuit 1 (front right and rear left)

Brake circuit 2 (front left and rear right)

Brake circuit 2 (front left and rear right)

Brake circuit 1 (front right and rear left)

Diagonal brake line split

Figure 6.23 A dual line system

Brake servo/booster

To help the driver operate the brakes, a **brake servo** is often used. It is usually mounted between the brake pedal and the master cylinder. It is a large vacuum-operated **diaphragm**. It helps the driver to apply pressure to the master cylinder when the brake pedal is operated.

- It consists of a large sealed container, with a rubber diaphragm mounted in the middle, separating the two halves of the servo. This diaphragm is connected to the input piston rod of the brake master cylinder.

- On a petrol engine, a vacuum created in the inlet manifold is directed to both sides of the servo diaphragm. Because vacuum exists on both sides, pressure is even, and the diaphragm stays still.

- When the driver operates the brake pedal, a valve at the back of the servo allows a higher atmospheric pressure to enter the rear chamber behind the diaphragm.

- The higher pressure acts against the surface of the diaphragm. It pushes it forward and assists with the operation of the brake master cylinder.

- When the brake pedal is released, the atmospheric pressure valve closes. Vacuum is restored to both sides of the diaphragm and a return spring helps move it back to its original position.

Key terms

Brake servo – a mechanism designed to provide vacuum-operated assistance to the driver's effort when the brake pedal is pressed.

Diaphragm – a rubber partition separating two chambers.

Did you know?

A diesel engine does not create a vacuum in the inlet manifold like a petrol engine. So that it can operate a vacuum servo, an engine-driven vacuum pump is used. This vacuum pump is sometimes called an 'exhauster'.

Safe working

The brake servo is only a form of power assistance. It is sometimes referred to as a 'brake booster'. It should be 'fail safe', meaning that if it stopped working the brakes would still operate but the pedal would have to be pushed much harder.

Testing the vacuum brake servo operation

Checklist			
PPE	**VPE**	**Tools and equipment**	**Source information**
• Steel toe-capped boots • Overalls • Latex gloves	• Wing covers • Steering wheel covers • Seat covers • Foot mat covers	• Exhaust extraction equipment	• Technical data

1. Sit inside the car and pump the brake pedal until it becomes hard. Remember to observe all health and safety, including the correct use of PPE and VPE.

2. Hold your foot on the brake pedal. Check that the car is not in gear and that any exhaust extraction equipment is attached. Start the engine.

3. If the servo is working correctly, you should be able to feel the pedal creep down slightly. The pedal should not go all the way to the floor.

4. If creep is felt, then vacuum from the engine is assisting the brakes with servo operation.

5. The servo should be able to keep its vacuum when the engine is switched off. To test this, keep your foot held down on the brake pedal and turn off the engine. If vacuum is leaking, you will feel the pedal push back against your foot.

Basic vehicle braking systems operation

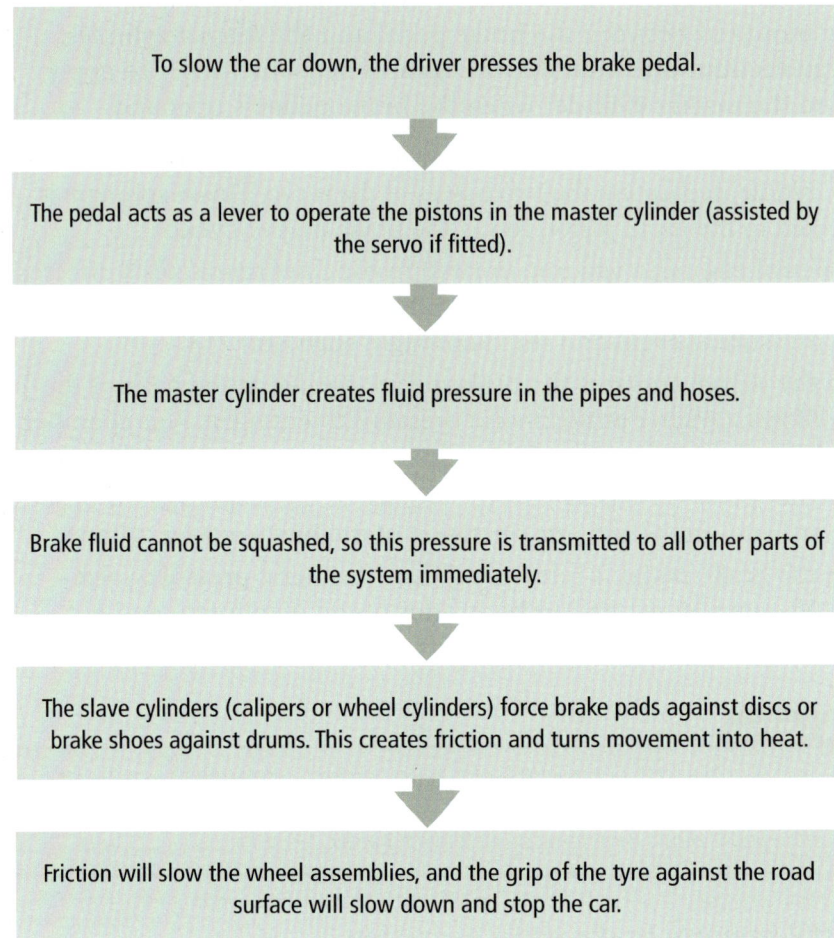

To slow the car down, the driver presses the brake pedal.

The pedal acts as a lever to operate the pistons in the master cylinder (assisted by the servo if fitted).

The master cylinder creates fluid pressure in the pipes and hoses.

Brake fluid cannot be squashed, so this pressure is transmitted to all other parts of the system immediately.

The slave cylinders (calipers or wheel cylinders) force brake pads against discs or brake shoes against drums. This creates friction and turns movement into heat.

Friction will slow the wheel assemblies, and the grip of the tyre against the road surface will slow down and stop the car.

Figure 6.24 A basic vehicle braking system

Key terms

Efficiency – how well something works.

ABS – an anti-lock braking system that uses computer controlled valves to reduce hydraulic pressure in a wheel that is about to skid.

Anti-lock braking systems (ABS)

Many modern cars are designed with **ABS** (an anti-lock braking system). This helps to reduce the possibility of a tyre skidding under braking.

- An ABS system uses electronics to detect the rotating speed of individual wheels.
- If one wheel is turning considerably slower than the others, a computer decides that it is about to skid.
- This computer then operates a series of valves to reduce or remove hydraulic pressure from that particular brake assembly. This allows the wheel to speed up again.
- When the wheel is once again turning at a speed similar to the others, normal braking is returned.

Figure 6.25 An anti-lock braking system electronic control unit

ABS provides more vehicle control in an emergency braking situation.

Right-hand (RH)
front wheel

Master cylinder

Right-hand (RH)
rear wheel

ABS actuator
assembly

Solenoid valve

P   P   M

Motor

Pumps

Reservoir

Left-hand (LH)

Left-hand (LH)

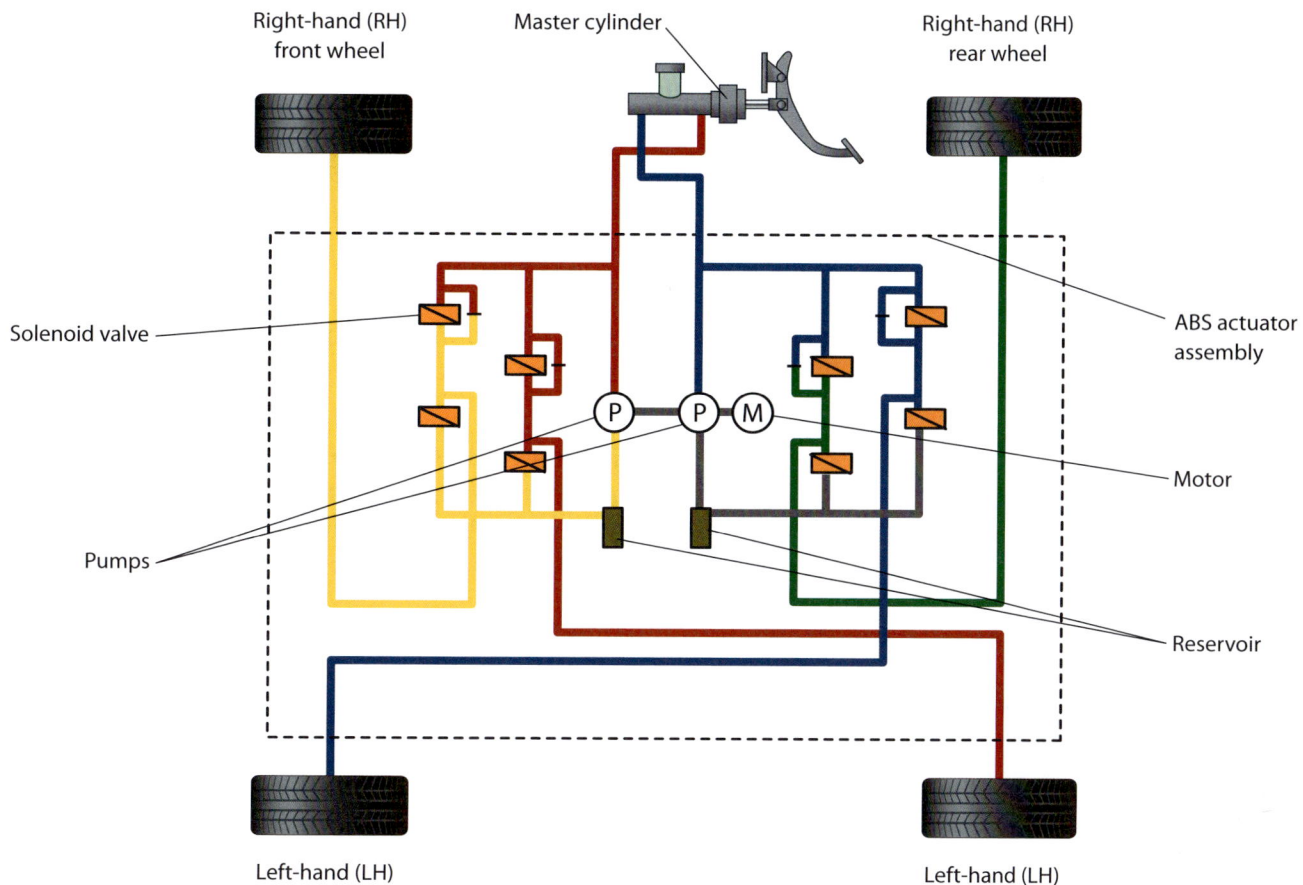

Figure 6.26 ABS layout

Weight transfer and proportioning valves

When the vehicle is moving in a forward direction and the brakes are applied, weight transfer normally puts most of the car's weight over the front axle. This places more load onto the front tyres and gives them additional grip. On the other hand, the rear tyres become unloaded and grip is reduced. If hydraulic brake pressure is allowed to be the same on all four wheels, the rear tyres could skid and cause an accident.

To reduce the possibility of this happening, some vehicles use a **proportioning valve** which splits the hydraulic forces. The valve will allow full pressure to go to the front brakes, letting them operate normally. At the same time, it reduces the brake hydraulic pressure to the rear wheels. This helps to prevent wheel lock-up.

Key term

Proportioning valve – a valve fitted in the braking system to reduce hydraulic pressure to the rear brakes. This is so that the rear wheels are less likely to skid during the weight transfer of braking.

CHECK YOUR PROGRESS

1 Name three components of a drum brake system.
2 Name three components of a disc brake system.
3 Give two advantages of disc brakes over drum brakes.

Safe working

Any work conducted on a critical safety system, such as brakes, requires that you take accurate measurements and follow correct settings. If you do not, damage, injury or even death could occur.

Routine maintenance on vehicle braking systems

It is important to have a good source of technical data when you are conducting routine maintenance on light vehicle braking systems. This way, you will have access to the correct maintenance schedules and procedures.

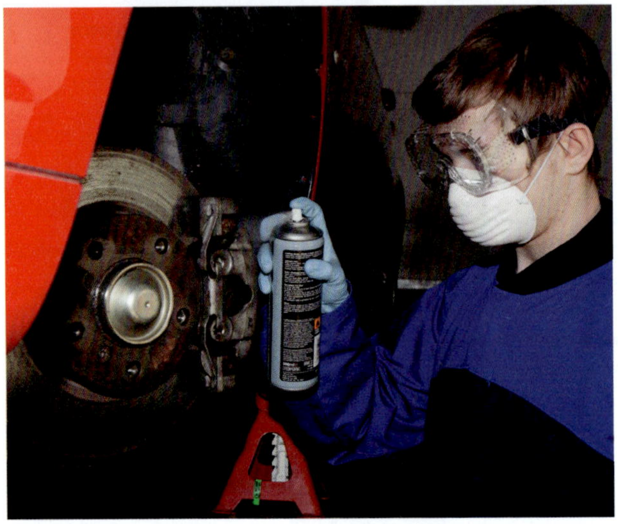

Figure 6.27 A mechanic spraying brake cleaner to remove dust

You must give special consideration to the selection of personal protection equipment (PPE) when you replace braking system components such as pads, discs or shoes.

As friction material wears down, it will create dust that can be hazardous to health. This dust should be properly controlled by the use of dedicated brake cleaning solvents. It is advisable to wear a particle mask too.

Older brake friction materials may contain asbestos. If breathed in, it can cause scarring of the lungs known as asbestosis. Asbestos dust can also cause cancer. Both of these conditions may not show symptoms for a number of years. However, you need to be aware of the risks involved and take appropriate precautions.

Safe working

Before you do any work on a light vehicle braking system, carry out a visual inspection for any component damage or fluid leaks. If you find a problem, report it immediately to your supervisor. They can contact the owner and gain authorisation for any extra work needed.

Conduct a visual inspection for fluid leaks

Checklist			
PPE	**VPE**	**Tools and equipment**	**Source information**
• Steel toe-capped boots • Overalls • Latex gloves	• Wing covers • Steering wheel covers • Seat covers • Foot mat covers	• Exhaust extraction equipment • Ramp or hoist • Inspection lamp	• Technical data

1. Raise the car with the engine running so that servo assistance is available. Remember to use exhaust extraction if working in a confined area.

2. Get an assistant to put their foot on the brake pedal, to place the system under pressure. Then use an inspection lamp to check all visible braking system components.

Replace brake pads

Checklist			
PPE	**VPE**	**Tools and equipment**	**Source information**
• Steel toe-capped boots • Overalls • Latex gloves • Particle mask	• Wing covers • Steering wheel covers • Seat covers • Foot mat covers	• Inspection lamp • Vehicle hoist or ramp • Brake cleaner • Spanners • Screwdrivers • Socket set • Brake hose clamp • Caliper piston retraction tool • Torque wrench	• Technical data • COSHH data

1. Gain access to the brake pad system components (remove the wheels, etc.). Remember to observe all health and safety requirements, including the correct use of PPE and VPE.

2. Clean any brake dust away using an appropriate brake cleaner.

3. Inspect braking system components (discs, calipers, pads, hoses and pipes, etc.) for damage or excessive wear.

4. Use an appropriate tool to clamp the brake hose and release the bleed nipple. It is advisable to connect a hose and bleed bottle to the nipple to catch any fluid.

5. Undo any brake pad retaining clips, push back the caliper pistons slightly and remove the pads.

6. Lay the removed components out in order. In this way they can be compared with any new parts to make sure they are the same. It also makes it easier to remember in what order to refit them.

7. Fully push back the caliper piston(s), close the bleed nipple and remove the brake hose clamp.

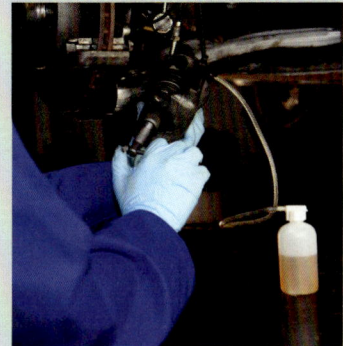

8. Make sure all friction surfaces are free from grease and oil.

9. Lubricate brake pad and caliper sliding surfaces. Specialist brake greases are designed for this purpose.

10. Reassemble. Make sure all components are correctly secured, including the correct use of torque settings for nuts and bolts.

11. Pump the brake pedal to reset calipers and pads.

12. Refit the wheels and tighten wheel nuts to correct torque.

13. Road test the car for safety and to check correct operation.

During the road test, new brake pads will require 'bedding in'. This involves careful but firm operation so that pad surface irregularities are worn away, giving good pad to disc contact.

Working life

Ahmed is working on the brakes of a customer's car with oil and grease on his hands. This oil gets transferred to the front brake pads, but Ahmed doesn't bother to clean it off properly. When he's finished working on the brakes, he puts the wheels back on the car and does the wheel nuts up without using a torque wrench.

1   What might happen when the car is driven?

2   Who could be affected by this?

Replace brake shoes

Checklist			
PPE	**VPE**	**Tools and equipment**	**Source information**
• Steel toe-capped boots • Overalls • Latex gloves • Particle mask	• Wing covers • Steering wheel covers • Seat covers • Foot mat covers	• Inspection lamp • Vehicle hoist or ramp • Brake cleaner • Spanners • Screwdrivers • Socket set • Return spring tools • Torque wrench	• Technical data • COSHH data

1. Gain access to the brake shoe system components (remove wheels, drums, etc.). Observe all health and safety requirements, including the correct use of PPE and VPE.

2. Clean any brake dust away using an appropriate brake cleaner.

3. Inspect braking system components (drum, shoes, wheel cylinder, hoses and pipes) for damage, excessive wear or fluid leakage.

4. Follow the manufacturer's instructions to strip out and remove the brake shoes. Specialist tools are available for the removal and refitting of shoes and springs.

5. Lay the components out in order. In this way they can be compared with any new parts to make sure they are the same. It also makes it easier to remember in what order to refit them.

6. Lubricate the contact points where the brake shoes slide against the backplate. Specialist brake greases are designed for this purpose.

7. Refit brake shoes, checking for correct positioning of any leading or trailing shoes.

8. Make sure all friction surfaces are free from grease and oil.

9. Reassemble and refit the brake drum. Make sure all components are correctly secured, including the correct use of torque settings for nuts and bolts.

10. You should now adjust the brake shoes following the manufacturer's procedures.

11. After the shoes have been adjusted, you can adjust the handbrake.

12. Refit the wheels and tighten the wheel nuts to correct torque.

13. Road test the car for safety and to check correct operation.

Safe working

Before carrying out any brake shoe adjustment, make sure that the handbrake is in the fully off position.

Always check the operation of the brakes before the car is started and moved when you have been working on braking systems. If you do not, it is possible that the first time the brakes are used they may not function correctly. This could lead to an accident.

Maintenance of the hydraulic system

The fluid in a hydraulic system must not contain any air bubbles or it will not work properly. If air has got into the hydraulic system, when the brakes are pressed, instead of the pressure acting through the fluid and operating the wheel cylinders or brake calipers, the air bubbles are squashed. This reduces the overall efficiency of the hydraulic system. When this happens you will normally feel too much brake pedal movement. It is often referred to as **spongy**.

Any air must be removed by a process called 'bleeding'. This involves forcing the air out of the braking system so that only fluid remains.

To remove air from the system, brake calipers and wheel cylinders are fitted with **bleed nipples**.

> **Key terms**
>
> **Spongy** – a term given to the soft feel of a brake pedal when there is air in the hydraulic fluid.
>
> **Bleed nipples** – a small manually operated valve, usually found on a hydraulic slave cylinder. It is designed to allow the removal of air from the braking system.

Bleeding the hydraulic braking system

Checklist			
PPE	**VPE**	**Tools and equipment**	**Source information**
• Steel toe-capped boots • Overalls • Latex gloves • Goggles	• Wing covers • Steering wheel covers • Seat covers • Foot mat covers	• Inspection lamp • Vehicle hoist or ramp • Brake cleaner • Spanners • Screwdrivers • Socket set • Bleed bottle • Brake fluid • Torque wrench	• Technical data • COSHH data

1. Locate and top up the clutch fluid reservoir with fresh clutch/brake fluid.

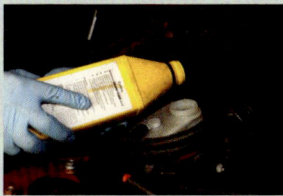

2. Connect a bleed bottle and pipe to the clutch slave cylinder.

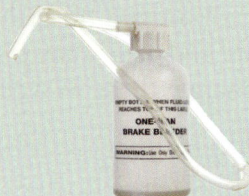

3. Open the bleed screw. Ask an assistant inside the car to push the clutch pedal down.

4. Close the bleed nipple and ask your assistant to release the clutch pedal.

5. Repeat steps 3 and 4 until all air has been removed from the system.

6. Top up the clutch fluid reservoir and clean up any spilt fluid.

7. Repeat the process for all bleed nipples, until all of the air has been removed.

8. Clean up any spilt brake fluid.

9. Refit the wheels and tighten the wheel nuts to correct torque.

10. Road test the car for safety and to check correct operation.

> **⚠ Safe working**
>
> Remember that you must dispose of the old brake fluid correctly in accordance with the Environmental Protection Act 1990.

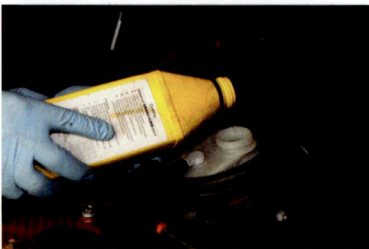

Figure 6.28 Topping up brake fluid

Brake fluid

The fluid designed to be used in braking systems must have certain properties. Two main types of brake fluid are in common use:

- glycol-based fluid
- silicone-based fluid

Glycol-based fluid

Glycol is a glycerine–alcohol-based fluid, normally labelled DOT 3 or DOT 4. It is designed to have a high boiling point. This is because the braking system components become very hot during operation. If the fluid reached the point where it could boil, then bubbles might appear, and bubbles can be compressed or squashed. This can lead to excessive movement at the brake pedal. If the fluid boils, it is possible that the brake pedal will go all the way to the floor. This is a situation called **vapour lock**.

Another problem with glycol-based brake fluid is that it is **hygroscopic**. This means it attracts moisture from the air. Over a period of time, this absorbed moisture can lead to problems.

- The first problem is that it lowers the boiling point of the brake fluid and can lead to vapour lock.
- The second problem is that it may corrode metal brake system parts from the inside out.

Silicone-based fluid

The other type of brake fluid (DOT 5) is a silicone-based liquid with a high boiling point. Unlike a glycol-based fluid it is not hygroscopic. This means it will not attract moisture from the atmosphere.

Glycol- and silicone-based brake fluids are not compatible and should not be mixed. If the brake fluid is to be changed from one type to another, you should fully drain and flush the system before refilling. Some rubber components and seals may also not be compatible. As a result you may need to change these as well.

Testing brake operation

Braking **efficiency** is a measurement of how well the brakes work. It is usually measured with a brake roller tester.

Key term

Brake efficiency – a measurement of how well brakes work at slowing the car down when compared with the weight of the vehicle.

Roller brake testing

Checklist			
PPE	**VPE**	**Tools and equipment**	**Source information**
• Steel toe-capped boots • Overalls • Latex gloves	• Wing covers • Steering wheel covers • Seat covers • Foot mat covers	• Brake roller tester	• Technical data • Vehicle weight

1. Place the front wheels of the car into the rollers of the brake tester. Remember to observe all health and safety requirements, including PPE and VPE.

2. Start both rollers and allow the car to centralise in the tester. Apply the handbrake.

3. Run each roller individually, while steadily applying the brakes. Record maximum effort from the gauges.

4. Run both rollers at the same time. Steadily apply the brakes and record any difference in performance between the two.

5. Place the rear wheels of the car into the rollers of the brake tester.

6. Run each roller individually, while steadily applying the brakes. Record maximum effort from the gauges. This is done for both hand and foot brakes.

7. Finally, apply the brakes while running both rollers to see that the gauges rise and fall at the same rate.

8. Remove the car from the rolling road and calculate performance. A chart is normally provided which will show the vehicle weight to be used when calculating performance.

To calculate performance:

$$\frac{\text{Total footbrake effort (kg)}}{\text{Vehicle weight (kg)}} \times 100 = \text{performance (\%)}$$

$$\frac{\text{Total handbrake effort (kg)}}{\text{Vehicle weight (kg)}} \times 100 = \text{performance (\%)}$$

Minimum braking efficiency

The current minimum braking efficiency required for a car to be used on a public highway is:

- 50% for the footbrake
- 16% for the handbrake if a dual line system is used
- 25% for the handbrake if a single line system is used.

The maximum imbalance allowed between both front brakes is 25%.

An example is shown below:

A car with a split braking system weighs 1000 kg		
Left-hand front	Right-hand front	Total
250 kg	170 kg	420 kg
Left-hand rear	Right-hand rear	Total
155 kg	160 kg	315 kg
Footbrake total		735 kg
Left-hand handbrake	Right-hand handbrake	Total
120 kg	90 kg	210 kg

Total footbrake effort	÷	Vehicle weight	x 100	= Performance
735 kg	÷	1000 kg	x 100	73.5%
Total handbrake effort	÷	Vehicle weight	x 100	= Performance
210 kg	÷	1000 kg	x 100	21%

Front brake imbalance

Highest effort	–	Lowest effort	÷ Highest effort	x 100	= Imbalance
250 kg	–	170 kg	÷ 250 kg	x 100	32%

Find out

Does the brake efficiency above meet the current legal requirements?

Completion of work and testing

A road test must be conducted to check for safety and correct operation following any work on the braking system.

You should record on a job card any work carried out on a car's braking system. In this way there is a record of this maintenance.

CHECK YOUR PROGRESS

1 What is a master cylinder with two pistons called?
2 How do you know if a servo is working?
3 What is brake efficiency?
4 What is meant by the term hygroscopic?

Steering and suspension systems – components and maintenance

The steering and suspension systems of cars are responsible for **manoeuvrability**, **handling** and comfort.

The steering system turns the movement of the steering wheel into a side-to-side movement that can pivot the front wheels.

The front wheel and brake assemblies are mounted on **ball joints**. The ball joints allow them to turn from side to side for steering, and move up and down to allow for suspension operation.

The vehicle's suspension system provides the car with good handling. It helps to keep the tyres in contact with the road as the car moves over an uneven surface. It also absorbs the road shocks to give the passengers a comfortable ride.

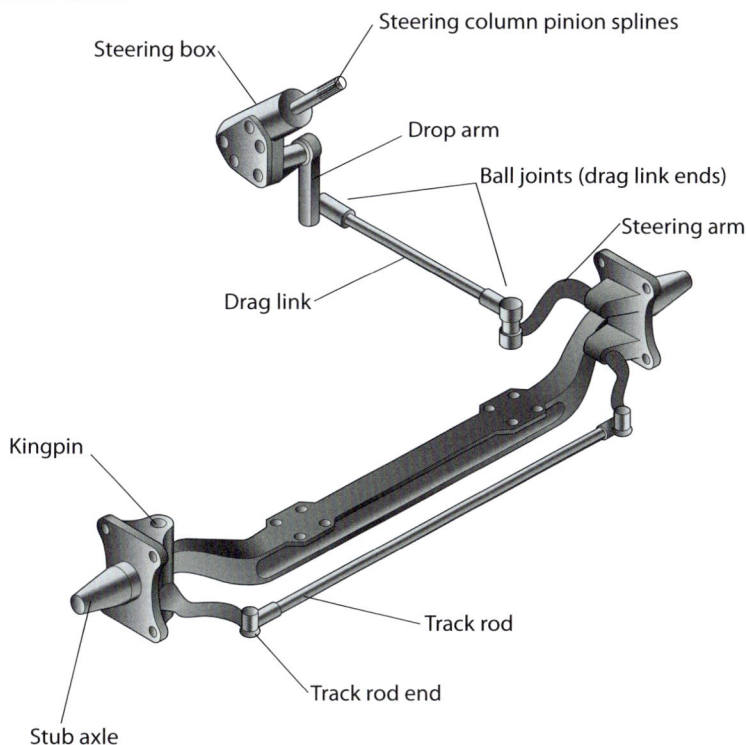

Figure 6.29 Steering box

Non-assisted steering and suspension components

The steering wheel steers the car. It is controlled by the driver, who decides which direction the car is going in. The steering wheel is connected, inside a supporting tube, to a metal shaft known as the steering column. As the driver turns the steering wheel, this movement is transferred down this column. It passes through a series of universal joints to a steering gearbox. A steering gearbox is needed because, if the wheels were connected directly to the steering wheel, the driver would struggle to be able to turn it.

Steering wheel

Steering column

Collapsible on impact

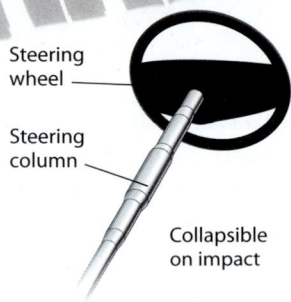

Figure 6.30 Steering wheel and column

Working life

Dave has to remove a steering wheel. He has undone the nut in the middle but it won't come off. He decides to use a hammer to hit the steering column to try and loosen it.

1 Why is this a bad idea?

The job of the steering gearbox is to turn the rotating movement of the steering wheel into side-to-side movements, to control the steering and multiply the driver's effort. This is done by a small gear connected to the steering column. This meshes with a larger gear inside the gearbox, increasing the turning effort.

The two main types of steering gearbox in common use are:

- the rack-and-pinion – it is lightweight and compact and used on many modern cars
- the steering box – this is an older type of system which is rarely used on small cars but can be found on expensive prestige models and off-road vehicles.

Rack-and-pinion steering

The most common steering gearbox is the rack-and-pinion. It is lightweight, and easier and cheaper to manufacture compared with a steering box system.

A rack-and-pinion has fewer parts than a steering box and during normal operation there are fewer items to wear out. This reduces the feeling of slack in the steering, making it more reliable and giving it a more direct feel.

The **rack** is a long metal shaft with gear teeth cut into its side. The **pinion** is a small gear which meshes with the rack at 90°. It turns the rotating movement of the steering wheel and column into a side-to-side movement at the rack.

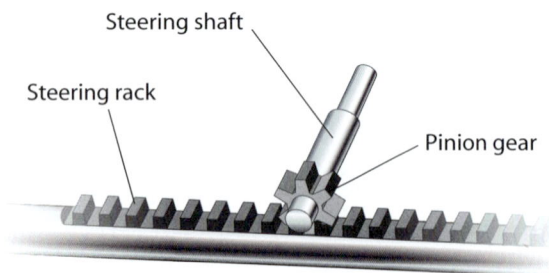

Steering shaft

Steering rack

Pinion gear

The size, shape and number of teeth on these two gears provide **torque** multiplication. This makes it easier for the driver to steer and control the car.

The rack is a solid metal shaft. If it was connected directly to the steered wheels, as the car went over a bump and the suspension moved up and down, it would try to bend. To prevent this, two short metal shafts called **track rods** are joined to either end of the steering rack with ball joints. They pivot up and down with suspension movement.

Figure 6.31 Rack and pinion

The outer ball joint is known as the **track rod end**. It is attached to the back of the **hub** by a **steering arm**.

The inner ball joint is usually covered with a rubber boot called the **rack gaiter**. This helps to keep lubricating oil and grease around the joint and in the rack. It also helps to keep dirt and moisture out.

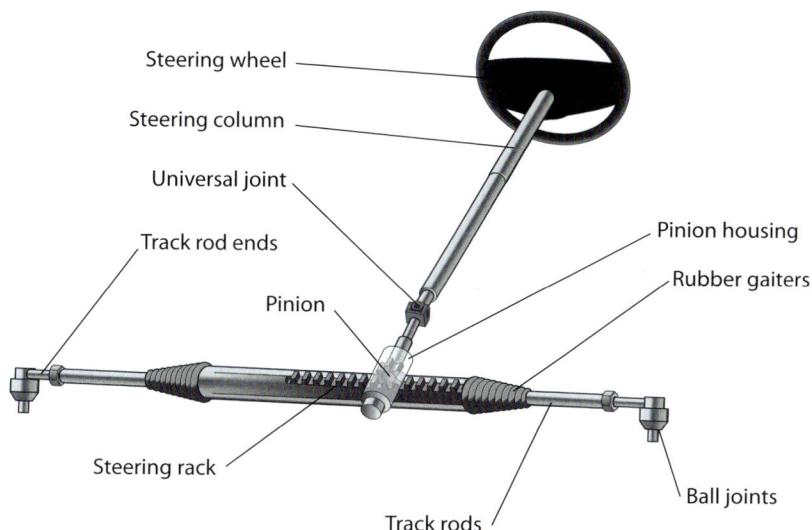

Labels: Steering wheel, Steering column, Universal joint, Track rod ends, Pinion, Steering rack, Track rods, Pinion housing, Rubber gaiters, Ball joints

Figure 6.32 Rack-and-pinion

Table 6.13 Rack-and-pinion steering components and their purposes

Rack-and-pinion steering component	Purpose
Steering wheel	Allows the driver to control the steering
Steering column	A metal shaft to connect the steering wheel to the rack-and-pinion
Rack-and-pinion	Turns rotating movement of the steering wheel into side-to-side movement at the wheels. Reduces driver effort by multiplying torque
Rack gaiters	Rubber boots to keep lubrication in the steering rack and dirt and moisture out
Track rods	Short shafts that allow the suspension to move up and down without bending the steering
Track rod ends	Ball joints at the ends of the steering arms which allow the wheels to pivot

Steering box

The second type of steering is known as a steering box, as in gearbox. This uses a **worm** gear, like the screw thread on a nut and bolt, to turn a rotating gear through 90°. This will then move a steering arm.

There are a number of different designs of steering boxes including worm and roller, worm and peg, and **recirculating** ball.

- In a worm and roller steering box, the worm gear moves the drop arm using a roller gear which meshes with the worm.

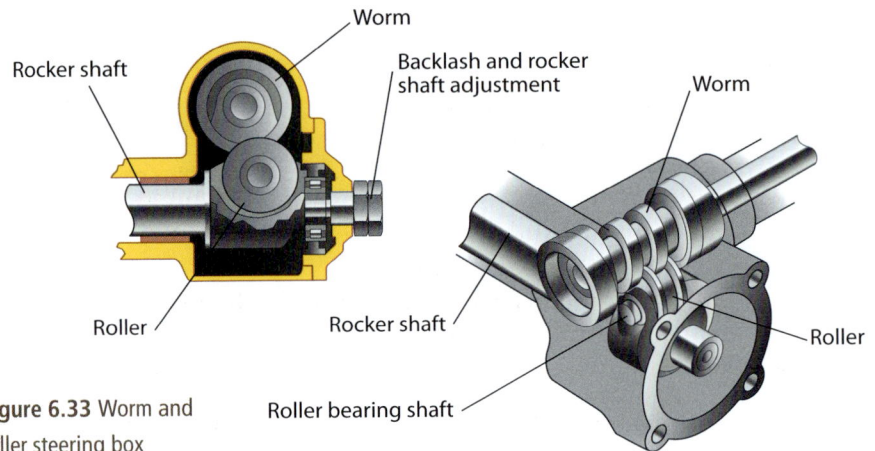

Figure 6.33 Worm and roller steering box

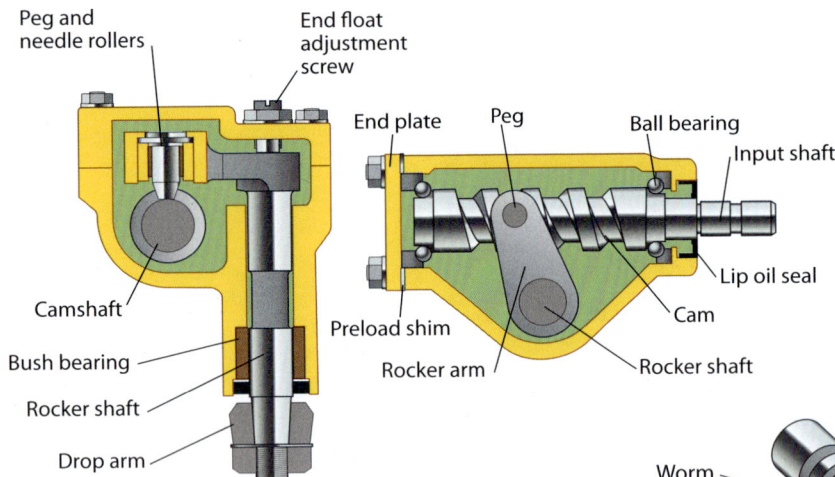

Figure 6.34 Worm and peg steering box

- In a worm and peg steering box, the worm gear moves the drop arm using a short peg connected to a shaft.
- In the recirculating ball steering gearbox, the thread between the worm and nut is filled with ball bearings to reduce friction.

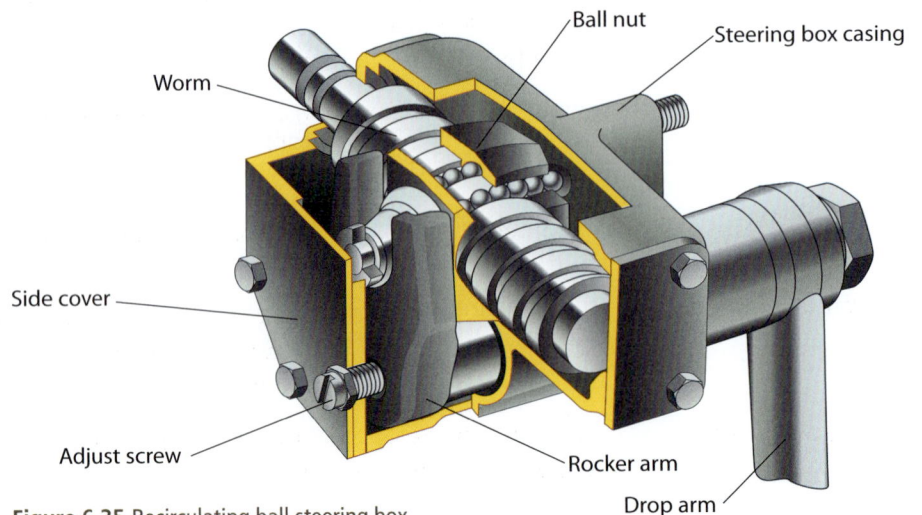

Figure 6.35 Recirculating ball steering box

The steering arm, sometimes called the **drop arm** or **Pitman arm**, is then connected to the rest of the steering linkage. This is done by a short rod known as a **drag link**. The movement is relayed to the wheels.

Two types of steering linkage are available for steering box construction. If the front axle of the vehicle is a **solid beam axle** then the steering arm is connected directly to a one-piece track rod. This is then joined to both front wheel pivots. As the steering is turned from side to side, this track rod relays the movement to the front wheels.

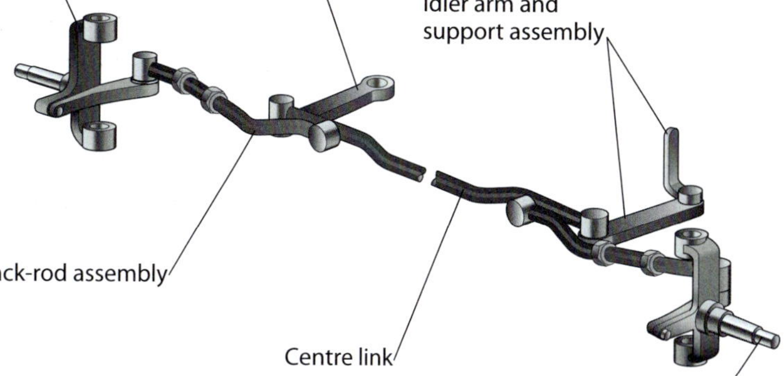

Figure 6.36 The steering linkage

A different type of linkage is needed if independent front suspension is used. If a one-piece track rod was used, as one wheel went over a bump, it would pull the steering to one side. However, the track rod itself is split up into three shorter sections, with a ball joint connecting each section. This way, as one wheel goes over a bump, it can move independently up and down and not affect the rest of the steering.

The steering box is mounted on one side of the car, so the track rod will need to be supported on the opposite side. This is done with an **idler arm**.

Table 6.14 Steering box system components and their purposes

Steering box system component	Purpose
Steering wheel	Allows the driver to control the steering
Steering column	A metal shaft to connect the steering wheel to the steering box
Steering box	Turns rotational movement of the steering wheel into side-to-side movement at the wheels. Reduces driver effort by multiplying torque
Idler arm	A movable arm designed to support the steering mechanism on the opposite side of the car to the steering box
Drop arm and drag link	Connects the output of the steering box to the track rods
Track rods	Shafts that connect drag link to the steered wheels and allow the suspension to move up and down without bending the steering
Track rod ends	Ball joints at the ends of the steering arms which allow the wheels to pivot

Key terms

Worm – a form of gear that is produced as a spiral on a shaft, similar to a screw thread.

Recirculating – going round and round.

Drop arm – the output lever of a steering gearbox.

Pitman arm – another name for the drop arm.

Drag link – a short linkage that connects the drop arm to the track rod assembly.

Solid beam axle – a solid metal axle shaft with the wheel assemblies at either end.

Idler arm – a component that supports the steering linkage and copies the movement of the drop arm.

1 What does the steering wheel do?
2 Name three types of steering box.
3 What is the purpose of an idler arm?

Power assisted steering (PAS)

Many cars are now fitted with power assisted steering (PAS). This makes steering even easier. Hydraulic pressure or electric motors can be used to assist the driver, especially during slow moving manoeuvres.

As the vehicle speeds up, power assistance is designed to reduce so that steering 'feel' is maintained.

All steering assistance must be **fail safe**. This means that if the power assistance goes wrong, the driver is still able to steer. However, it will take more effort to turn the wheel and the steering will feel heavier.

Suspension

A suspension system is needed so that the car can move smoothly over bumps.

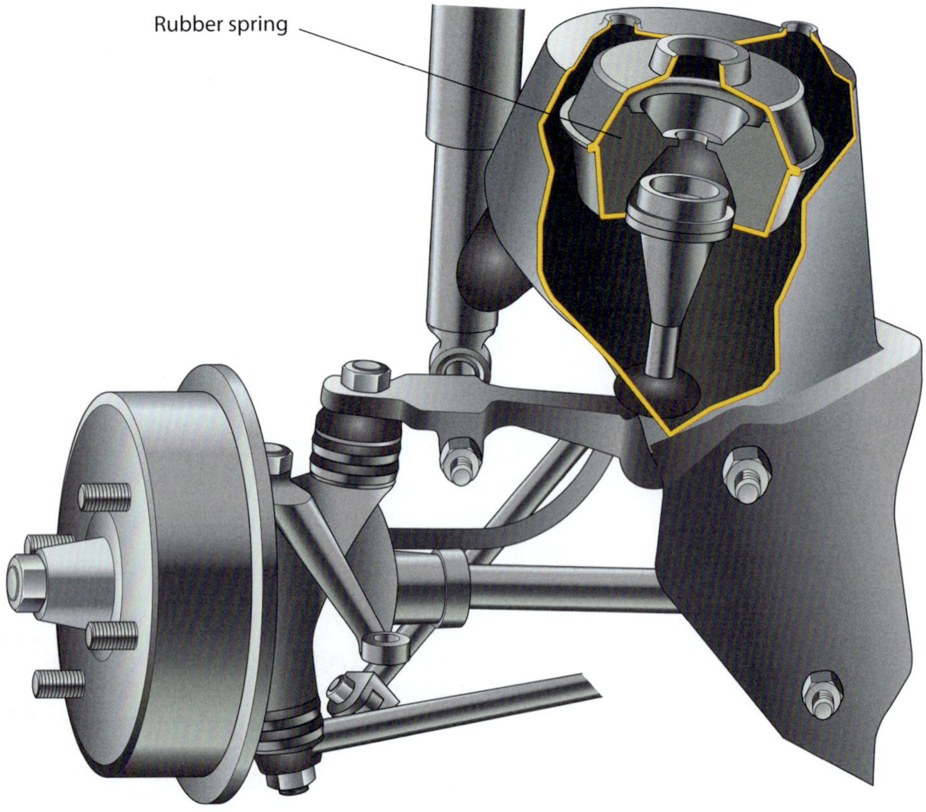

> **Key term**
>
> **Fail safe** – if the power assistance fails, then the driver must still be able to operate the steering safely.

Rubber spring

Figure 6.37 Rubber springs

The suspension system consists of two main elements:

- springs which absorb the impact from bumps and potholes
- dampers which take the bounce out of the spring (**oscillation**).

Most car suspension springs are made from sprung metal but other materials can also be used.

- Rubber is commonly used as an additional springing component in many suspension systems, bushes, mountings and bump stops. It has also been used as the main spring members in some older cars.
- Pneumatics (gases) can also be used as springs because they are compressible. If a gas is sealed inside a collapsible container it can be squashed and will try to return to its original shape. This is just like the air in a tyre which gives bounce.
- If air is used in suspensions, it is normally topped up by an on-board compressor.
- Nitrogen is another type of gas that is used in suspensions. This is because it is a stable non-flammable gas (inert).
- If nitrogen is sealed behind a rubber diaphragm, hydraulic fluid can be pushed through a series of pipes to transfer movement from the suspension to the gas. The gas is squashed and then returns to its original shape. This forces fluid back the other way to operate against the suspension and road bumps.
- A suspension system that uses nitrogen as the spring, and hydraulic fluid combined, is called **hydro-pneumatic**.
- Some vehicle manufacturers are starting to use fibre-reinforced plastics, like glass-reinforced plastic (GRP), as their suspension springs.

> **Key terms**
>
> **Oscillation** – bouncing of a suspension spring.
>
> **Hydro-pneumatic** – a suspension system that uses nitrogen gas under pressure to act as a spring, and a liquid to act as a hydraulic operating mechanism between the suspension and the gas container.

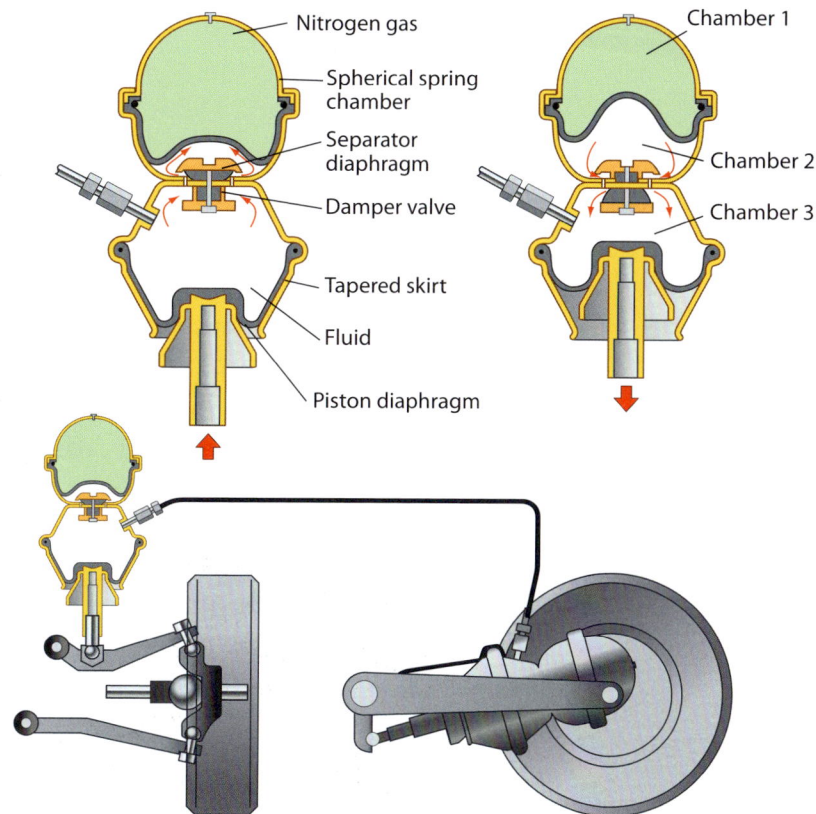

Figure 6.38 Hydragas unit

Three types of metal spring are used on cars:

- leaf spring
- coil spring
- torsion bar.

Leaf spring

A leaf spring is a slightly curved steel strip. It is made thicker in the middle or supported by extra strips for strength.

It is curved when not under load. However, when it is placed under tension, because of weight or going over a bump, it is forced to flatten. Then, because of its sprung nature, it tries to return to its original shape.

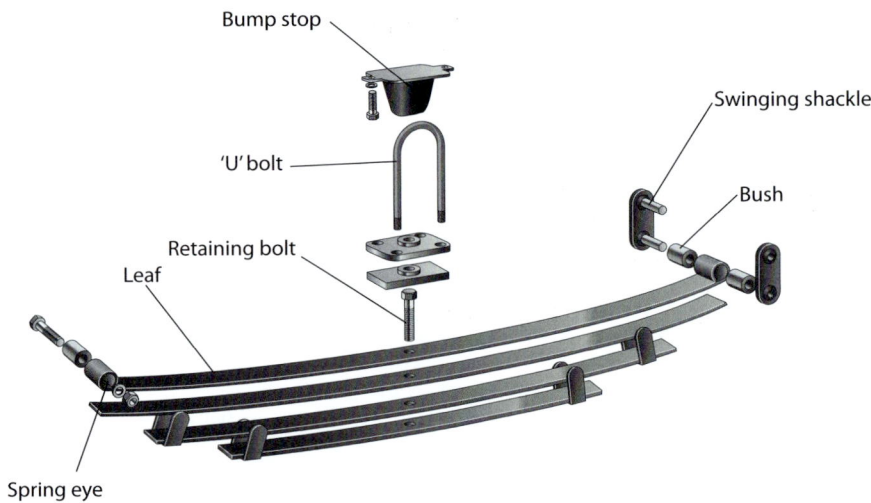

A leaf spring is attached at each end of the vehicle chassis by a nut and bolt. If it was mounted solidly at both ends, it wouldn't be able to flex or change shape and act as a spring.

A linkage known as a **swinging shackle** is used at one end to allow the spring to lengthen under load.

The centre of the spring is clamped to the car's axle. As it goes over a bump, the axle moves upwards. This flattens the spring and absorbs the shock.

Figure 6.39 Leaf spring assembly

Coil spring

A coil spring is a sprung steel rod wound in a spiral known as a **helix**. It is mounted between the suspension components and the car's bodywork.

As the car goes over a bump, the suspension moves up and down. This squashes and stretches the spring. The spring is wound in a spiral so, as it is **compressed**, it actually twists slightly and the overall coil gets slightly fatter.

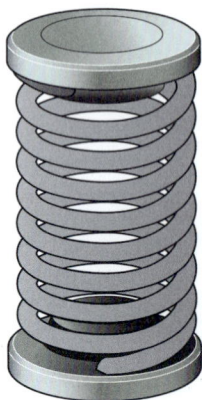

Constant rate coil spring

Variable rate coil spring

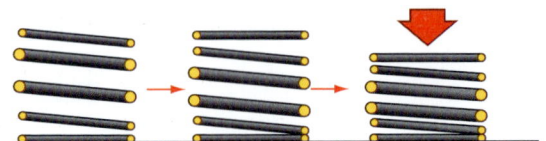

Taper coil spring

Figure 6.40 Coil spring and taper coil spring

Torsion bar

A **torsion** bar is a sprung metal rod that is anchored securely to the vehicle body at one end. It is connected to a suspension arm at the other end. As the car goes over a bump, the suspension arm is levered upwards and the torsion bar is twisted.

The torsion bar then tries to unwind itself. This forces the suspension arm back the other way.

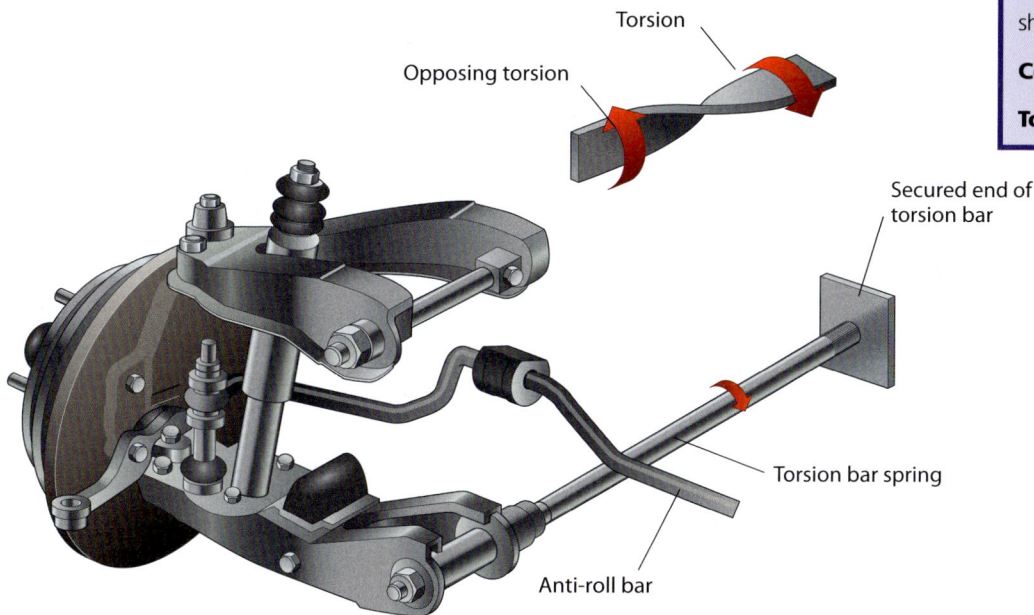

Figure 6.41 Torsional bar and elasticity

This design can be used on both front and rear suspensions. Due to its compact nature, it has become very popular with manufacturers of small hatchback cars as a rear suspension set-up.

The rods are usually mounted longitudinally (front to rear) if used for front suspension. If used on the rear, it is common to mount them transversely (side to side).

Table 6.15 Spring types and their actions

Spring type	Action
Leaf spring	A curved metal bar which tries to return to its original shape when straightened
Coil spring	A sprung steel rod wound in a spiral. When compressed it twists and expands, then uncoils back to its original position
Torsion bar	A sprung metal rod that tries to unwind itself when twisted
Rubber	Compresses and then the elastic nature returns it to its original state
Air	When sealed inside a collapsible container, air will squash and then return to its original state
Nitrogen	Sealed behind a diaphragm and acted on by hydraulic force, nitrogen will squash and then return to its original position
GRP	Glass-reinforced plastics that can be used to make leaf springs

Suspension dampers

When a vehicle's wheel goes over a bump in the road, this creates **bump** and the suspension and spring move upwards. Suspension dampers, sometimes called shock absorbers, are designed to reduce this oscillation or bounce.

Without dampers, the springs would continue to bounce uncontrollably. This would make the ride extremely uncomfortable and the handling very dangerous.

Most suspension dampers are telescopic. That means they collapse in on themselves like a telescope. They consist of a sealed, oil-filled cylinder and a piston attached to a rod.

One end of the damper is attached to the suspension components and the other end is attached to the car's chassis.

- When the wheel goes over a bump, the suspension and spring move upwards. This causes the damper to compress and push the piston along the cylinder. This forces oil through small holes in the piston.
- The piston movement will try to force oil from the lower part of the cylinder to the upper part.
- As the spring returns the suspension to its original position (**rebounds**), the damper extends. This moves the piston back the other way.
- This movement forces oil through a series of valves, slowing down the motion and absorbing most of the spring's rebound energy and converting it into heat.

Upper mounting
Compression stop
Piston rod
Integrated seal
Valve
Hydraulic oil
Floating piston
Gas chamber

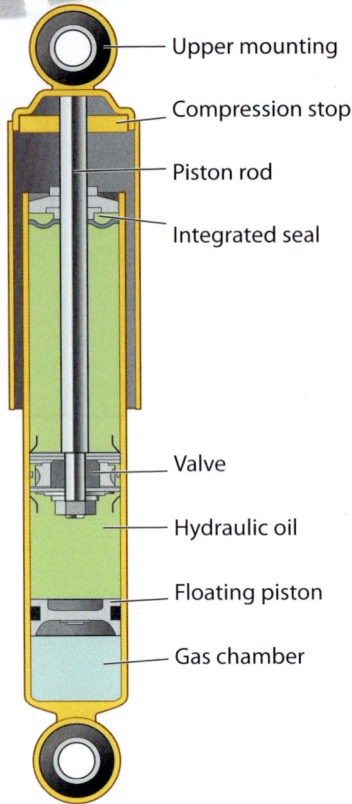

Figure 6.42 Telescopic damper

Key terms

Bump – the upward movement of the suspension.

Rebound – the downward movement of the suspension.

Find out

Test the operation of a suspension damper and make recommendations about its function. You can do this by choosing a vehicle and pushing downwards on one corner. (Be careful not to push on a place that will cause damage to the car.) Record how many times the suspension bounces. Is this acceptable?

CHECK YOUR PROGRESS

1 Name three types of metallic spring.
2 Name two types of non-metallic spring.
3 What is the purpose of a suspension damper?

Independent and non-independent suspension systems

On older cars, the suspension was connected to a beam axle. This is a solid metal beam that runs from one wheel assembly to the other across the car. With this type of suspension, as one wheel goes over a bump, it will also affect the angle of the other wheel, at the end of the axle on the other side of the car. This reduces handling and performance and can make the body of the car **roll** or tip to one side.

The answer to this problem is to give each wheel its own spring and suspension attachment. As any particular wheel goes over a bump, it has no effect on the others. This type of set up is called 'independent suspension'.

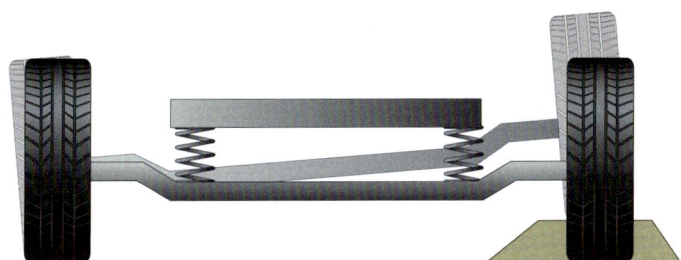

Figure 6.43 Non-independent suspension

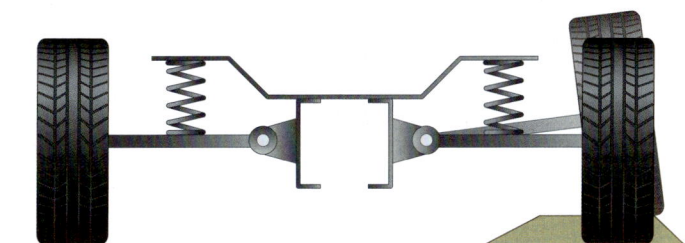

Figure 6.44 Independent suspension

Table 6.16 Independent suspension system components

Component	Description
Macpherson struts	Macpherson struts are a form of combined suspension assembly. They contain spring, damper and attachments, all in one unit. The telescopic damper is mounted inside a tube and is often referred to as the 'strut'. At the top of the strut assembly, a coil spring is fitted between two spring mountings. This surrounds the upper part of the damper. The top of a Macpherson strut is attached to the vehicle body in a specially shaped and reinforced housing called a turret. The lower part is connected to a suspension arm. The suspension arm can be a wishbone or track control arm design. The upper mounting of the strut assembly, usually where it is bolted to the **turret**, contains a rotational bearing assembly. This allows it to pivot when the steering is turned.

Component	Description
Wishbones 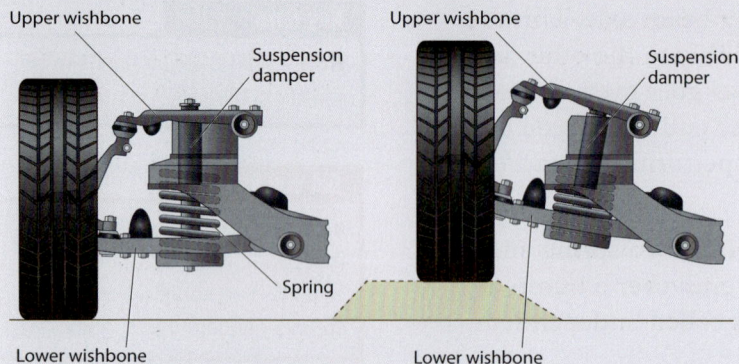 Upper wishbone, Suspension damper, Spring, Lower wishbone	A wishbone is a triangular-shaped suspension arm with one mounting point at the outer end by the wheel pivot. It is normally attached by a ball joint to allow for steering and suspension movement. It has two mountings at the inner end where it attaches to the car's chassis or **subframe**. This double mounting point, where it joins the chassis, acts to stabilise the arm from front and rear movement, created when accelerating and braking.
Track control arms 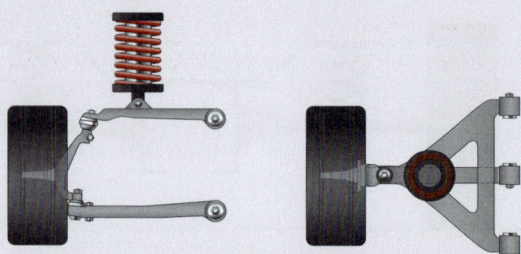	A track control arm is similar to a wishbone assembly. However, it only has one mounting point at the inner end where it joins the car's chassis or subframe. To stop it moving backwards and forwards during acceleration and braking, it needs a 'tie rod' attached to it and the chassis. In some designs, car manufacturers use the anti-roll bar ends to act as the tie rod.
Trailing arms 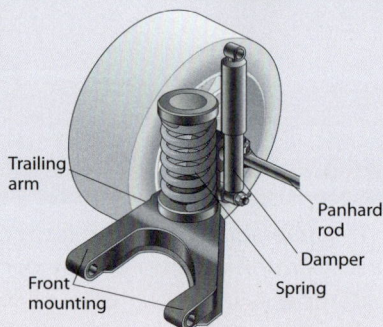 Trailing arm, Front mounting, Panhard rod, Damper, Spring	On a rear suspension assembly, some manufacturers use a suspension component called a trailing arm. This suspension arm is mounted front to rear under the car with the front end attached to the vehicle frame. This is so that it can pivot up and down with suspension movement. Because it is dragged behind during the normal forward motion of the car, it has been given the name 'trailing arm'.
Anti-roll bars 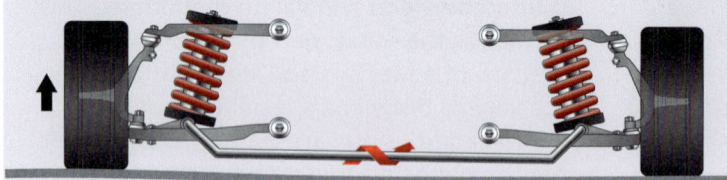	Body roll is a problem that occurs on vehicles fitted with independent suspension systems. Roll occurs when cornering forces act on the vehicle body, trying to tilt it to one side. The force of roll has the effect of unloading (taking the weight off) vehicle suspension units and tyres. This can reduce grip during manoeuvring operations and cause handling issues that can lead to an accident. To help reduce body roll, anti-roll bars are sometimes fitted. An anti-roll bar is a form of torsion bar spring, attached at either end to the right- and left-hand suspension units across the car. As the car goes around a corner and roll begins, the springing action of the bar reacts against some of the loads placed on the suspension. This helps to reduce body movement.

Non-assisted steering and suspension systems operation

So that the steered wheels of the car can follow the correct line under any driving or manoeuvring condition, a series of steering angles are used called **steering geometry**.

Some typical steering angles or geometry are outlined below.

Tracking

When travelling in a straight line, all four wheels of the car should run parallel. The steered wheels at the front should not point inwards or outwards when viewed from above. This steering setting is known as **tracking**.

The tracking is determined by the length of the steering rack, track rods and other linkages. The steering rack and linkages can be made longer or shorter so that tracking can be adjusted during normal maintenance procedures.

When viewed from above, the front of the wheels and tyres is referred to as the 'toe'.

- If the fronts of the wheels are pointing inwards slightly, the steering is said to be 'toeing-in'.
- If they are pointing outwards slightly, the steering is said to be 'toeing-out'.

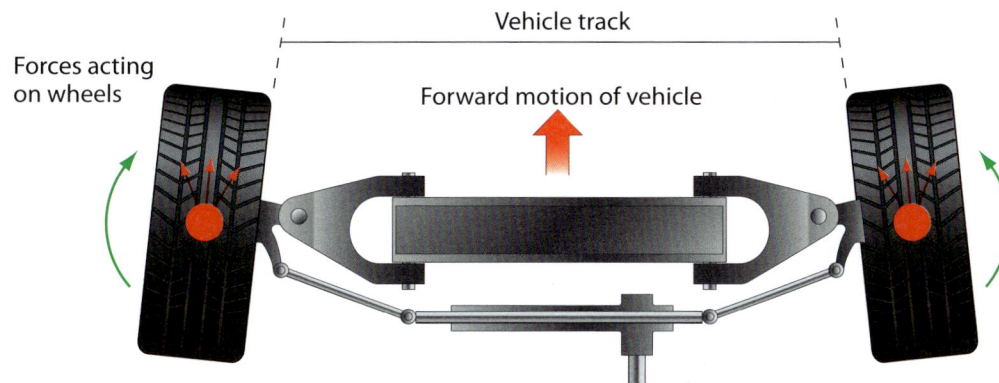

Figure 6.45 Toe-in wheel alignment

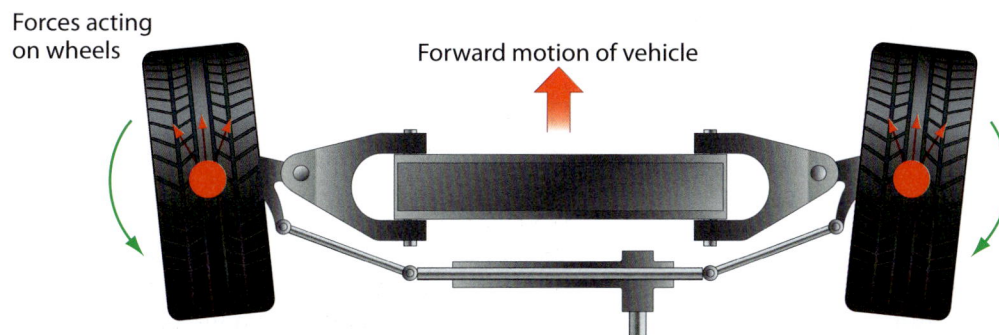

Figure 6.46 Toe-out wheel alignment

Key terms

Turret – a specially reinforced section of an inner wing.

Subframe – a chassis member bolted to the vehicle body. It is designed to support engine, transmission, steering and suspension components.

Steering geometry – a series of angles created between the main steering components. They are designed to allow the wheels and tyres to follow the best possible course when manoeuvring.

Tracking – where both wheels run parallel as they roll along the road.

Figure 6.47 Feathered tyre wear

Figure 6.48 Shoulder wear

Key term

Feathering – a form of tyre tread wear which feels smooth to the touch in one direction and rough in the other.

Safe working

Take care when inspecting a tyre tread for feathering. It is possible that sharp steel bracing wires may have been exposed due to tyre wear. These could cut your hand. Also, foreign objects and contaminants may be on the surface of the tyre, so PPE such as latex gloves can help to protect your hands.

If the wheels are not kept parallel when the vehicle is moving, the tyres will be forced to slip across the road surface. This causes rapid wear to the tread.

The effects of incorrect tracking are feathering and shoulder wear.

• **Feathering** is a common indication of incorrect tracking. When feathering occurs, the blocks to the tyre tread get worn in one direction. This can usually be felt by rubbing your hand over the surface of the tyre tread. If you rub your hand in one direction, the tyre tread will feel relatively smooth. When you rub your hand in the other direction, it will feel rough.

• Shoulder wear is another symptom of incorrect tracking. It could be rapid wear on the inside or outside of the tyre tread. If the tracking is toeing-out too much, wear can occur on the inside edges of the tyres. If the tracking is toeing-in too much, wear can occur on the outside edges of the tyres.

Did you know?

When the car starts to roll, the front wheels can toe-in or toe-out slightly. Manufacturers often include an initial toe setting in the technical specifications, to be used when adjusting the tracking. With a rear-wheel drive car, a slight toe-in is often included. With a front-wheel drive car, a slight toe-out is often included.

True rolling and toe-out on turns (TOOT)

When a car turns a corner, the steered wheels follow two different paths. The inner wheel of the bend travels on a much tighter curve than the outer wheel.

If both steered wheels were turned at exactly the same angle when the vehicle went round the bend, one of the tyres would struggle to grip and slip across the road surface. This would lead to rapid tyre wear and loss of control.

To help prevent this, a series of steering angles that create toe-out on turns are used. This means the inner wheel turns at a sharper angle than the outer wheel. This makes the front of the tyres 'splay outwards'.

Now as the vehicle goes around the corner, the steered wheels follow the correct path and curve. This is called 'true rolling motion'.

Ackerman principle

Toe-out on turns is achieved by a process called the Ackerman principle. This makes the inner wheel turn at a sharper angle than the outer wheel.

The Ackerman principle uses geometry (triangles) to produce toe-out on turns when the steering is moved from the straight ahead position.

Looking at Figure 6.49, you can see that the steering linkage, the steering rack or the track rod, is shorter than the pivoting point of the front of steered hubs. They are connected to these hubs by angled steering arms.

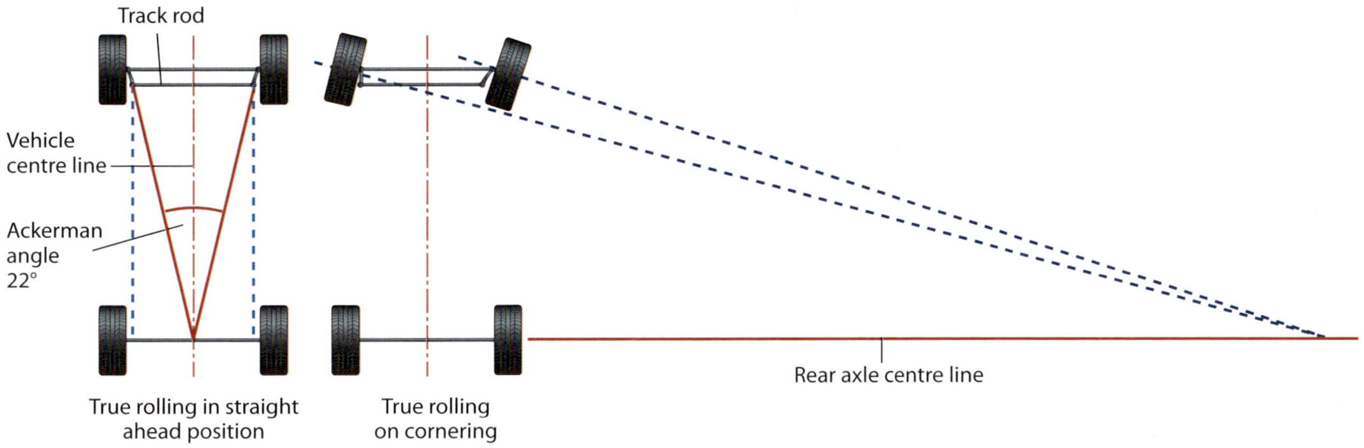

Figure 6.49 Ackerman system

An imaginary triangle is created at each end of the track arm where it meets the pivoting point of the steering hubs.

As the steering is moved from the straight ahead position, and the rack and track rods are moved to one side, one triangle is 'squashed' while the other is 'stretched'. This creates a sharper angle on the inner steered wheel, giving a true rolling motion.

The angle of the steering arms connected at the pivot is calculated by drawing two imaginary lines to the centre of the back axle, as shown. Now, when the steering is turned to one side, the Ackerman principle will create different wheel angles. If two lines are drawn at right angles to the steered wheels, they will meet up on an imaginary line that is at right angles to the rear wheels, as shown.

Camber angle

Camber angle is a steering geometry setting that will allow the top of the wheel to lean outwards or inwards when viewed from the front.

- If the tops of the wheels lean outwards, this is known as 'positive camber'.
- If the tops of the wheels lean inwards, this is known as 'negative camber'.
- If the wheels are completely vertical when viewed from the front, this is called 'neutral camber'.

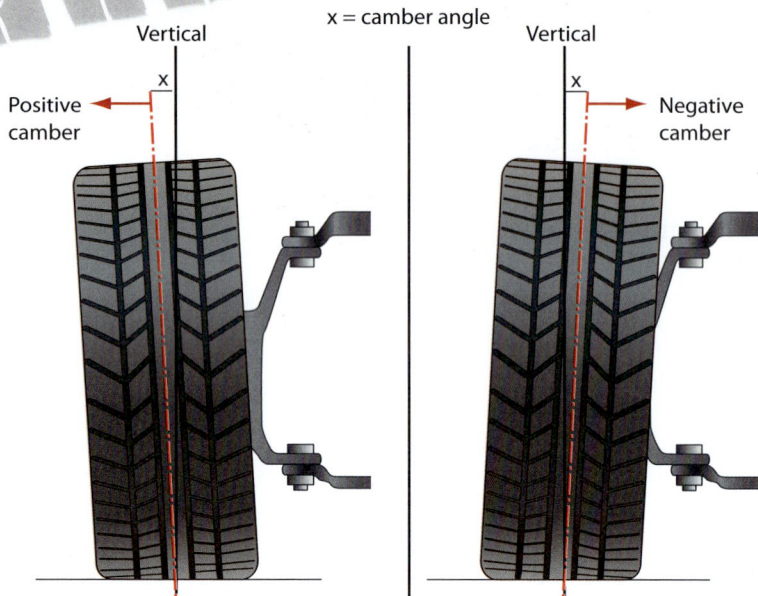

Figure 6.50 Positive and negative camber

- A positive camber angle can give light nimble steering. This is because the contact patch on which the tyre is moved, moves inwards under the pivot point of the front steered hubs. However, this can produce a tendency for the steered wheels to follow road irregularities, such as lumps and bumps.

- A negative camber angle can give very good road holding because the bottom of the tyres are 'splayed outwards' and react against cornering forces. However, this will produce heavy steering, making it feel hard to turn.

Did you know?

Excessive camber angle can lead to rapid tyre wear. This would normally be seen as a line worn on either the inner or outer edge of the tyre around the entire circumference.

Caster angle

Caster angle is a steering geometry setting that will help keep the wheels and tyres in a straight line when rolling along the road. It can help produce a self-centring action. When a corner has been turned, the steering wheel tries to pull itself back to the straight ahead position. This is done by angling the suspension of the steered wheels either forward or backwards when viewed from the side. This is just like the angle of the front forks on a bicycle or motorbike.

The weight of the vehicle is projected forward of the contact point of the tyre where it touches the road surface. This produces drag that will try to pull the wheel straight after a corner.

Did you know?

You can see caster action in operation by looking at the wheels of a shopping trolley. The pivot point of shopping trolley wheels is offset forwards of the centreline of the wheel where it touches the ground. As it is pushed forwards, drag pulls the wheel straight in the direction of motion.

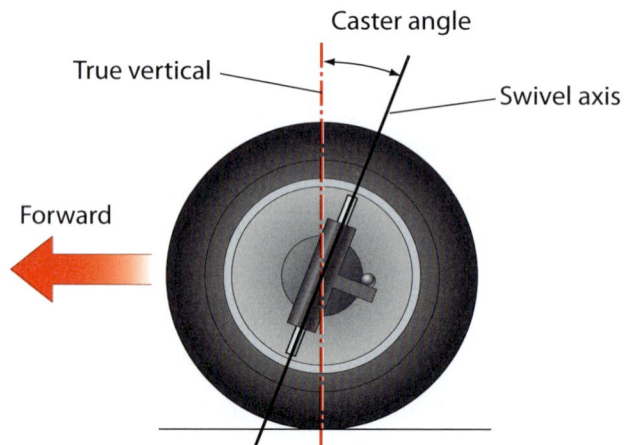

Figure 6.51 Caster angle

Swivel axis inclination (SAI) or kingpin inclination (KPI)

These allow the front suspension to pivot and the wheels to steer. They are called kingpins on older beam axle type vehicles or front suspension swivels, and usually ball joints on more modern cars.

These ball joints or swivels are normally offset with the lower swivel being mounted further out towards the wheel and tyre than the upper swivel. This produces an inclination (leaning inwards).

If an imaginary line is drawn down through the swivel axis inclination, and an imaginary line is drawn down through the camber angle of the wheel, they will meet at a point either below the road surface, at the road surface, or above the road surface. This is known as scrub radius.

The scrub radius produces a 'turning moment' which will help the steering when turned from the straight ahead position.

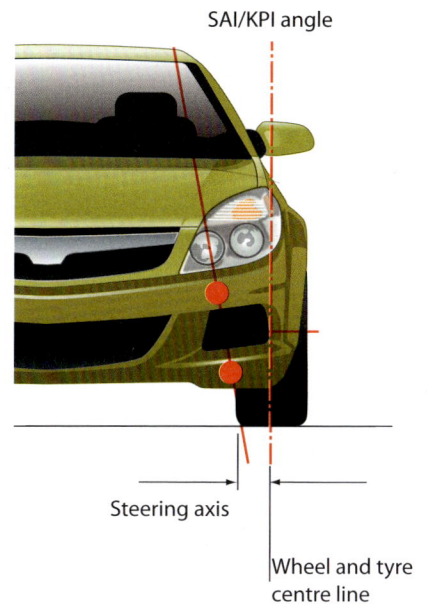

Figure 6.52 Swivel axis inclination

Table 6.17 Steering geometry

Steering angle	Description
Tracking	A steering setting designed to describe if the steered wheels are parallel
TOOT	Toe-out on turns lets the steered wheels splay outwards when turning a corner
Ackerman principle	A series of steering angles designed to allow the wheels to travel with true rolling motion when going round the corner
Camber	The outward or inward tilt of the road wheels when viewed from the front
Caster	The angle created by the tilt of the front suspension from the vertical when viewed from the side
SAI (or KPI)	The inwards tilt of the front suspension ball joints (swivels)
Thrust line	An imaginary line at right angles to the rear axle centre line that does not point straight ahead due to movement in the rear wheel suspension joints during forward motion of the vehicle

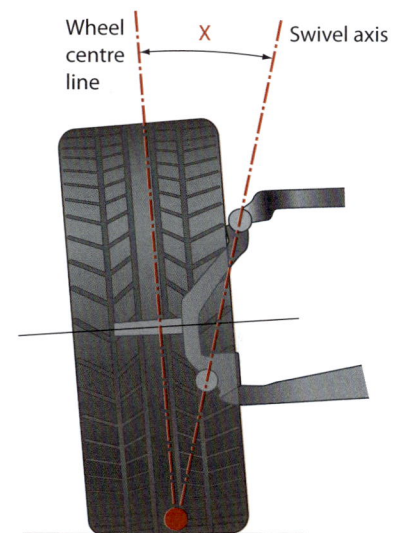

Figure 6.53 Negative offset (negative scrub radius)

CHECK YOUR PROGRESS

1   What steering setting helps the wheels return to the straight ahead position after a corner?
2   Why do wheels need to toe-out on turns?
3   What sort of tyre wear might occur if tracking is incorrect?

Routine maintenance on steering and suspension systems

Routine maintenance conducted on a car's steering system will usually include:

An inspection for wear and excessive movement

An inspection for physical damage to any steering components

The checking and adjustment of steered wheel alignment (particularly tracking)

Figure 6.54 A flow chart to show routine maintenance of a steering system

As with other types of maintenance, it is very important that you refer to the manufacturer's technical data. If incorrect steering adjustment settings are used, handling issues can occur. These could lead to rapid tyre wear and make the car unstable.

Track rod ends

If you need to replace a track rod end it is important that certain procedures are followed. If you don't do this, it is possible that steering alignment might be incorrect and toe-out on turns could be affected. Always make sure that the adjustment of the track rod ends is the same on both sides of the car.

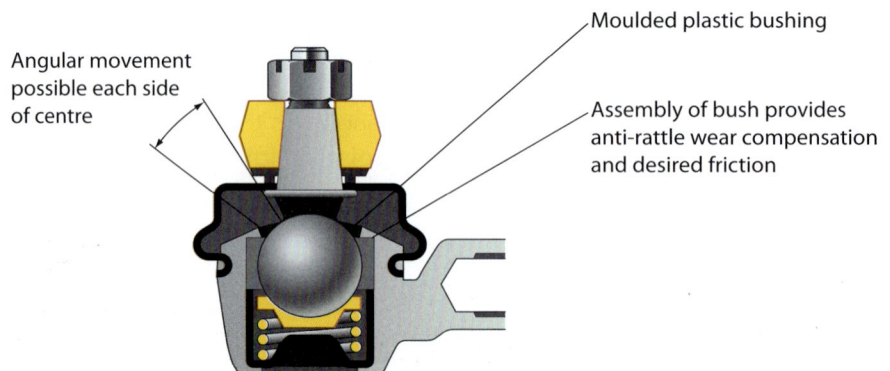

Angular movement possible each side of centre

Moulded plastic bushing

Assembly of bush provides anti-rattle wear compensation and desired friction

Figure 6.55 Ball joints (track rod end)

Replace a track rod end

Checklist			
PPE	**VPE**	**Tools and equipment**	**Source information**
• Steel toe-capped boots • Overalls • Latex gloves	• Wing covers • Steering wheel covers • Seat covers • Foot mat covers	• Socket set • Spanners • Ball joint splitter • Jack • Axle stand • Wheel alignment equipment • Torque wrench	• Technical data • Wheel alignment data

1. Support the vehicle safely and remove the road wheel.

2. Clean around the track rod thread, using a wire brush.

3. Slacken the track rod lock nut half a turn.

4. Remove the nut on the ball joint.

5. Use the ball joint splitter to release the ball joint taper from the hub.

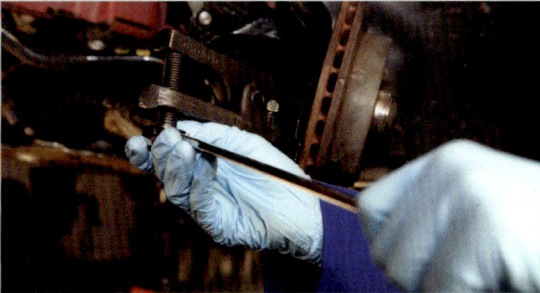

6. Unscrew the joint from the track rod.

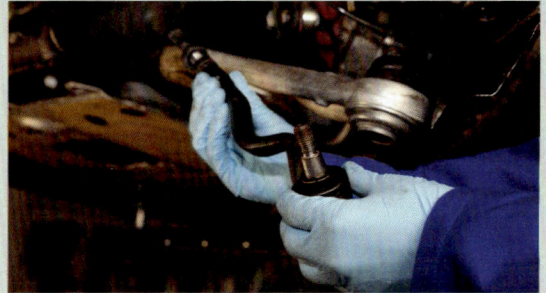

7. Screw the new track rod end on to the track rod up to the lock nut and back off half a turn.

8. Press the ball joint taper into the steering hub and fit the retaining nut, using a torque wrench.

9. Tighten the lock nut to the track rod end.

10. Replace the wheel and torque the wheel nuts.

11. Lower the vehicle and roll it backwards and forwards for the suspension to settle.

12. Check the wheel alignment.

Did you know?

It is important before you do any steering alignment checks that:

- the car is placed on a flat and level surface
- the handbrake should be applied and all the wheels chocked to prevent the car moving during any adjustments
- the steering wheel should be in the straight ahead position
- tyre pressures should be correct
- suspension ride heights should be within the manufacturer's specification
- the car is correctly loaded (for example, no heavy weights in the boot).

CHECK YOUR PROGRESS

1 List three important factors that should be correct before you check steering alignment.
2 When setting the tracking, why should you make the same adjustment to both track rod ends?
3 What special tool is used to remove the track rod end from the steering arm?

FINAL CHECK

1 What does the word 'hygroscopic' mean?

   a brake fluid can damage paint
   b brake fluid can boil
   c brake fluid can absorb water
   d brake fluid can contain air

2 Which of the following is not a type of steering box?

   a ball and pinion
   b recirculating ball
   c rack-and-pinion
   d worm and peg

3 A tyre is worn around the shoulders of the tread. What could have caused this?

   a over inflation
   b under inflation
   c incorrect camber
   d incorrect caster

4 A tyre marked 185/70R14 H will fit a road wheel with what diameter?

   a 14 cm
   b 185 mm
   c 14 inches
   d 185 cm

5 What should be used to tighten wheel nuts?

   a an air gun
   b a wheel brace
   c a torque wrench
   d all of the above

6 Which one of the following is not a type of spring?

   a hydraulic
   b pneumatic
   c coil
   d torsion

7 When should tyre pressures be checked?

   a it doesn't matter
   b when the tyres are hot
   c when the tyres are cold
   d after a long journey

8 What reduces suspension oscillation (bounce)?

   a coil springs
   b leaf springs
   c dampers
   d torsion bars

9 When does steering receive the most power assistance?

   a when moving slowly
   b when moving quickly
   c when the engine is started
   d it is the same all the time

10 Most brake servos are operated by what?

   a vacuum
   b hydraulics
   c motors
   d none of the above answers

GETTING READY FOR ASSESSMENT

The information contained in this chapter, as well as continued practical assignments in your centre or workplace, will help you to prepare for both the end-of-unit tests and diploma multiple-choice tests. This chapter will also help to prepare you for working on light vehicle chassis systems safely.

You will need to be familiar with:

- Tyre construction methods (both cross-ply and radial)
- Tyre inflation methods (both tubed and tubeless)
- Tyre sidewall markings and their meanings
- Minimum legal requirements for tyre tread wear
- Tyre tread wear patterns and the faults that cause them
- Tyre removal and replacement methods
- Wheel and tyre balancing procedures
- Correct methods for removing and refitting road wheels
- The names and functions of braking system components
- The operation of drum brakes
- The difference between leading and trailing brake shoes and their effect on self-servo operation
- The operation of disc brakes
- The operation of hydraulic braking systems
- The minimum legal braking efficiency required for both footbrake and handbrake mechanisms
- The meanings of the key terms hygroscopic, vapour lock and brake fade
- The components used in both steering rack and steering box mechanisms
- The operation of steering rack and steering box mechanisms
- The meaning of the key steering geometry terms camber angle, caster angle, Ackerman principle, toe-out on turns, tracking and swivel axis inclination (SAI)
- Suspension spring types to include coil springs, leaf springs, torsion bars, rubber and pneumatic
- The advantages and disadvantages of independent and non-independent suspension systems
- The function and operation of suspension damper units

Before trying a theory end-of-unit test or diploma multiple-choice test, make sure you have reviewed and revised any key terms that relate to the topics in that unit. Be sure to read all questions fully and take time to digest the information so that you are confident about what the question is asking you. With multiple-choice tests, it is very important that you read all of the answers carefully, as it is common for the answers to be very similar and this may lead to confusion.

For practical assessments, it is important that you have had sufficient practice and that you feel that you are capable of passing. It is best to have a plan of action and work method that will help you. Make sure that you have sufficient technical information, in the way of vehicle data, and appropriate tools and equipment. It is also wise to check your work at regular intervals. This will help you to be sure that you are working correctly and to avoid problems developing as you work.

When undertaking any practical assessment, always make sure that you are working safely throughout the test. Ensure that all health and safety requirements are observed and that you use the recommended personal protective equipment (PPE) and vehicle protection equipment (VPE) at all times. When using tools, make sure you are using them correctly and safely.

Good luck!

7 Light vehicle drivelines

This chapter will give you an introduction to light vehicle driveline systems, components and operation. It provides the basic knowledge that will help you with both theory and practical assessments.

It will introduce the basic operating principles of light vehicle manual and automatic transmission system components.

Finally, it will help you plan a systematic approach to transmission inspection and maintenance.

This chapter covers:

- Safe working on light vehicle drivelines
- Light vehicle driveline layouts
- Light vehicle clutch systems
- Light vehicle gearboxes
- Routine maintenance on light vehicle drivelines

WORKING PRACTICE

The light vehicle driveline system includes all of the components that will take the turning effort from the engine crankshaft to the road wheels. These components include:

- clutch
- gearbox
- final drive
- differential
- drive shafts.

You must always use personal protective equipment (PPE) and you also need to think about the possibility of crush or bump injury. You will also come into contact with chemicals such as lubrication oils. Make sure that your selection of PPE will protect you from these hazards.

Personal Protective Equipment (PPE)

Safety helmet protects the head from bump injuries when working under cars.

Safety gloves provide protection from oils and chemicals. They also protect the hands when handling objects with sharp edges.

Safety mask protects against clutch dust inhalation.

Overalls provide protection from coming into contact with oils and chemicals.

Barrier cream protects the skin from old transmission oil, which can cause dermatitis and may be carcinogenic (a substance that can cause cancer).

Safety boots protect the feet from a crush injury and often have oil- and chemical-resistant soles. Safety boots should have a steel toe-cap and steel mid-sole.

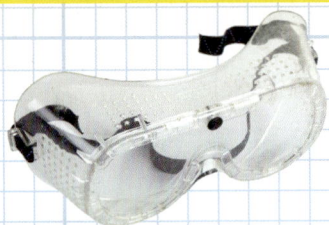

Safety goggles/glasses reduce the risk of small objects or chemicals coming into contact with the eyes.

To reduce the possibility of damage to the car, always use the appropriate vehicle protection equipment (VPE):

Wing covers

Seat covers

Steering wheel covers

Floor mats

If appropriate, safely remove and store the owner's property before you work on the vehicle. Before returning the vehicle to the customer, reinstate the vehicle owner's property. Always check the interior and exterior to make sure that it hasn't become dirty or damaged during the repair operations. This will help promote good customer relations and maintain a professional company image.

Vehicle Protective Equipment (VPE)

Safe Environment

During the repair or maintenance of light vehicle driveline systems you may need to dispose of certain waste materials such as lubrication oil. Under the Environmental Protection Act 1990 (EPA), you must dispose of them in the correct manner. They should be safely stored in a clearly marked container until they are collected by a licensed recycling company. This company should provide you with a waste transfer note as the receipt of collection.

To further reduce the risks involved with hazards, always use safe working practices including:

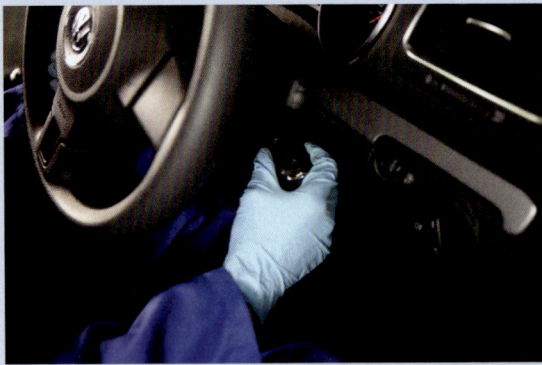

1. Immobilise vehicle (by removing the ignition key). If possible, allow the engine to cool before starting work.

2. Prevent the vehicle moving during maintenance by applying the handbrake or chocking the wheels.

3. Follow a logical sequence when working. This reduces the possibility of missing things out and of accidents occurring. Work safely at all times.

4. Always use the correct tools and equipment. (Damage to components, tools or personal injury could occur if the wrong tool is used or a tool is misused.) Check tools and equipment before each use.

5. Following the replacement of any vehicle components, thoroughly road test the vehicle to ensure safe and correct operation. Make sure that all work is correctly recorded on the job card and vehicle's service history, to ensure that any maintenance work can be tracked.

6. If components need replacing, always check that the parts are the correct quality and type for the vehicle if it is still under guarantee. Inferior parts or deliberate modification might make the warranty invalid. Also, if parts of an inferior quality are fitted, this might affect vehicle performance and safety.)

Preparing the car

Tools

Socket set

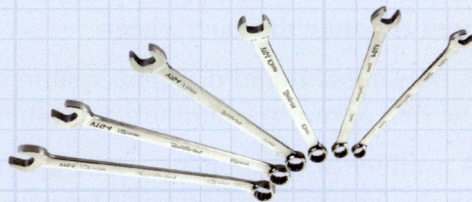

Spanners

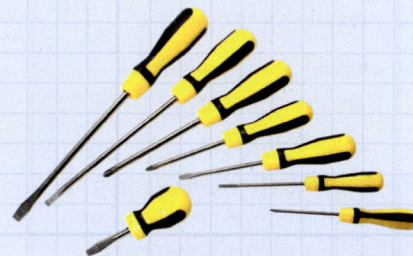

Screwdrivers

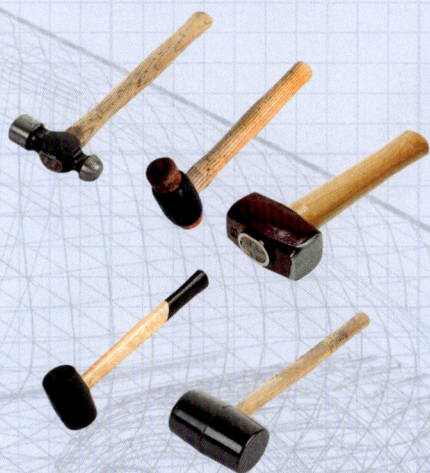

Hammers

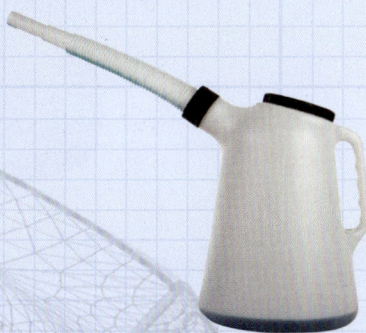

Measuring jug

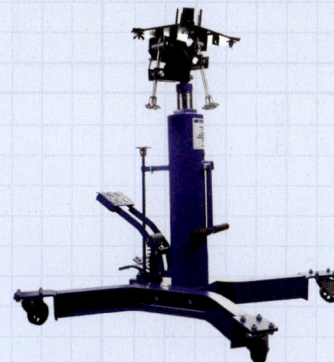

Transmission jack

Safe Working

- If you are using a ramp or vehicle hoist, always check that the car is evenly positioned, secure and its weight does not exceed any safe working loads (SWL).

- Always clean up any fluid spills (especially gearbox lubrication oil) immediately to avoid slips, trips and falls.

- Always use exhaust extraction when running engines in the workshop.

- Always wear gloves to protect your hands from transmission lubricating oil.

- Always wear a particle mask and do not allow clutch dust to become airborne.

- Always use mechanical lifting equipment where possible when handling heavy transmission components.

- Always take great care when using compressed air. There is a possibility that it can penetrate the skin causing severe injury or even death.

Power – the rate at which work is done.

Torque – turning effort.

Drive – the turning of the transmission components and the car wheels.

Transmission – this is made up of the clutch, gearbox and drive shafts. It enables power to be transmitted from the engine to the wheels so that the car will move.

Manual – where the driver selects the gear to use.

Automatic – where the car selects the gear to use.

Light vehicle driveline layouts

The engine is the car's **power** plant. It provides the energy needed for movement.

As the engine operates, it creates a turning effort called **torque** which is used to rotate the car's wheels.

A transmission and driveline system is needed to connect and transfer the **drive** from the engine's crankshaft to the wheels.

A number of components make up the vehicle's driveline, depending on the layout and the type of **transmission** used (i.e. **manual** or **automatic**).

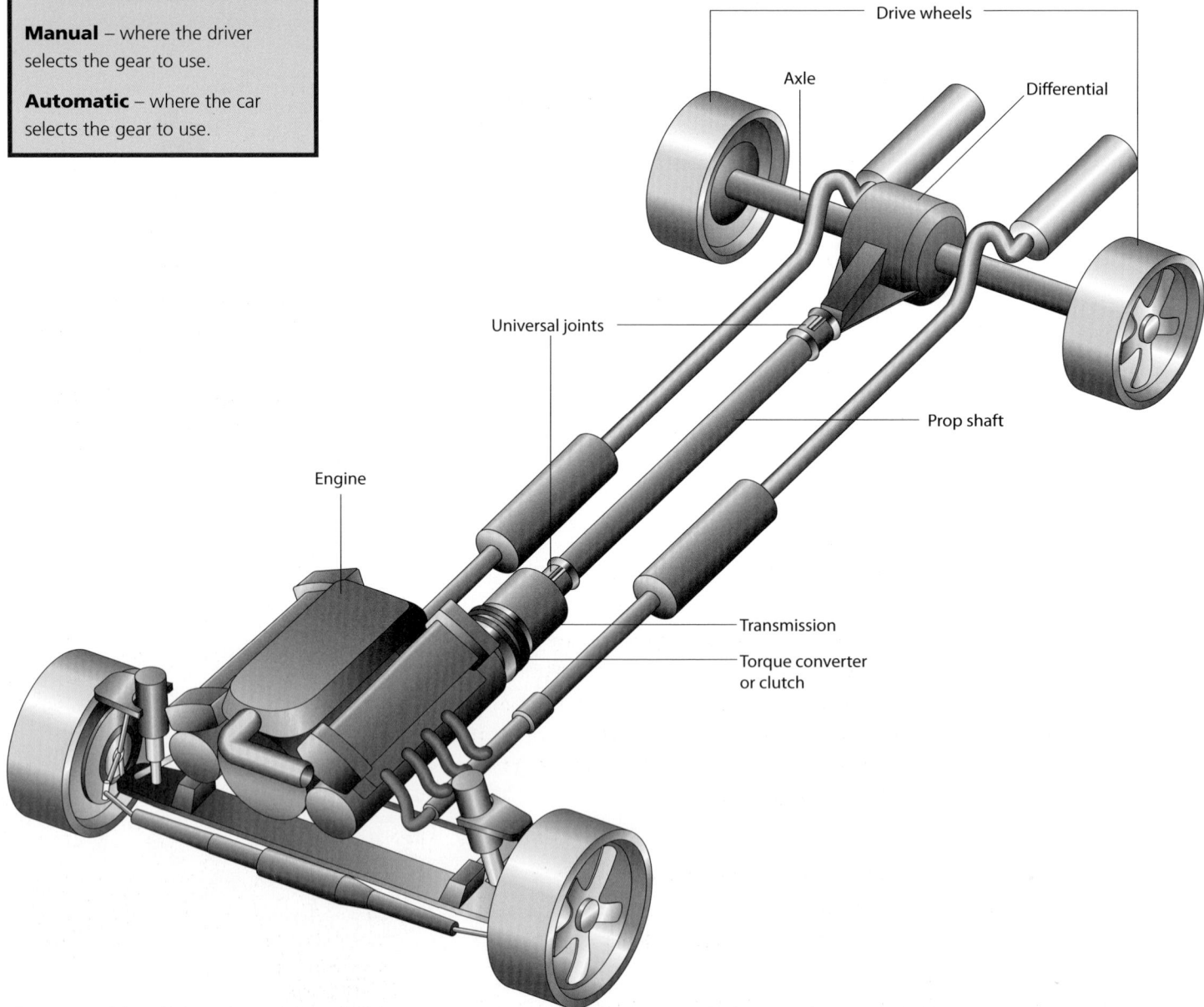

Figure 7.1 Light vehicle engine and transmission

Table 7.1 Standard manual transmission components and their purposes

Standard manual transmission component	Purpose
Clutch	Connects and disconnects the engine from the gearbox
Gearbox	Increases the turning effort from the engine
Prop shaft	Used in a rear-wheel drive car to connect the gearbox to the rear axle
Drive shaft	Used in a front-wheel drive car to send drive to the wheels
Final drive	Also helps increase turning effort and in a rear-wheel drive car turns the drive through a right angle
Differential	Allows one wheel to travel faster than the other when turning a corner

Light vehicle clutch systems

A **clutch** in the car either links up or separates the power from the engine to the driving shafts that turn the wheels. This means the movement of the car can be stopped and started.

> ### Did you know?
>
> A clutch is needed for three main reasons:
>
> 1 It provides a smooth take-up of drive (going from stationary to moving).
>
> 2 It provides a 'temporary position of **neutral**' (allows the car to come to a stop without taking it out of gear or stalling the engine).
>
> 3 It will allow the engine to be disconnected from the gearbox so that gear change can take place (without the gear teeth of the transmission hitting each other, making a lot of noise and causing a lot of damage).

> ### Key terms
>
> **Clutch** – a mechanism for connecting and disconnecting two moving components.
>
> **Neutral** – when no drive is being transmitted.
>
> **Engaged** – when all parts of the clutch are held together and drive is being transmitted.
>
> **Disengaged** – when the clutch components are separated and no drive is being transmitted.

A clutch is made up of three main sections, built like a sandwich (see Figure 7.2). All of the component parts are squashed together to provide grip.

- When the three sections are held tightly together, they all spin round at the same time. The clutch is said to be **engaged**.
- When the driver presses down on the clutch pedal, the parts separate. When this happens there is no connection between the engine flywheel and the input shaft of the gearbox. The clutch is now said to be **disengaged**.

The friction clutch

- A friction clutch is fitted at one end of the engine. It is bolted to the engine's flywheel.
- The surfaces of the clutch components are held together by very strong spring pressure.
- The grip in the clutch (to transmit drive) is provided by the friction from the surfaces in contact and the pressure with which they are held together.

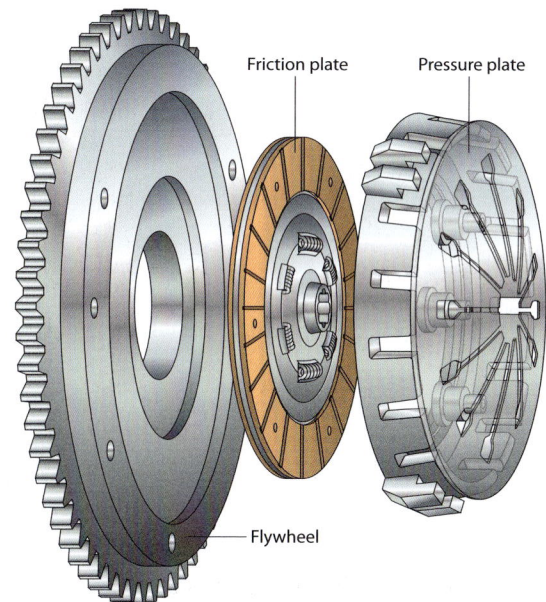

Figure 7.2 Clutch components

Friction plate   Pressure plate

Flywheel

A friction clutch is what is known as a gradual engagement type. To engage the clutch:

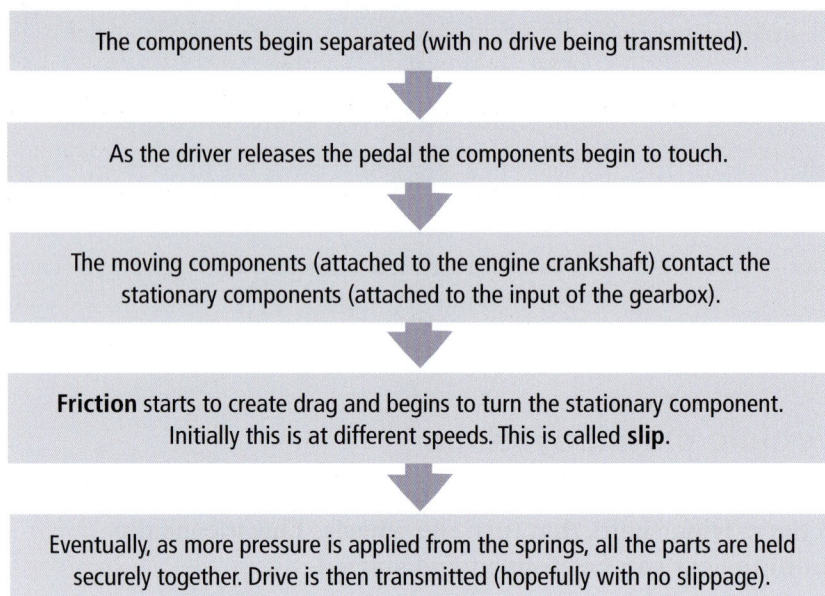

The components begin separated (with no drive being transmitted).

⬇

As the driver releases the pedal the components begin to touch.

⬇

The moving components (attached to the engine crankshaft) contact the stationary components (attached to the input of the gearbox).

⬇

Friction starts to create drag and begins to turn the stationary component. Initially this is at different speeds. This is called **slip**.

⬇

Eventually, as more pressure is applied from the springs, all the parts are held securely together. Drive is then transmitted (hopefully with no slippage).

Figure 7.3 What happens when the clutch is engaged

Requirements of a clutch

- A clutch must transmit drive smoothly with no **drag**. Drag is where the clutch is not fully disengaged and rubs together, still transmitting some **drive**.
- A clutch must transmit drive smoothly with no slip once it is fully engaged. Slip is where drive is not fully transmitted and the friction surfaces slide over each other, producing a loss of turning effort.

The amount of **torque** produced by an engine can have a considerable effect on how the clutch must operate. A friction clutch must be able to transmit all of the engine torque to the rest of the transmission system without slip.

Three methods are used that increase the torque transmitting ability of a friction clutch. These are outlined in Table 7.2.

Table 7.2 Three methods that increase the torque transmitting ability of a friction clutch

Method	Advantage or disadvantage
Increase the diameter or mean radius of the clutch (i.e. make it bigger and increase the size of the surface area and the amount of friction gained).	The disadvantage of this is the more torque an engine produces, the larger the clutch must be. This might not be practical for design purposes.
Increase the forces squeezing the surfaces together. This is normally done by using stronger springs in the clamping mechanism.	The disadvantage of this is the stronger the spring force, the more effort that is required to operate it. This can mean that the clutch pedal feels hard to push down.
Increase the number of surfaces in contact. These are often referred to as multi-plate clutches. A multi-plate clutch can still be thought of like a sandwich, but with more layers of filling in the middle.	The advantage of this is by increasing the number of surfaces in contact, the overall size of the clutch can be kept fairly small. The surface area is increased and this provides more friction. More drive can then be transmitted.

Table 7.3 The main components of a clutch system

Friction clutch component	Purpose
Flywheel	Rotates with the crankshaft and forms a flat surface to drive the clutch
Clutch cover	Houses the clutch. It is bolted to the flywheel and rotates at crankshaft speed
Pressure plate	Provides the clamping surface (operated by springs) to drive the friction plate
Friction plate	Clamped between the pressure plate and the flywheel. Transfers drive to the input shaft of the gearbox
Release bearing	Operates against the clutch springs while the engine is turning
Release fork	Pushes against the release bearing when the pedal is pushed to operate the clutch mechanism

> **Did you know?**
>
> A flywheel is a heavy rotating disc. It is mounted on the end of the engine's crankshaft. Its purpose is to store the spinning motion from the crankshaft and keep the engine turning smoothly.

Friction plate construction

The friction plate is a disc with a lining material attached to it. The lining material is similar to that found on brake pads and shoes.

A sprung metal plate is sandwiched between the friction materials. It acts as a shock absorber when it is engaged.

A set of **torsion** springs help prevent **juddering** as drive is taken up. The centre of the friction plate is **splined** to the input shaft of the gearbox.

Release bearing construction

The clutch release bearing is normally a sealed ball bearing type unit. The outer part of the bearing usually touches the fingers at the centre of a diaphragm spring and rotates with it. The centre part of the bearing is connected to a lever arm mechanism called the 'release fork'. When operated, this arm pushes the release bearing against the diaphragm spring fingers. This releases the clamping force of the clutch pressure plate.

Freeplay

When the clutch is engaged, no force should be placed on the release bearing where it touches the diaphragm springs of the pressure plate. A small clearance or gap is normally needed at the pedal end. This is called 'clutch pedal **freeplay**'.

Clutch pedal freeplay is needed for two reasons:

1 It makes sure that the release bearing is not continuously pressed against the rotating fingers of the diaphragm spring. This will help reduce wear on both the diaphragm spring fingers and the clutch release bearing itself.

2 It ensures that the clutch mechanism is fully engaged.

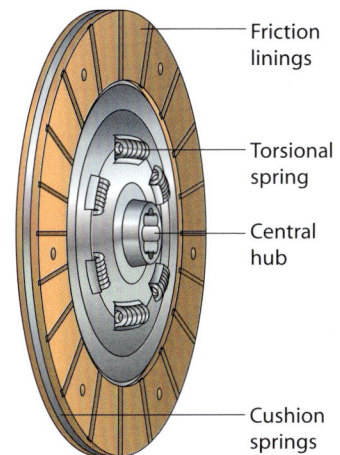

Figure 7.4 A clutch friction plate assembly, showing a friction material (facing) with radial grooves and rivet holes and clutch hub with torsion damping springs

> **Key terms**
>
> **Torsion** – twisting.
>
> **Juddering** – a vibration felt in the transmission during take-up of drive.
>
> **Spline** – grooves machined into a shaft, which are matched to grooves machined within a hub. When mounted together, relative rotational movement is prevented.
>
> **Freeplay** – a small gap or movement allowance within the adjustment of a mechanism.

Correct amounts of freeplay

If freeplay didn't exist, then the bearing might be pushing on the clutch release springs. This would partly disengage the clutch friction plate. This would then cause the drive plate to slip and to wear out early.

If too much freeplay exists, then when the clutch pedal is pressed by the driver, the clutch might not fully disengage. This would mean that the friction plate is still being gripped slightly between the pressure plate and the flywheel. If this happens, the engine could not be fully disconnected from the gearbox. This can make changing gear difficult. If the clutch friction plate does not fully disengage, this is referred to as **clutch drag**.

Clutch operation

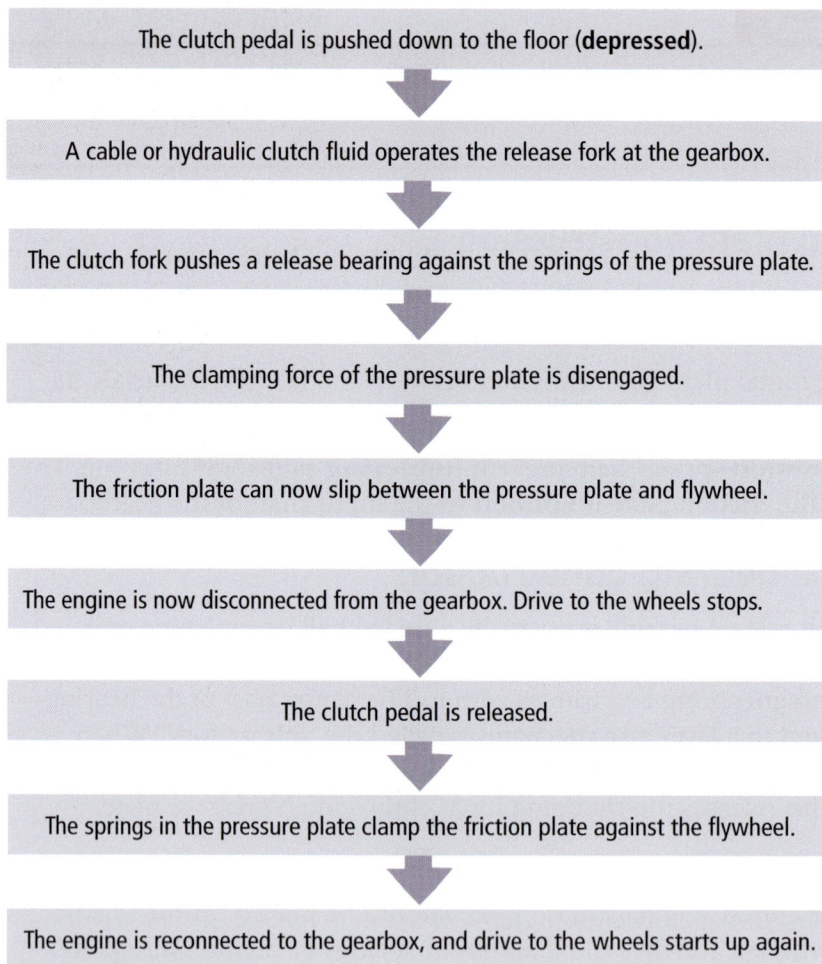

The clutch pedal is pushed down to the floor (**depressed**).

⬇

A cable or hydraulic clutch fluid operates the release fork at the gearbox.

⬇

The clutch fork pushes a release bearing against the springs of the pressure plate.

⬇

The clamping force of the pressure plate is disengaged.

⬇

The friction plate can now slip between the pressure plate and flywheel.

⬇

The engine is now disconnected from the gearbox. Drive to the wheels stops.

⬇

The clutch pedal is released.

⬇

The springs in the pressure plate clamp the friction plate against the flywheel.

⬇

The engine is reconnected to the gearbox, and drive to the wheels starts up again.

Figure 7.5 How a clutch works

Working life

While servicing a customer's car, Naseem adjusts the clutch cable so that there is no slack.

1   What is slack in the clutch cable called?

2   Why is it needed?

3   What might happen to the clutch now?

Coil and diaphragm spring clutches

The main differences in construction of light vehicle clutches are found in the design of the springing mechanism that clamps the surfaces together.

Coil spring clutches

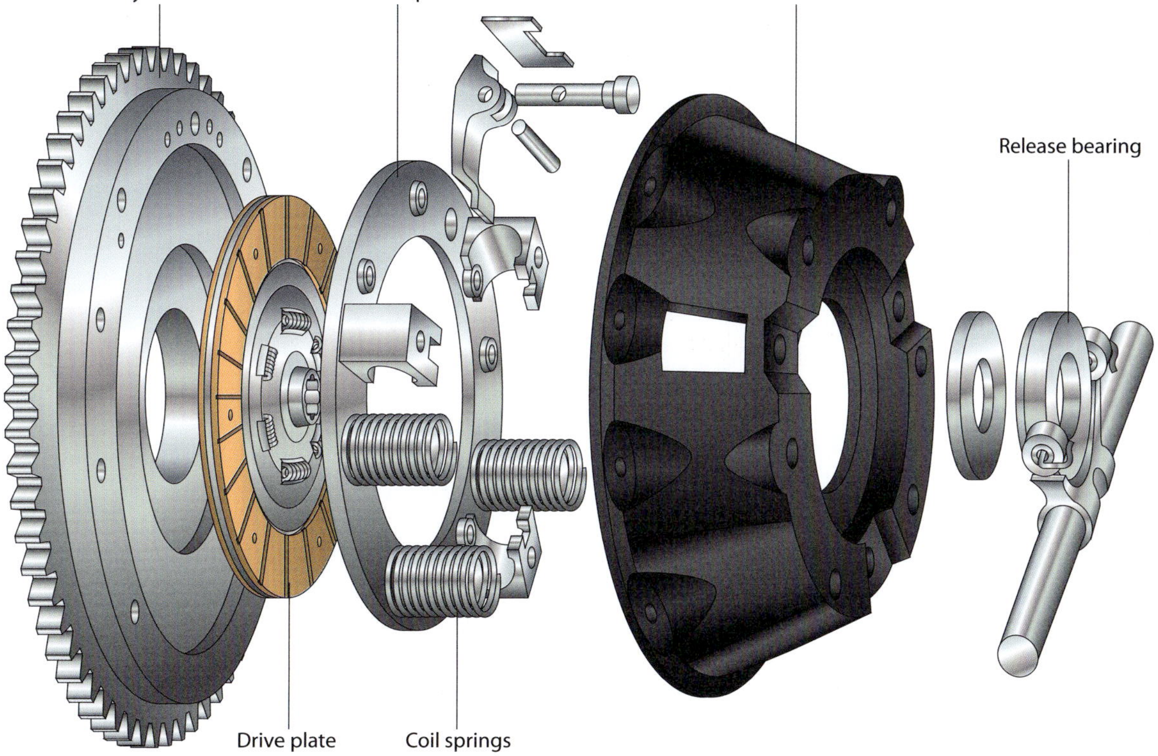

Figure 7.6 Coil spring clutch

A number of early clutches used coil springs to provide the clamping effort. Over a period of time, coil spring tension and length would change due to wear and tear. When this happened, an uneven clamping force could be produced at the pressure plate. This uneven clamping force might lead to clutch drag, slip or vibrations caused as take-up of drive is required.

Diaphragm spring clutches

To overcome these problems a spring mechanism called 'a diaphragm' was developed. This diaphragm spring is a single metal plate, made into a series of sprung steel fingers. It is slightly dished in shape. When one end of the fingers is pressed by the clutch release bearing, the fingers pivot about a **fulcrum** (like a seesaw). This moves the opposite end of the diaphragm fingers in the other direction. When this happens, the pressure plate is moved away from the friction plate and disengages the clutch.

Key term

Fulcrum – a pivot point.

When the driver lifts his foot off the clutch pedal, the ends of the fingers of the diaphragm spring are released. Because the steel fingers are sprung, they return to their original position and reapply pressure to the friction plate. This then reconnects the drive. Because the spring diaphragm fingers are made from a single piece of metal, an even clamping force can be produced. This overcomes many of the problems created when using coil spring mechanisms.

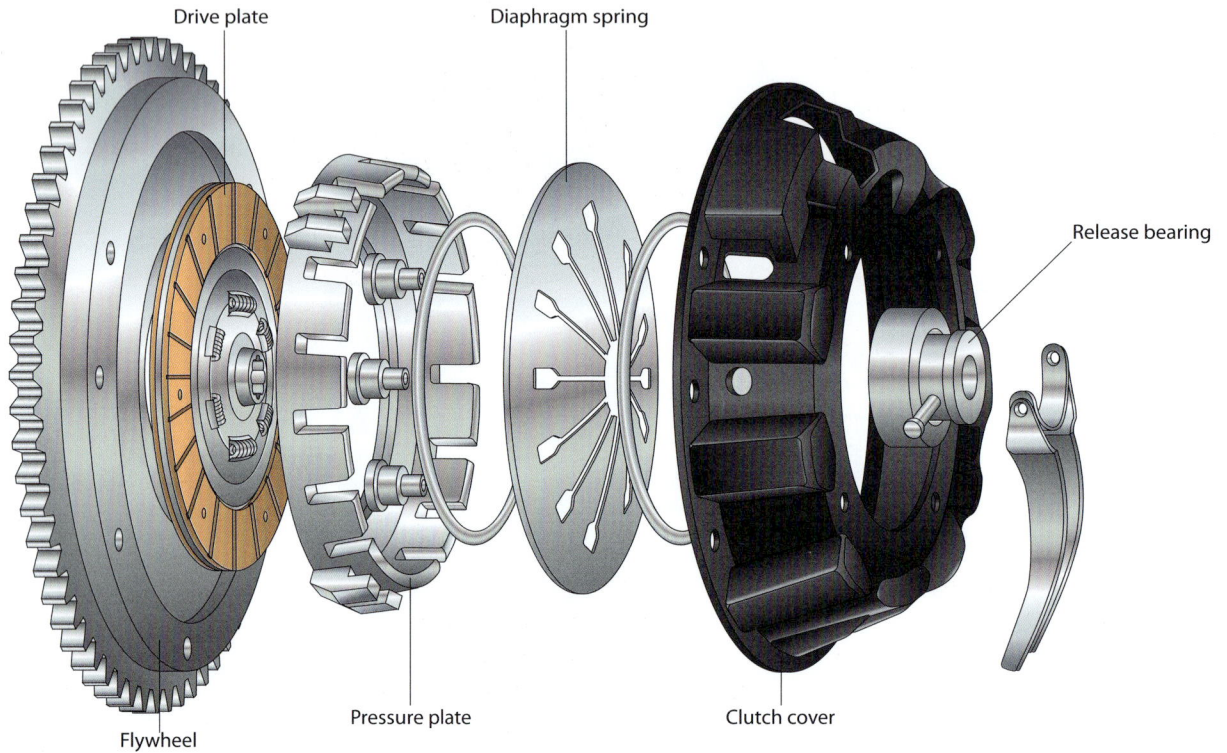

Drive plate

Diaphragm spring

Release bearing

Flywheel

Pressure plate

Clutch cover

Figure 7.7 Diaphragm spring clutch assembly

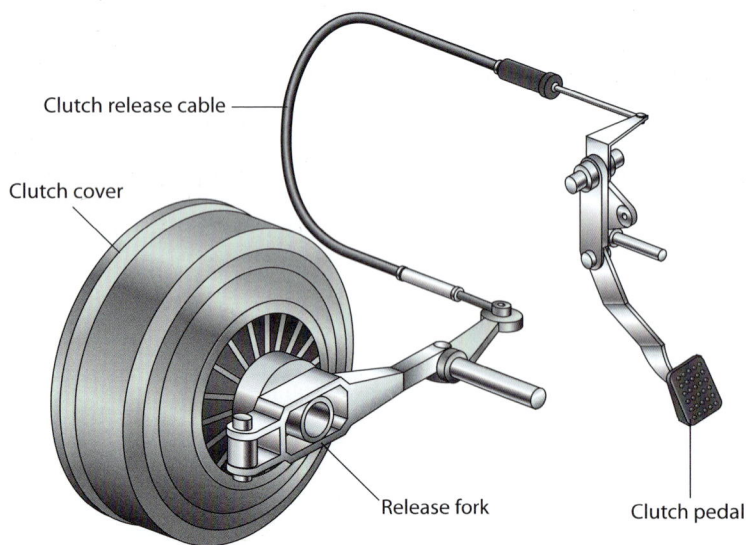

Clutch release cable

Clutch cover

Release fork

Clutch pedal

Figure 7.8 A cable-operated clutch assembly

Cable or hydraulic?

When the clutch pedal is pushed down by the driver, the movement is transmitted to the clutch. This is done in one of two ways, either mechanically (cable) or hydraulically.

Cable

In a mechanical system, a clutch cable is attached to one end of a clutch pedal. When the pedal is pressed, it pulls on this cable to operate the lever at the gearbox end. This lever is called the clutch fork. The clutch fork then moves and presses the release bearing against the fingers of the diaphragm spring. It pushes it inwards towards the flywheel.

The disadvantage of cable operation is that, as the clutch wears out, the clutch cable becomes slack and needs adjusting. This can be done with a nut and bolt manually, or some systems use an automatic adjuster.

Hydraulic

In this system the clutch pedal is attached to a clutch master cylinder. When the pedal is depressed (pushed down) a piston pushes fluid through a series of pipes and hoses. **Hydraulic** fluid will then transfer this movement to a slave cylinder piston at the clutch release end and operate the clutch.

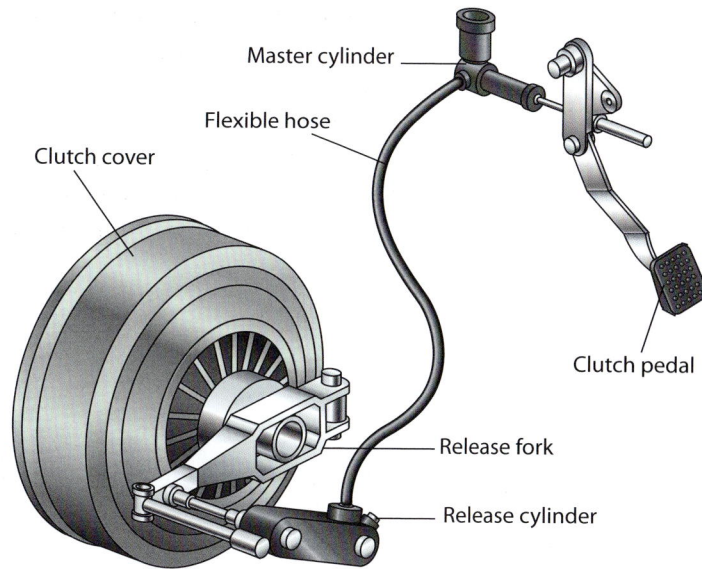

Figure 7.9 A hydraulically-operated clutch assembly

Table 7.4 The advantages and disadvantages of hydraulic operation

Advantages	Disadvantages
As the clutch friction plate wears, hydraulic fluid takes up the space created. This makes the system self-adjusting.	As with a hydraulic system found on brakes, air bubbles must be avoided. This is because air is a gas and therefore compressible. The system should be bled so that no air exists, otherwise it might not function correctly. (See braking systems on page 209).
Hydraulics produces smooth efficient operation no matter how long the fluid pipes and hoses are or whether they go around bends.	Because a hydraulic system relies on fluid, any leaks in the system could cause loss of operation. If this hydraulic fluid contaminates friction surfaces, grip might be lost and clutch slip could occur.

Key term

Hydraulics – using fluid to operate.

Did you know?

The hydraulic fluid used in a clutch system is often the same as that used in a brake system.

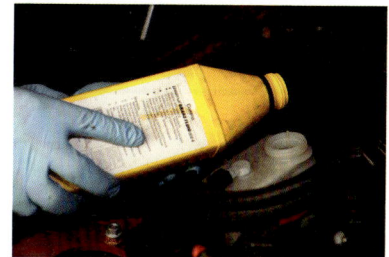

Figure 7.10 Topping up a clutch fluid reservoir

CHECK YOUR PROGRESS

1 Name the three main components of a friction clutch.
2 List the three main functions of a clutch.
3 What is torque?

Light vehicle gearboxes

To allow a car to be used under different conditions and road speeds, a gearbox is needed. The engine will only produce its greatest turning effort at certain speeds. Without a **gearbox** the engine would struggle and find it very hard to pull away, carry heavy weights, tow a caravan or go uphill.

A manual transmission uses gears which will multiply torque using leverage. **Leverage** is able to increase the effort supplied. This makes it easier for the engine to turn the wheels.

Increased torque

Torque conversion mechanism

Figure 7.11 A transmission system is needed to increase the torque (turning effect) produced by the engine to help the car go up a hill

Key terms

Gearbox – the housing which contains gears, shafts, bearings and selectors.

Leverage – a mechanical advantage gained by using a lever.

Inertia – a force that holds something steady or keeps it moving at a constant speed and direction.

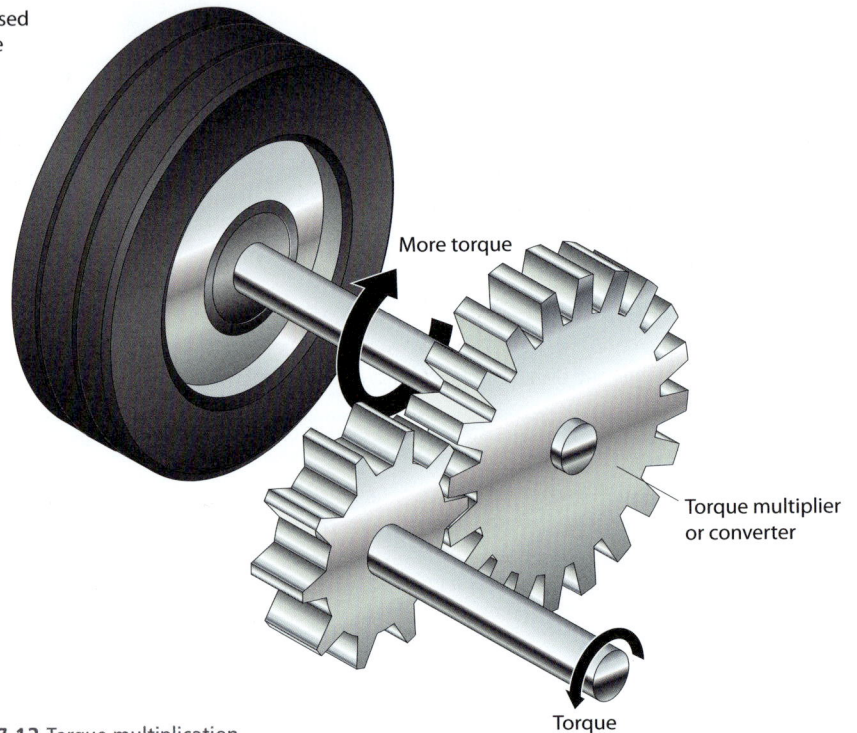

More torque

Torque multiplier or converter

Torque

Figure 7.12 Torque multiplication

Leverage can be explained using the following example.

Imagine two spanners – one short and one long. Which would make it easier to undo a wheel nut?

The longer spanner will make it easier because you have a greater amount of leverage. This will multiply your turning effort at the nut.

If you look at the two gears in Figure 7.14, one is larger than the other. Imagine that the outside edge (circumference) of the gear was split and it was laid out flat. You can see that you would have a longer line from the larger gear and a shorter line from the smaller gear.

Figure 7.13 Six different-sized combination spanners – the longer spanners will have a greater amount of leverage and make it easier to undo a wheel nut

These can now be compared to the two spanners to demonstrate how a larger gear can multiply turning effort or torque.

Unfortunately you don't get something for nothing.

When you use a longer spanner to multiply the turning effort at the nut, the opposite end of the spanner has to move a far greater distance.

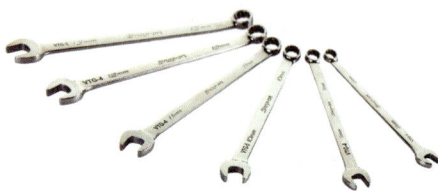

This is the same when it comes to gears. While it multiplies torque, the outer part of the gear has to move a lot further. Because of this it travels slower. The trade-off for torque multiplication is speed.

- As torque goes up, speed goes down.
- As speed goes up, torque goes down.

Once the car is moving, it has **inertia** and because of this it needs less torque. This means that other gears can now be used to multiply (increase) the speed of the vehicle and reduce torque.

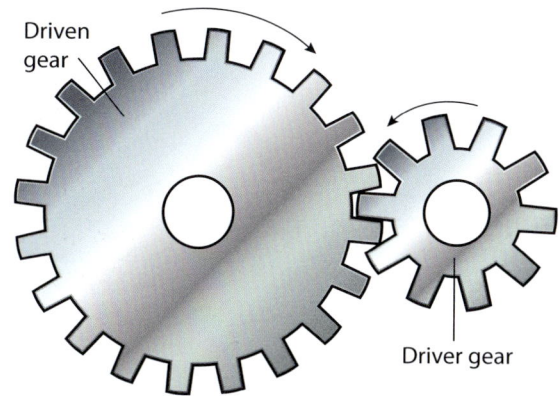

Figure 7.14 Rotating gears

Gear ratios

Gear ratios show how much the torque is multiplied and how much the speed is reduced in the gearbox. It compares the number of teeth between the **driver gear** and the **driven gear**.

Here is an example:

If the input gear has 10 teeth, and the output gear has 20 teeth, then the input gear will need to turn two complete revolutions for the output gear to turn once. This would be known as a gear ratio of 2:1.

Key terms

Gear ratio – the difference between the number of teeth on the driven gear and the number of teeth on the driver gear.

Driver gear – the input gear.

Driven gear – the output gear.

Did you know?

A 2:1 gear ratio means that speed is reduced by half. For example, an input turning at 200 rpm will give an output of 100 rpm.

The output torque or turning effort will double. For example, an input torque of 200 Newton metres will provide an output torque of 400 Newton metres.

To calculate a gear ratio, you can use a simple sum:

$$\text{Ratio} = \frac{\text{Driven}}{\text{Driver}}$$

That is, the number of teeth on the output gear divided by the number of teeth on the input gear).

As in the previous explanation, this is 20 divided by 10 equals 2 (2:1).

Find out

Calculate the following gear ratios:

- Input gear 10 teeth – output gear 80 teeth
- Output gear 224 teeth – input gear 14 teeth
- Input gear 15 teeth – output gear 180 teeth
- Output gear 64 teeth – input gear 13 teeth

(You may need a calculator for the last one.)

Figure 7.15 Gear ratios

A useable range of gear ratios can be achieved by varying the number of teeth on the input and output gears.

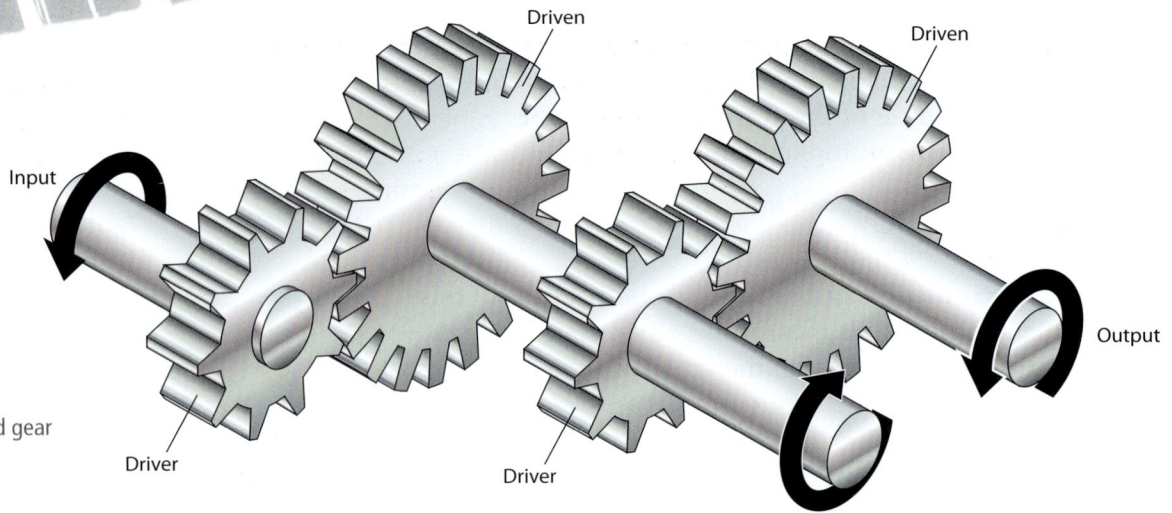

Figure 7.16 Compound gear train ratio calculations

Compound gears

If drive is transmitted through a number of different gears, this is known as a compound gear set.

To work out overall gear ratio, each set of gears should be calculated separately and then multiplied together for a total.

$$\text{Ratio} = \frac{\text{Driven}}{\text{Driver}} \times \frac{\text{Driven}}{\text{Driver}}$$

Other functions of the gearbox

- The gearbox is able to give a permanent position of neutral. This means the engine is completely disconnected from the road wheels. Now the engine can be run while the vehicle is stationary.
- The gearbox can change the direction of rotation between the crankshaft and the driven road wheels. By doing this the car is able to reverse.

The main components of a manual gearbox

Table 7.5 Manual gearbox components and their purposes

Manual gearbox component	Purpose
Casing	An outer housing to hold the gears
Bell housing	The gearbox clutch housing
Spur gear	Increases turning effort (straight cut teeth)
Helical gear	Increases turning effort (teeth cut at an angle)
Selector hub	Connects the selected gear to the output shaft of the gearbox
Selector fork	Operates the selector hub
Selector rods	Move the selector forks
Baulk rings	Makes sure that gears are turning at the same speed before selection
Detent	Helps hold gears in position
Interlock	Stops two different gears being selected at the same time

The gearbox is the second stage in the transmission system after the clutch. It is usually bolted to the back of the engine, with the clutch mounted between the two units. The driver operates the gear lever, which is connected to a series of selector rods or cables inside the gearbox. These selector rods engage and disengage appropriate gear wheels. This provides a ratio that gives the most efficient torque transmission to the road wheels.

Gear mesh and direction of rotation

Two circular wheels with teeth cut around the edge are brought into contact with each other. If one of them is provided with torque as an input, the other will be driven as an output. When you connect them together, if the first gear turns in a clockwise direction, the second gear will turn in an anticlockwise direction.

If an intermediate gear, known as an **idler gear**, is introduced then the direction of the output is reversed. So the input gear turns clockwise, the idler gear turns anticlockwise and the output gear turns clockwise.

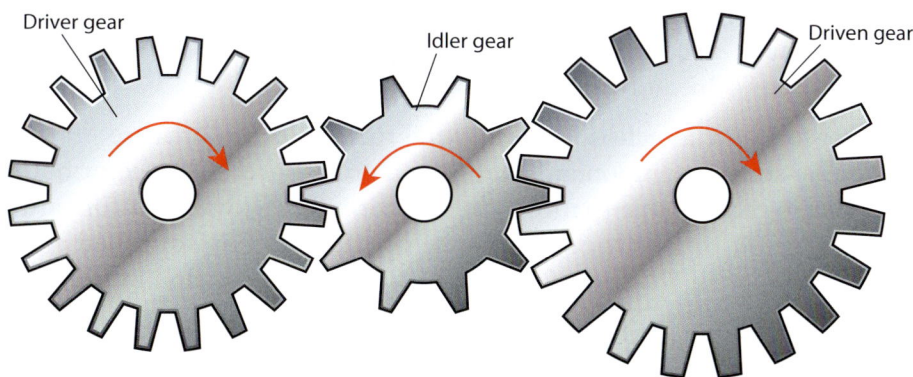

> **Did you know?**
>
> The main casing of the car's transmission can often be separated into two parts. These are the bell housing which encloses the clutch mechanism, and the gearbox, which holds the gears.

Figure 7.17 Idler gear, used to provide reverse

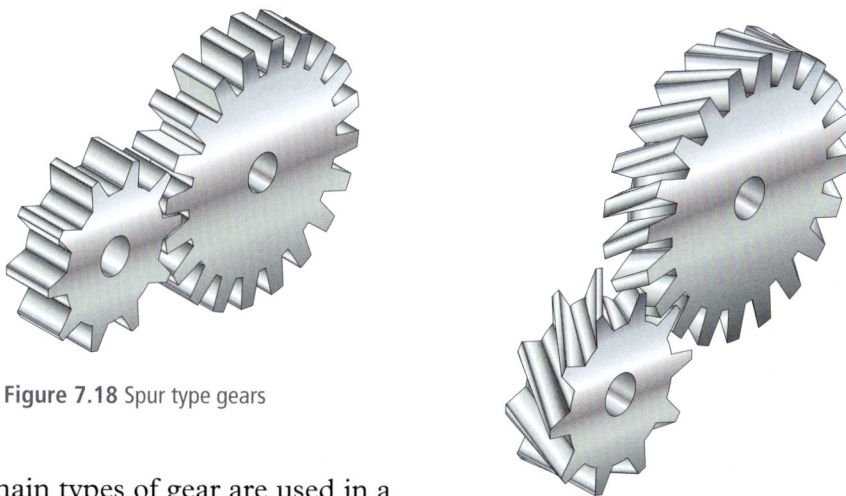

> **Did you know?**
>
> The idler gear has no overall effect on the gear ratio between the input and the output but simply changes the direction of rotation. This is how reverse gear is achieved. This means it can be ignored if you are calculating gear ratio.

Figure 7.18 Spur type gears

> **Find out**
>
> Draw and label a spur gear set and a helical gear set. Show the direction of the teeth.

Two main types of gear are used in a standard manual transmission system. These are **spur gears** and **helical gears**.

Figure 7.19 Helical cut gears

> **Key terms**
>
> **Idler gear** – a gear inserted between two other gears which will reverse the direction of rotation.
>
> **Spur gears** – a gear wheel with straight cut teeth.
>
> **Helical gears** – a gear wheel with teeth cut at an angle. If extended it would form into a spiral.

Spur gears

Spur gears are direct acting and create low amounts of drag. This will help improve overall performance but can make them noisy in operation. A spur cut gear can be slid in and out of **mesh** with another spur cut gear. They are ideal for use as a reverse gear idler.

Helical gears

A helical cut gear has teeth on an angle, as shown in Figure 7.19. The teeth are not just cut on a diagonal – if they were extended around a cylinder they would actually be shaped like the coils of a spring. Helical cut gear teeth provide a large surface area, which makes them very strong. This means that they last longer and are much quieter than a spur gear. Because of its design and shape, a helical cut gear cannot be slid in and out of mesh. Helical gears have to be used in a gearbox known as a **constant-mesh**.

Gear selection

In a constant-mesh type gearbox, a method of selecting a gear is needed (other than sliding them in and out of engagement). A selector hub is used, which is splined to the output shaft of the gearbox. When the driver requires a certain gear, they push or pull a gear selector lever inside the car. This moves rods inside the gearbox.

Selector forks are attached to these metal rods. Selector forks are horseshoe-shaped components that sit in a groove over the selector hub. The selector fork is able to move the selector hub backwards or forwards.

Once a gear is selected, a mechanism called a **detent** helps to hold the transmission in gear.

The selector hub sits between the gears. When slid towards an appropriate gear setting, **dog teeth** on the side of the selector hub meet corresponding dog teeth on the side of the gear wheel. The two components are then locked together:

- The selected gear is positively attached (through the **dog clutch**) to the selector hub.
- The selector hub is positively attached (through the splines) to the output shaft of the gearbox.

All the components now turn together and because of this engine torque is transmitted. The selector hub usually has a dog clutch on each face. This means that when it is slid in one direction a gear is selected, and when it is slid the other way it engages with another gear. When the selector hub is not engaged with either gear wheel, both can run freely without transmitting any power (neutral).

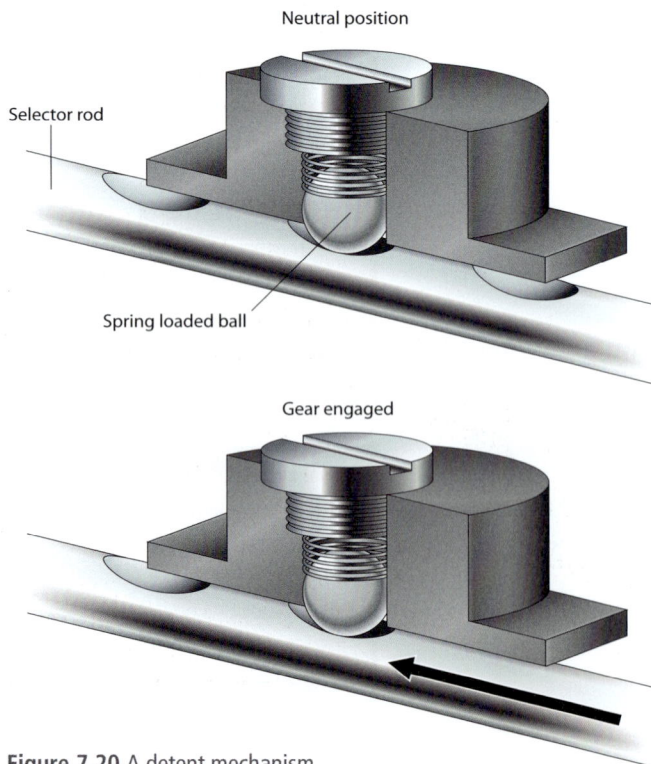

Neutral position

Selector rod

Spring loaded ball

Gear engaged

Figure 7.20 A detent mechanism

Synchronisation

Because gears of different sizes are used inside the gearbox, they will all be travelling at different speeds. So that they can be selected without damage to the teeth their speed has to be the same. The process of speeding up or slowing down the gears is called **synchronisation**.

A system is used on the selector hub. It is sometimes referred to as the **synchromesh**. When the driver selects a gear, two surfaces in the selector hub act like a friction clutch. This equalises the speed of both the gear and the selector hub. The gears can now be selected without damage.

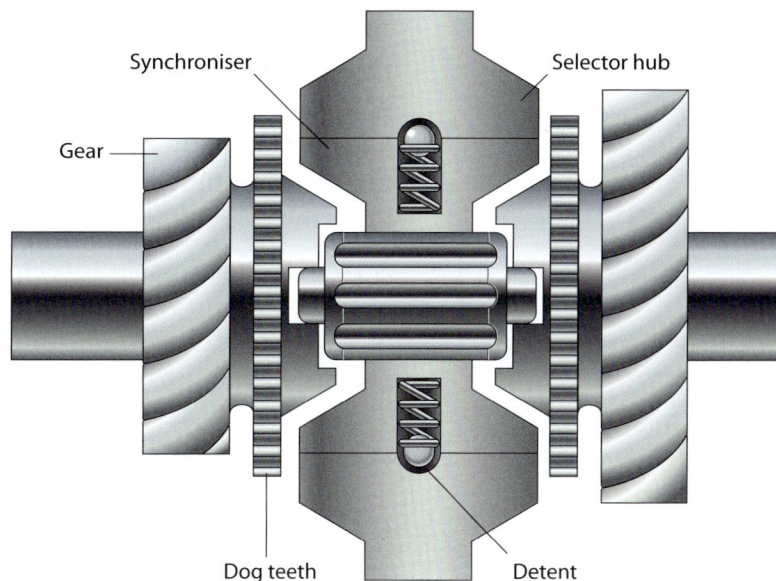

Figure 7.21 A synchromesh gear selector hub

Did you know?

The lubrication of a manual gearbox is achieved by oil-splashed feed. This means that as the gears turn, they scoop up gear oil and drag it around the gears. This lubricates the gearbox components.

Baulk ring

It is possible to 'out run' the speed of the synchronisation (by forcing the gear selector lever too quickly). Because of this **baulk rings** are used (see Figure 7.22). These block the movement of the selector hub until it is moving at the same speed as the gear.

Baulk rings need friction to work correctly. Small grooves are cut around the friction surface of the baulk ring. They act like the tread on a tyre and cut through the oil to provide grip.

Figure 7.22 A synchromesh baulk ring

Key terms

Mesh – gear teeth connected to each other.

Constant-mesh – this is where all of the gear teeth are continually in contact with each other.

Detent – a mechanism designed to help lock the selected gear in place.

Dog teeth – a series of square cut teeth which provide positive engagement.

Dog clutch – a positive engagement type clutch which uses dog teeth to lock the rotating components together.

Synchronisation – two components moving at the same speed.

Synchromesh – a mechanism designed to synchronise gear speeds.

Baulk ring – a blocking mechanism designed to prevent the engagement of gears before their speeds are synchronised.

Interlock mechanism

It is possible that when the gear stick is moved by the driver, two selector rods are moved at the same time. If this happens the selector forks inside the gearbox may try to lock two gears to the output of the transmission simultaneously. If both gears were selected at the same time, the gearbox would lock solid.

A system known as an **interlock** is provided in the selection mechanism to try to prevent this. The interlock will commonly use a series of balls or rods that lock the selector shafts when operated. In this way only one shaft is able to move at any one time.

Key term

Interlock – a mechanism designed to prevent the selection of two gears at the same time.

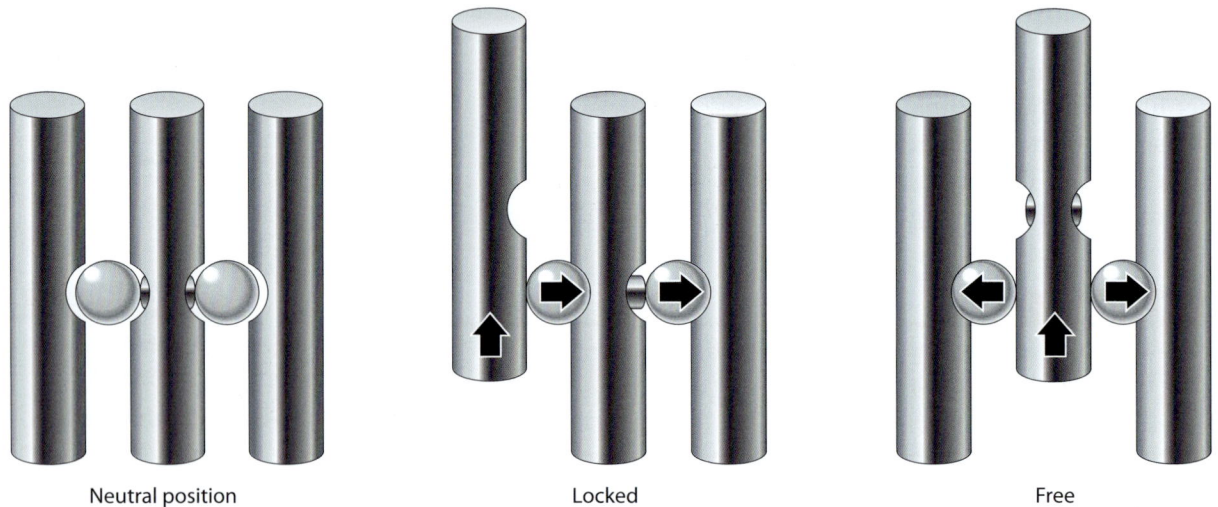

Neutral position   Locked   Free

Figure 7.23 An interlock mechanism

CHECK YOUR PROGRESS

1 What is the difference between a spur gear and a helical gear?
2 How is reverse achieved in a gearbox?
3 What component helps to hold the transmission in gear?

Drivelines

Drive shaft/prop shaft

Once the gearbox has multiplied the torque from the engine, the turning effort must now be transferred to the wheels. Drive shafts or prop shafts are used. The one which is used depends on transmission layout (i.e. front-wheel drive, rear-wheel drive or four-wheel drive).

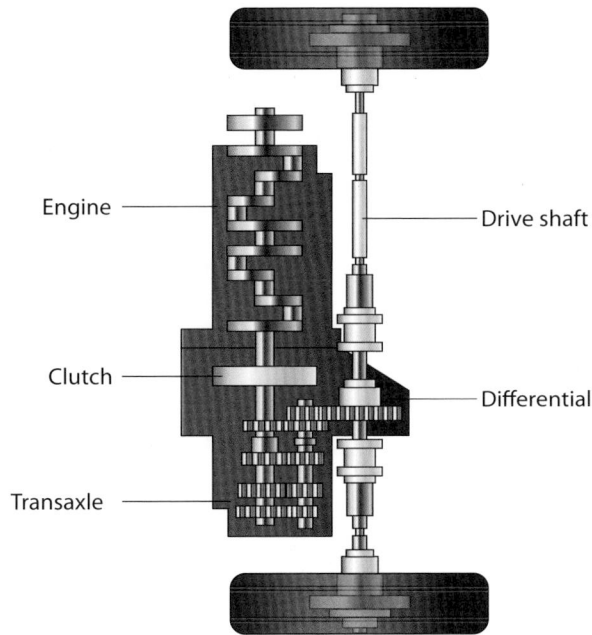

Figure 7.24 Front-wheel drive

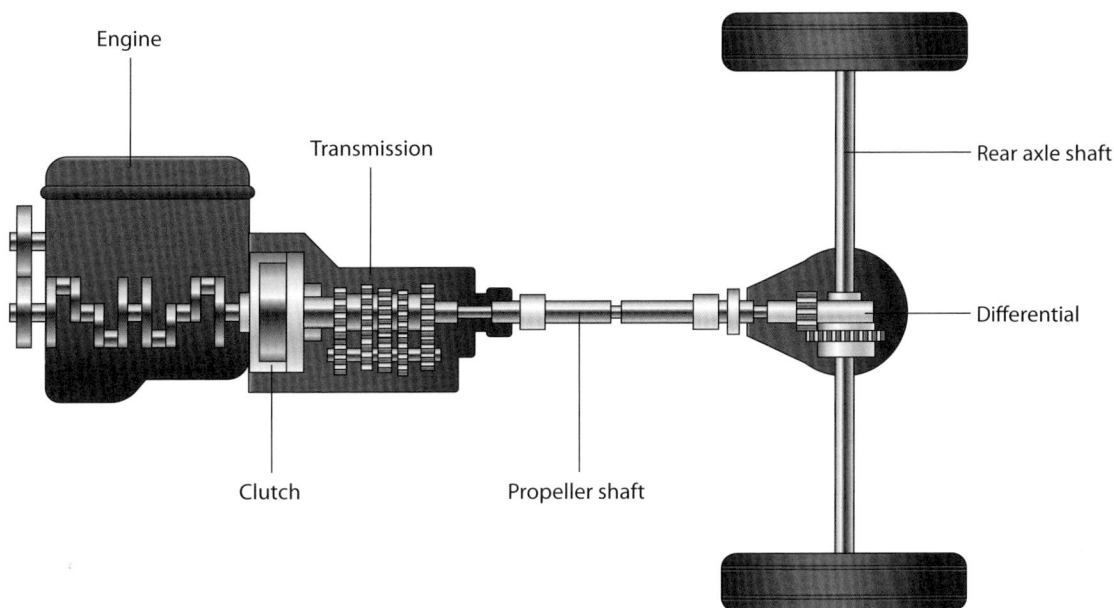

Figure 7.25 Rear-wheel drive

Prop shafts

A prop shaft (or propeller shaft to give it its full name) is used with front engine rear-wheel drive vehicles.

The prop shaft is simply a metal tube which is strong enough to transmit the full power of the engine and torque multiplied by the gearbox. The shaft is connected to the back of the gearbox. It runs beneath the floor to join it to the back axle.

A **universal joint (UJ)** is usually needed at either end of a prop shaft. This is because, as the suspension moves up and down, a difference in height exists between the rear axle and the gearbox. Without universal joints, as the car goes over a bump, suspension movement will try to bend the prop shaft.

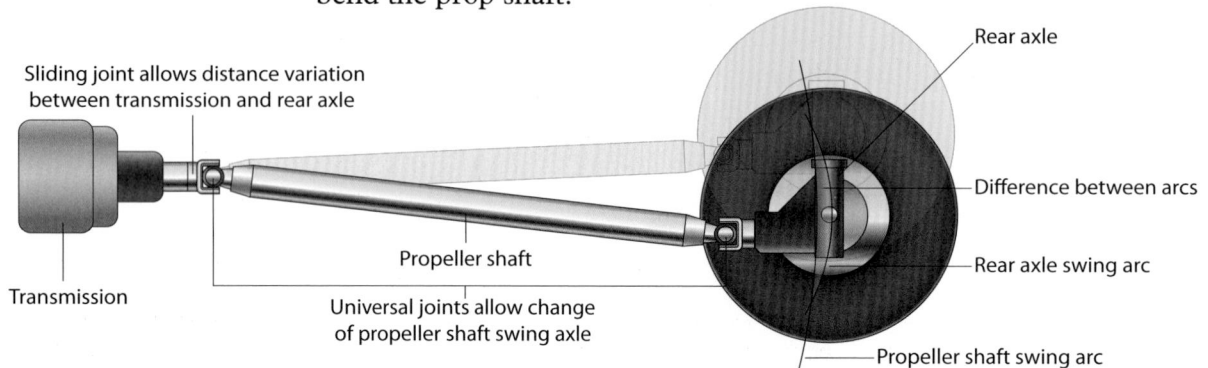

Figure 7.26 Universal joints stop suspension movement bending the prop shaft

Universal joints (UJ)

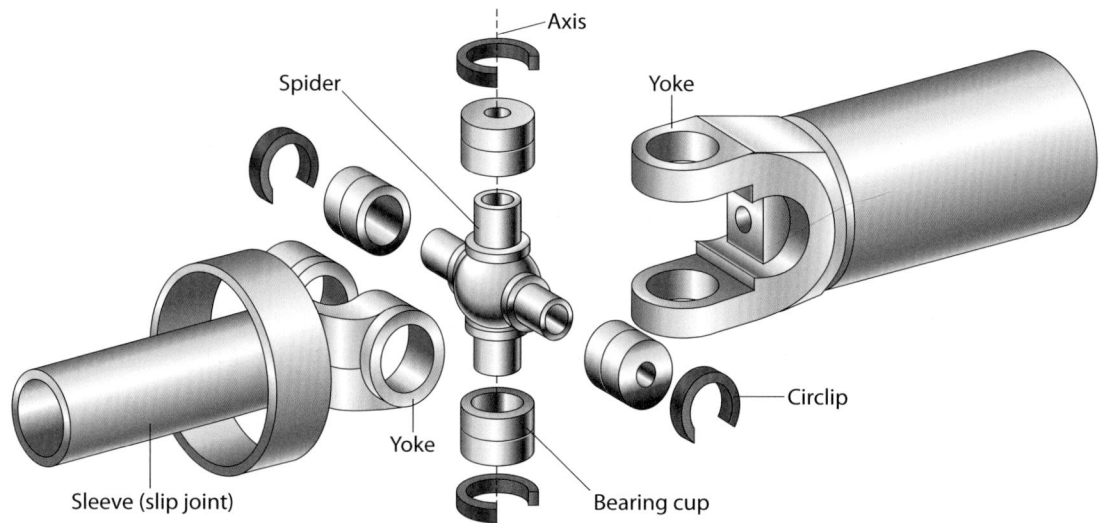

Figure 7.27 Hookes type universal joint

The most common type of universal joint is the Hookes UJ, as shown in Figure 7.27.

It is made up of two **yokes** pivoted on a central crosspiece. This is sometimes called a spider. The spider is formed by two pins crossing over each other at right angles. The yokes, one on the input shaft and the other on the output shaft, are connected to the spider so that they are at right angles to each other.

This arrangement allows the input and output shafts to rotate together even when they are at different angles.

A universal joint will speed up and slow down as it turns. This is because of the way it is made. The speeding up and slowing down will cause a vibration in the transmission system. To prevent this, the universal joints at either end of the prop shaft are 'synchronised'. This means that as one speeds up the other slows down, and the difference in speed is cancelled out.

Rear universal joint
SYNTHESIS
Front universal joint

Front universal joint

Parallel

Figure 7.28 Synchronised universal joints

Rear universal joint

Sliding joints

As the rear suspension travels up and down, this movement tries to stretch or **compress** the prop shaft. Because of this, a sliding joint is often included in the design. This means that the length of the prop shaft is able to vary (it can get longer or shorter).

Drive shafts

A drive shaft can be used with front engine front-wheel drive cars. The shafts leave the transmission system **transversely** (across the width of the car) and connect to the hub of the driving wheels.

As with a prop shaft, these shafts must be able to cope with suspension movement. In the case of front-wheel drive they must be able to deal with steering movement as well. Because the steering and suspension produce large angular movements, a universal joint (UJ) isn't suitable to use with a drive shaft. This is because a universal joint speeds up and slows down causing a vibration. To replace the universal joint, a special type of coupling known as a **constant velocity joint** (CV joint) is used.

A number of different designs and styles can be used depending on the manufacturer. However, the job of a constant velocity joint is to transmit drive at a constant speed, regardless of steering and suspension movement. These joints can wear out very easily and are usually covered with a rubber boot. In this way grease can be used to provide lubrication. This helps them to last longer.

Key terms

Universal joint (UJ) – a coupling that connects rotating shafts. It allows freedom of movement in all directions.

Yoke – a connection between two things so that they move together.

Compress – squash.

Transverse – sideways, across the car.

Constant velocity (CV) joint – the joint at the end of the drive shaft on a front-wheel drive vehicle. It is able to transmit drive with no variation in speed.

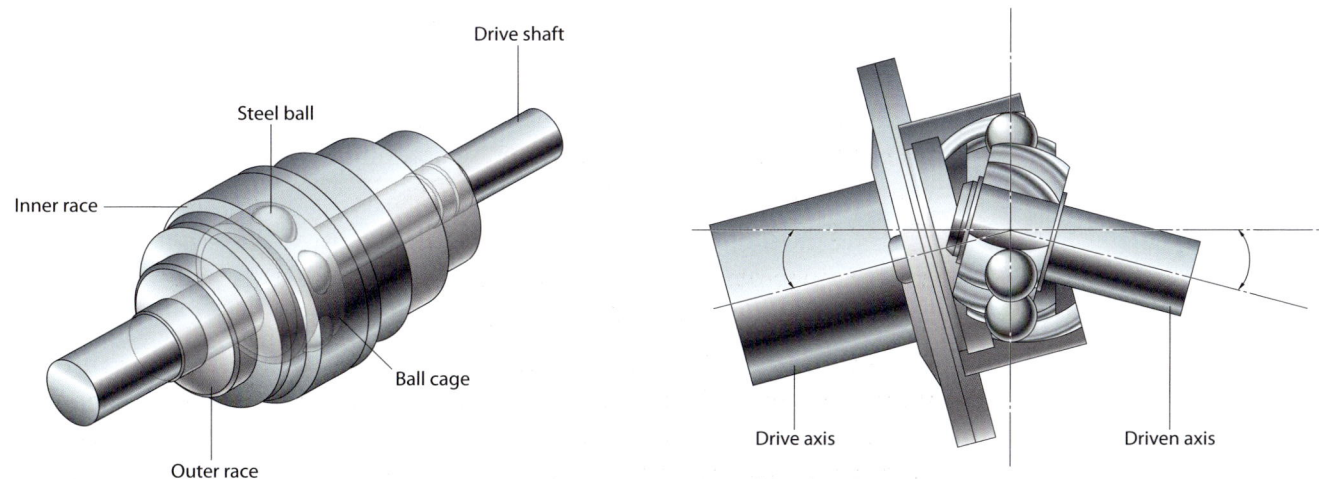

Drive shaft

Steel ball

Inner race

Ball cage

Outer race

Drive axis

Driven axis

Figure 7.29 A constant velocity (CV) joint

Final drive

Having completed its journey through the gearbox, the turning effort is now transmitted to the **final drive unit**. The final drive is the last stage in the transmission of power from the engine to the road wheels.

A final drive unit normally consists of two gears called the 'crown wheel and pinion'. This will provide a fixed final gear ratio to increase torque. With a rear-wheel drive, it will also turn the rotation from the prop shaft through right angles to turn the wheels.

The final drive will also include a differential unit which allows one wheel to travel faster than the other when turning a corner.

On front-wheel drive cars, the final drive is similar to the ones found on rear-wheel drive. The main difference is that there is no need for a prop shaft to take drive from the gearbox to a rear axle.

On cars with a transverse engine (this is where the engine is mounted across the car from left to right) the final drive doesn't need to turn the drive through a right angle.

Final drive gear ratios

As with the gears in a gearbox, the final drive reduction depends on the number of teeth on the crown wheel and pinion. A typical figure for final gear ratio is approximately 4:1.

If the crown wheel and pinion are used on a rear-wheel drive, a special type of gearing known as **bevel** is used. Most rear-wheel drive axles will combine this bevel gear with a spiral **hypoid** design, as shown in Figure 7.30.

In a hypoid design, the gear teeth are curved in such a way that the pinion can be fitted below the centreline of the crown wheel. This means that the prop shaft can be set lower down. Because of this the transmission tunnel in the floor that houses the prop shaft can be made lower or may disappear altogether. The flatter floor area inside the car gives more space for passengers. It also lowers the centre of gravity of the vehicle and makes it more stable on the road.

Did you know?
If the final drive and gearbox are incorporated into a single unit, this is often called a **transaxle**.

Key terms

Final drive unit – a transmission unit containing the differential, crown wheel and pinion.

Transaxle – a unit in which the gearbox and final drive are contained within one casing.

Bevel – gears that mesh at an angle.

Hypoid – this describes the pinion mounted in an offset position against the crown wheel in a final drive unit (normally set below the centreline).

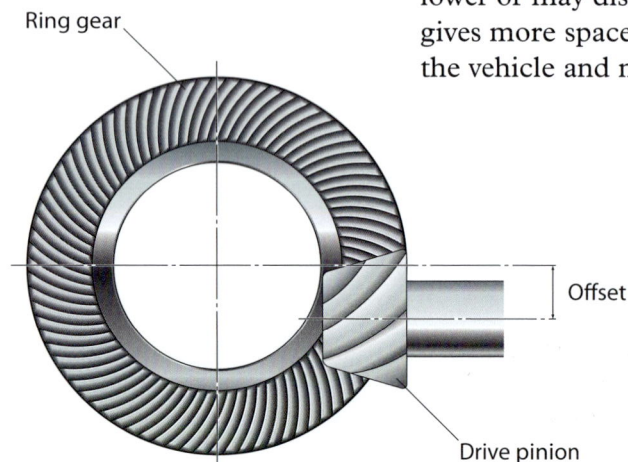

Ring gear / Offset / Drive pinion

Figure 7.30 A hypoid rear-wheel drive crown wheel and pinion

Did you know?
The fixed final gear reduction reduces the output speed of the gearbox. This is because even when a car is travelling at 70 mph, the road wheels are only turning between 700 and 1200 rpm (depending on their size). Compare this with the speed of the engine crankshaft which may be doing around 4000 rpm.

Differentials

When a car is travelling along a straight piece of road, both the driven wheels cover the same amount of ground at the same speed.

When it comes to take a bend, the inner driven wheel doesn't have to travel as far as the outer one, and so it needs to travel more slowly. (If both wheels turned at the same speed when trying to turn the corner, the inner wheel would be forced into a skid.)

On a bend, drive must be transmitted at different speeds. This is done using a **differential** unit. This is housed inside the final drive casing. When needed, some of the driving force from the inner wheel is transferred to the outer wheel. This will speed the outer wheel up and slow down the inner wheel.

A differential unit allows this because of a gearing system. The turning effort taken from the **crown wheel** is transmitted to the differential casing, where a metal pin is fixed. As the differential casing turns, the drive pin moves end over end. Two small gears are mounted on the drive pin. They are often called 'planet gears'. They are in constant mesh with two side gears, often called 'sun gears'. (See Figure 7.33.)

When the car is travelling in a straight ahead direction, the drive pin turns end over end. It locks the planet gears directly to the side gears and drives them all at the same speed, as described in Figure 7.32 below.

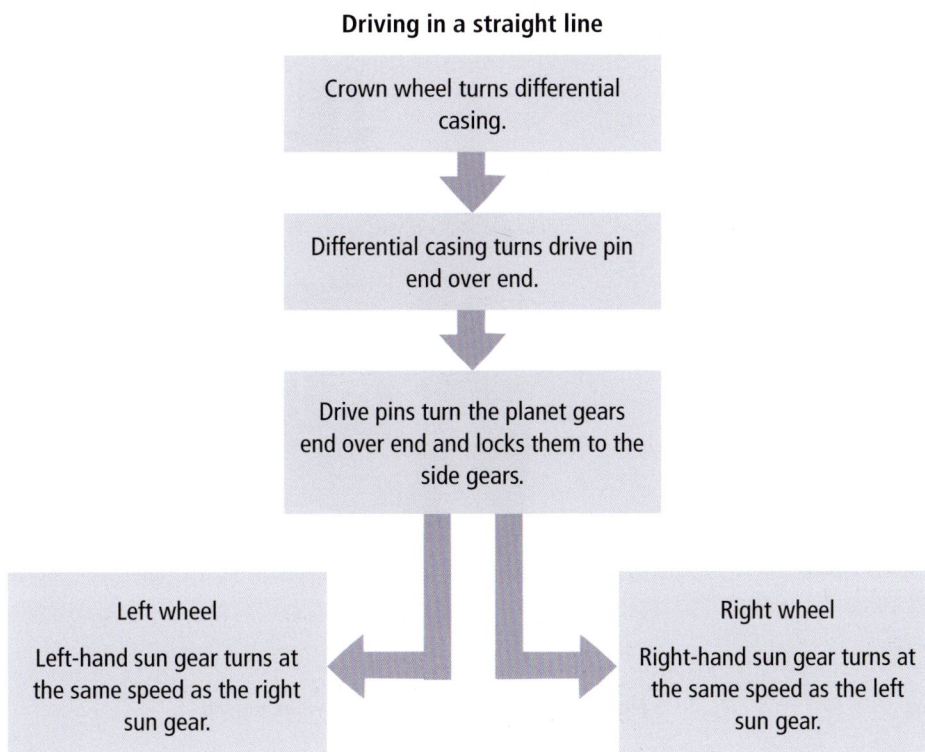

Distance A < Distance B

RPM of inside wheel < RPM of outside wheel

Figure 7.31 Driven wheels move at different speeds when cornering

> **Key terms**
>
> **Differential** – a mechanism that allows one driven wheel to travel faster than the other when the car goes around a bend.
>
> **Crown wheel** – a large metal gear wheel with teeth around the outer edge.

Driving in a straight line

Crown wheel turns differential casing.

↓

Differential casing turns drive pin end over end.

↓

Drive pins turn the planet gears end over end and locks them to the side gears.

Left wheel Left-hand sun gear turns at the same speed as the right sun gear.	← →	**Right wheel** Right-hand sun gear turns at the same speed as the left sun gear.

Figure 7.32 What happens when you drive in a straight line

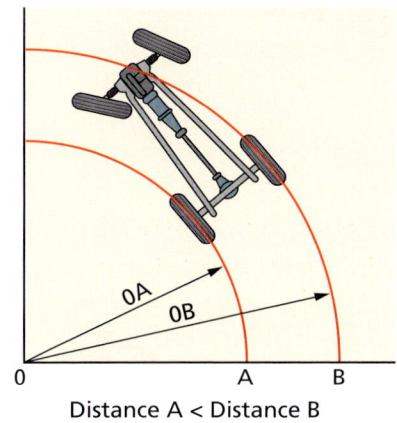

As the vehicle turns a corner, the extra load will try to slow down one wheel and reduces the speed at one sun gear. The drive pin still turns end over end. This provides torque or turning effort to the sun gears. However, the planet gears will now rotate on the pin, allowing more drive to be transmitted to one wheel than the other. This will let one wheel travel faster than the other but still transmit drive with the same torque. (See the flow chart in Figure 7.35.)

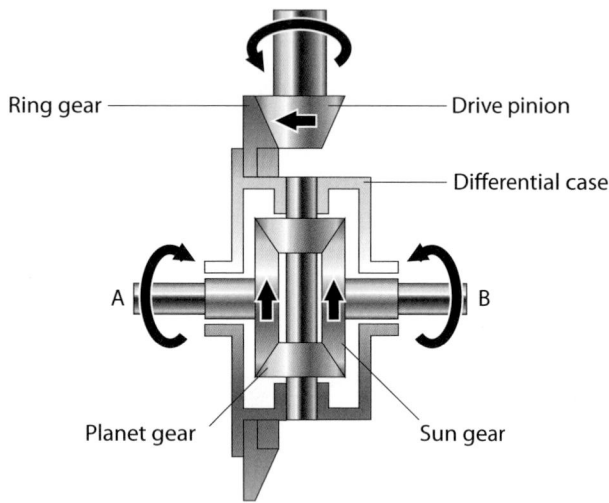

Ring gear — Drive pinion

Differential case

A — B

Planet gear — Sun gear

Straight ahead travel
RPM A = B

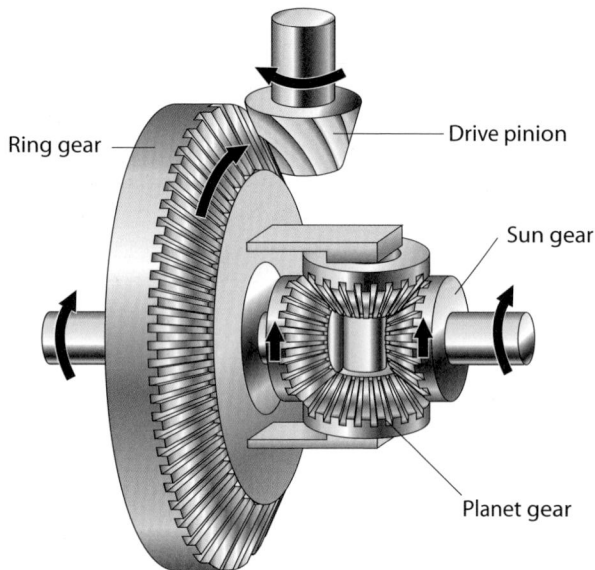

Ring gear — Drive pinion

Sun gear

Planet gear

(Larger resistance) A — B (Smaller resistance)

Turning
RPM A < B

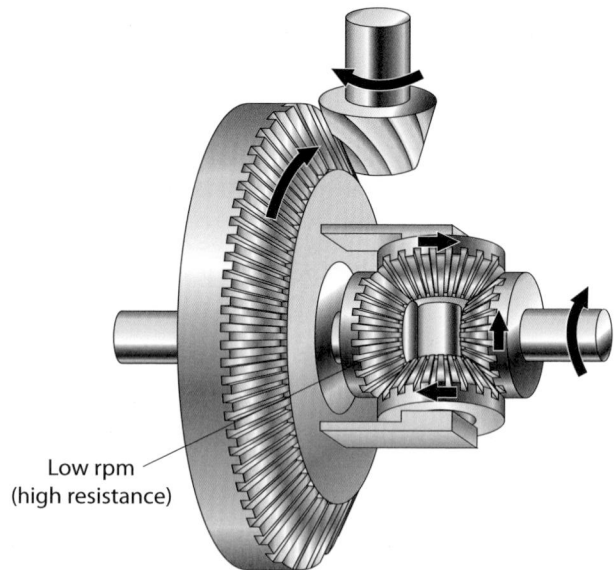

Low rpm
(high resistance)

Figure 7.33 Differential operation when travelling straight ahead

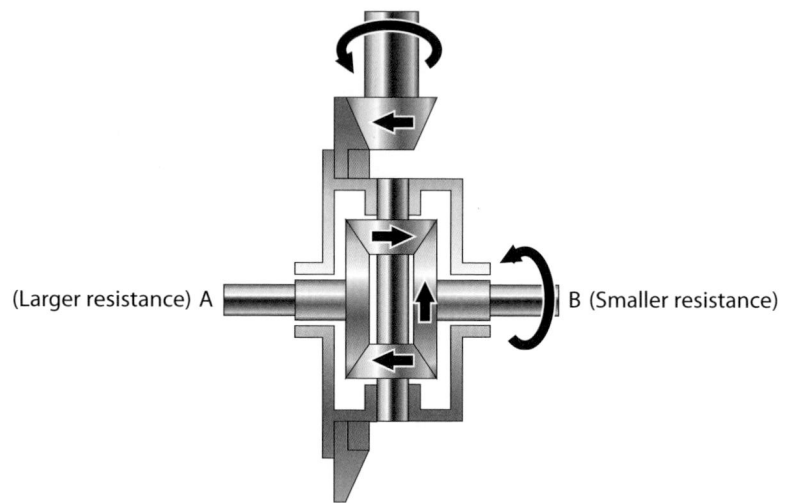

Figure 7.34 Differential operation when turning

Turning a left-hand bend

```
┌─────────────────────────────┐
│ Crown wheel turns differential │
│           casing.             │
└─────────────────────────────┘
              │
              ▼
┌─────────────────────────────┐
│ Differential casing turns drive pin │
│          end over end.        │
└─────────────────────────────┘
              │
              ▼
┌─────────────────────────────┐
│ Drive pins turn the planet gears │
│ end over end but the planet gears │
│  rotate against the side gears.  │
└─────────────────────────────┘
       │              │
       ▼              ▼
┌──────────────┐  ┌──────────────┐
│ Left wheel   │  │ Right wheel  │
│ Left-hand sun │  │ Right-hand sun │
│ gear turns at a │  │ gear turns at │
│ slower speed than │  │ a faster speed than │
│ the right sun gear. │  │ the left sun gear. │
└──────────────┘  └──────────────┘
```

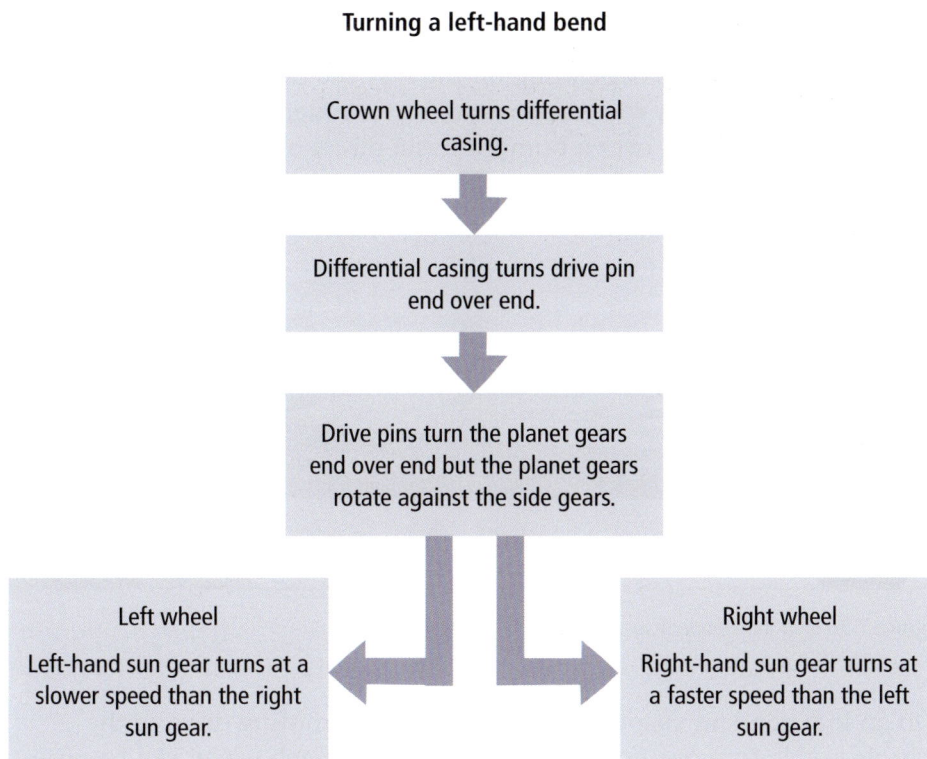

Figure 7.35 What happens when you turn round a left-hand bend

Centre diff/transfer box

In a four-wheel drive car, the engine can be mounted transversely (sideways) or longitudinally (in a straight line) depending on the manufacturer's design.

As drive leaves the gearbox, an extra unit is often used called a **transfer box**. This splits the drive so that it can be used by the front and rear axles.

A four-wheel drive car has at least two differential assemblies – one for the front wheels and one for the back. In addition to this, cars with permanent four-wheel drive sometimes have a centre differential that splits the drive from front to rear.

When a car is travelling in a forward direction, the front axle reaches a corner first. As the front axle starts to go around the bend, it will need to be travelling at a different speed from the back axle. The centre differential which is sometimes used operates in the same way as a standard differential. However, instead of allowing a difference in speed between the right-hand and left-hand wheels, it allows a difference in speed between the front and rear axle.

Key term

Transfer box – a mechanism used on four-wheel drive cars to split the drive between the front and rear axles.

Axle types

Axles are given names depending on the way the drive shafts and wheel hubs are supported by the suspension. A non-independent axle is rigid. When one wheel goes over a bump it has a direct effect on the opposite wheel, as shown in Figure 7.36.

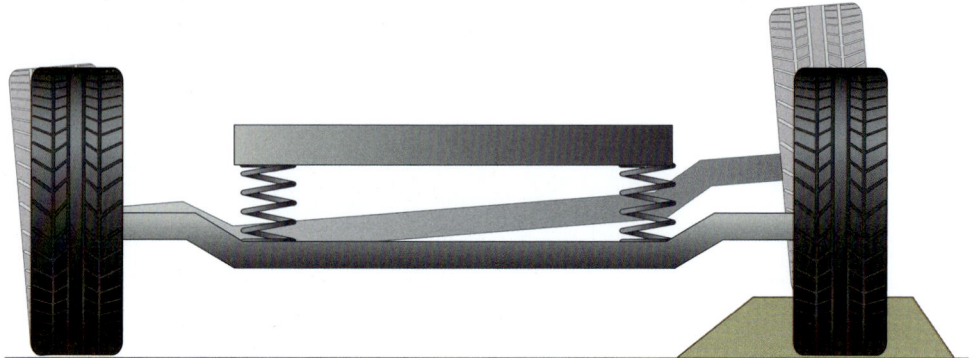

Figure 7.36 Non-independent suspension

On an independent suspension, each shaft transmitting drive to the wheels is able to move up and down separately. Each wheel can be given its own spring. The driveshafts will have universal joints at each end to allow for the change in angle as they go over bumps in the road, and won't affect the operation of each other, as shown in Figure 7.37.

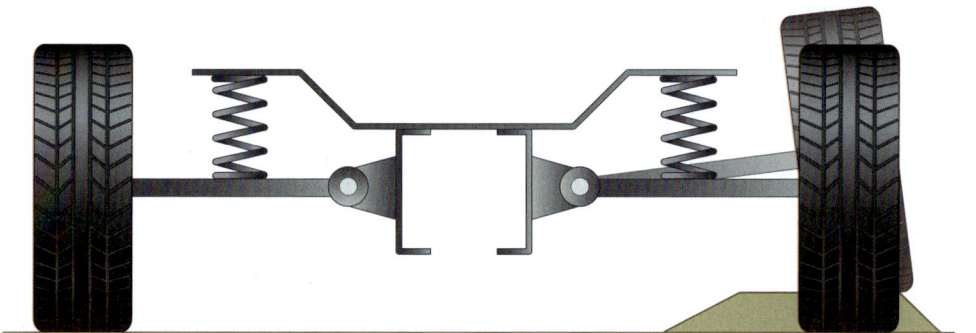

Find out

Look at the vehicles in your workshop and state whether their driven axles use independent or non-independent suspension.

Figure 7.37 Independent suspension

CHECK YOUR PROGRESS

1 Name three different types of driveline layout.
2 What do the letters CV stand for (in connection with drive shafts)?
3 What is it called when the transmission and final drive are combined in one unit?

Light vehicle automatic gearboxes

Some cars are fitted with an automatic transmission system. Its purpose is the same as a manual transmission system: to multiply the torque created by the engine. However, instead of the driver choosing which gear to use, the gearbox selects an appropriate ratio depending on the engine speed and load. This is usually done through a system of hydraulics (see pages 246–247) with mechanical or electronic control.

Table 7.6 The main automatic transmission and torque converter components and their purposes

Automatic transmission component	Purpose
Torque converter	Replaces the clutch from a manual gearbox. Fluid is used to drive turbine vanes and the input shaft of the gearbox
Gearbox	Multiplies the torque (turning effort) from the engine. Gears are automatically selected by the gearbox depending on vehicle speed and load
Prop shaft	Used in a rear-wheel drive car to connect the gearbox to the rear axle
Drive shaft	Used in a front-wheel drive car to send drive to the wheels
Final drive	Also helps to increase turning effort, and in a rear-wheel drive car, it turns the drive through a right angle
Differential	Allows one wheel to travel faster than the other when turning a corner

Torque converters

An automatic transmission still needs a way to provide a temporary neutral position and to give a smooth take-up of drive. This is done with the use of a component known as a **torque converter**. It is mounted in a similar position to a standard clutch. However, instead of using friction surfaces clamped together to provide drive, fluid is forced between turbine blades. This creates drag which rotates the input shaft of the gearbox.

A torque converter consists of three main components:

- the impeller
- the turbine
- the stator.

They are all sealed inside a casing, as shown in Figure 7.38.

A torque converter casing is pressurised with automatic transmission fluid (ATF). When spun by the engine crankshaft, fluid is taken into the impeller blades and thrown outwards by centrifugal force. The fluid strikes the blades of the turbine, making it spin. The spinning turbine is connected to the input shaft of the gearbox, which now turns.

Key term

Torque converter – a form of fluid flywheel designed to connect drive to the gearbox in place of a standard clutch.

Did you know?

When the brake pedal is released, and fluid from the impeller begins to push against the turbine, drive to the gearbox begins. This means that the car will start to move. This movement is often referred to as 'creep'.

The hydraulic fluid now leaves the turbine and strikes the blades of the stator. The stator directs it back into the impeller at high speed. This force helps to multiply the torque provided by the crankshaft.

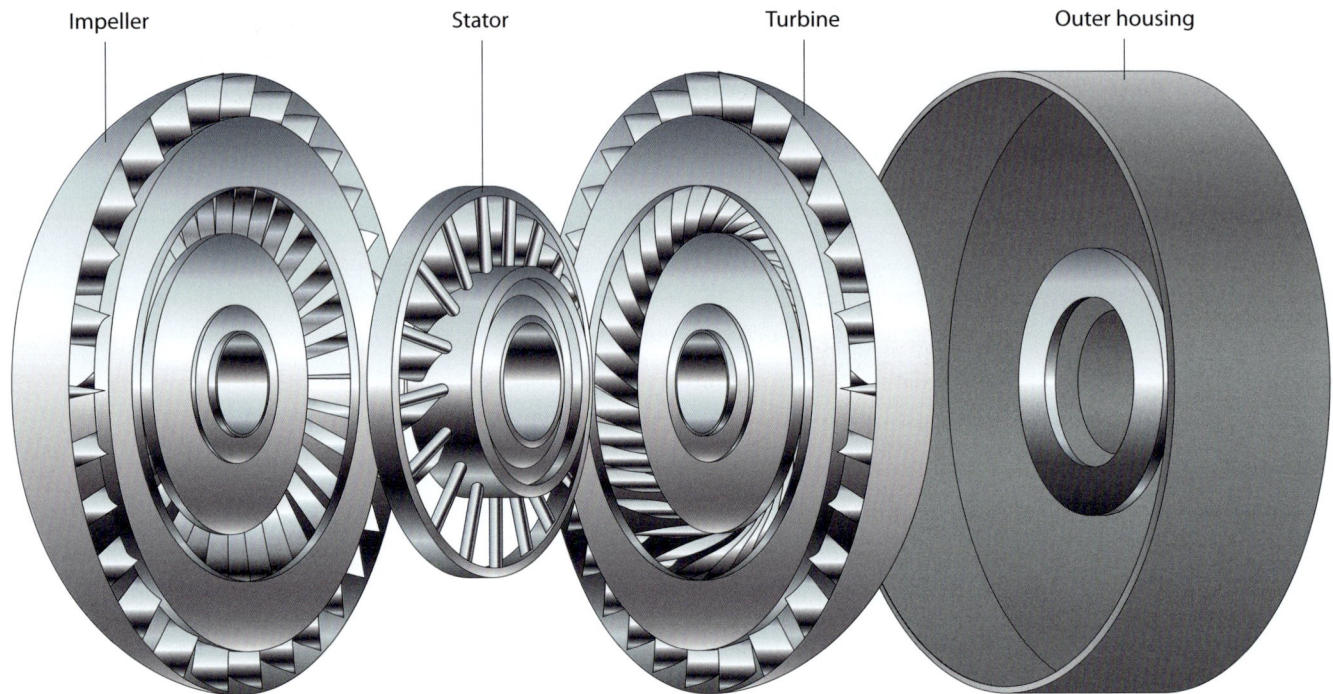

Impeller   Stator   Turbine   Outer housing

Figure 7.38 A torque converter

Key term

Inhibitor switch – a starter motor cut-out. It is designed to prevent the driver starting the car while in gear.

Epicyclic gears

Unlike a manual gearbox, standard gears are not used in an automatic transmission system. A component known as an **epicyclic**, as shown in Figure 7.39, is able to provide three forward speeds and reverse.

- The outer gear is called the annulus or ring gear.
- The inner gear is called the sun gear.
- The intermediate gears are called the planets.

A combination of any two gears operating together provides the driving action. This means that the third gear set is locked in position, becoming an idler.

This locking-in position can be achieved by the use of multi-plate clutch sets, one-way clutches or a mechanism known as a **brake band**.

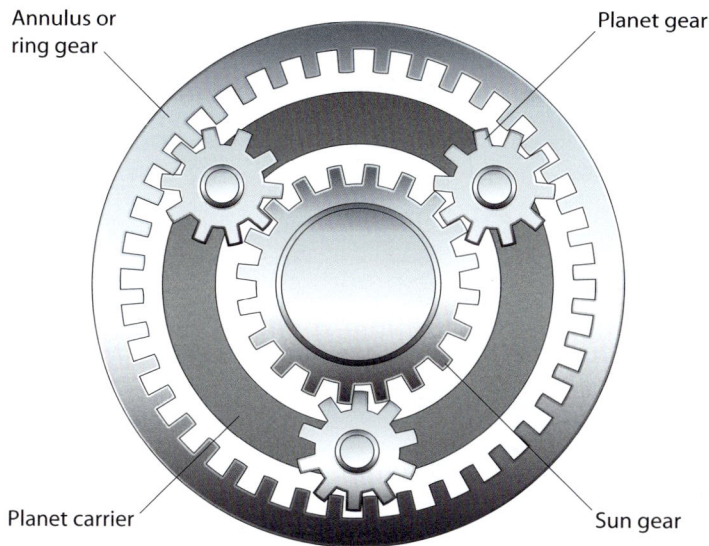

Figure 7.39 An epicyclic gear mechanism

Labels: Annulus or ring gear; Planet gear; Planet carrier; Sun gear

Brake bands

A brake band is a thin strip of metal surrounding one of the gear sets. One end of the brake band is anchored securely to the casing of the gearbox. The other is attached to the end of a hydraulic piston, as shown in Figure 7.40.

When gear selection is needed, the hydraulic piston squeezes the ends of the brake band together. This grips the gear set and holds it stationary.

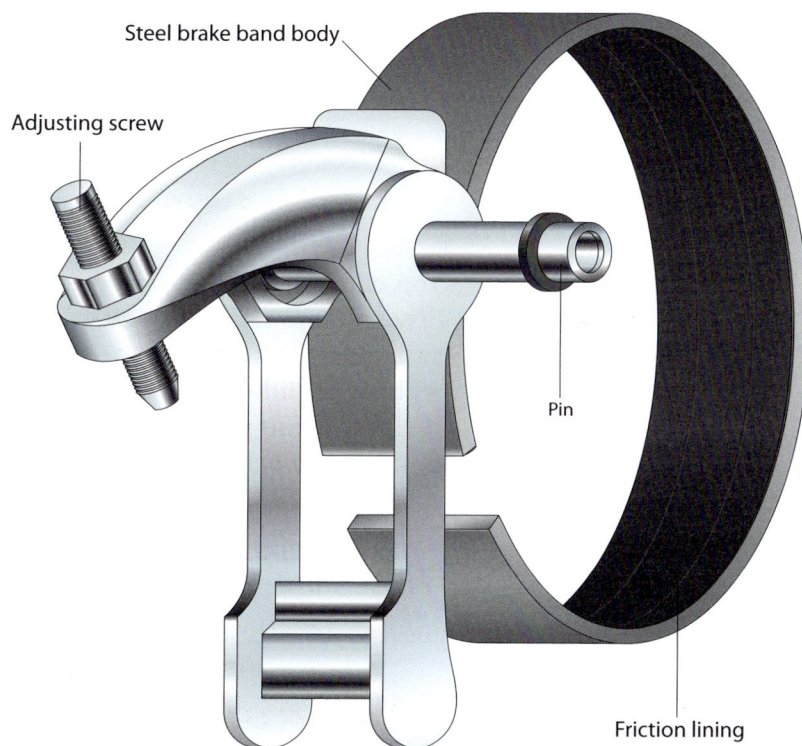

Figure 7.40 A brake band

Labels: Steel brake band body; Adjusting screw; Pin; Friction lining

Continuously variable transmission (CVT)

A **CVT** gearbox is a form of automatic transmission.

Instead of using mechanical gear sets, such as an epicycle, a drive belt is held between two pulleys, as shown in Figure 7.41. This is done in a similar way to the chain and sprockets on a bicycle.

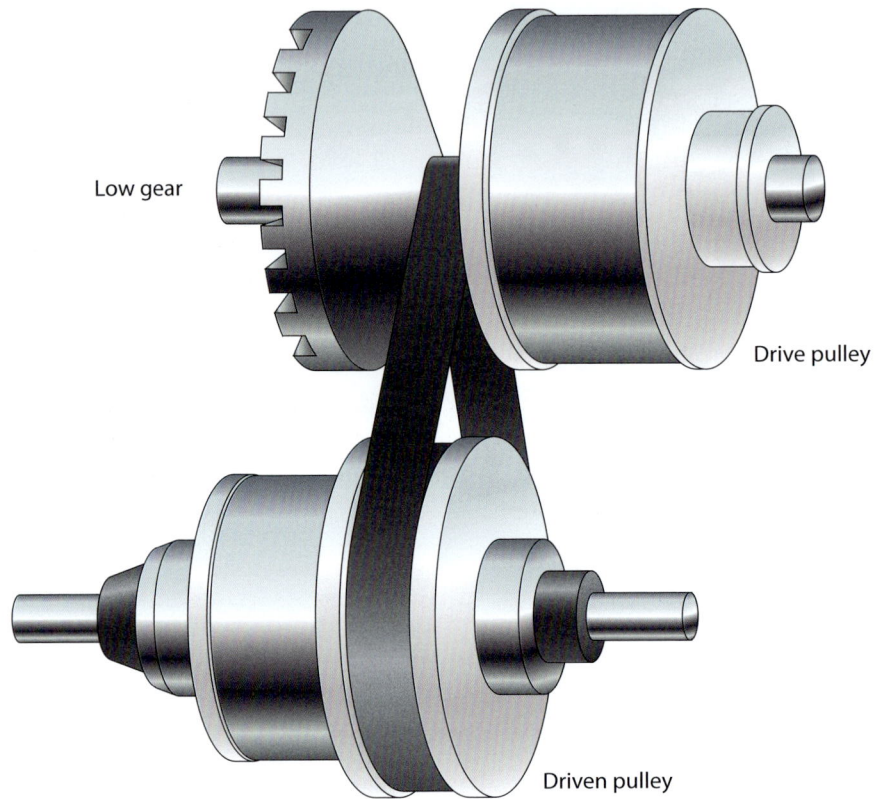

Low gear

Drive pulley

Driven pulley

Figure 7.41 Continuously variable transmission (CVT)

You can achieve different gear ratios by changing the size of the drive pulleys. This is done by allowing the drive pulleys to expand and contract. This way the drive belt is able to ride up and down within the pulleys, varying their size and therefore the gear ratio. Because these pulleys do not rely on fixed gear sizes, a stepless gear ratio can be achieved, maintaining the best efficiency for any engine speed or load.

CHECK YOUR PROGRESS

1 Name the three main sections of a torque converter.

2 Name the three main gears found in an epicyclic.

2 What do the letters CVT stand for?

Routine maintenance on light vehicle drivelines

Regular maintenance of transmission and driveline systems is very important for a long and fault-free service life. Lubrication and freeplay adjustments should be assessed at recommended service intervals. Damage or excessive wear will need to be reported to the owner, so that authorisation for any extra work can be given.

When carrying out routine maintenance on light vehicle driveline systems, it is important that you have a good source of technical data. This will provide you with correct measurements and specifications. Sources of technical data can include:

* paper-based workshop manuals
* technical data books
* manufacturers' technical bulletins
* electronic-based technical data and Internet-sourced information.

Working life

Anita has put a car on a four-post ramp, ready to be worked on. Instead of placing the gearbox in neutral, she has left it in gear, to help to stop it rolling while it is on the ramp.

1 Is this OK?

Anita comes back to the car later after finishing another job. She forgets the car is still in gear and starts the engine.

2 What might happen when the engine starts?

3 What could she have done differently?

Checking and topping up manual gearbox oil

Checklist			
PPE	VPE	Tools and equipment	Source information
• Steel toe-capped boots • Overalls • Latex gloves	• Wing covers • Steering wheel covers • Seat covers • Foot mat covers	• Socket set • Spanners • Vehicle hoist or ramp	• Technical data • Gear oil specification

1. Raise the car on a ramp, making sure that it stays level at all times. Remember to observe all health and safety requirements, including the correct use of PPE and VPE.

2. With the help of a workshop manual, locate the transmission level bung.

3. Remove the gearbox bung.

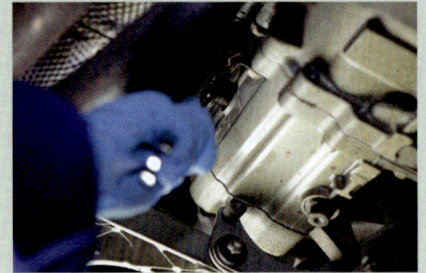

4. The gearbox oil should be level with the opening in the side of the casing (i.e. either dripping out slightly or just below the edge. This can often be checked using an improvised dipstick).

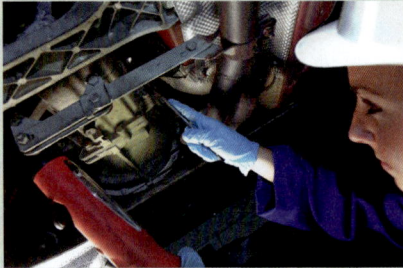

5. To top up the gearbox, use an oil that is graded EP (extreme pressure). The type and quantity of oil should always be that recommended in the manufacturer's technical specifications.

6. When oil starts to drip from the level hole, refit the bung.

⚠️ **Safe working**

Any oil spills should be cleaned up immediately to avoid slips, trips and falls.

Find out

Choose a car (manual transmission) and look up the transmission oil type/grade and quantity.

Checking and topping up automatic gearbox oil

Checklist			
PPE	VPE	Tools and equipment	Source information
• Steel toe-capped boots • Overalls • Latex gloves	• Wing covers • Steering wheel covers • Seat covers • Foot mat covers	• Lint-free cloth	• Technical data • Gear oil specification

1. Place the car in neutral or park and start the engine. Remember to observe all health and safety, including the correct use of PPE and VPE.

2. Open the bonnet and remove the transmission dipstick.

3. Wipe the dipstick with a lint-free cloth.

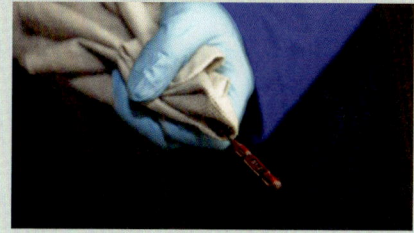

4. Reinsert the dipstick, remove it and check the oil level against minimum and maximum marks.

5. Top up the gear oil with automatic transmission fluid (ATF) by pouring down the dipstick hole.

6. Make sure that you don't exceed the maximum level shown on the dipstick.

Manual transmission system clutch replacement

Over a period of time, a manual transmission system clutch may wear out or be damaged by use. When this happens a replacement clutch assembly will be needed.

Checklist			
PPE	**VPE**	**Tools and equipment**	**Source information**
• Steel toe-capped boots • Overalls • Latex gloves	• Wing covers • Steering wheel covers • Seat covers • Foot mat covers	• Vehicle hoist or ramp • Socket set • Spanners • Screwdrivers • Clutch alignment equipment	• Technical data • Torque settings

1. Raise the car on a ramp and remove the gearbox. Remember to observe all health and safety requirements, including the correct use of PPE and VPE.

2. You can now unbolt the clutch assembly from the flywheel and remove it.

3. It is recommended that the three main components are all replaced at the same time. These are the clutch cover/pressure plate, the friction or drive plate, and the clutch release bearing.

4. Before reassembling the clutch, all friction surfaces should be degreased with a suitable brake/clutch cleaner.

5. When you refit the clutch assembly, it is important that the friction plate is aligned in the middle of the flywheel. This way the input shaft of the gearbox can be correctly located when the transmission assembly is remounted.

6. Specialist tools can be used to clamp the friction plate in the middle of the flywheel as it is assembled.

7. Tighten the mounting bolts in a diagonal sequence. In this way, you can make sure that an even clamping force is applied to the pressure plate as it is attached to the flywheel.

8. Refit the gearbox. Reassemble and road test to check for correct function and operation.

Drive shaft gaiter replacement

A constant velocity joint requires lubrication at all times. To keep lubricating grease in, and dirt out, a rubber gaiter is used to enclose the joint. This drive shaft gaiter must be replaced if it becomes split or damaged.

Checklist			
PPE	VPE	Tools and equipment	Source information
• Steel toe-capped boots • Overalls • Latex gloves	• Wing covers • Steering wheel covers • Seat covers • Foot mat covers	• Vehicle hoist or ramp • Socket set • Spanners • Screwdrivers • Circlip pliers	• Technical data • Torque settings

1. Strip out and remove the drive shaft. Remember to observe all health and safety requirements, including the correct use of PPE and VPE.

2. Once removed, the old gaiter can be cut away and the CV joint disconnected (refer to the manufacturer's procedures).

3. The gaiter can now be replaced.

4. You should clean the CV joint and re-lubricate with fresh grease.

5. Refit the joint and the drive shaft assembly.

6. Make sure that all the items are correctly aligned and any nuts and bolts tightened to the manufacturer's recommended torque settings.

7. Road test and check for correct function and operation.

Figure 7.42 A split constant velocity boot

1 Which of the following is not a function of the clutch?

  a to multiply torque
  b to provide a smooth take-up of drive
  c to disengage the engine for gear change
  d to provide a temporary position of neutral

2 A gearbox is fitted to a car to:

  a vary the speed and torque output from the engine
  b allow speed and torque to be equalised
  c ensure high speed operation is possible
  d change the direction of engine drive

3 As torque in the gearbox is increased:

  a turning effort is reduced
  b speed is increased
  c speed is reduced
  d all of the above answers

4 Clutch slip can be caused by:

  a worn friction material
  b clutch springs that are too tight
  c excessive free play at the clutch pedal
  d an overheated engine

5 Clutch drag can be caused by:

  a worn friction material
  b clutch springs that are too loose
  c excessive free play at the clutch pedal
  d an overheated engine

6 Gearbox lubrication is usually achieved by:

  a dry sump lubrication
  b total loss
  c splash feed
  d pressure feed

7 Torque is best described as:

  a power
  b speed
  c pressure
  d turning effort

8 If an input speed of 3000 rpm is put through a gear ratio of 2 : 1, the output speed will be:

  a 1500 rpm
  b 6000 rpm
  c 150 rpm
  d 300 rpm

9 If an input torque of 200 Nm is put through a gear ratio of 4 : 1, the output torque will be:

  a 100 Nm
  b 200 Nm
  c 400 Nm
  d 800 Nm

10 An interlock stops:

  a torque
  b speed
  c synchronisation
  d two gears being selected

GETTING READY FOR ASSESSMENT

The information contained in this chapter, as well as continued practical assignments in your centre or workplace, will help you to prepare for both the end-of-unit tests and diploma multiple-choice tests. This chapter will also help to prepare you for working on vehicle driveline systems safely.

You will need to be familiar with:

- The operation of the clutch
- The operation of a manual gearbox
- How to calculate gear ratios
- The power flow of drive through the transmission system from the engine to the road wheels
- The common differences between a manual and automatic transmission system
- How to check the oil level in both a manual and automatic gearbox

This chapter has given you an introduction and overview to light vehicle driveline systems and components. It has provided you with the basic knowledge that will help you with both theory and practical assessments.

Before trying a theory end-of-unit test or diploma multiple-choice test, make sure you have reviewed and revised any key terms that relate to the topics in that unit. Be sure to read all questions fully and take time to digest the information so that you are confident about what the question is asking you. With multiple-choice tests, it is very important that you read all of the answers carefully, as it is common for the answers to be very similar and this may lead to confusion.

For practical assessments, it is important that you have had sufficient practice and that you feel that you are capable of passing. It is best to have a plan of action and work method that will help you. Make sure that you have sufficient technical information, in the way of vehicle data, and appropriate tools and equipment. It is also wise to check your work at regular intervals. This will help you to be sure that you are working correctly and to avoid problems developing as you work.

When undertaking any practical assessment, always make sure that you are working safely throughout the test. Ensure that all health and safety requirements are observed and that you use the recommended personal protective equipment (PPE) and vehicle protection equipment (VPE) at all times. When using tools, make sure you are using them correctly and safely.

Good luck!

8 Light vehicle electrical systems

This chapter will give you an introduction and overview to light vehicle electrical and lighting systems. It provides the basic knowledge that will help you with both theory and practical assessments.

It will introduce the basic operating principles of car electrical and lighting systems components. It will also help you to identify the main components used in vehicle electrical and lighting systems, including the main electrical principles and key terms.

Finally, it will help you to interpret and create your own simple electrical circuits, remove and replace electrical lighting components and align headlamp units.

This chapter covers:

- Safe working on vehicle electrical and lighting systems
- Vehicle electrical systems and principles
- Simple electrical circuits
- Vehicle lighting system components and operation
- Electrical component replacement

WORKING PRACTICE

Personal Protective Equipment (PPE)

Electricity can be extremely hazardous if not handled correctly. Working with car electrical and electronic systems presents a number of dangers which can involve personal injury or vehicle and component damage. You need to be aware of these hazards and risks as you work on the electrical components of cars. You must always use personal protective equipment (PPE) and you also need to think about the possibility of crush or bump injury. You will also come into contact with chemicals such as battery acid. Make sure that your selection of PPE will protect you from these hazards.

Safety helmet protects the head from bump injuries when working under cars.

Overalls provide protection from coming into contact with oils and chemicals.

Safety gloves provide protection from oils and chemicals. They also protect the hands when handling objects with sharp edges.

Barrier cream protects the skin from old engine oil, which can cause dermatitis and may be carcinogenic (a substance that can cause cancer).

Safety boots protect the feet from a crush injury and often have oil- and chemical-resistant soles. Safety boots should have a steel toe-cap and steel mid-sole.

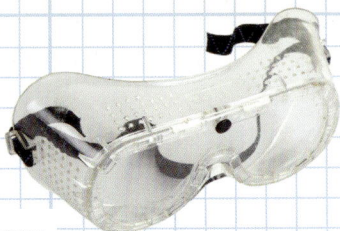

Safety goggles reduce the risk of small objects or chemicals coming into contact with the eyes.

To reduce the possibility of damage to the car, always use the appropriate vehicle protection equipment (VPE):

Wing covers

Seat covers

Steering wheel covers

Floor mats

If appropriate, safely remove and store the owner's property before you work on the vehicle. Before you return the vehicle to the customer, reinstate the vehicle owner's property. Always check the interior and exterior to make sure that it hasn't become dirty or damaged during the repair operations. This will help promote good customer relations and maintain a professional company image.

Vehicle Protective Equipment (VPE)

Safe Environment

During the repair or maintenance of light vehicle electrical systems you may need to dispose of certain waste materials such as car batteries. Under the Environmental Protection Act 1990 (EPA), you must dispose of them in the correct manner. They should be safely stored in a clearly marked container until they are collected by a licensed recycling company. This company should provide you with a waste transfer note as the receipt of collection.

To further reduce the risks involved with hazards, always use safe working practices including:

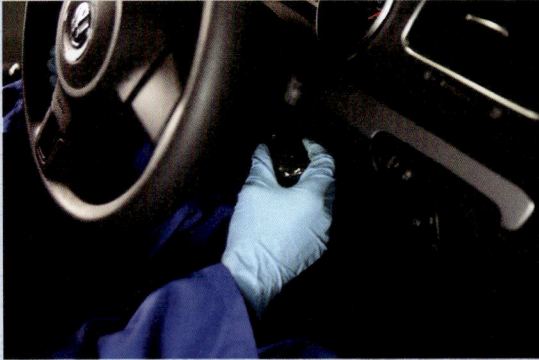

1. Immobilise vehicle (by removing the ignition key). If possible, allow the engine to cool before starting work.

2. Prevent the vehicle moving during maintenance by applying the handbrake or chocking the wheels.

3. Follow a logical sequence when working. This reduces the possibility of missing things out and of accidents occurring. Work safely at all times.

4. Always use the correct tools and equipment. (Damage to components, tools or personal injury could occur if the wrong tool is used or a tool is misused.) Check tools and equipment before each use.

5. Following the replacement of any vehicle components, thoroughly road test the vehicle to ensure safe and correct operation. Make sure that all work is correctly recorded on the job card and vehicle's service history, to ensure that any maintenance work can be tracked.

6. If components need replacing, always check that the parts are the correct quality and type for the vehicle if it is still under guarantee. Inferior parts or deliberate modification might make the warranty invalid. Also, if parts of an inferior quality are fitted, this might affect vehicle performance and safety.)

Preparing the car

Tools

Screwdrivers

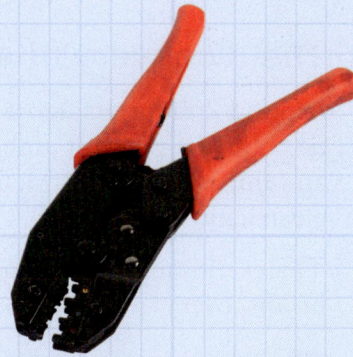

Crimping tool

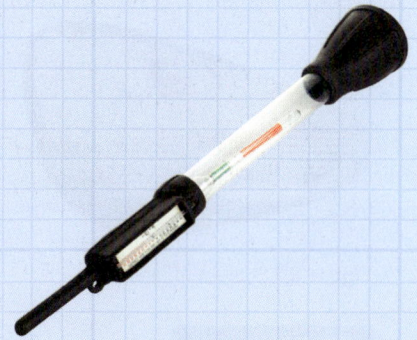

Battery hydrometer

Battery charger

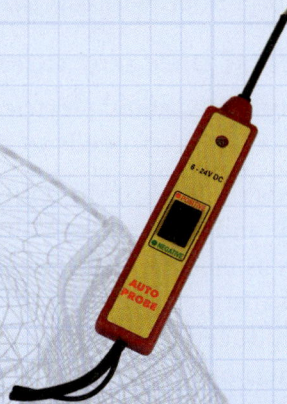

Test light/power probe

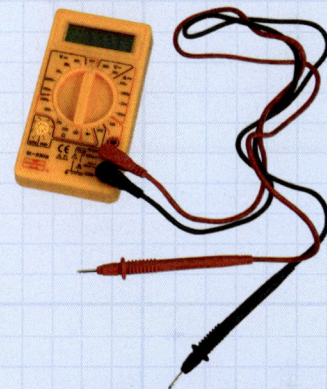

Multimeter

Safe Working

- If you are using a ramp or vehicle hoist, always check that the car is evenly positioned, secure and its weight does not exceed any safe working loads (SWL).

- Always use exhaust extraction when running engines in the workshop.

- Isolate electrical circuits by disconnecting the battery where possible.

- Make sure that multimeters are set to the appropriate scale/unit and correctly connected to the circuit that is tested.

- When replacing headlamp bulbs, do not touch the glass/quartz with your hands.

- When checking battery electrolyte, wear the correct PPE (goggles, gloves, etc.) to protect you from the sulphuric acid.

- Always remove any metal jewellery when you work on electrical circuits, to reduce the risk of accidental electrical discharge and injury.

- Always make sure that electrical repairs are of a good standard, to reduce the possibility of overheating and fire.

Figure 8.1 A piece of amber

Key terms

Static – stationary, not moving.

Current – moving electricity.

Alternating current (AC) – electricity that moves in two directions (backwards and forwards).

Direct current (DC) – electricity that only moves in one direction.

Figure 8.2 Every substance is made up of molecules and every molecule is made up of atoms

Vehicle electrical systems and principles

Electricity has existed since the beginning of time. It is contained in every substance known to man.

The discovery of electricity

Around 2500 years ago a Greek scientist named Thales found that if he rubbed a piece of amber on a piece of cloth, small particles of dust and fluff would be attracted to it. What he had discovered was **static** electricity.

Thales did not understand what was happening, but he did write down his discovery.

The naming of electricity

Around 1550, William Gilbert, Queen Elizabeth I's doctor, found that if he rubbed a silk cloth on a glass rod it would attract even heavier objects such as feathers. He named this phenomenon electricity after the Greek word for amber, which is *elektron*.

The first electric current

Unfortunately, static electricity is difficult to turn into a usable source of energy. Electricity needs to move to make it useful.

Towards the end of the 18th century, two Italian scientists, Luigi Galvini and Alessandro Volta, created the first moving electricity known as electric **current**. There are two types of current – **alternating current (AC)** and **direct current (DC)**. This current was produced from a chemical reaction and eventually led to the invention of the battery.

Electrical principles and terminology

What is electricity?

Every substance known to humankind is made of molecules. The molecules of a substance are made up from **atoms**. For example, if the substance is water, the molecule is H_2O. This means that the molecule is made up of two hydrogen (H) atoms joined to one oxygen (O) atom.

The reason why it can be difficult to understand electricity is because it is contained within atoms. Atoms are very small and hard to imagine.

- The easiest way to imagine an atom is like a miniature solar system, with a sun in the middle and planets orbiting around the outside.
- In the case of an atom, the **nucleus** represents the sun. The nucleus is made of positively charged particles known as **protons**. It also contains particles with no charge known as **neutrons**.
- Orbiting around this nucleus (in a similar way to the planets) are negatively charged particles known as **electrons**. As the name suggests, it is the electrons that produce electric current.

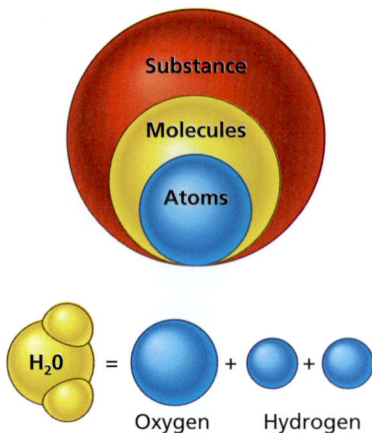

Different atoms have different numbers of protons and electrons, as shown in Table 8.1.

Table 8.1 The numbers of protons and electrons in different atoms

Type of atom	Number of protons/electrons
Hydrogen	1
Copper	29
Lead	82

To make the electric current, you need to move electrons from one atom to the next. To do this they need to be given a push by an external force or pressure.

The pressure used to move electrons can be created by:

* magnets
* a chemical reaction.

Conductors and insulators

Orbiting electrons are held in place in a similar way to the gravity acting on the planets circling around the sun.

Because the hydrogen atom is so simple, the attraction between the nucleus and the electron is very strong. This makes it very hard to move the electron. When electrons don't move easily the element is known as an **insulator**.

A copper atom contains 29 electrons and 29 protons. The electrons orbit in circles that get bigger and bigger.

The electrons in the farthest orbit have a far weaker bond/attraction to the nucleus than those in a hydrogen atom. These outer electrons are known as 'free electrons'. If an external pressure is applied, electrons can be moved from one atom to the next. This movement of electrons is electric current. When electrons do move easily the element is known as a **conductor**.

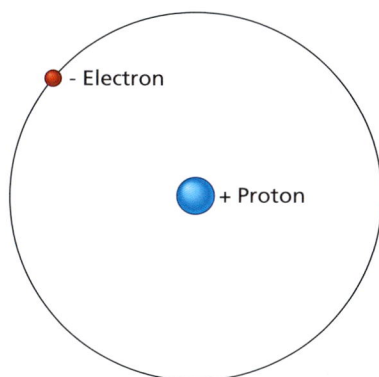

Figure 8.3 A helium atom – the nucleus is made up of protons and neutrons while the electrons spin around the outside

Key terms

Atom – the smallest component of any chemical element.

Nucleus – the centre of an atom.

Protons – positively charged particles.

Neutrons – particles with no charge.

Electrons – negatively charged particles.

Insulator – a chemical element which does not allow the easy movement of electrons.

Conductor – a chemical element that allows the easy movement of electrons.

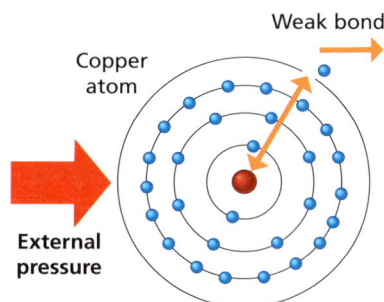

Figure 8.4 A hydrogen atom – one electron is in orbit around the nucleus

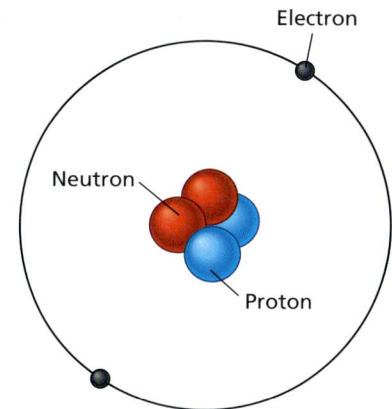

Figure 8.5 A copper atom with 29 electrons in orbit around the nucleus – if external pressure is applied, the free electron in the outer orbit can be dislodged

Find out

Make a list of materials used on cars that are good conductors and insulators of electricity.

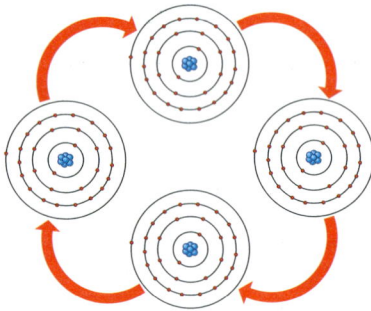

- Conductors are used on cars where we want electricity to flow easily, such as wiring.
- Insulators are used on cars to reduce the movement of electricity, such as the coating on the outside of a spark plug lead.

Circuits

For electrons to move from one atom to another, the conductor must be connected in a **circuit**. This means that as one electron leaves it can be replaced by one from behind.

If not connected in a circuit, the electrons cannot flow (move), as the last electron in the conductor has nowhere to go. If the circuit is broken it is said to have lost **continuity**.

Magnets

Electricity and magnetism are very closely linked. Both electricity and magnets have positive and negative, or North and South, poles. Both attract and repel.

- If a copper conductor (wire) is passed by a magnet, the magnetic attraction will move electrons through that copper conductor and create electric current.
- If an electric current is passed through a copper conductor then it will generate an invisible magnetic field.

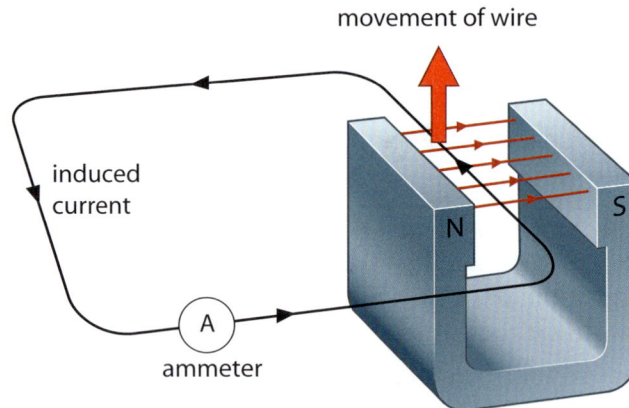

Figure 8.6 These atoms are in a circuit – the arrows represent how electrons will move from one to the next

Figure 8.7 The wire is moving in a magnetic field and creating a current

The magnetic effect of electrical current can be used to make things move (by magnetic **attraction** or **repulsion**). That movement can be used to make a motor.

The movement of magnets past a conductor can be used to make electric current. This is the principle of a **generator**.

- Motors turn electrical energy into mechanical energy.
- Generators turn mechanical energy into electrical energy.

Heat

As electric current flows, some of the energy can be turned into heat. This heat can be used by vehicle systems such as lighting and screen demisting.

Chemical reaction

Electrical energy can be produced by or converted into chemical energy. Because of this it is possible to store electricity and take it with you in the form of a battery. If you keep a battery **charged**, it provides a portable source of electricity that can be used when needed.

The principle of a direct current circuit

As it is very hard to imagine electrons moving from one atom to the other, the process is often best described using water.

Imagine a simple water tower containing:

- a reservoir of water at the top (to represent a battery)
- a pipe leading from the bottom of the reservoir (to represent the wire)
- a tap on the end of the pipe (to represent a switch)
- a small water wheel at the end (to represent a motor).

This can be used to show the operation of a simple electrical circuit.

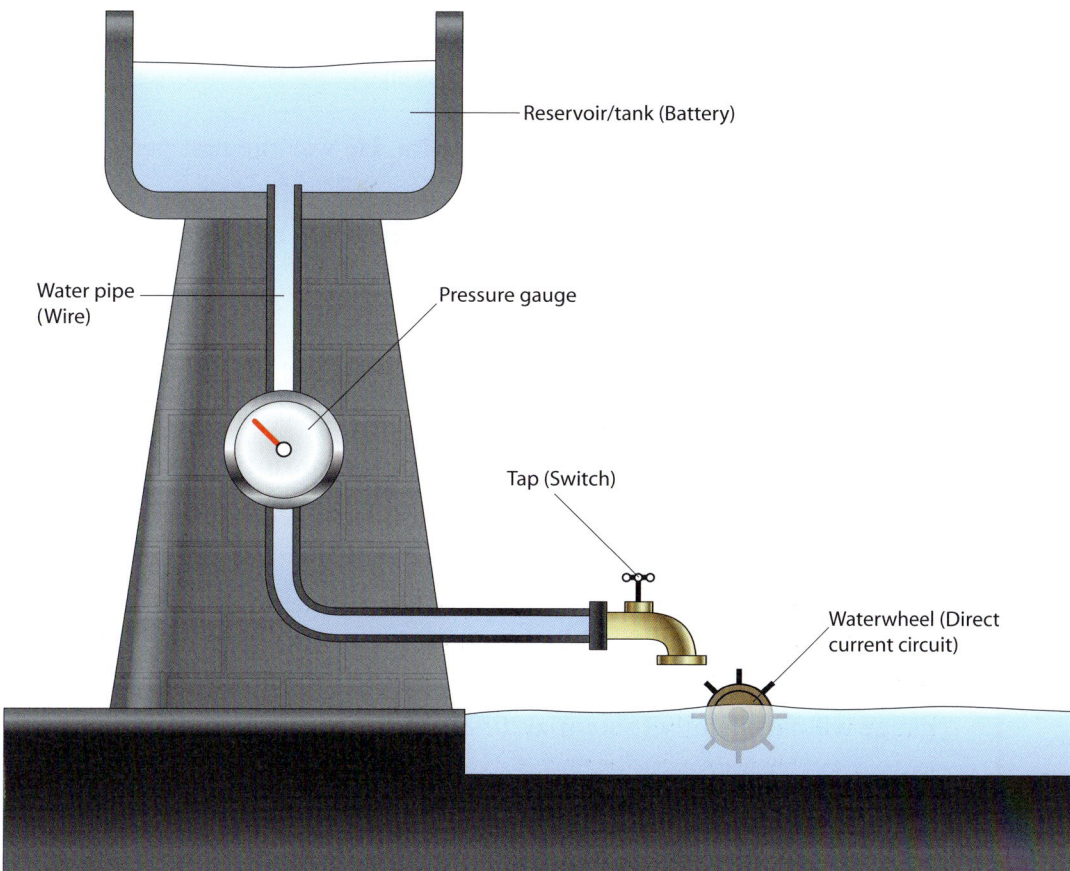

Figure 8.8 This water tower represents the operation of a simple direct current electric circuit

Key terms

Circuit – an unbroken loop.

Continuity – refers to an electrical conductor (something that allows electricity to move easily) which is unbroken or complete (i.e. continuous).

Attraction – bringing together.

Repulsion – pushing away.

Generator – a mechanical component that makes electricity.

Charged – the storage of electricity, in a battery for example.

When the tap is turned on, gravity pushes water down through the pipe, under pressure, out through the tap and on to turn the water wheel. This is similar to the way electric current flows through a circuit and turns a motor when the switch is turned on.

Electrical units

The four main electrical units that you will be using are:

- volts
- amps
- ohms
- watts.

They are each named after the person who first described their function.

Volt

Voltage was named after Alessandro Volta. He was the first person to produce moving electricity.

Voltage is used to describe the force or pressure in any part of an electrical circuit.

There are two main types of electrical voltage:

- the stored pressure, when everything is switched off
- the system pressure, when the circuit is switched on.

The stored pressure is known as **EMF** or **electromotive force**.

The pressure found in the circuit when it is switched on is known as **potential difference** (**Pd**).

Just as in the water diagram (Figure 8.8), when the tap is switched off the pressure gauge will read high (electromotive force).

When the tap is switched on and water can flow, the pressure on the gauge will fall slightly (potential difference)

Amp

Amps are the quantity of electricity. The **amp** was named after a Frenchman, André-Marie Ampère.

⚠ **Safe working**

Although voltage is electrical pressure and can lead to electric shock, it is the quantity or amps that are likely to kill.

Amps are used to describe the amount of electric current (moving electricity) in any part of an electrical system. If you compare them to the water diagram in Figure 8.8, for example, amps are the amount of water flowing down the pipe and coming out the end.

If you think of the water tank as the system's battery, its capacity, or the amount of electricity that it can hold, is measured in **amp hours** (**Ah**). Normally, the larger the battery, the more electricity it can hold.

Ohm

An **ohm** is the **resistance** to electrical flow. Ohms are named after a German mathematician called Georg Ohm.

Ohms can be compared to the water diagram in Figure 8.8. The tap at the end of the pipe can increase or reduce the flow of water if you change how far it is opened. If you open it a long way, a large amount of water can flow and this turns the water wheel fast. As the tap is slowly closed, the amount of water coming out of the end of the pipe is reduced. The water wheel turns slower until eventually it stops altogether. The tap is used to control the amount of water flowing in this circuit and it is similar to a 'dimmer switch' found on lighting circuits (a tap for electricity).

Watt

The **watt** is a measurement of electrical **power** made or used. It is named after a Scottish engineer called James Watt.

Compare watts to the water diagram in Figure 8.8. As the water wheel is turned, the energy given up by the water flowing in the circuit to turn the water wheel is measured in watts.

If you could turn the water wheel mechanically in the opposite direction, it would be able to scoop up water, forcing the energy backwards and generating current (this would be like the charging circuit). This generation of energy could also be measured in watts.

Ohm's law

If any one of the units within a circuit (volts, amps, ohms or watts) is changed (i.e. increased or decreased), this will affect all the other units. For example:

- If the voltage (or pressure) in the water system was increased, more water would flow and the amperage (or quantity) would also increase.
- If the resistance to flow was increased (if the tap was partially closed, for example) then less water would flow and the amperage (or quantity) would also decrease.

This was explained by Georg Ohm with the following mathematical calculations:

$$\text{amps} = \text{volts} \div \text{resistance}$$
$$\text{resistance} = \text{volts} \div \text{amps}$$
$$\text{volts} = \text{amps} \times \text{resistance}$$

Key terms

Volts – unit of electrical pressure.

Electromotive force (EMF) – measurement of electrical voltage with nothing switched on.

Potential difference (Pd) – the difference in electrical voltage when a circuit is switched on and energy is being used.

Amp – unit of electric current which describes quantity.

Amp hours (Ah) – the units used to measure the capacity of the battery.

Ohm – unit of electrical resistance.

Resistance – something that slows down movement.

Watt – unit of electrical power.

Power – the rate at which work is done.

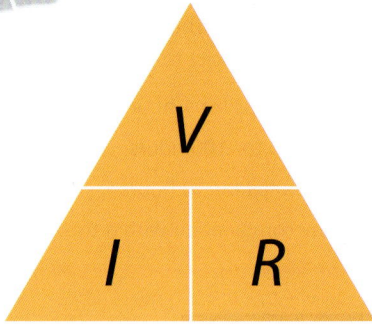

Figure 8.9 The Ohm's law triangle

With Ohm's law, if you know two of the electrical measurements, you can calculate the third.

The Ohm's law triangle is a good method for calculating the missing unit. It is laid out as shown in Figure 8.9.

In Figure 8.9:

- V = volts (this is sometimes shown as the letter 'E' to represent EMF, but still means volts)
- I = amps (the letter 'I' is used to represent instantaneous current flow)
- R = ohms (the letter 'R' is used for resistance because an 'O' could be confused for a zero).

How to use the triangle

Cover up the unknown unit with your thumb and you are left with the calculation required. For example, amperage is unknown, so cover the 'I' and you are left with V/R (i.e. volts divided by resistance).

Find out

Calculate the following examples.

- A car with a 12V battery is connected to a heated rear window element with a resistance of 0.5 ohms. How much current will flow?
- A starter motor with a resistance of 0.05 ohms is turning over very slowly. When checked it is drawing a current of 100 amps. How many volts are there in the battery?
- A headlamp bulb of a 12V car is drawing 8 amps. What is the resistance of the bulb?

The power triangle

Watts or power can be calculated in a similar way as:

amps = watts ÷ volts

volts = watts ÷ amps

watts = amps × volts

A power triangle can be used in the same way as Ohm's law. It is laid out as shown in Figure 8.10.

In Figure 8.10:

- P = power (in watts – this is sometimes shown as the letter 'W' to represent watts, but still means power)
- V = volts (this is sometimes shown as the letter 'E' to represent EMF, but still means volts)
- I = amps (the letter 'I' is used to represent instantaneous current flow).

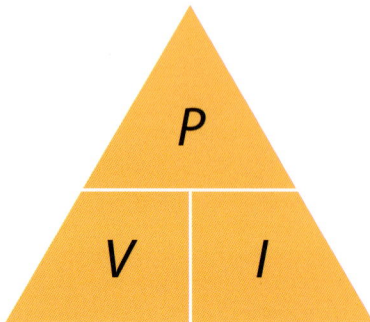

Figure 8.10 The power triangle

How to use the triangle

Cover up the unknown unit with your thumb, and you are left with the calculation required. For eample, amperage is unknown, so cover the 'I' and you are left with W/V (i.e. watts divided by volts).

Find out

Calculate the following examples.

- A 60 watt bulb is connected to a 12V supply. What is the current drawn? What size fuse would you use to protect the circuit? (Choose a fuse slightly larger than the current flowing in the circuit.)
- What is the power of a 12V starter motor, drawing a current of 50 amps?
- What is the voltage supply to a 21 watt bulb, which is drawing 1.5 amps?

Types of circuit – series and parallel

Series circuits

A **series** circuit is one where the **consumers** are connected in a line one after another. Because they are all in the same circuit, they share the electricity (like Christmas tree lights). Because of this, each consumer will only get part of the voltage available.

A series circuit is broken if any one of the consumers fails within the circuit. No electricity can flow and the rest of the consumers stop working (like Christmas tree lights).

Parallel circuits

A **parallel** circuit is one where the consumers are connected next to each other. Each consumer has its own power supply and **earth** route back to the battery. Because each consumer has its own power supply and earth, they will all receive the full voltage available, and work at full power. If one consumer in the circuit fails, then all the others keep working.

Key terms

Series – connected one after another.

Consumer – a component that uses electrical energy (bulbs, motors, etc.).

Parallel – connected side by side.

Earth – the negative part of an electrical circuit.

CHECK YOUR PROGRESS

1 List the four main electrical units of measurement.

2 How are components connected in a series circuit?

3 How are components connected in a parallel circuit?

The electrical systems

Battery system

Key terms

Cell – a device that delivers electrical current as a result of a chemical reaction.

Lead – used as the negative plate of a lead acid battery.

Lead peroxide – used as the positive plate in a lead acid battery.

Because vehicles are mobile, the electrical energy source must be portable (be able to move around with the vehicle).

Electrical energy can be carried in a chemical container known as a battery.

A standard lead acid battery contains a number of **cells**. Each cell is capable of producing approximately 2.1V. If you link six cells together, in series, you get a battery with a voltage of 12.6V (or rounded down to 12V, as it is more commonly known).

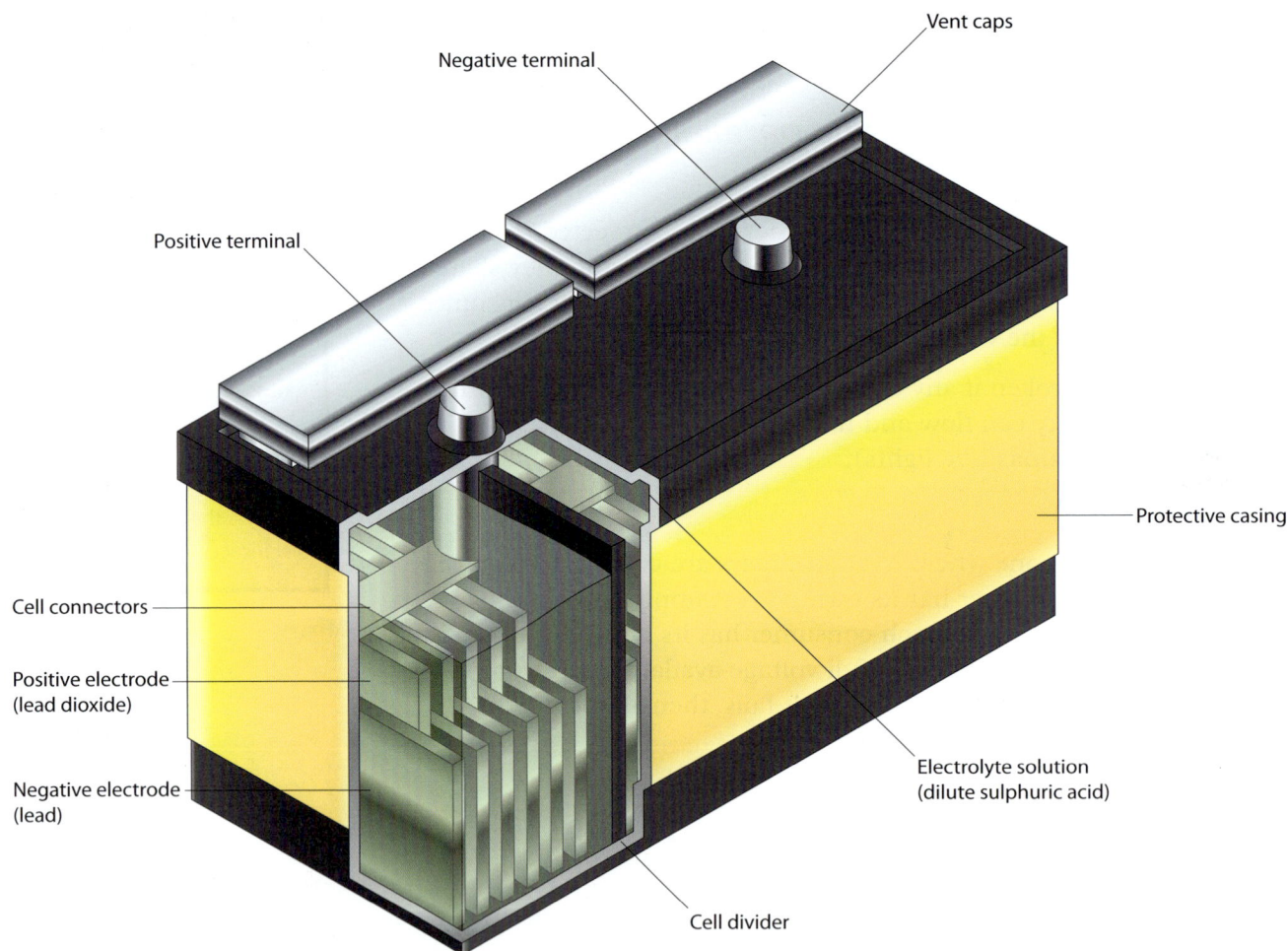

Vent caps

Negative terminal

Positive terminal

Protective casing

Cell connectors

Positive electrode (lead dioxide)

Negative electrode (lead)

Electrolyte solution (dilute sulphuric acid)

Cell divider

Figure 8.11 The components of a car's lead acid battery

Each cell contains a number of lead plates which are chemically different. The negative plate is made of **lead** and the positive plate is made of **lead peroxide**.

Each plate contains a different number of electrons. If connected in a circuit the positive plate would like to share the extra electrons with the negative plate. As the electrons move from plate to plate, electric current is made.

Making the circuit

The first half of the circuit is made with an **electrolyte**. This is a liquid made from **sulphuric acid** and **deionised water**. This electrolyte covers the plates and will allow electrons to move from one plate to another (just like swimming). The top of each plate is then connected to the rest of the circuit, which contains a consumer to use up the electrical energy. When the circuit is complete the electrons combine with the electrolyte and move from one plate to another as a chemical reaction.

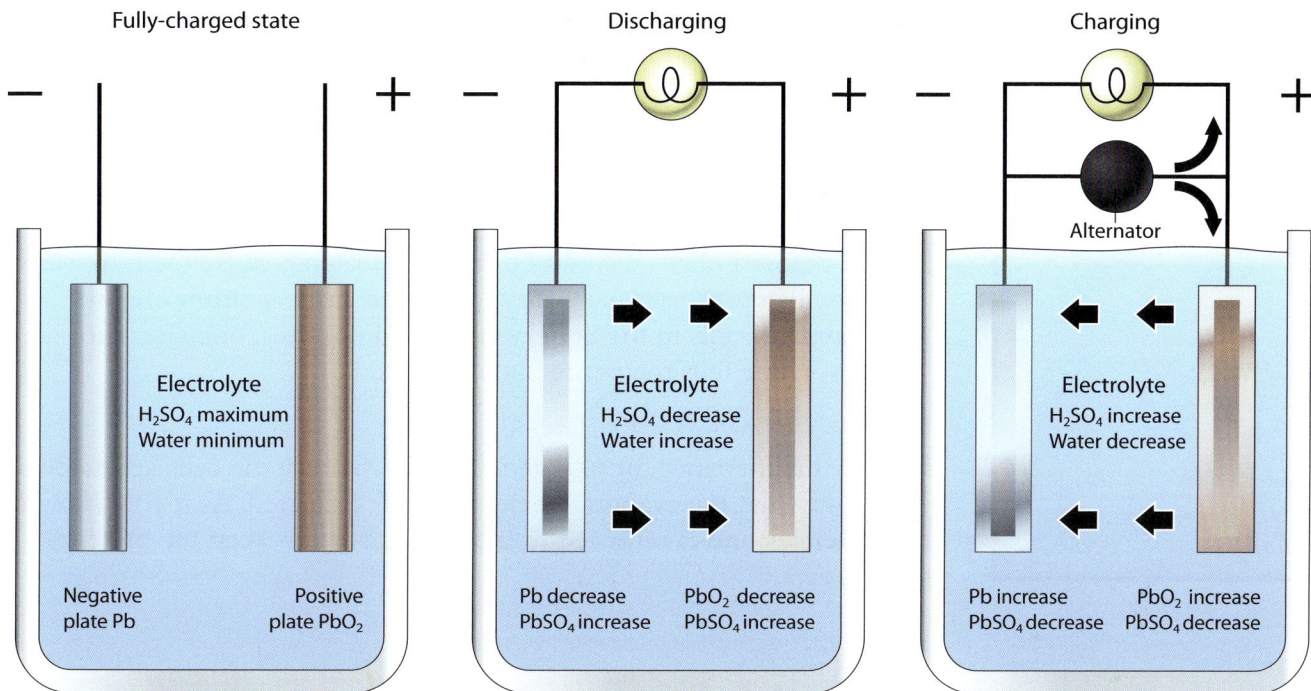

Fully-charged state	Discharging	Charging
− +	− +	− +
		Alternator
Electrolyte H_2SO_4 maximum Water minimum	Electrolyte H_2SO_4 decrease Water increase	Electrolyte H_2SO_4 increase Water decrease
Negative plate Pb Positive plate PbO_2	Pb decrease $PbSO_4$ increase PbO_2 decrease $PbSO_4$ increase	Pb increase $PbSO_4$ decrease PbO_2 increase $PbSO_4$ decrease

Figure 8.12 Chemical reactions inside a lead acid battery (Pb = lead, PbO_2 = lead peroxide, $PbSO_4$ = lead sulphate, H_2SO_4 = sulphuric acid)

The movement of electrons creates electric current (amps). As current moves through the system it gives up its energy to be changed into either a source of heat, magnetism or chemical reaction.

✚ Emergency

The electrolyte used in the lead acid battery is made up from deionised water and sulphuric acid. Sulphuric acid is highly **corrosive** and can cause severe burns if you splash it on your skin or in your eyes. You must always use personal protective equipment when handling electrolyte.

A battery uses a one-way circuit (direct current). As electrical current is created by a chemical reaction, both plates will eventually become the same chemically (known as **lead sulphate**) and the current will stop. When this happens no more electricity flows through the circuit and the battery is described as 'flat'.

Key terms

Electrolyte – a liquid that an electric current can pass through. Sulphuric acid and deionised water are used in a car's lead acid battery.

Sulphuric acid – used as an electrolyte base in a lead acid battery.

Deionised water – water with no electrical charge.

Corrosive – worn away by chemical reaction (can melt or burn).

Lead sulphate – a chemically depleted plate from a discharged battery.

Charging the battery

A pressure/voltage higher than that coming out of the battery can be created. This is done by adding an electricity pump known as a generator to the circuit. It forces electrons from one plate back to the other and recharges the battery. This process happens over and over, with the lead acid battery discharging and recharging.

Charging and discharging creates heat which will slowly lead to **evaporation** of the electrolyte. As evaporation takes place, the level of electrolyte within the cell will drop.

If the level of electrolyte is not kept above the plates, two things will happen:

1. The exposed part of the plates will not be able to be used during the charge and discharge process. This means that the capacity or amount of electricity in the cell that can be used is reduced.
2. The exposed plates will not be cooled. Heat can cause them to warp or bend so that eventually they touch and cause a **short circuit**. Because all cells in the battery are linked in series, one after the other, if one fails the entire battery is dead.

Maintenance free

In a low maintenance, or maintenance free, battery the cells are made in such a way that the evaporated electrolyte is collected. As it cools down it **condenses** and is returned to the cell. This helps keep the battery topped up.

Key terms

Evaporation – the change of state from liquid to vapour.

Short circuit – the electricity takes a short cut and bypasses the circuit consumer.

Condensation – the change of state of a substance from vapour to liquid.

Table 8.2 Battery components and their purposes

Battery component	Purpose
Plates	Lead plates containing different amounts of electrons. Used to create electric current
Electrolyte	Sulphuric acid and deionised water, used to create the chemical reaction inside the battery
Terminals	The positive and negative electrical connections on the battery

⚠️ **Safe working**

When a lead acid battery is charged, hydrogen sulphide gas is given off. This gas is extremely flammable. If ignited, it could cause the battery to explode. When charging a battery, remove all sources of ignition and work in a well-ventilated area.

Replacing batteries

Checklist			
PPE	**VPE**	**Tools and equipment**	**Source information**
• Steel toe-capped boots • Overalls • Latex gloves • Goggles	• Wing covers • Steering wheel covers • Seat covers • Foot mat covers	• Spanners • Screwdrivers • Socket set	• Technical data

1. Be aware of health and safety issues, including PPE and VPE.

2. Turn off all electrical circuits.

3. Record radio security codes, etc.

4. Remove the key from the ignition.

5. **Disconnect** the battery leads – **negative first, positive last**.

6. Remove battery hold down clamps and change the battery.

7. **Reconnect** the battery leads – **positive first, negative last**.

8. Reinstall electrical security codes.

9. Dispose of the old battery in accordance with environmental requirements.

Working life

Simon is disconnecting a battery. He has forgotten to remove his jewellery. As he is disconnecting one of the terminals on the battery, he accidentally touches a spanner to the ring on his hand, and against the battery terminal. Before he has time to react, a short circuit has been created. The heat from the electrical discharge melts the ring into his finger. When he arrives at hospital, the emergency department has to cut his finger open to remove the ring.

1   Can you think of a reason why Simon did not remove his ring before he started work?

2   Was this accident the fault of Simon or his employer?

Safe working

When you connect or disconnect the battery from a vehicle, the terminals should be removed and refitted in a certain order. This will reduce the possibility of a short circuit which can cause damage or injury.

Battery testing

When charging and discharging, the chemical make-up of the electrolyte changes as some of the sulphur in the acid is converted. This creates lead sulphate on the surface of each plate. As the battery discharges, the strength of the sulphuric acid falls and leaves more water behind. The strength of electrolyte can be measured using a **hydrometer**. A hydrometer is a tool that detects the **specific gravity** (**SG**) of a liquid. The specific gravity of a liquid is the density when compared with water.

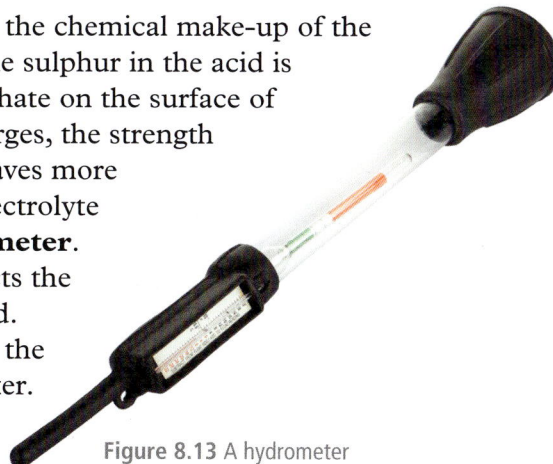

Figure 8.13 A hydrometer

Key terms

Hydrometer – a special tool designed to measure the density of a liquid.

Specific gravity (SG) – a measurement of the density of a liquid when compared to water.

A hydrometer uses a floating indicator which shows the specific gravity (or density) of the liquid being tested.

On the scale of specific gravity, water is used as a base and given a figure of 1. If the liquid is denser than water, then its specific gravity will be higher than 1.

- A discharged battery cell will have a specific gravity of around 1.060 (SG), which is almost water.
- A fully charged battery cell will have a specific gravity of around 1.280 (SG), which is denser than water.

Checking specific gravity

Checklist			
PPE	**VPE**	**Tools and equipment**	**Source information**
• Steel toe-capped boots • Overalls • Latex gloves • Goggles	• Wing covers • Steering wheel covers • Seat covers • Foot mat covers	• Hydrometer • Battery charger	• Technical data • Specific gravity chart

1. The battery must be at least 75% charged to get an accurate reading of specific gravity.

2. Undo the cell caps. Remember to observe all health and safety requirements, including PPE and VPE.

3. Insert the hydrometer and use it to suck up a sample of electrolyte.

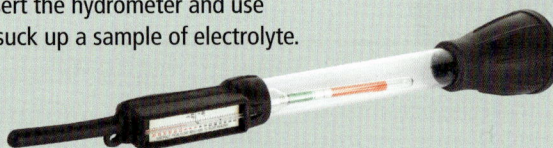

4. Take the specific gravity reading. Be careful to avoid splashes.

5. Return the electrolyte to the cell and repeat for the other cells.

6. Compare the readings recorded.

7. If you find a large difference in specific gravity between one cell and the others, the battery condition is poor and you should recommend that the battery is replaced.

Working life

A customer complains that their car is hard to start on cold mornings (the engine turns over slowly). Andy decides to check the battery with a hydrometer. After comparing the readings from each cell:

- four are found to have a specific gravity of 1.26
- one is found to have a specific gravity of 1.20
- one is found to have a specific gravity of 1.15.

1 Why has this happened?

2 What should Andy recommend to the customer?

The operation of the vehicle charging system

In many cars an **alternator** recharges the battery. As you learned earlier, if a magnet is moved in a coil of copper wire, or a coil of copper wire is moved within a magnetic field, an electric current is generated. The mechanical spinning of an engine can be used to spin a magnet or coil of copper wire and create an electricity pump.

An alternator produces alternating current (AC). This is electricity that moves in one direction then the other. This is because a magnetic field has North and South poles and as it spins electricity is created first in one direction and then the other.

Figure 8.14 An alternator

Find out

Use two magnets to explain the difference in polarity. What happens when you bring opposite poles together (North and South)? What happens when you bring the same poles together (North and North or South and South)?

Construction of an alternator

Rotor

The component that spins (rotates) in an alternator is usually referred to as a rotor. In many cases this is a combined **electromagnet** with a number of North and South poles. For an electromagnet to operate, it must be supplied with electric current from the battery. Because the rotor moves, a way of supplying electric current is needed. This is provided by a **slip ring** and **brush** assembly. This way, as the rotor spins, an electrical connection can be maintained.

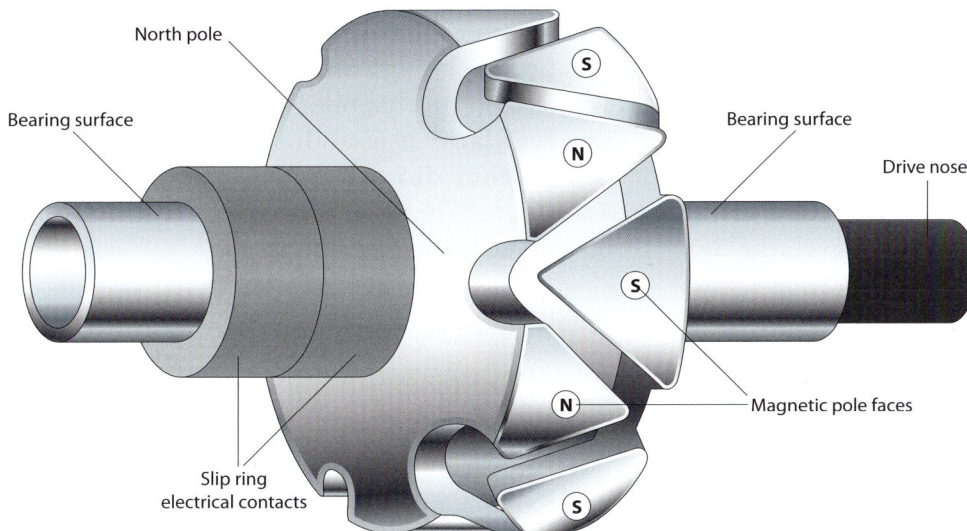

Key terms

Alternator – a modern type of electrical generator used to charge the car battery.

Electromagnet – a metal core made into a magnet by passing electric current through a surrounding coil.

Slip rings – a pair of copper tracks connected to the alternator's rotor. They help to provide a movable electrical connection.

Brush – pieces of carbon or metal ending in wires or strips. They are used to make electrical connections inside generators or motors.

Figure 8.15 The rotor of an alternator

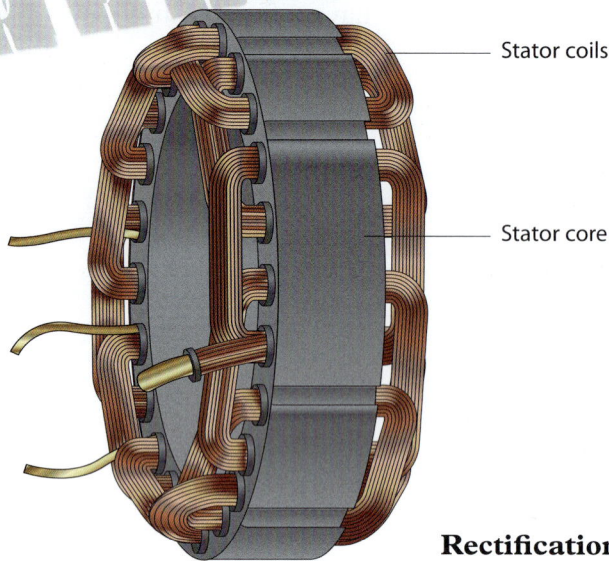

Figure 8.16 An alternator stator unit

Stator

The electromagnet needs to spin inside a coil of copper wire. This wire coil is normally stationary. It is known as the stator. An electrical current is created as the electromagnetic rotor spins in this coil of wire. It has a voltage high enough to charge the battery.

More than one coil of copper wire is used to increase the output of the alternator. In fact many systems use three. Each coil of copper wire is known as a **phase**, leading to the term **three phase**. Each phase will produce an alternating current, giving three times the output of a single phase generator.

Rectification

A battery cannot accept alternating current, and if left unchanged only half of the generated electricity could be used. A component known as a **rectifier** is used to change alternating current to direct current (AC to DC).

A series of **diodes** are connected to the output of the alternator phases. Diodes only allow electricity to flow in one direction. They can be arranged so that by the time it reaches the battery all the electricity has been converted into direct current.

Here is a simple example. Imagine a traffic policeman standing on a T-junction with cars coming at him from the left and from the right. The traffic policeman is able to redirect them so that they all travel down the road that is directly in front of him. It doesn't matter which way the cars arrive, they all leave down the same road.

Regulation

The faster the magnet spins, the greater the voltage or pressure produced. As the engine speeds up, a very high voltage can be created that would damage the battery. To stop this happening, an electrical pressure relief valve known as a **regulator** is used. Inside the regulator, an electronic component called a **Zener diode** is used to 'dump' excess electricity when voltage rises too high. When the voltage falls to the required level, the diode resets and charging continues.

Table 8.3 Alternator components and their purposes

Alternator component	Purpose
Rotor	An electromagnet turned by the vehicle engine.
Stator	A series of three coils of copper wire, where electric current is generated.
Rectifier	A set of diodes, used to convert AC to DC
Regulator	A component used to prevent electrical voltage going too high

Replacing an alternator

Checklist			
PPE	**VPE**	**Tools and equipment**	**Source information**
• Steel toe-capped boots • Overalls • Latex gloves • Goggles	• Wing covers • Steering wheel covers • Seat covers • Foot mat covers	• Spanners • Socket set • Lever bar • Belt tension gauge • Multimeter	• Technical data • Radio security codes

1. Be aware of health and safety issues, including PPE and VPE.

2. Turn off all electrical circuits.

3. Record radio security codes, etc.

4. Remove the key from the ignition.

5. Isolate the electrical system by disconnecting the battery negative lead.

6. Slacken the alternator adjustor and remove the auxiliary drive belt. Disconnect all electrical components and wiring from the alternator.

7. Unbolt the alternator mountings and replace the alternator. Reconnect all wiring to the correct alternator terminals.

8. Refit and adjust the auxiliary drive belt.

9. Reconnect the battery negative lead.

10. Reinstall the electrical security codes.

11. Start the engine and check the charge rate with a voltmeter.

Starting an engine

For the engine to start the crankshaft must be turned at speed, normally faster than 180 rpm. An electric motor can be used to do this.

Simple motors can be made by passing an electric current through a coiled wire. This is called a field coil. It is wound around a central shaft known as the **armature**. The electric current produces an invisible magnetic field. This is repelled or attracted by magnets which enclose it, causing the armature to turn. The crankshaft can be turned by attaching a gear to the end of the armature and connecting it with a gear on the engine.

You need a lot of energy to turn a starter motor. The small, lightweight switches, such as the starter/ignition switch that is found in the controls of a car, cannot cope with the amperage required. Because of this, heavy duty switches are needed to connect the starter motor to the battery.

Some starter motors include the heavy duty switch as part of their design. Figure 8.17 shows a pre-engaged starter motor with a **solenoid** attached. A solenoid can be used to act as a heavy duty switch. It is operated by the ignition switch and connects the main electrical wires from the battery to the starter motor. The large amounts of electrical current needed to operate the motor can now power the starter.

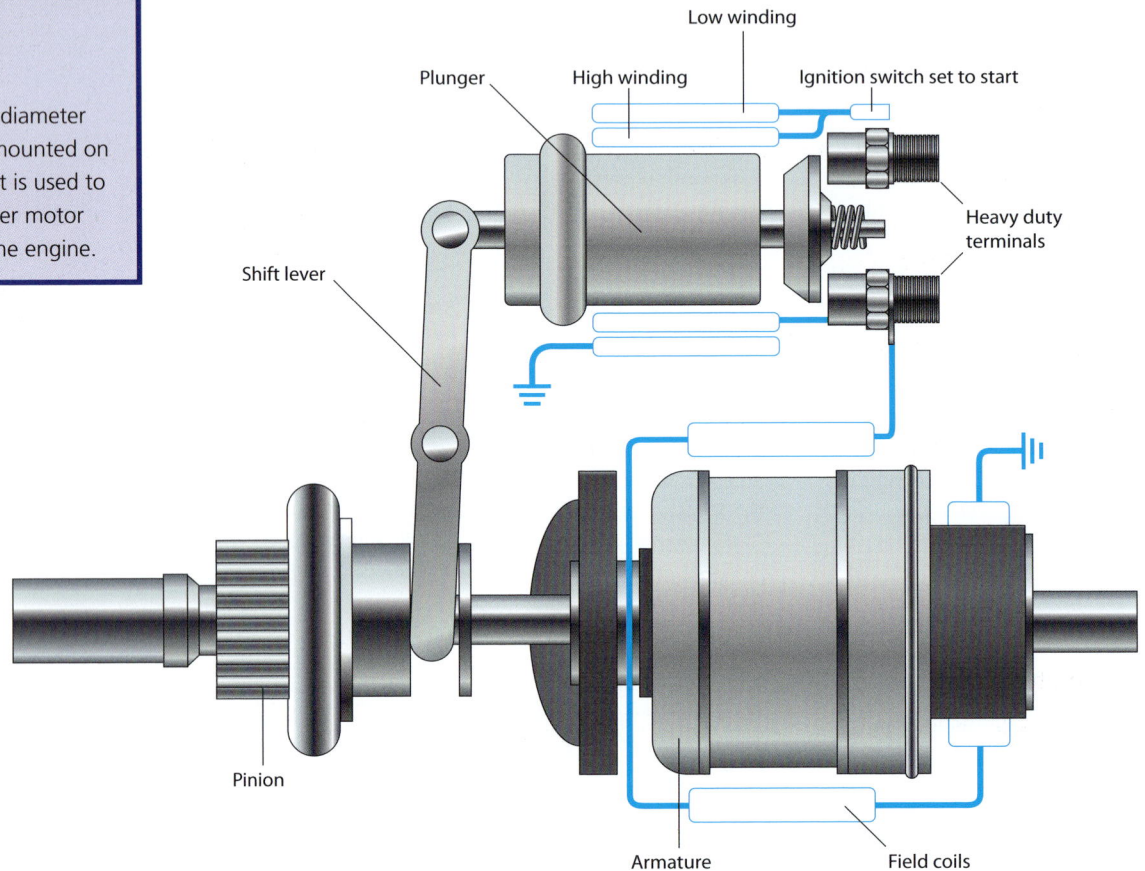

Figure 8.17 A pre-engaged starter motor including the solenoid and pinion

Figure 8.18 An inertia starter motor

Inertia starter motors

Some vehicles have an old-fashioned form of starter motor known as an inertia starter motor. It uses the difference in speed of the armature and the **pinion gear** to spin it along a **helix** known as a 'Bendix'.

- As voltage is applied to the armature of the starter motor, it turns so fast that it 'outruns' the speed of the pinion gear. This moves it along the helix and allows it to connect with the **ring gear** on the flywheel.

- When the engine starts, the speed of the flywheel is far quicker than the pinion. The gear is thrown back down the helix towards the armature of the starter motor (out of engagement).

Pre-engaged starter motors

Another type of starter motor is the pre-engaged type. It is more commonly used.

A pre-engaged starter motor can normally be recognised as it has a solenoid mounted on top. In this type of starter motor the solenoid has two functions:

1. It acts as the heavy duty switch to connect the power to the motor.
2. One end of the solenoid shaft is connected to a pivoting rod. The rod pushes the pinion gear along the armature shaft and into engagement with the ring gear on the flywheel.

When the starter button or key is released, a large spring pushes the solenoid back in the opposite direction. This disconnects the heavy duty power to the starter motor and pulls the pinion gear out of mesh from the flywheel.

Figure 8.19 A pre-engaged starter motor

Table 8.4 Starter motor components and their purposes

Starter motor component	Purpose
Armature	The rotating central shaft of a starter motor
Field coils	Creates the magnetic field that is used to turn the armature
Solenoid	Can act as a heavy duty switch. Sometimes used to move the pinion gear into engagement
Pinion gear	A small gear attached to the end of the armature

CHECK YOUR PROGRESS

1 How many volts does a single cell of a battery produce?
2 What are the two liquids used to make electrolyte?
3 In an alternator, what component is used to convert alternating current (AC) to direct current (DC)?

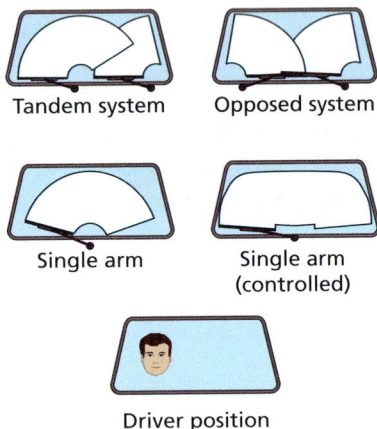

Tandem system

Opposed system

Single arm

Single arm (controlled)

Driver position

Figure 8.20 Windscreen wiper layouts

Auxiliary systems

Other electrical systems on cars, apart from starting, charging, ignition or engine management, are normally grouped under the term 'auxiliary systems'. These can include:

- lighting
- windscreen wipers
- windscreen heaters
- horn (see page 307–308)
- central locking
- electric windows.
- electric mirrors
- alarm
- immobiliser
- in-car entertainment
- satellite navigation

Windscreen wipers

Windscreen wipers are designed to give the driver a clear view of the road during rainy conditions. They are needed for safe driving. Wiper systems can be fitted to both front and rear windscreens. The front wipers are laid out in a number of different formats as shown in Figure 8.20. Two electrical motor systems are in common use.

Front windscreen wipers often use a standard rotating motor, with a crank arm fitted. As it rotates, the wipers are pulled backwards and forwards.

In rear wiper motor systems, it is common to reverse the **polarity** of electric current. In this way, the motor will change direction.

Windscreen wiper motors will need a system to 'park' the wipers (i.e. to stop them, when switched off, at the bottom of the screen). To do this, power to the motor continues even after the wiper switch has been turned off. When the wipers reach the bottom of the screen, an automatic system cuts the power.

Working life

Dave is an experienced technician. He has asked Kim, a trainee, to replace the windscreen wiper arms on a car. Dave has told Kim to turn on the ignition and wiper motor, switch off the motor and wait for it to stop, and then turn off the ignition before she fits the new wiper arms.

1   What might happen if Kim ignores Dave's advice?

Windscreen heater

During cold and damp weather, many windscreens mist up. This is due to condensed water moisture from the air. A screen demister is used to clear the windscreen so that the driver can see clearly.

A wire is placed in the windscreen and when it heats up it clears the glass. The heating elements of both the front and rear windscreens can use large quantities of current. A timer circuit is often used to reduce the possibility of damage caused by overheating.

Central door locking

Central locking uses solenoids or actuators in each door. When they are operated from a master control, they will automatically lock all of the other doors. This includes the boot and petrol flap.

As the key is turned in the master door lock, a **micro switch** will send a signal to the control unit. This will then operate the solenoids or actuators on all of the other doors.

All central locking solenoids or actuators will be connected in parallel. This way, if one fails, the others will continue to operate.

Lots of modern central locking systems operate by remote control. This is normally from the key fob. An advancement of this system is known as 'lazy lock'. In this system there is a main control unit. When it is operated, it not only shuts the central locking, but may also close electric windows, sunroofs and in some cases convertible hoods.

Alarm and immobiliser

A large proportion of all reported crime includes theft of, and theft from, motor vehicles. Many car manufacturers are now including anti-theft systems on their cars when they are built. This is in an attempt to reduce this crime statistic. However, a determined thief will always find a way around these systems.

- An active anti-theft system is one that attempts to deter theft in the first place, for example an alarm.
- A passive anti-theft system is one which uses an immobiliser, for example to stop the car being taken away.

A number of systems on a car can be immobilised including:

- fuel supply
- spark ignition
- starter motor system
- engine management electronic control unit (ECU) operation.

These systems are connected so that the electrical circuit is broken if someone tries to steal the vehicle. This makes it difficult for a potential thief to get the vehicle started.

In-car entertainment

In car entertainment, or ICE, is an electrical comfort/convenience system which may include:

- radio
- CD player
- DVD player
- television
- satellite navigation.

Common electrical components and symbols

Table 8.5 Common electrical components – their purposes and symbols

Symbol	Component
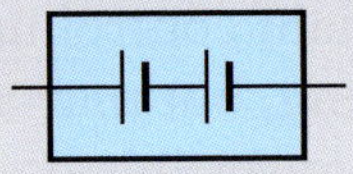	**Battery** – source of portable electrical power
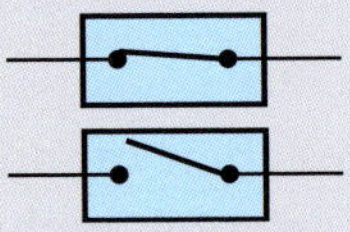	**Switch** – component that controls electrical current in a circuit
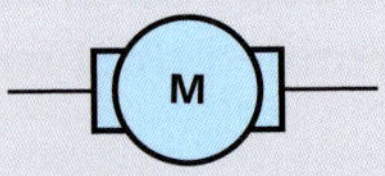	**Motor** – component that uses electromagnetism to provide mechanical movement
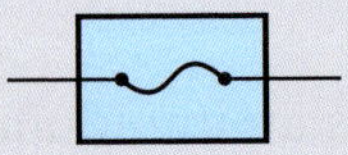	**Fuse** – component that protects an electric circuit from excess current
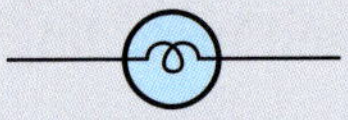	**Bulb** – component that converts electrical heat into light
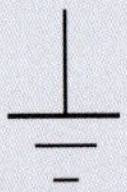	**Earth** – electrical ground connection designed to complete a circuit
	**Diode** – component that acts as a one-way valve for electricity
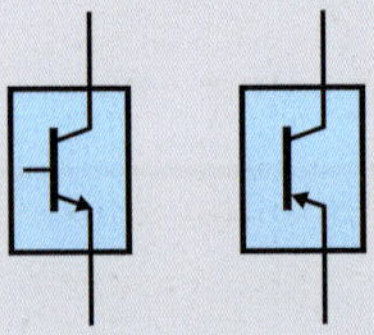	**Transistor** – component that acts as a switch with no moving parts
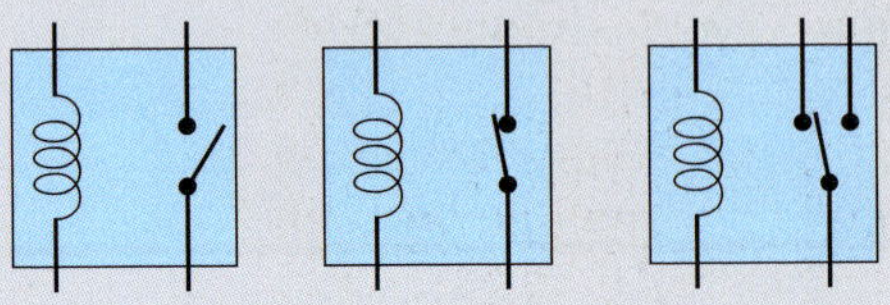	**Relay** – remote, electromagnetic switch. The three different types of electric relay symbol are M4, B4 and double throw.

CHECK YOUR PROGRESS

1 List three electrical auxiliary systems.
2 Name two methods that manufacturers use to try to reduce motor vehicle theft.
3 What do the letters ICE stand for?

Simple electrical circuits

An electric circuit is one in which electric current can flow in an unbroken loop.

A complete circuit will usually consist of:

- a source of energy, such as a battery
- connection, such as wires
- control switches
- a consumer; something to do useful work (motor, bulb, etc.)
- a protection device (fuse or circuit breaker).

Wires and cables

Wire colours

Because lots of wires are used in a car's electrical system, the external **insulated** coating is usually colour-coded. These colour codes can be shown on a wiring diagram. This will help you to identify the correct wires when diagnosing electrical circuit faults.

Wire sizes

Electrical wires come in different sizes. The fatter the wire the more electricity it can carry.

> **Key term**
>
> **Insulation** – covering material (as found on the outside of wires). It is designed to prevent the conduction of electricity.

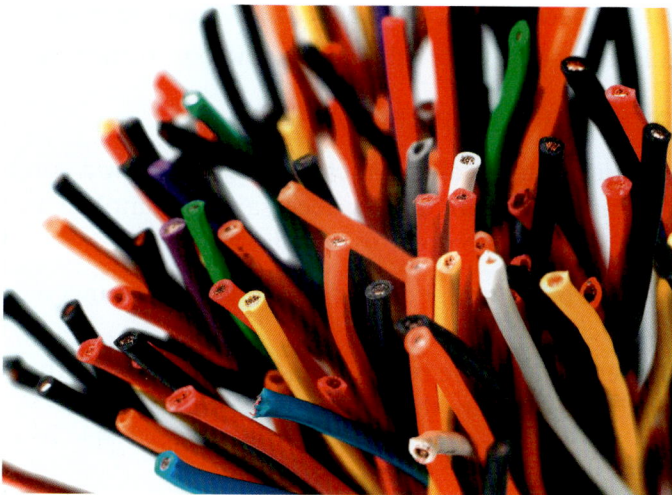

Figure 8.21 Electrical wires

Some typical wire size grades and uses are shown in Table 8.6.

Table 8.6 Typical wire size grades and their uses

Number of strands/ wire diameter	Continuous current rating/uses
9/0.30 mm	5.75 amps – side lamps, tail lamps, reversing lamps, horns
14/0.30 mm	8.75 amps – side lamps, tail lamps, reversing lamps, horns, general wiring
28/0.30 mm	17.5 amps – headlamps, fog/driving lamps, windscreen wiper motor
44/0.30 mm	27.5 amps – charging cable, battery feed
65/0.30 mm	35 amps – charging cable for alternator, dynamo
84/0.30 mm	42 amps – charging cable for alternator, dynamo
97/0.30 mm	50 amps – heavy duty alternator
120/0.30 mm	60 amps – heavy duty alternator
80/0.40 mm	70 amps – heavy duty alternator
37/0.71 mm	105 amps – starter/battery cable
37/0.90 mm	170 amps – starter/battery cable
61/0.90 mm	300 amps – starter/battery cable

You should choose wire for an electrical circuit so that:

- it is suitable for the amount of electricity to be carried
- its internal electrical resistance is not too great
- its overall size is not too large which will add to weight and cost.

Terminals and connectors

- An electrical connector joins two parts of a circuit together.
- An electrical terminal is where the circuit ends or 'terminates'.

Insulated and earth return systems

Most cars are made from metal and metals tend to be good conductors of electricity. Therefore it is not always necessary to complete an electrical circuit back to the battery using wire alone.

- The negative end of electrical circuit wiring can be connected to the vehicle body or chassis.
- The negative terminal of the battery can be connected to the vehicle body or chassis to complete the circuit.

This is known as an earth return.

Figure 8.22 A battery earthing point

Vehicles that may have an insulated return system include:

- fuel tankers
- ambulances
- emergency vehicles.

In these types of vehicles, the wiring of electric circuits terminates at an electrical connector known as a 'bus bar'. This is insulated from the body or chassis. A single insulated earth return wire can now connect the 'bus bar' back to the negative side of the battery.

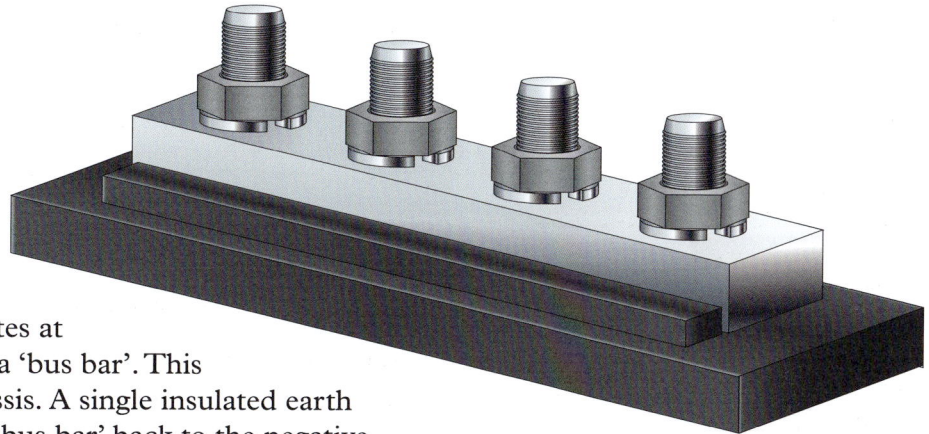

Figure 8.23 A bus bar

Fuses

If electrical current is allowed to flow in a circuit, and there is no device such as a bulb or a motor to do some useful work, then the electrical energy will be converted into heat. This can lead to rapid and expensive damage to the electrical circuit. For example, wiring can be burnt out. Fuses or circuit breakers are normally used to protect the circuit.

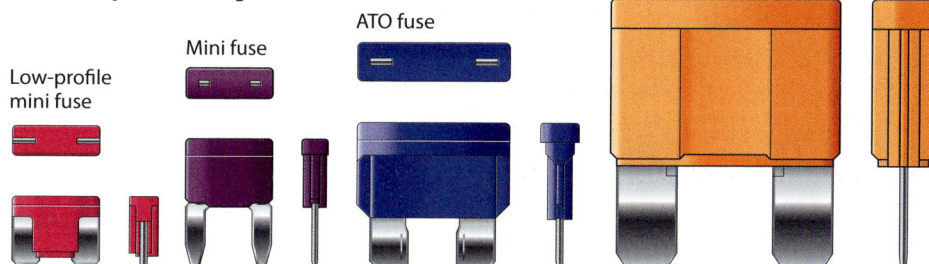

Maxi fuse

ATO fuse

Mini fuse

Low-profile mini fuse

Figure 8.24 Blade fuses in different sizes: mini, standard and maxi

outer casing of fuse

fuse wire

When the current is too high the wire gets hot....

....and melts.

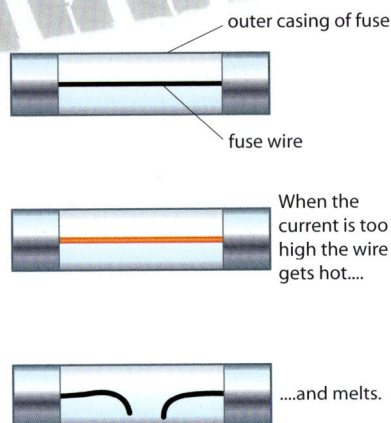

Figure 8.25 A high electrical current melting fuse wire

A fuse is a weak link placed in the circuit. It is designed to fail if a rapid increase in current flow occurs.

A thin piece of wire is fitted in line (series). The wire is rated in amps just above that of the current designed to flow in the system. This relatively cheap component is designed to burn out before any further damage can occur to the rest of the circuit. Once the fuse has burnt out an open circuit exists and no further current can flow (the electricity stops). In this way a fuse can save more expensive parts from damage.

Fuses come in different shapes and designs but are often colour-coded to show their **rating** in amps.

Key term

Rating – the value or size of a fuse.

Replacing electrical fuses

Checklist			
PPE	**VPE**	**Tools and equipment**	**Source information**
• Steel toe-capped boots • Overalls • Latex gloves • Goggles	• Wing covers • Steering wheel covers • Seat covers • Foot mat covers	• Multimeter	• Technical data • Wiring diagrams

Find out

Choose a vehicle and investigate fuse colour codes and ratings for the following circuits: headlamp, horn, heated rear window.

Safe working

When you remove or refit electrical components, be careful not to damage the component. You must also make sure that it is secure. A poor connection, for example, could cause a high resistance that might make the component break or even catch fire when switched on.

1. Be aware of health and safety regulations, including correct use of PPE and VPE.

2. Check for voltage at both ends of the fuses using a voltmeter.

3. Identify the blown fuse/inoperative circuit.

4. Identify the circuit and fuse rating from the technical data shown on the fuse box cover.

5. Switch off the circuit.

6. Replace the fuse with one of the correct rating.

7. Check the operation of the circuit.

Working life

Leroy is fixing a headlight on a customer's car. He finds that the 10 amp fuse designed to protect the circuit has blown. He replaces the fuse with one of the same type, which immediately blows. He wonders if he should use a fuse with a higher amperage instead.

1   Why is this not a good idea?

2   What further problems might this cause?

1 What is the purpose of a fuse?
2 What is the difference between a terminal and a connector?
3 Why are wires different colours?

Correct procedures to make circuits

When working on electric circuits, it is important that you choose the correct tools and use them safely.

Connections

- Electrical connections should always be of the best type and design. This is because a poor connection increases circuit resistance and reduces the overall efficiency.
- You should always connect wiring and components with the proper polarity (meaning correctly fitted to positive and negative).
- If you use electrical 'crimp' terminals, the correct size and type should be chosen according to the wiring and current designed for that circuit. They are often colour-coded. These colours correspond with markings on the crimping pliers so that a secure connection is made when operated.
- Wiring and components should be chosen that are suitable for the circuit. They should also be rated for voltage and amperage within the range of operation.

Multimeters

The **multimeter** is a piece of electrical test equipment designed to allow you to measure a number of different electrical units in an electrical circuit.

There are two types of multimeter:

- **analogue**
- **digital**.

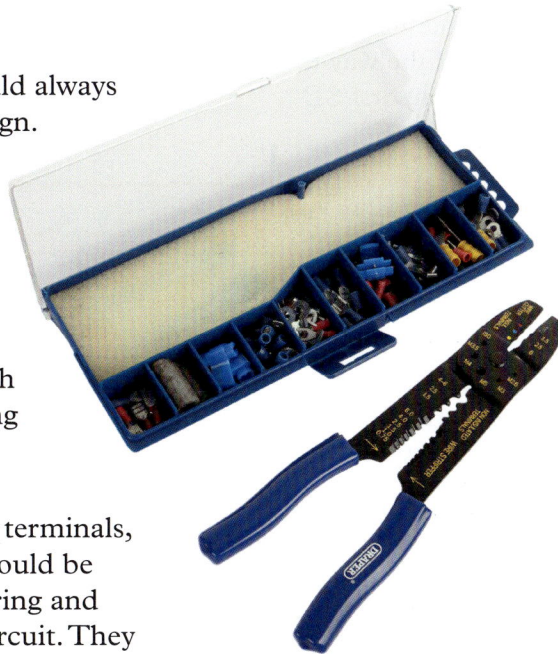
Figure 8.26 A crimping kit with crimping pliers

Safe working

There are various risks and hazards that you should assess when you are using electrical tooling including:

- sharp edges for cutting and probing
- hot work, such as soldering
- the risk of electric shock (use insulated tooling to reduce the risk).

Key terms

Multimeter – an electrical measuring tool with functions for volts, amps and ohms.

Analogue – an instrument readout that displays using a moving needle or similar method.

Digital – instruments which show numbers on a screen.

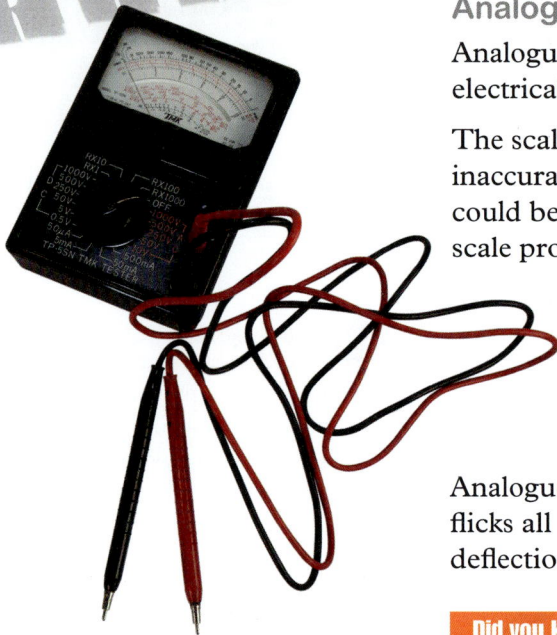

Figure 8.27 An analogue multimeter

Analogue

Analogue multimeters use a needle that moves across a scale to record electrical readings in a circuit.

The scale can be difficult to read and because of this you might take inaccurate readings. A needle that points somewhere between two units could be reading any fraction available. It will depend on the range of scale provided by the manufacturer.

Did you know?

The old-fashioned name for this type of unit was an 'AVO meter'. This stood for amps, volts and ohms.

Analogue multimeters also have an upper range limit. If the needle flicks all the way to the end of the scale, it is known as full-scale deflection, or FSD.

Did you know?

It is quite normal for the last digit (on the far right of the screen) to continuously change. This figure can often be ignored.

Digital

Digital multimeters display digits or numbers on a liquid crystal display screen (LCD). These numbers are clearly displayed and are easy to read accurately.

Two types of digital multimeter are common. These are either manually operated or **autoranging**.

Key term

Autoranging – a multimeter that automatically selects the scale of readings to be used.

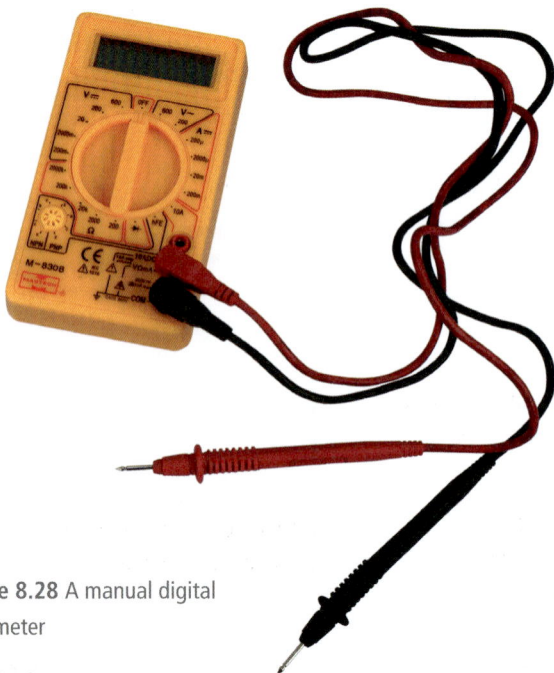

Figure 8.28 A manual digital multimeter

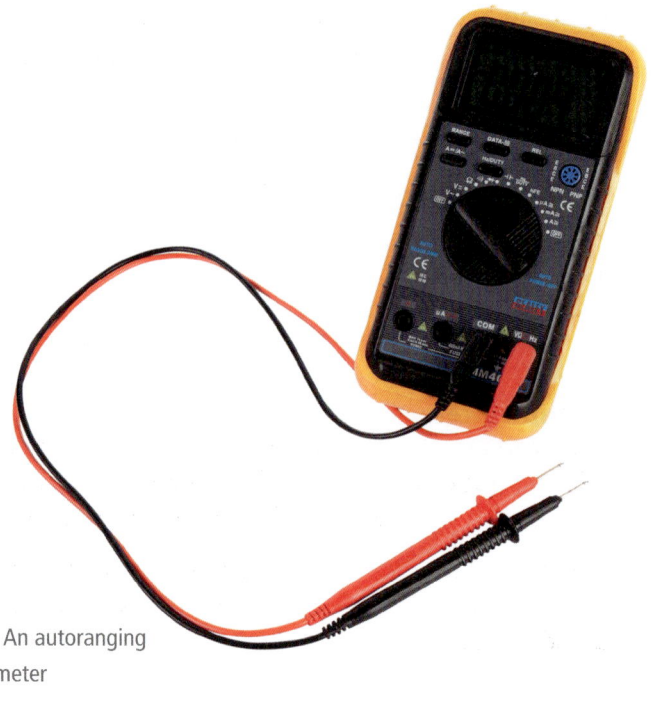

Figure 8.29 An autoranging digital multimeter

With a manual multimeter, you select the unit (volts, amps ohms) and the scale to be measured. You normally do this by turning a dial on the front of the unit.

With an autoranging multimeter, you select the unit (volts, amps ohms) but the scale is automatically selected by the multimeter.

> **! Safe working**
>
> When using an autoranging multimeter be sure that you know the size of the reading (scale) being measured (i.e. millivolts, volts, kilovolts and megavolts). This will be displayed on the screen but can be in very small writing.
>
> When using a manual multimeter, if you don't know the scale to be used, follow this procedure:
>
> - For the testing of volts and amps, the highest scale on the dial should be selected first. Then rotate the dial slowly down through the scales until an accurate reading is seen.
> - For the testing of ohms, the lowest scale on the dial should be selected first. Then rotate the dial slowly up through the scales until an accurate reading is seen.

> **⊘ Safe working**
>
> Never use a multimeter if the leads or meter are damaged.

The electrical units of volts and amps are often broken down into two further areas – direct current and alternating current.

- The direct current scale is normally shown on the meter as a straight line with a number of dots underneath it. $\overline{}\ \overline{}\ \overline{}$ This symbol is designed to prevent confusion. If just a single line was used it might be mistaken for a minus and if two lines were used it might be mistaken for an equals sign.
- The alternating current scale is normally shown on the meter as a wavy line. $\sim$

Voltages higher than 60V DC and 30V AC can cause electric shock. Always make sure that you are holding onto the insulated plastic part of the test probes when carrying out electrical measurement.

Make sure that the test leads are correctly connected on the multimeter and that the selector dial is in the correct position.

If the low battery symbol appears on the screen of a digital multimeter, replace the battery straight away.

> **Did you know?**
>
> Under normal circumstances electricity is invisible and has no taste or smell. Because of this, the risk of an accident is high. If electricity is released (discharged) by mistake it can have serious consequences.
>
> Large amounts of electricity (amperage) can cause serious burns. High electrical pressures (voltage) may affect human organ operation, such as heart and brain function.

Checking voltage, amperage and resistance

Voltage

When you are checking for electrical voltage in a circuit, make sure the probes are connected with the black in the common or ground socket. The red lead should be in the socket indicating voltage.

A voltmeter will measure the pressure difference in an electric circuit between where you put the black probe and where you put the red probe.

> **! Safe working**
>
> You must always connect a voltmeter in parallel.

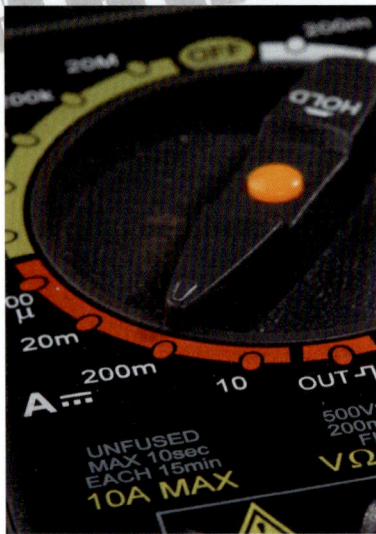

Figure 8.30 An ammeter

For example, if a battery was being tested, the black probe would normally be put on the negative terminal and the red probe would be placed on the positive terminal. The difference in electrical pressure between these two points can now be measured. (For example, a fully charged battery would show a reading on the voltmeter of approximately 12.6V.)

> **Did you know?**
>
> Most voltage that is going to be measured on a car is direct current, so you should choose the scale with the straight and dotted lines.

Amperage

When you are measuring the electrical current in a circuit, you should use an ammeter.

You need to take care when using an ammeter because if it is connected wrongly, it can be damaged.

Using an ammeter

Checklist			
PPE	**VPE**	**Tools and equipment**	**Source information**
• Steel toe-capped boots • Overalls • Latex gloves • Goggles	• Wing covers • Steering wheel covers • Seat covers • Foot mat covers	• Multimeter set to measure amps	• Technical data

1. Connect the probe leads so that the black lead is in the common or ground socket and the red lead is connected to the socket used for measuring amps. This socket is normally separate from the one that would be used to measure volts or ohms.

2. The selector dial should be turned to amps measurement.

3. You must connect an ammeter in series to the electrical circuit being tested. This means breaking into and using the ammeter as part of the circuit. This can be done by disconnecting a wire and putting the black probe on one end of the connector and the red probe on the other, so that electric current will flow through the meter completing the circuit. (A good place to connect an ammeter is at the fuse box. Remove the fuse completely and replace it with the ammeter.)

4. If the ammeter is connected correctly, current will flow and the circuit will operate normally.

5. The reading can now be taken from the screen.

> **Safe working**
>
> An ammeter must be connected in series. Never connect an ammeter in parallel (across a circuit). A good ammeter has a very low internal resistance. As a result, if the ammeter is connected in parallel, excessive current will flow and the ammeter will be damaged.

Resistance

When checking for electrical resistance, the power to the circuit should always be switched off first. The component being tested must also be disconnected from the circuit.

Using an ohmmeter

Checklist			
PPE	VPE	Tools and equipment	Source information
• Steel toe-capped boots • Overalls • Latex gloves • Goggles	• Wing covers • Steering wheel covers • Seat covers • Foot mat covers	• Multimeter set to measure ohms	• Technical data

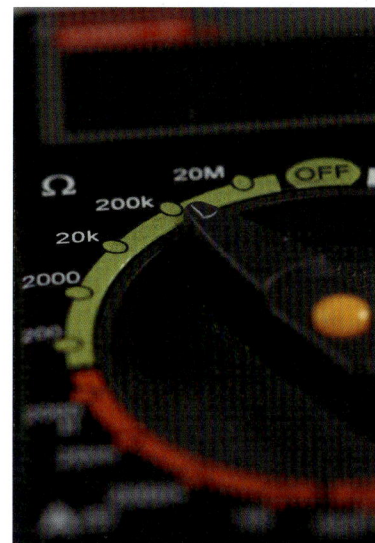

Figure 8.31 An ohmmeter

1. You should connect the probe leads so that the black lead is in the common or ground socket and the red lead is connected to the socket marked with the omega symbol (Ω).

2. The selector dial should be turned to the lowest ohms setting and the tips of the probes connected together, so that the meter can be calibrated.

3. When the leads are connected the readout should show zero or as near to zero as possible.

4. Once calibrated the component may now be tested for resistance by connecting it in parallel to the ohmmeter. A bulb, for example, would be connected with the red and black probes across the two terminals and the resistance can be read from the screen.

Safe working

When checking resistance, the circuit should be switched off and the component disconnected.

Read and interpret simple wiring diagrams

Electric circuits are designed using wiring diagrams. To help with the design and reading of wiring diagrams, many manufacturers use pictures or symbols to represent electrical system components. Examples of some common electrical and electronic symbols can be found on page 298.

The following are simple examples of basic wiring diagrams.

Horn

In this simple horn circuit (see Figure 8.32), power supply from the battery goes through a fuse and then on to the horn unit. To complete the circuit, it is then connected to a horn switch. This can be mounted on the steering wheel or on one of the multifunction (headlamp, indicator, etc.) stalks behind the steering wheel. This switch is then connected to earth to complete the circuit.

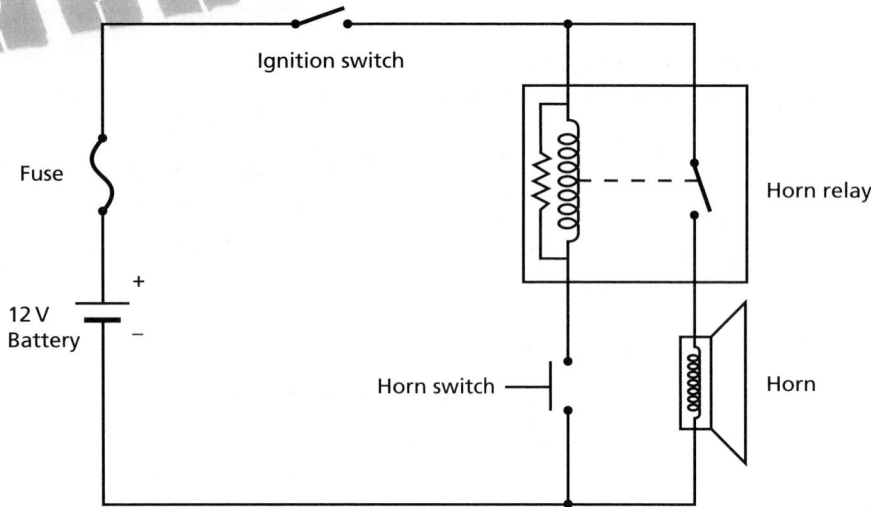

Figure 8.32 A basic horn wiring circuit

The horn switch is usually in the off position (open circuit) until it is required. When operated, the switch is closed and current can flow from the battery. It flows through the fuse, onto the horn unit (where the energy is converted into sound), through the switch and down to earth, where it returns to the battery completing the circuit.

Brake lights

In the simple brake light circuit shown in Figure 8.33, power supply from the battery goes through a fuse and then onto the brake light switch. This is usually mounted at the brake pedal. After the brake light switch, the electric current is split into a parallel circuit, to feed each brake light bulb individually. Once the current has passed through the filament in the bulb, it then travels to an earth connection, completing the circuit.

The brake light switch is usually in the off position (open circuit) until it is required. When operated, the brake light switch is closed, and current can flow from the battery. It flows through the fuse, through the brake light switch and on to the light bulbs, which convert the electrical energy into heat making them glow.

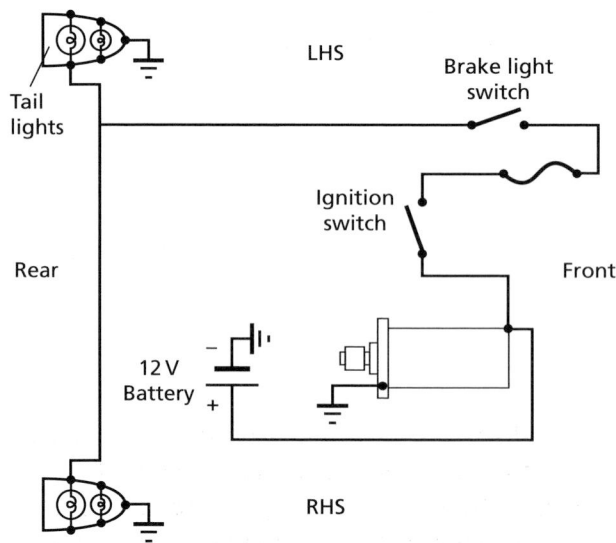

Figure 8.33 A rear brake light circuit

Headlights

The headlamp circuit shown in Figure 8.34 is more complicated in design. A lighting circuit of this type will normally also operate tail lights.

The power supply from the battery is usually split through two fuses. In this way the left- and right-hand side lighting circuits are separate in case one should fail. (If all the lights were connected through one fuse, and that fuse blew, all the lights would go out and this would be dangerous.)

After the fuse the circuit is connected to the main lighting control switch. Each bulb will receive its own feed, and because of this, if one bulb fails, the others keep working.

The headlamp bulbs are normally a twin **filament** type. Each filament has a different power rating and glows with a different brightness (main beam is usually brighter than dip beam). A dip/main beam relay is often used to control the power switching between these two filaments.

The output side of all bulbs in this headlamp lighting circuit are connected to earth to complete the circuit.

> **Key term**
>
> **Filament** – a very thin piece of wire which gets hot and glows when electric current passes through it.

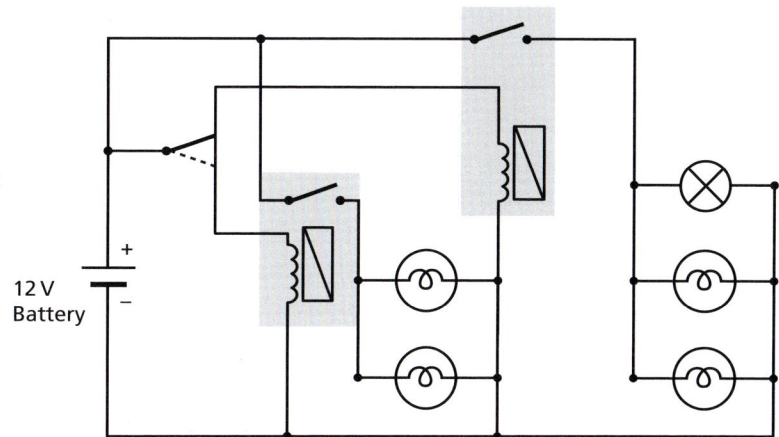

Figure 8.34 A simple headlamp wiring circuit

CHECK YOUR PROGRESS

1 Name two types of multimeter.
2 What do the letters AVO mean on an old-fashioned multimeter?
3 What do the following symbols mean when found on a multimeter?
   a = – – =
   b ~~

Vehicle lighting system components and operation

Types of bulbs

Car electric light bulbs are usually made of a thin **tungsten** wire known as a filament. Electric current is then forced through the filament. The resistance of the wire causes the electrical energy to be converted into heat. This makes the filament glow. When the filament reaches a temperature of 2300°C, it gives off white light.

> **Key term**
>
> **Tungsten** – a heavy grey metallic element.

If the filament was exposed to oxygen, this extreme heat would be enough to burn the wire away. Therefore it is normally enclosed in a glass bulb with all of the air removed.

The main types of bulbs used on modern cars include:

- tungsten filament
- quartz-halogen
- high intensity discharge (xenon)
- light emitting diodes (LED).

Some bulbs may contain an **inert gas**, such as argon under pressure. This will allow the bulb to run slightly hotter and produce a brighter light.

Quartz-halogen headlight bulbs run very hot. They are normally enclosed in a bulb made from the rock crystal quartz rather than glass. This is because quartz can withstand much higher temperatures. They are pressurised with a gas made from **halogens** (non-metallic elements, such as iodine). This means they glow much brighter than a standard bulb.

High intensity discharge (HID) lights are a new type of headlamp being used on modern cars. They contain no electrical filament to get hot and glow. Instead they use an electrical spark to react with a gas, usually **xenon**, to produce light. An igniter unit similar to an ignition coil creates the spark. Then a **ballast** unit takes over to maintain the spark and keep the gas glowing.

Light emitting diodes (LED) use the movement of electrons through a diode to create light. This type of bulb uses a **semi-conductor** material, so that when current is passed through it, light is given out. This bulb has no filament to get hot. As it uses the movement of electrons to create light, its illumination is extremely quick.

Light emitting diodes are very small so in order to produce the same size of light, LEDs are often grouped together in clusters.

Lighting regulations

A number of lighting systems are mandatory (each vehicle must have them) in the UK including:

- side and rear lamps
- headlamps
- stop lamps
- rear fog lamps
- indicators
- hazard lamps.

Key terms

Inert gas – a non-reactive gas element.

Halogen – a description of a series of gases such as argon.

Xenon – a gas used in some discharge type light bulbs.

Ballast – a resistor used in HID lighting circuits to control electrical output.

Semi-conductor – a material that has the properties of both a conductor and an insulator.

Find out

Investigate and find images of the following headlamp types:

- low beam units/ high beam units
- combined units
- European lens
- American lens
- projector type
- HID (xenon).

Find out

Investigate a car in your workshop and identify all lights and note their colour.

The lighting system regulations also state the colour of light that is to be used on cars. The colour of lighting has been specifically designed. Common colours include:

- white light to show that the vehicle is coming towards you
- red light to show that the vehicle is travelling away from you
- amber light to show a change in direction.

Table 8.7 shows the colours of lamps that are required on roads in the UK.

Table 8.7 Light types and colours

Light type	Colour
Headlamp	White light (may sometimes be yellow but this colour must not affect the operation of any other lamp)
Front side lamp	White light (may sometimes be yellow if built into the yellow headlamp lighting system)
Direction indicator	Directional indicators are used to help other road users know which way the car is intending to move. They are normally amber in colour
Brake light	Red (and they must not affect the operation of any other lamp)
Rear lamp	Red (and sometimes these rear lamps will incorporate reflectors, so that a red reflection can be seen if illuminated by headlights, even when switched off)
Reverse lamps	White light (remember, white light helps show direction of travel)
Fog lamps	White or yellow light (a yellow light is better at penetrating fog, while white light can produce a reflective glare)

Replacing a rear light

Checklist			
PPE	**VPE**	**Tools and equipment**	**Source information**
• Steel toe-capped boots • Overalls • Latex gloves • Goggles	• Wing covers • Steering wheel covers • Seat covers • Foot mat covers	• Screwdrivers • Spanners • Socket set	• Technical data • Workshop manual

1. Be aware of health and safety issues, including PPE and VPE.	2. Turn off the switch to the light circuit.	3. Remove the key from the ignition and inform others of the work being carried out.	4. Disconnect the bulbs from the lamp unit.
5. Following the manufacturer's instructions, remove and replace the rear lamp unit.	6. Reconnect the bulbs.	7. Test the operation of all lights.	

Bulb

Filament above
focal point

Dip beam

Parabolic
reflector

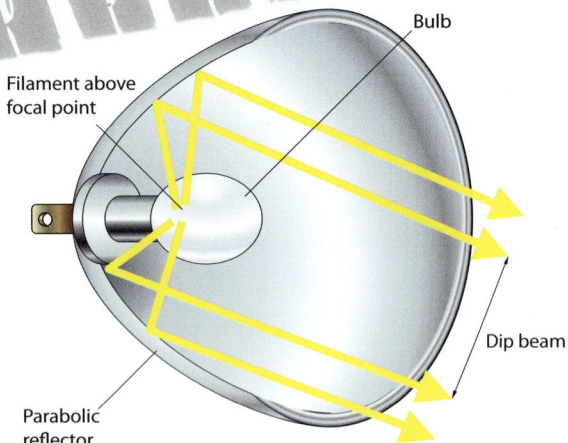

Figure 8.35 A pre-focus bulb and a reflector can direct light in a concentrated beam

Headlamps

Headlamps use a bulb that is usually called **pre-focus**. This type of bulb uses a curved reflecting lens to direct light forward in a concentrated beam. This gives vision in the dark (it illuminates where you point it like a torch).

This beam is then directed through a shaped lens covering the front of the lamp unit. **Prisms** moulded into this lens can further bend the light to achieve a precisely directed headlamp beam pattern.

Indicators

Direction indictors and hazard warning lights use a flasher unit to turn the bulbs on and off.

In a **bimetallic strip** type flasher unit, the electrical contacts are mounted on two strips of different metal. The indicator bulbs light up when they are switched on. Electrical current heats up the bimetallic strip, which bends and breaks the circuit. As soon as the circuit is broken, the bulb goes out. The bimetallic strip cools and reconnects the circuit, so starting the whole process over again.

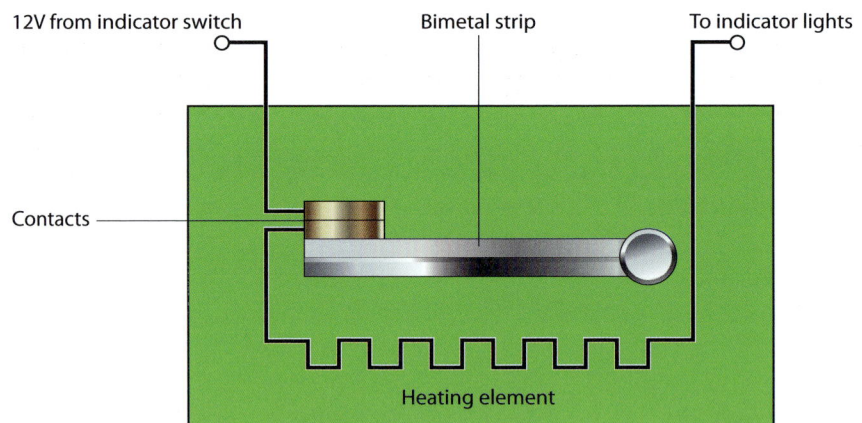

Key terms

Pre-focus – a bulb mounted in front of a reflector to create a beam of light.

Prism – shaped transparent glass which is designed to bend light.

Bimetallic strip – two different strips of metal that will expand at different rates when heated.

Did you know?

Indicators must flash at the rate of between 60 and 120 flashes per minute.

12V from indicator switch

Bimetal strip

To indicator lights

Contacts

Heating element

Figure 8.36 A bimetallic strip indicator flasher unit

CHECK YOUR PROGRESS

1 List three different types of light used on a car.
2 How fast should indicators flash?
3 Why do you think lights have different colours?

Electrical component replacement

Whenever you replace an electrical component on a car, you must make sure that:

- the component is switched off
- the electrical supply is isolated (disconnected) where possible
- the component is cool.

Remember that, during use, some of the energy of an electrical component will be turned into heat.

Replacing a halogen headlamp bulb

Checklist			
PPE	**VPE**	**Tools and equipment**	**Source information**
• Steel toe-capped boots • Overalls • Latex gloves • Goggles	• Wing covers • Steering wheel covers • Seat covers • Foot mat covers	• Screwdrivers	• Technical data • Bulb wattage requirements

1. Switch off the headlamps and allow them to cool. Remember to observe all health and safety requirements, including PPE and VPE.

2. Open the bonnet and remove the bulb cover on the back of the headlamp.

3. Remove the wiring connector and unclip and remove the bulb.

4. Be careful not to touch the quartz globe when you replace a quartz-halogen headlamp bulb. The natural oils found on the skin of your hands will contaminate the quartz. When switched on, the extremely high temperatures created will boil these oils. This will damage the bulb and lead to it failing sooner than it should. Instead, you should hold the bulb by the metal base.

5. When you fit the bulb into the headlamp unit, it is important that you mount the locating lugs in the correct position.

6. Refit all covers and connections.

7. Switch on the lamps and test operation and connections.

Replacing a headlamp unit

Checklist			
PPE	**VPE**	**Tools and equipment**	**Source information**
• Steel toe-capped boots • Overalls • Latex gloves • Goggles	• Wing covers • Steering wheel covers • Seat covers • Foot mat covers	• Screwdrivers • Spanners • Socket set	• Technical data • Headlamp alignment instructions

1. Switch off the headlamps and allow them to cool. Remember to observe all health and safety requirements, including PPE and VPE.

2. Follow the manufacturer's instructions for correct removal as it is easy to damage clips and mounting positions.

3. After you have safely removed the headlamp unit, you must use the correct type of headlamp to replace the original. If you don't, the type of light emitted, the beam pattern and the direction of beam may be affected.

4. Once fitted, check the headlamp for alignment to make sure that it meets the UK regulations.

Headlamp alignment

Checklist			
PPE	VPE	Tools and equipment	Source information
• Steel toe-capped boots • Overalls • Latex gloves • Goggles	• Wing covers • Steering wheel covers • Seat covers • Foot mat covers	• Screwdrivers • Spanners • Socket set • Headlamp alignment tool	• Technical data • Headlamp alignment instructions

1. Remember to observe all health and safety, including PPE and VPE.

2. Before you make any headlamp alignment checks, it is important that certain conditions are met (see the 'Did you know?' box below).

3. You will need to set up a headlamp aligning tool in accordance with the manufacturer's instructions for the headlamp to be tested and set.

4. When switched on, a beam pattern will be cast on a scale in the alignment tool. This will depend on the headlamp type and operation.

5. The beam pattern must be appropriate for the headlamp type. It must also fall within scale **tolerances** laid down by lighting regulations.

6. Adjusters are normally included in the design of the headlamp unit so that corrections can be made if headlamp alignment is incorrect.

Examples of headlamp patterns and styles are shown in Figure 8.37.

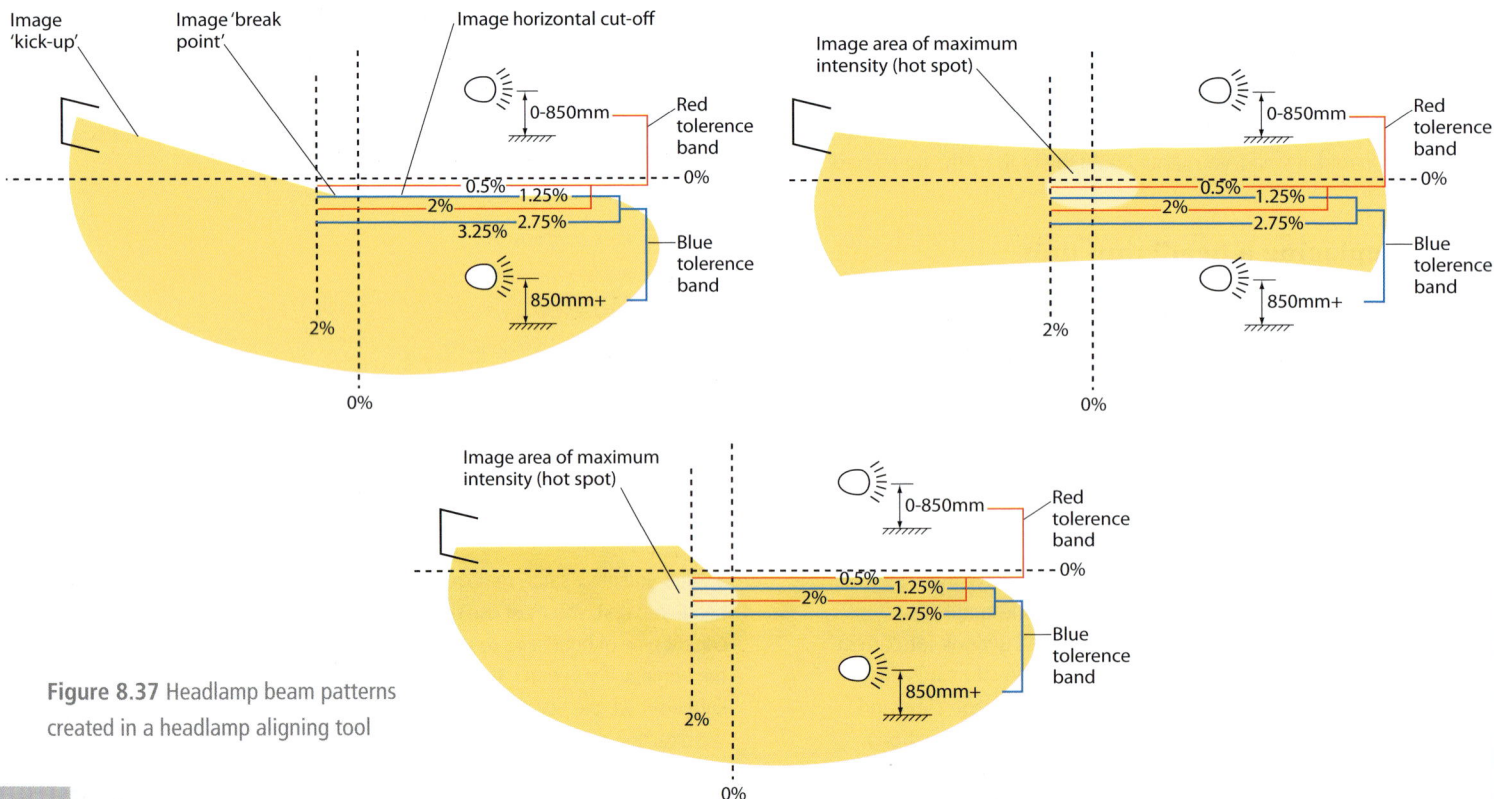

Figure 8.37 Headlamp beam patterns created in a headlamp aligning tool

FINAL CHECK

1 A hydrometer can be used to check a battery's:

   a state of charge
   b electrolyte level
   c voltage
   d capacity

2 When compared to a standard filament bulb, LEDs have the advantage of:

   a being cheaper
   b lighting up quicker
   c having a brighter filament
   d being bigger

3 A circuit passing 12 volts through a resistance of 3 ohms will produce a current of:

   a 40 watts
   b 40 amps
   c 4 amps
   d 4 watts

4 Voltage can be described as:

   a electrical pressure
   b electrical capacity
   c electrical current
   d electrical power

5 A battery converts:

   a mechanical power into electrical power
   b chemical power into mechanical power
   c electrical power into mechanical power
   d chemical energy into electrical energy

6 An alternator converts:

   a mechanical energy into electrical energy
   b chemical power into mechanical power
   c electrical power into mechanical power
   d chemical energy into electrical energy

7 Indicators should flash approximately:

   a 90 times a second
   b 90 times a minute
   c 180 times a minute
   d once every 2 seconds

8 To measure electrical resistance a digital multimeter should be set to:

   a voltage AC
   b ohms
   c voltage DC
   d continuity

9 A starter motor converts:

   a mechanical energy into electrical power
   b chemical power into mechanical power
   c electrical energy into mechanical energy
   d chemical power into electrical energy

10 What is a typical specific gravity reading for a battery in good condition?

   a 1.100
   b 1.150
   c 1.200
   d 1.280

GETTING READY FOR ASSESSMENT

The information contained in this chapter, as well as continued practical assignments in your centre or workplace, will help you to prepare for both the end-of-unit tests and diploma multiple-choice tests. This chapter will also help to prepare you for working on light vehicle electrical and lighting systems safely.

You will need to be familiar with:

- Electrical safety precautions
- The electrical units – volts, amps, ohms and watts
- How to calculate Ohm's law
- Types of electrical: power supply, wiring, consumer device, switching and circuit protection
- Common electrical symbols
- Electrical wiring diagrams
- The basic use of multimeters
- How to replace electrical and lighting components

This chapter has given you an introduction and overview to light vehicle electrical and lighting systems. It has provided you with the basic knowledge that will help you with both theory and practical assessments.

Before trying a theory end-of-unit test or diploma multiple-choice test, make sure you have reviewed and revised any key terms that relate to the topics in that unit. Be sure to read all questions fully and take time to digest the information so that you are confident about what the question is asking you. With multiple-choice tests, it is very important that you read all of the answers carefully, as it is common for the answers to be very similar and this may lead to confusion.

For practical assessments, it is important that you have had sufficient practice and that you feel that you are capable of passing. It is best to have a plan of action and work method that will help you. Make sure that you have sufficient technical information, in the way of vehicle data, and appropriate tools and equipment. It is also wise to check your work at regular intervals. This will help you to be sure that you are working correctly and to avoid problems developing as you work.

When undertaking any practical assessment, always make sure that you are working safely throughout the test. Ensure that all health and safety requirements are observed and that you use the recommended personal protective equipment (PPE) and vehicle protection equipment (VPE) at all times. When using tools, make sure you are using them correctly and safely.

Good luck!

Index

Key terms are indicated by **bold type**.

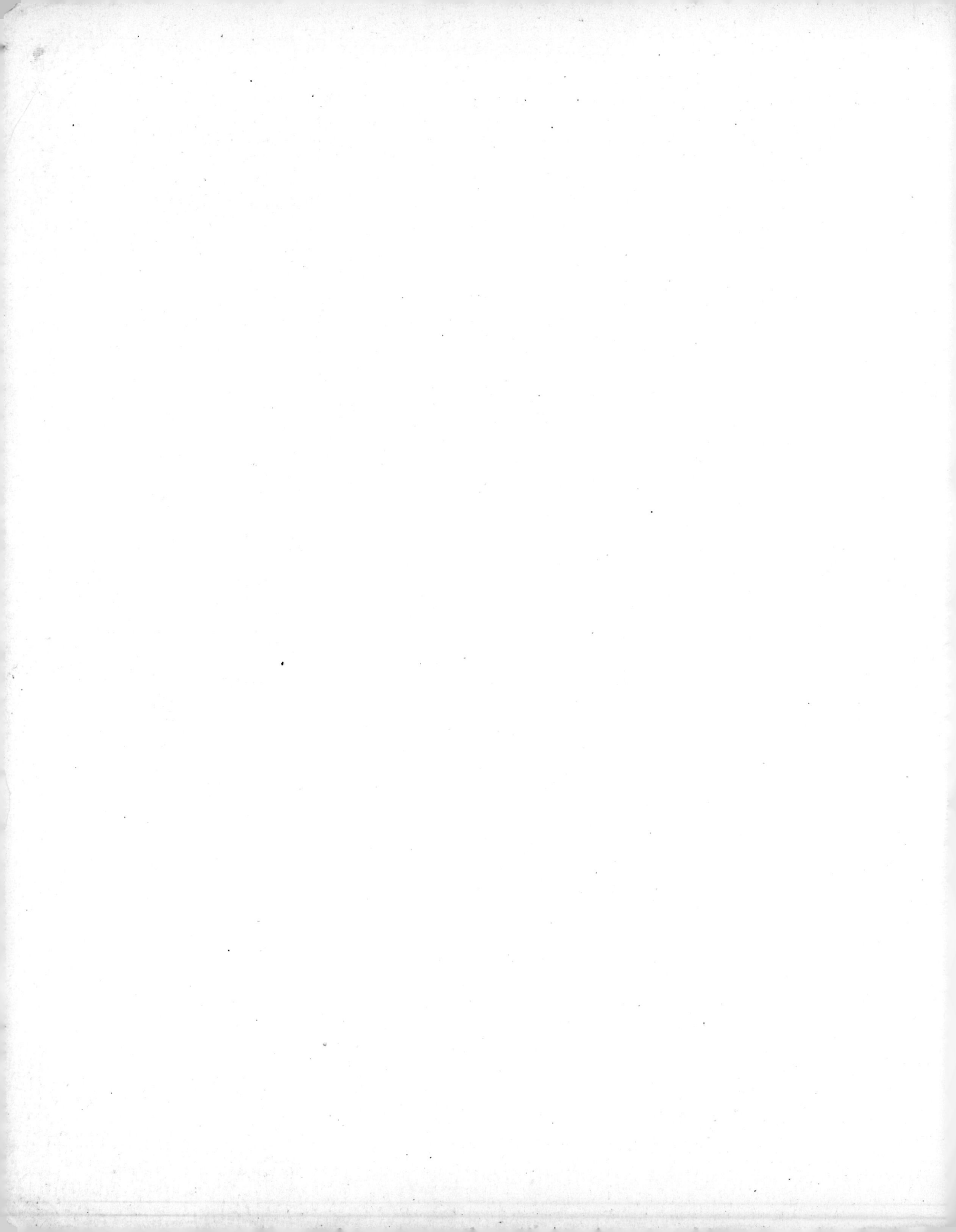

In a Straits-Born Kitchen

LEE GEOK BOI

mc Marshall Cavendish
Cuisine

Some of the recipes in this collection were first published in other titles by Lee Geok Boi: *Classic Asian Noodles,* Marshall Cavendish, 2007; *Classic Asian Salads,* Marshall Cavendish, 2009; *Classic Asian Rice,* Marshall Cavendish, 2010; *Asian Soups, Stews & Curries,* Marshall Cavendish, 2014; *Asian Seafood,* Marshall Cavendish, 2017

© 2021 Marshall Cavendish International (Asia) Private Limited

Published by Marshall Cavendish Cuisine
An imprint of Marshall Cavendish International

A member of the
Times Publishing Group

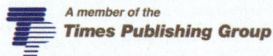

Other Marshall Cavendish Offices:
Marshall Cavendish Corporation, 800 Westchester Ave, Suite N-641, Rye Brook, NY 10573, USA • Marshall Cavendish International (Thailand) Co Ltd, 253 Asoke, 16th Floor, Sukhumvit 21 Road, Klongtoey Nua, Wattana, Bangkok 10110, Thailand • Marshall Cavendish (Malaysia) Sdn Bhd, Times Subang, Lot 46, Subang Hi-Tech Industrial Park, Batu Tiga, 40000 Shah Alam, Selangor Darul Ehsan, Malaysia

Marshall Cavendish is a registered trademark of Times Publishing Limited

National Library Board, Singapore Cataloguing in Publication Data

Name(s): Lee, Geok Boi.
Title: In a Straits-born kitchen / Lee Geok Boi.
Description: Singapore : Marshall Cavendish Cuisine, [2021] | Includes bibliographical references.
Identifier(s): OCN 1232438545 | ISBN 978-981-49-2876-2 (paperback)
Subject(s): LCSH: Cooking, Peranakan. | Cooking, Malaysian. | Cooking, Singaporean. | Cooking, Indonesian. | LCGFT: Cookbooks.
Classification: DDC 641.5959–dc23

Printed in Singapore

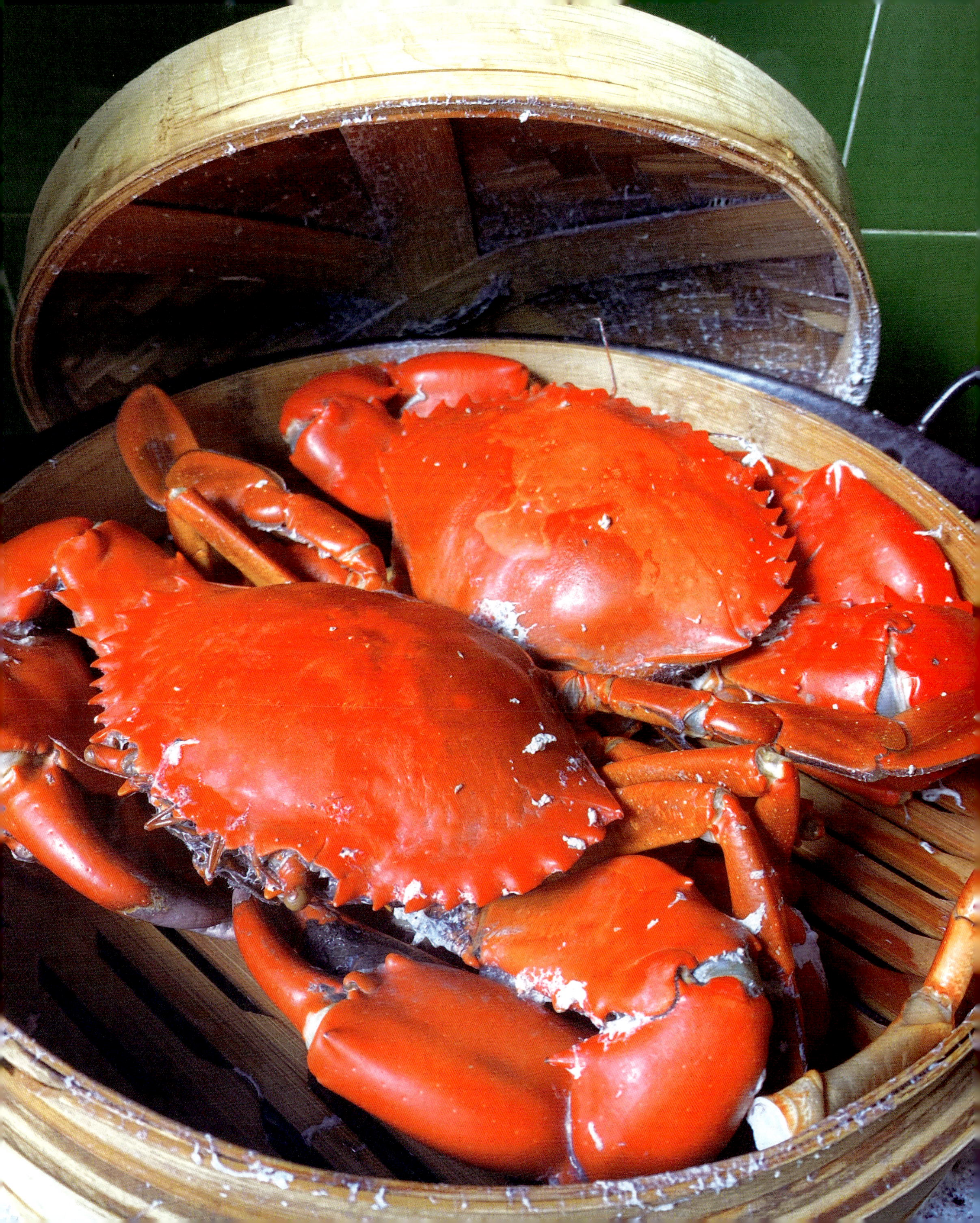

Contents

The Straits-Born: An Introduction

Straits-born cuisine shows the histories of exploration, international trade, migration and colonisation. It is the story of unique communities formed at the cross-roads of the fabled Maritime Silk Road. Although many think of Straits-born as "Peranakan Chinese", the Malay word "Peranakan" means "local born", and the Peranakan community consists of Penang Chinese, Melaka Chinese, Eurasians (Portuguese, Dutch and English), Chetti Melakans aka Chetti/Chitty, Singapore Straits Chinese and Indonesian Chinese, when spoken in the context of colonial era Singapore. The communities were defined partly by their cultural practices such as their womenfolk wearing the *sarong kebaya* on formal occasions and the use of a Creole dialect but not necessarily. Today, they are defined by their heritage recipes. Straits-born cuisine could only have come about because of where the cooks were from, the essential ingredients that were on hand easily, and where there were already established ancient culinary traditions in the use of these ingredients. It is a cuisine that grew out of the multitude of natural resources and plenty enjoyed by ancient Island South East Asia of which Singapore is a part. Straits-born cuisine bloomed in Singapore because it became the meeting place of the Straits-born diaspora whose cooking was further shaped by the prosperity created on the island.

When modern Singapore was founded by the English East India Company in 1819, it was the last of the three Straits Settlements and it attracted migrants from the older established Straits-born communities. Melaka had been a fabled trading emporium in the 15th century and Penang had been founded in 1786. The hotly contested Spice Islands (once Moluccas, today Maluku) and the rest of the Indonesian Archipelago were coming together as part of the Dutch colonial empire. The bazaars of Melaka saw silks from China, cottons from India and edible and non-edible products from all over the region — what the history books called "Straits produce" — were the trade goods of the ships that flocked to the port. The arrival of the Europeans in South East Asia starting with the Portuguese and Spanish in the 16th century brought produce native to Central and South America, the most significant being Central American chillies. Introduced by the Portuguese to South India when they captured Goa, and Melaka, chillies changed the culinary flavours of not just South Indians and South East Asians, but also far-flung Korea. Imagine South Indian curries, kimchi or *sambal belacan* without chilli?

The people of Island South East Asia were not devoid of fabulous flavours. There was pepper (*Piper nigrum*) originally native to India, but which was already being grown in the Indonesian islands since the days of the Buddhist-Hindu kingdoms in Java and Sumatra in the 12th and 13th centuries. Turmeric (*Curcuma longa*), cardamoms (*Elettaria Cardamomum*), two members of the ginger family, were originally native to India, as were tamarind (*Tamarindus indica*) and mangoes (*Mangifera indica*). The trade connections that brought Buddhism, Islam and Hinduism to South East Asia also introduced these essential ingredients to Straits-born cooking. The Indian and Arab trading ships also carried the seed spices from further afield: coriander (*Coriander sativum*), Italy; cumin (*Cuminum cyminum*), West Asia; fennel (*Foeniculum vulgare*), southern Europe and West Asia; fenugreek (*Trigonella foenum-graecum*), southern Europe and West Asia; cinnamon (*Cinnamomum zeylanicum*), Sri Lanka; Chinese cassia (*Cinnamomum cassia*), southern China. The fabled Spice Islands had nutmeg (*Myristica fragrans*), candlenuts aka *buah keras* (*Aleurites moluccana*) and cloves (*Syzygium aromaticum*). South East Asia itself had several native gingers and numerous fragrant plants and flowers. Lemongrass (*Cymbopogon*)/*serai* is a South East Asian native, as is galangal (*Alpinia galanga*) aka *lengkuas* aka *lam keor* (Hokkien meaning blue ginger) aka greater galangal, both almost de rigueur in Straits-born cooking. Ginger (*Zingiber officinale*) was common from India across South East Asia to China. Two examples of fragrant leaves used in Straits-born cooking are kaffir lime aka *makrut* lime aka *limau perut* (*Citrus hystrix*) and pandan leaves (*Pandanus amaryllifolius*)/screwpine. Flowers are used too. There is the blue of the pea flower aka butterfly pea flower (*Clitoria ternatea*) aka *bunga talang* in rice desserts and the superb fragrance of the torch ginger bud (*Etlingera elatior*) in salads and seafood curries. Coconut palms as well as several species of palms including the sago palm were native to South and South East Asia and coconut-enriched dishes go back centuries.

It was trade that brought the South Indians, Hokkiens and Europeans to the ports of South East Asia. Indians from the southern coasts of the subcontinent were amongst the very earliest traders starting circa 4th century, bringing with them more than trade goods. They brought Hindu and Buddhist cultures to South East Asia. Bali is Hindu to this day, and Java's Prambanan and Borobodur, Cambodia's Angkor Wat complex testify to these faiths. The two great Hindu epics, the Ramayana and the Mahabharata, are embedded in traditional Javanese, Thai and Cambodian cultural iconography and performing arts. The Srivijaya trading emporium founded circa 11th century to which the ancient history of Singapore is linked, was a Hindu-Buddhist polity. Thus was formed the community of Chetti Melakans who originated from these early Tamil traders and who are today the smallest of the Straits-born communities. The largest Straits-born community, the Straits Chinese, evolved from the Hokkien traders who were part of the growing maritime networks between southern China and South Asia from as early as the 11th century. Fujian on the South China Sea coast was once described as "eight parts mountains, one part water and one part farmland". Looking towards the sea came naturally. Hokkien was the lingua franca in the early overseas Chinese communities in this region. In the 16th and 17th centuries, the Portuguese and Dutch were the next groups of men to descend on South East Asia. Like the Tamils and Hokkiens, the men married local women who, by then could also be Chetti, Melaka Chinese or Malay women, thus creating the Portuguese and Dutch Eurasian community. These women kept their traditional South East Asian cooking, but tweaked it to suit their menfolk and their religious leanings. Thus evolved Straits-born cuisine with its roots in regional culinary essentials, but also incorporating the diverse range of new ingredients from international trade and the arrival of the Europeans. After 1819, the Straits-born diaspora from Penang, Melaka and Semarang to Singapore saw the development of a Singapore Straits-born cuisine that brought together their heritage recipes — as seen in the growing range of Straits-born cookbooks.

Tracing Straits-Born Cooking

The codification of Straits-born cooking is a fairly recent phonemenon. The first local cookbook to say "Nonya Recipes" on the title page was *Mrs Lee's Cookbook: Nonya Recipes and Other Favourite Recipes* self-published in 1974. However, it is not the first published mention of Straits Chinese cooking. The very first local cookbook was published in 1931, of which the 9th and last edition came out in 1962. Edited by Mrs AE Llewellyn, *The YWCA of Malaya Cookery Book* subtitled *A Book of Culinary Information and Recipes Compiled in Malaya* referred to the 1931 edition which was edited by Mrs RE Holttum, wife of the Director of the Singapore Botanic Gardens, and Mrs TW Hinch, wife of the principal of Anglo-Chinese School. In the preface of the 1962 edition, Mrs Llewellyn said that the cookbooks were compiled from recipes contributed by people of all races from Malaya and Singapore and went on to ask for criticism from the different communities in order to confirm the authenticity of the recipes as well as acknowledgement of recipes adapted from other communities. She said, "By receiving these criticisms I hope to build sections of authentic, readily acceptable recipes, clearly distinguishable in origin, but which will become popular with all Malayans."

In the light of Mrs Llewellyn's request for confirmation of ethnic authenticity in the recipes, the way the recipes are presented says something about their history. There is no Straits-born section, but a small handful of recipes in the Chinese section were labelled as "Straits Chinese": chicken satay, *otak-otak*, *popiah goreng* (fried spring roll). Pork Sambal was labelled "Possibly Straits Chinese", but with a spice paste consisting of onion, fennel, dried chillies and garlic, I would say "definitely Straits Chinese" because of the combination of spices, but also

because traditional Chinese cooking did not use fennel. One recipe not labelled as Straits Chinese, but which looked distinctly Straits Chinese, was Pig Liver Balls. With *ketumbar* (coriander) and *asam* (tamarind) water in it, it was certainly not Chinese. In fact, Pig Liver Balls looked like *Hati Babi* (Liver and Pork Balls) in *Mrs Lee's Cookbook* (1974) or *Hati Babi Bungkus* (Meat and Liver Balls) in Mrs Leong Yee Soo's *Singaporean Cooking* (1976). Many of the recipes in the Malay section of this first local cookbook would be seen as Straits-born today: rendang, laksa, *opor*, *agar-agar*, "*sarikauja*" (Malay custard). The recipe for *sarikauja* was much like the one my mother made: eggs, sugar and santan (coconut cream) steamed together to form a "custard cake" that was solid enough to cut with a knife. Could *sarikauja* be *serikaya* today? Or simply *kaya* (coconut egg jam)? My mother did call her custard cake *kaya* and we ate slices of it on bread.

In the very short Indonesian section in the 1962 edition, I found a recipe for chicken or mutton vindaloo. Such a recipe in the Indonesian section was puzzling until I remembered that Flores in the Indonesian Archipelago was once a Portuguese colony. Could this recipe have come from a Portuguese Eurasian or Goan, once resident in Flores? Vindaloos are classic South Indian vinegary dishes that came to be part of Portuguese Eurasian cuisine via the link between Melaka and Goa. Two cookbooks traced vindaloo to its origins. EP Veerasawmy in his *Indian Cookery for Use in All Countries* (1956) described vindaloos as "hot" and "it is therefore customary to use more chillies than in ordinary curries". In her very first tip in *100 Easy-To-Make Goan Dishes* (1977), Jennifer Fernandes wrote, "Goan food is pungent and I have used the required amount of chillies to keep the recipes as authentic as possible." Given that the Portuguese introduced chillies to its colonies, this was no surprise. South Indian food today is much spicier than North Indian food for example. A classic Portuguese Eurasian dish is Devil Curry, so-called because of its fiery appearance from lots of ground chillies.

Another dish of South Indian origin, but now very much a staple in many Singapore homes is *kurmah*. The dominant taste in *kurmah* is white pepper combined with other seed spices. Pepper was the tongue-tingler in pre-colonial food. It was the search for pepper that brought on European exploration. Apart from the *YWCA Cookery Book* and its few "Straits Chinese" recipes, the landmark *My Favourite Recipes* by Ellice Handy made almost no mention of "Straits Chinese" anything despite recipes today seen as Straits-born. This slim publication of 94 pages first came out in 1952 as a fundraiser for the Methodist Girls' School building fund. It has since become iconic of Singapore cooking and contributed to the spread of Straits-born cooking in Singapore and not only because Handy was a domestic science teacher. Another domestic science teacher who must have spread knowledge about Straits-born cooking was Lilian Lane whose 1964 *Malayan Cookery Recipes* had typical Straits-born recipes along with colonial era favourites. One of her recipes was for *rojak*, described as "Javanese fruit salad". This link between *rojak* and Java traces the fruit *rojak* of Penang back to 10th century Javanese copper plate inscriptions that list "*rujak* [as] a salad of raw green fruit, mixed with sugar or seasoning". Lane was an English woman who taught domestic science at the Malayan Teachers' College in Penang between 1956 and 1959. Her *rojak* recipe, however, did not have *hae ko* (black prawn paste) although it did have *teecheo* (Chinese sweet black sauce). In appearance then, Lane's *rojak* would have looked like today's Penang *rojak* even if its flavour was not. So when did Penang start producing and exporting *hae ko*? It seems to be Singapore's only source of *hae ko*.

One rare early local cookbook that linked Java to Singapore Straits-born cooking was Mrs Susie Hing's *In A Malayan Kitchen*. Mrs Hing was from Semarang in central Java, which in the Dutch colonial era was the hub of the Indonesian Chinese community. Tentatively dated as self-published in "1954", Mrs Hing's book had typical Indonesian recipes such as *opor ayam*, rendang Padang, *sate bakso* and *dendeng manis*. This last is a spiced savoury meat rather like Singapore *bak kuah*, but dried in the sun rather than grilled. (Could this be the inspiration for Singapore *bak kuah*?) Dried meat or *dendeng* dates back to 10th century Java, like *rujak*. Mrs Hing also had typical Malayan-Singapore recipes. (The concept of "Singaporean" is post-1965 when Singapore became independent.) Among such typical recipes as *roti jala* and Hokkien mee was a recipe for pineapple tarts, both the open and the closed pineapple-shaped tarts with spikes, and one for *Kroket Tjanker* or Java *Kwei Patti*. Unlike today's *kueh pie tee* with its bamboo shoots or *bangkuang* (yam bean) fillings, this Java *Kwei Patti* had a very meaty filling. So did *kueh pie tee* start as a Dutch colonial dish? *Mrs Lee's Cookbook* (1974) did not have a recipe for *pie tee*, but the 2nd edition of *My Favourite Recipes* (1960) had one with a bamboo shoot filling. And those closed spiky pineapple tarts were once considered Indonesian.

These colonial era cookbooks had substantial sections on European cakes, pastries and biscuits. Butter was a European addition to South East Asian diets. In the 1970s, sugee cake and Indonesian *kueh lapis* (aka *spekkoek*) became the forte of Eurasian bakers for the former, and Indonesian Chinese bakers for the latter.

Before the spread of refrigeration imported butter came in tins. Wheat flour for baking was another colonial era introduction. The wheat produced in northern India was a different kind of wheat and English-style baked goods was a colonial era development. (So Hokkien mee, longevity noodles and *mee sua* (wheat vermicelli) were colonial era creations?) Pre-European era sweets were made with rice, sago, rice and bean flours. Tapioca or cassava, originally from Brazil, was introduced by the Portuguese. One typical recipe found in the old cookbooks was for *kueh koya* which used mung bean flour. Without any kind of leavening — a European ingredient — these very pretty moulded biscuits were hard as rocks. (The Hakka nuns at my grandmother's Buddhist temple always gave us a tin of *kueh koya* for Chinese New Year and the children only ate it out of desperation. It was invariably the last of the goodies to be finished!) Almost nobody makes *kueh koya* these days. Another dessert ingredient made of mung bean flour was *tepong hoen kwe* often called green pea flour. This packaged flour has always been imported from Indonesia. The early cookbooks featured recipes for *kueh pisang* which was made of sliced bananas cooked in *tepong hoen kwe*, wrapped in banana leaves and steamed. This *kueh* is still popular.

Today's very popular Chinese New Year goodie that we know as love letters was once called *kueh belanda* (meaning Dutch cake in Malay). The *YWCA Cookery Book* recipe for "*Kueh Blanda* (Love Letters)" was flavoured with ground cardamom. As a child, I knew *kueh belanda* as *kueh kapit* because my Penang-born mother folded these biscuits into quarters Penang-style. In Singapore, the same biscuits were and still are rolled up into a cigarillo shape. Looking into *Ceylon Daily News Cookery Book*, a colonial era cookbook from Sri Lanka that dates back to 1929, I found an old Dutch recipe for *Ijzer Koekes* which looked like *kueh belanda* by another name. *Ijzer Koekes* had powdered cinnamon and cloves in the batter made of flour, sugar, eggs and coconut milk and they were baked in thin sheets with an iron held over a hot fire. The hot *koek* then had to be rolled up exactly like a love letter or *kueh belanda*. *Ijzer Koekes* was translated as "Iron (Thin) Cakes".

My Favourite Recipes (1960) did have a few old Straits-born sweets among the English cakes and biscuits, one of which was *tapeh* (fermented glutinous rice). *Tapeh* still enjoyed today for its mildly alcoholic flavour goes back to 10th century pre-Islamic Java when several types of fermented alcoholic beverages such as arack and palm wine aka toddy were listed as popular beverages. Handy's sago pudding recipe had a rare description: "Sometimes called Singapore, Penang, Malacca Pudding, Straits Settlement Pudding, or Palm Pudding." Sago palm flour was once a staple in parts of Island South East Asia where rice did not grow well or was not easily available.

These early cookbooks were aimed at teaching women how to prepare local dishes whether in the context of a domestic science class or housewives' cooking classes. Pre-Internet and cooking shows, many Straits-born women gave cooking classes to earn some pocket money. Mrs Lee said in the introduction to her cookbook, "(I)t has been one of my ambitions to write a book about Straits Chinese food so that the younger generation, including my grandchildren and later their children, will have access to these recipes which were usually kept within families as guarded secrets."

Favourite recipes were actually rarely kept secret, but passed on by word of mouth to favoured family and friends, and others through cooking classes and, today, through cookbooks and packaged spice pastes. My aunt from Alor Star, with the phenomenal taste buds, passed on her recipes to my mother, and I have been passing them on through my cookbooks. Like my mother and aunt, I was a collector and experimenter of recipes from my teenage days. I still am. Today, Internet videos and TV cooking shows have made learning about unfamiliar ingredients and dishes simplicity itself. Adopted and tweaked recipes become family favourites and eventually heritage recipes over time. Which was how Singapore Straits-born cuisine evolved. And why they overlap and show as many similarities as differences. The earliest cookbooks did not, could not or never thought to define Straits-born cooking, but over time, it defined itself. It grew out of the recipes of ordinary housewives sharing favourite recipes that were based on what was best in their environment, recipes that were enhanced and which expanded their culinary range, thus creating some of the most delectable food in the world. The only way to define Straits-born cooking is to look at what a Straits-born cook would dish out from her kitchen. Which is what this cookbook is about.

Basics

Cooking Tips and Other Essentials

Time-Saving Tips for Preparing Straits-Born Meals

Straits-born dishes have this reputation of being time-consuming to prepare.
It is not if you organise your time well.

- A couple of hours in front of the TV is also a good time to seed dried chillies, peel garlic or shallots. Keep seeded dried chillies in the fridge. They will get mouldy if left out. Peeled shallots or garlic can be bagged and kept refrigerated for a day or two, or ground and frozen. Once the food processer is out, take the chance to do some advance preparation for other recipes too.

- Chunks, slices and shreds of galangal, lemongrass, fresh turmeric and ginger freeze well. Buy extra and skin, rinse, cut into useful chunks or slices and freeze these aromatics for another day.

- Shredded or sliced ginger, lemongrass and galangal are best stored in between layers of plastic wrap in usable quantities to ease separation when required.

- Some of these whole aromatics do get watery once defrosted. To keep the flavours in the liquid, cut, slice or blend when the root is starting to defrost and not after.

- Seed fresh chillies before freezing in a box.

- Frozen ginger grates easily. If you use grated ginger often, always freeze a large piece for grating.

- Spice pastes freeze well. Always treble the amount of your favourite spice paste. If possible, freeze it first in an ice cube tray. Once frozen, repack the frozen cubes into a clearly labelled bag/box.

- Freeze ground garlic or shallots separately in clearly labelled boxes or in the same box separated by a small piece of plastic wrap. It is rare for a recipe to require equal amounts of both aromatics.

- Freeze raw or cooked dried chilli paste in a glass jar. Defrost for 5 minutes, then using a small spoon, scrape out what you need. Refreeze the jar. Refreezing does not affect the quality of the paste.

- Keep *sambal belacan* in a jar in the freezer. It does not freeze solid and it is easy to scoop out the needed spoonful or two. If kept in the chiller section, the *sambal* can get mouldy if not used quickly.

- If you stir-fry vegetables or noodles with garlic frequently, peel and chop a head or two of garlic and store in a glass jar with a tight lid. Top up with a layer of cooking oil to prevent the odour from spreading into everything else. The oil also prevents chopped garlic from turning green.

- Chopped spring onions or chives may be frozen and the frozen herb stirred directly into the hot dish/pot.

- Coriander and Chinese celery (*kin chye*) leaves do not freeze well, but the roots and stems do keep their flavour when frozen. Chop the frozen stems and add to the dish directly. Coriander roots may be smashed to get the juice for adding to a salad dressing. Chinese celery roots are never used, unless you want to try planting it.

- Try to keep related ingredients near each other in the freezer or in the same bag to reduce searching for the needed item.

- Always label anything that goes into the freezer. Frozen stuff can look very alike and be hard to sniff out or identify.

Sambal Belacan

Makes about 150 g (5⅓ oz)

Sambal belacan is an essential ingredient in Straits-born cuisine. It is not the same as *belacan* (dried shrimp paste). The former is a condiment and the latter is an essential ingredient in the condiment, but also in spice pastes (*rempah*) in Straits-born cuisine. The other essential part of *sambal belacan* is red chillies. The condiment is made by pounding fresh red chillies with toasted *belacan* in a mortar and pestle, but may also be blended in a food processor. More traditional cooks will insist that pounded *sambal belacan* tastes better, but the popularity of commercially-prepared bottled *sambal belacan* proves that convenience is more important today. *Belacan* comes in several forms ranging from liquid and pink, to blocks of dark brown paste firm enough to cut. Toasted *belacan* powder is also available. *Sambal belacan* has to be made with the block variety because the *belacan* must first be toasted. Large red chillies tend to lack bite, but give bulk. Combine with bird's eye chillies for bite and bulk.

150 g (5⅓ oz) fresh red chillies, seeded

45 g (1½ oz / 3 Tbsp) dried shrimp paste (*belacan*)

Ringing in Changes
- Add lime juice to taste
- Add finely shredded kaffir lime leaves
- Add finely chopped torch ginger bud
- Use this to make *sambal belacan* with shallots (page 76)

1. Chop seeded chillies coarsely by hand/in a food processor.

2. Toast dried shrimp paste. If toasting on a grill or oven-toaster, press into a thin flat piece and place on a metal plate or piece of foil. Toast both sides until slightly brown. Do not burn it or the *sambal* will taste bitter. Alternatively, crumble the shrimp paste into a frying pan and dry-fry until slightly brown. Stir constantly when dry-frying.

3. Pound chopped chillies and toasted shrimp paste in a mortar and pestle to your preferred texture. Some like the *sambal* to be coarse. Store in a clean glass jar in the freezer

Vindaloo Spice Mix

Makes about 250 g (9 oz)

Vindaloo spice mix is handy if you cook these vinegary spicy curries or Devil Curry often. I use Colman's English mustard powder in this spice mix because Indian mustard seeds never come ground. Although Colman's is a hot English mustard, its flavour is fine in vindaloo.

100 g (3½ oz) ground coriander

30 g (1 oz) ground turmeric

2 Tbsp ground cumin

1 Tbsp ground mustard

1 Tbsp ground black/white pepper

2 Tbsp ground ginger

75 g (2⅔ oz) red chilli powder

1. Combine in a large bowl or large plastic bag and stir or shake well to blend the spices. Pack in a ziplock bag or bottle with a label. Refrigerate for extended shelf life.

My Grandmother's Curry Powder

Makes about 500 g (1 lb 1½ oz)

In the pre-war days when few women were employed outside the home, Straits-born housewives often made foodstuff for sale to earn some money. My paternal grandmother made curry powder for sale. My mother continued making this curry powder to my grandmother's recipe for her own use until the last of the Little India spice millers disappeared in the 1980s. She would wash the whole seed spices and spread them out to dry in the sun on wicker trays. Once well-dried, the spices had to be lightly dry-fried to bring out the oils. The spices were then taken to Little India in Serangoon Road to be ground into a super fragrant curry powder. No commercial curry powder can match the fragrance of home-made curry powder. If you have a coffee grinder, try making small batches of curry powder to enjoy that awesome fragrance in your curries. Stick to the three top spices which have the strongest smells — coriander, cumin and fennel — and use ready ground spices for the rest.

250 g (9 oz) ground coriander

50 g (1¾ oz) ground cumin

50 g (1¾ oz) ground fennel

150 g (5⅓ oz) red chilli powder

2 Tbsp ground turmeric

1 Tbsp ground white pepper

1 tsp ground green cardamom

1 tsp ground cinnamon

2 tsp ground cloves

1. Combine all the ingredients in a bowl or large plastic bag and mix well. Pack into a ziplock bag and label. Refrigerate or freeze to extend shelf life.

Kurmah Spice Mix

Makes about 300 g (11 oz)

Even though *kurmah* spice mixes and pastes are available from supermarkets today, the home-made spice mix is much cheaper and as good. Use it to cook any meat.

100 g (3½ oz) ground coriander

50 g (1¾ oz) ground fennel

50 g (1¾ oz) ground cumin

100 g (3½ oz) ground white pepper

25 g (⅘ oz) ground turmeric

½ tsp ground cinnamon

½ tsp ground cloves

½ tsp ground cardamoms

1. Combine in a large bowl or plastic bag and stir or shake well to blend the spices. Pack in a ziplock bag or bottle with a label. Refrigerate or freeze for extended shelf life.

Asam Rempah

Makes about 500 g (1 lb 1½ oz)

Although the word "*asam*" is attached to this spice paste, it may also be used for dishes such as *sambals* that you want to enrich with coconut milk/coconut oil lees. It is the spice paste for *asam* fish curry and *buah keluak* curry. The proportions and ingredients can be tweaked to get varying flavours.

200 g (7 oz) shallots, peeled and sliced

50 g (1¾ oz) garlic, peeled and sliced

5 stalks lemongrass, finely sliced

100 g (3½ oz) dried red chillies, seeded

5-cm (2-in) knob galangal, scraped and sliced

5-cm (2-in) knob turmeric, scraped and chopped

½ Tbsp dried shrimp paste (*belacan*)

6 candlenuts (*buah keras*)

4 Tbsp water

1. Blend all the ingredients together in a food processor until smooth.

2. Pack into a box or ice cube tray and freeze until needed.

Ringing in Changes
- Increase, decrease or omit the lemongrass/galangal.
- Combine fresh red chillies with dried ones or substitute dried red chillies with fresh red chillies.
- Combine another ginger with the galangal or substitute galangal with a different ginger such as *krachai/chekok/*common ginger.
- Increase or decrease the amount of dried shrimp paste (but always have some).
- Omit fresh turmeric or use dried ground turmeric.
- Omit the candlenuts or add more for a richer flavour.

Asam Pedas/Lauk Pindang/Singgang

Makes about 600 g (1 lb 5⅓ oz)

This spice paste is the base of a light seafood curry that many Straits-born call *gulai*. Fresh pineapple, belimbing or tomatoes are often added to expand the dish, but also to increase the fruity flavour. *Asam pedas* is similar to a Portuguese Eurasian *singgang* or a Chetti *lauk pindang*. All are light sourish and spicy curries of seafood. It is an oil-free curry as the spice paste is simply boiled in tamarind juice to which seafood is added. Although tamarind paste is the traditional souring ingredient, the curry may also be soured with dried sour fruit slices (*asam gelugor*) or lime or lemon juice.

5-cm (2-in) knob turmeric, scraped

250 g (9 oz) fresh red chillies, seeded

3 stalks lemongrass, finely sliced

250 g (9 oz) shallots, peeled and sliced

2 tsp dried shrimp paste (*belacan*)

4 Tbsp water

1. Blend all the ingredients together in a food processor to a smooth paste.

2. Pack into a box or ice cube tray and freeze until needed.

Raw Dried Red Chilli Paste

Makes about 300 g (11 oz / 2$\frac{1}{2}$ cups)

Ground dried red chillies are useful in several ways. Uncooked, it is the base for a spice paste. Cooked with oil, it becomes useful for spicing up fried noodles like *char kuay tiao* (fried flat rice noodles) or a noodle soup or to ginger up a curry that's too mild.

100 g (3$\frac{1}{2}$ oz) dried red chillies, seeded, rinsed and softened in water

125 ml (4 fl oz / $\frac{1}{2}$ cup) water

1. Blend the ingredients together in a food processor until smooth.

2. Pack into a box or spoon into an ice cube tray and freeze until needed.

Cooked Dried Red Chilli Paste

Makes about 300 g (11 oz / 2$\frac{1}{2}$ cups)

100 g (3$\frac{1}{2}$ oz) dried red chillies, seeded, rinsed and softened in water, then drained and blended into a paste

125 ml (4 fl oz / $\frac{1}{2}$ cup) cooking oil

1. In a microwave-safe bowl, combine raw blended chilli paste and cooking oil. Cook in a microwave oven on High for 20 minutes, stirring after 10 minutes. Add 1–2 Tbsp water if chilli paste is very dry.

2. Leave to cool before bottling and freezing until needed.

How Much Chillies?

Quantifying the exact amount of chillies needed in a spice paste is difficult because the spiciness differs even with similar-looking chillies, what more that between little bird's eye chillies and the more common large variety. It is even difficult to tell how spicy a particular batch of chillies may be. (Apparently the spiciness is linked to the amount of sunshine that that particular chilli gets!) Erring on the side of caution is sensible as a dish that is too spicy is not edible. Feel free to use less or more than indicated in the recipes. While the recipes indicate the quantity for chillies in tablespoonfuls of ground chilli paste, you might find this conversion useful if you are working with whole chillies. Note that the conversions are approximations because ground chilli paste will have water in it.

1 Tbsp (20 g / $\frac{2}{3}$ oz) ground chilli paste	=	15 g ($\frac{1}{2}$ oz) seeded dried red chillies

Making Fresh Coconut Cream (Santan)

Before the availability of freshly ground coconut from the market and later packaged coconut cream, grating coconut on an old-fashioned scraper was dangerous work. The scraper was a piece of thick aluminium with spikes gouged out of it and injuries were not unknown. There is no substitute for freshly grated coconut and fresh *santan* in certain Straits-born delectables. Snacks and desserts like *ondeh-ondeh* must be coated with fresh grated coconut and *chendol*, sago pudding or *pulot hitam* must be topped with fresh *santan*, not packaged coconut cream. The same goes for some *kerabu-kerabu* (salads).

Grated coconut may be squeezed without any water to get a thick, rich and very tasty cream called *patti* although only the most traditional cooks insist on making *patti* today. The word is even seldom heard nowadays. Two squeezes are the norm, with the first squeeze made with a small amount of water to get *santan* or first milk. The second squeezing is simply coconut milk or second milk, and very thin third milk is also possible and useful for hard boiling. While *santan* is usually added at or near the end of cooking, the second or third milk is added early in the cooking.

Wet market grated coconut is convenient, but may sometimes be a little off because of Singapore's hot humid climate. One way around this problem is to buy the grated coconut early in the morning and refrigerate it as soon as you get home. Another way is to extract the *santan* as soon as you get home, add some salt to it and refrigerate until needed. If it's only grated coconut that you need, some cooks suggest mixing in a bit of salt and steaming the coconut lightly. Cool, then refrigerate until needed. I find that freezing grated coconut is another way to keep the coconut fresh. To get a good milk from frozen coconut, add warm water to the cold or defrosted grated coconut. To thicken the first milk, extract the second or even third milk, combine the two and refrigerate it. The cream will rise to the top and form a layer that is easily scooped out to thicken the first milk.

If you have a powerful blender that can crush ice cubes, make your own fresh *santan*. Get a whole coconut with the skin still on from a wet market. If the coconut has no cracks and the water inside won't leak out, the coconut will stay fresh outside the fridge. If there are cracks, it is best to either refrigerate the nut whole or in pieces. Whichever way, first rinse the nut clean. To break it open, place the cleaned nut in a large bowl and pierce one of the cracks with a sharp knife. The water can be enjoyed right off. Trim the brown skin before storing or just before processing. Either way, immerse the pieces of coconut in a bag of fresh water with a bit of salt added and refrigerate until needed. It will keep fresh for up to a week, but change the soaking water every couple of days.

Coconut Cream and Coconut Milk from Grated Coconut

Makes about 200 ml (7 fl oz / $^4/_5$ cup) coconut cream and 250 ml (8 fl oz / 1 cup) coconut milk

The richness of the *santan* depends on the age of the coconut as well as the amount of water added. Older nuts have a richer cream and stronger flavour. If it is easier, squeeze the *santan* out through a piece of cheesecloth (on right).

300 g (11 oz) grated coconut

Coconut Cream
150 ml (5 fl oz / $^2/_3$ cup) warm water

Coconut Milk
250 ml (8 fl oz / 1 cup) warm water

1. Combine grated coconut and 150 ml (5 fl oz / $^2/_3$ cup) warm water in a bowl. Mix and squeeze the coconut a few times.

2. Place a sieve over another bowl. Using both hands, take a large handful of coconut and squeeze the milk into the bowl. Set aside the squeezed handful of coconut in a separate bowl for the second milk. Repeat until all the coconut cream has been extracted.

3. To get the second milk, add 250 ml (8 fl oz / 1 cup) water to the coconut and repeat squeezing over a sieve.

Coconut Oil Lees

Preparation and cooking: 30 minutes

There was a time when Straits-born housewives made their own coconut oil from freshly squeezed coconut cream. Freshly made coconut oil is super fragrant as is the lees at the bottom of the pan once the oil has been extracted. Coconut oil lees adds instant richness to a dry sambal, curry or *kerabu*. Better yet, it can be made with packaged coconut cream.

Packaged coconut cream

1. Place packaged coconut cream in a saucepan and bring to the boil. Turn down the heat and simmer gently, stirring and scraping up the bottom of the pot until the oil comes out and the lees starts turning brown. Take care not to burn the lees. It should be a nice shade of brown.

2. Keep the lees together with the oil in a box. Refrigerate until needed or freeze it.

Krisek (Toasted Grated Coconut)

Makes about 100 g (3¹/₂ oz) • Cooking time: 30 minutes

Krisek is plain dry-fried grated coconut that is a classic thickener for *rendang* and *otak-otak* and equally tasty as a light topping for a Straits-born salad. *Serunding* (coconut *sambal*) is *krisek* with flavouring. It is also a good salad topping or as a dressing for plain rice. It is commonly added to a dish of *sayur lodeh* (rich vegetable stew) or *lontong* (rich vegetable stew with rice cubes). A small quantity of minced beef, chicken, dried shrimps or dried krill may be included for more flavour or to ring changes in the *sambal*. *Krisek* and *serunding* are usually prepared in advance, bottled and kept in the fridge until needed.

250 g (9 oz) fresh grated skinned coconut

1. Dry-fry grated coconut in a wok over low heat, stir-frying continuously until nicely brown.

Serunding (Coconut Sambal)

Makes about 300 g (11 oz)

250 g (9 oz) fresh grated skinned coconut

6 dried red chillies, wiped clean, seeded and thinly sliced

5 kaffir lime leaves, crushed

5-cm (2-in) knob galangal, scraped and thinly sliced

1 large pinch ground turmeric

1 tsp salt

2 Tbsp sugar

25 g (1 oz) shallots, peeled and thinly sliced

1 Tbsp cooking oil

2 Tbsp minced beef/minced chicken/dried shrimps (*hae bee*)/ dried krill (*grago*) (optional)

1. If using dried shrimps/dried krill, rinse clean and pat dry. Pick out any debris. If dried shrimps are large, chop to preferred texture.

2. Except for the optional ingredient, shallots and oil, mix together grated coconut with the rest of the ingredients. Set aside.

3. Heat oil in a wok and fry sliced shallots until pale brown and fragrant. Add the optional ingredient and stir-fry until fragrant.

4. Add coconut mixture and continue stir-frying until coconut is dry and nicely brown and the flavours well-blended. Keep the heat low so as not to burn the coconut.

5. Leave to cool before bottling and refrigerating. It will keep for at least 2 weeks. When using the *serunding*, leave the lime leaves and galangal in the bottle.

MSG, Seasoning Powders, Bouillon Cubes

Glutamic acid is a protein salt that occurs naturally in certain foods such as meats and seafood, but also in cheeses, mushrooms, tomatoes and seaweed. Glutamic acid produces umami, the fifth basic flavour after bitter, sweet, sour and salty. It is a savoury flavour that creates a moreish quality when combined with salt. Umami is a Japanese word and it was a Japanese scientist who first extracted monosodium glutamate from seaweed in 1908, after which Ajinomoto went on the market worldwide as a flavour enhancer. Several other brands popped up in China particularly and monosodium glutamate better known as MSG became an essential in the larder of Asian cooks. In the 1950s, MSG was even put on the table in salt shakers as a condiment not only in Japan, but also in a number of other Asian countries, or mixed into the traditional seasonings such as chilli powders and soy sauces to give more oomph to the food. (I once came across the suggestion in a cookbook that a dish of MSG powder be part of the array of condiments on the dining table for the convenience of guests who might want more flavour in their food!)

It was US consumers who first came up with the "Chinese restaurant syndrome" in the 1970s. The syndrome — extreme thirst, palpitations, headaches etc. — was said to be the result of consuming MSG in Chinese food. MSG became stigmatised as something harmful and to be banned from all food products. (It never happened and there are still lots of food products that contain MSG although its presence may be disguised as flavourings, flavour enhancers or the like. It was only in recent years that MSG was removed from canned soups made in the US.) The stigmatisation of MSG spread all over the world and sadly, food writers everywhere including myself began talking bad about MSG in the 1980s. Since then, and with more experience as a cook, I now see that stigmatising MSG was not only uncalled for, but may even be an act of stupidity because many cookbook writers urged substituting MSG with sugar. Sweetness is not umami and this substitution may be why the food in Singapore now seems sweet rather than savoury.

A close study of the older cookbooks compared to more recent ones (like my cookbooks for example) will find that MSG gets mentioned in the older books, but is routinely ignored in the more recent publications. Compare the recipe for *bakwan kepiting* (crabmeat meatball soup) in *Mrs Lee's Cookbook* (1974) to that in Wee Eng Hwa's *Cooking For The President* (2010). Mrs Lee's recipe had 2 teaspoons of MSG to about 2 litres of soup. In Wee's book, the *bakwan kepiting* recipe had 2 teaspoons of sugar in the soup and no mention of MSG. Terry Tan's 1983 *Straits Chinese Cookbook* listed bouillon cube and MSG in a recipe that is credited to his mother. His soups have varying amounts of MSG, quite often just a pinch. Mrs Leong Yee Soo's 1988 *The Best of Singapore Cooking* also lists MSG as an ingredient particularly in the soups, and if not MSG, chicken cube (aka chicken bouillon cubes). A 2017 cookbook, *Daily Nonya Dishes* by Lloyd Mathew Tan does not list MSG or bouillon cubes in any of his recipes. Instead, sugar is commonly listed with salt. Sugar is also the MSG substitute in Sharon Wee's *Growing Up In A Nonya Kitchen* published in 2012. For example, her Hokkien mee soup lists 1½ tablespoons of sugar. On the other hand, *Mrs Lee's Cookbook* lists both 2 teaspoons of MSG as well as 2 teaspoons of sugar for Hokkien mee soup. What about the landmark *My Favourite Recipes* by Ellice Handy? My 1960 edition which also includes all the recipes in the original 1952 edition has one or two scattered mentions of "seasoning powder" and "¼–½ Knorr-Swiss Clear Chicken Broth cube", but for most of the recipes, the cook has to make stock from scratch.

Before the availability of bottled MSG, cooks obtained the umami flavour by hours of boiling meat bones and meat scraps to get a tasty stock. Apart from meat bones being expensive, many cooks today are short of time. The answer is a food additive. (I once ate a steamboat dinner that started with just a chicken bouillon cube in a pot of boiling water. At the end of the meal of boiled seafood and meatballs, the resulting soup was delicious.) Like all food additives, while a little makes a huge difference, double that does not necessarily make it better. (It won't kill you, but too much MSG may make you thirsty although sometimes I think this is the result of autosuggestion.) Think chillies and how too much can ruin the dish. While light soy sauce and fish sauce have naturally occurring glutamic acid, there is a limit to how much of either sauce you can add to the food without over-salting it. Certain dishes — noodles, vegetables and soups — definitely taste better with a little umami boost in the form of a pinch of MSG, seasoning powder aka stock powder or bouillon cube. All brands of seasoning or stock powder or bouillon cubes may say "no MSG added", but the operative ingredient — what gives the flavour — is still glutamic acid extracted from meat stock.

In Singapore, the commonest flavours in seasoning powders or bouillon cubes are chicken or *ikan bilis* (dried anchovies). However, in West Asia, I have also seen beef and mutton bouillon cubes as well as seasoning powders in these flavours. Bouillon cubes are made from condensing rich stock into a cube. Presumably the same may be said for seasoning powders? Bouillon cubes are convenient only if you use the whole cube, but not if all you want is a pinch. Seasoning powders are more convenient for Asian cooking, but outside Asian stores, you have to hunt for this. If all you can find are bouillon cubes, mash a couple of cubes into a small jar and keep a little handy spoon in there.

As with my other cookbooks, I have not listed MSG or seasoning powder as an ingredient in any of the recipes here, but note that when I cook soups, fry noodles or vegetables, I always add a pinch of seasoning powder to the dish. The choice is yours, but don't substitute with sugar.

The Agak-Agak Method: Measurements and the Taste Test

The conversions from metric to ounces in the recipes are not exact, except in the cakes and recipes where *agak-agak*, or "guesstimate", does not work. Cakes and biscuits do need more exact measurements. Where they do not matter, the $^1/_2$ or $^3/_4$ oz may be rounded up. Such miniscule amounts are really insignificant when it comes to a curry or soup. In fact, my mother never owned a kitchen scale of any kind, although my aunt must have had a measuring cup because she did bake. (When I started baking as a teenager, the first kitchen tool I bought was a measuring cup.) The cooks of old used a hand, spoon, rice bowl, teacup and whatever else, and then they eyeballed the ingredients. Old cookbooks even had quantities of ingredients listed by the price: "10 cents of this" or "5 cents of that"!

This *agak-agak* method was how women traditionally cooked. Recipes were passed down verbally and the wannabe cook had the benefit of seeing how much a handful should be. My mother gave me recipes as "some onions, some chillies…"

and would gesture to signify the amount she was thinking of. Or she would say "*agak-agak*". The final judgement was the taste test; when the curry or stew was on the boil, the cook then added more of this or that to taste. Thus, treat the measurements as recommendations and adjust by doing the taste test. Do this near the end of cooking when the meat is the right texture and no more water is needed for further cooking. The flavour will be more accurate if you adjust the seasoning at the end rather than in the middle of cooking.

Taste-testing also applies to texture. Vegetables and cuts of meats may vary in toughness, so adjust cooking times according to how you want them. To check the texture of meat, use a fork. If the fork goes in easily, the meat is tender. If younger folks prefer firmer vegetables, remove half while the vegetables are still firm, and cook the rest further for the old folks. There is a lot of "guesstimating" in cooking, which is why no two cooks using the same recipe will come up with exactly the same taste. (But I do try!)

Slow Cooking Methods

Many Straits-born stews, soups and pork stock in particular require long cooking. Apart from boiling on a stove-top, the two other easier slow cooking methods are in a slow cooker or in a covered casserole dish in an oven set to between 90°C (195°F) and 150°C (300°F). The lower temperature takes more time and is excellent for overnight cooking when you don't want to get the meat too tender. Make sure there is head room in the casserole for the food to boil up. Note that evaparation is minimal in the oven method. Adjust the liquid accordingly. If doing stocks, these two methods yield a clear stock, unlike boiling on the stove.

Cooking Oil

In this book, cooking oil refers to canola (rapeseed)/soy/peanut (groundnut)/sunflower/corn/vegetable oil. Do not use an oil with a strong flavour, such as olive oil/butter/ghee/lard/mutton fat unless so specified. When deep-frying something, use peanut (groundnut) oil if available. This oil is more stable at high temperatures and the deep-fried food will have better keeping qualities.

Rendering Lard

To render lard or any animal fat, get a piece of fat from the butcher. Using a sharp knife, cut the fat into small cubes. Put into a pot and set it over low-medium heat to render the lard. Take care not to over-brown the bits of crispy fat. Use them as a topping for fried noodles.

Serving Sizes

The quantities of each recipe should serve at least four because most Straits-born meals are communal, consisting of several dishes to be eaten with rice and shared with others. With the exception of *kerabu* (salad) and stir-fried vegetables, the curries and stews keep well and some, like the meat ones, freeze well too. When freezing cooked food, freeze in serving portions.

Kerabu, Acar and Vegetables

From Garden to Table

Kerabu and *urap* are Malay-Indonesian words for salads dressed Straits-born style. It depends on who is talking about it. *Kerabu* is the word used in Penang and northern Peninsular Malaya while *urap* is more common southwards in Melaka, Singapore or Indonesia. Straits-born salads show many similarities with Thai salads. There are always raw or lightly blanched vegetables, some kind of protein, shallots or onions, fresh herbs and a dressing usually of lime juice and *sambal belacan* in the case of Straits-born salads. In place of lime juice, sour fruit may be the tart ingredient when these fruits are in season. Popular would be green mangoes, pineapple or belimbing. The gardens of many Straits-born homes often had a mango or belimbing tree, or the families knew someone who did. Foraging or popping into the garden to collect some leaves or pick fruit for a *kerabu* was common in days when most Straits-born families lived in houses with gardens.

A *kerabu* can be simple or elaborate depending on what there is or what the cook wants to start with. Pineapple *kerabu* may simply be coarsely chopped pineapple mixed with sliced fresh chillies, sugar and dark soy sauce or the fresh chillies replaced with *sambal belacan* because this condiment is always in any Straits-born kitchen. Add some boiled prawns or that slice of leftover asam pork — chopped or slivered so that a little goes a longer way. Add *serunding* or *krisek* to get a larger or different-tasting dish. *Kerabu*-making relies on imagination, the taste-test and what is on hand. Always keep the dressing ingredients nearby until the *kerabu* tastes just right.

A little more exacting is preparing an *acar* (pickles) or chutney. Straits-born cooks pickle more than just vegetables or fruit. Some will pickle fish and many have a bottle of pickled krill called *cinchalok* on standby for an easy appetiser. Although my mother regularly used the *agak-agak* method even with pickles, I found that following a quantified recipe made for more consistent results when it came to pickles. *Acar* needs time to develop its true flavour and taste-testing and adjusting the flavours once the pickling has started is not an option. About the only adjustment possible is the amount of sweetness or tartness. *Acar* and chutneys are good standby dishes for days when the master chef is too tired or lazy to prepare anything other than boil a pot of rice. They are also excellent as side dishes with simpler fare like a fried fish or steamed prawns.

Vinegar Acar

Preparation and cooking time: 2 days; advance preparations required

This is a Melaka-style pickle that requires no refrigeration, unlike *acar awar*, a Penang pickle with ground peanuts in it. The pickles get more sour over time, but the quick fix is to make a thick sugar syrup with white sugar, cool and stir the syrup into the pickles. Leave overnight before consuming. Always spoon out any pickles or chutney with a clean dry spoon, and never return any uneaten portion to the jar. The pickled chillies in the picture is made with the same pickling mix as this vinegar acar.

Assorted vegetables to make up 4 kg (8 lb 12 oz) of trimmed vegetables

- Cucumber, cored and julienned
- Carrots, peeled and julienned
- Cauliflower, cut into florets
- White cabbage, cut into 2-cm (1-in) squares
- Green chillies stuffed with shredded green papaya/carrot
- Green beans, stringed and cut into 2-cm (1-in) pieces
- Sugar snap peas, stringed and cut into 2-cm (1-in) pieces
- Small garlic cloves, peeled
- Small shallots, peeled

Pickling Mix

150 g (5^1/$_3$ oz) dried shrimps (*hae bee*), softened in water

2 Tbsp dried shrimp paste (*belacan*)

500 g (1 lb 1^1/$_2$ oz) old turmeric, scraped and chopped

1.5 litres (48 fl oz / 6 cups) Chinese white rice vinegar

500 ml (16 fl oz / 2 cups) cooking oil

100 g (3^1/$_2$ oz) garlic, peeled and chopped

2.5 kg (5 lb 8 oz) white sugar

2^1/$_2$ tsp salt

125 g (4^1/$_2$ oz) white sesame seeds, toasted and coarsely pounded

1. The pickling mix must be prepared a day ahead and be stone cold before the prepared vegetables are added.

2. Chop dried shrimps and *belacan* with some vinegar in a food processor until fine. Set aside.

3. Blend turmeric with vinegar in a food processor to get a fine pulp. Let turmeric soak for 30 minutes or so until richly coloured. Strain vinegar through a sieve and set aside. Discard the turmeric pulp.

4. Heat oil in a 5-litre (4-quart / 20-cup) stainless steel saucepan and fry chopped garlic until fragrant and pale gold. Add chopped dried shrimps and *belacan*, taking care not to over-brown garlic.

5. Add turmeric-coloured vinegar, sugar and salt and bring to a vigorous boil. Turn off heat and cool pickling mix overnight. Cover with a thick kitchen towel rather than the pot lid, to prevent moisture from dripping into the mix.

6. Prepare cucumbers first. It's the only one that needs to be salted so that the juices can be squeezed out. The more juices you squeeze out, the more crisp the pickles. Rub trimmed cucumber with ^{1}/$_2$ tsp salt. Stand for 30 minutes or longer, then squeeze out juices.

7. There are three ways to prepare the vegetables for pickling. One is to spread them out on rattan trays and dry in the sun till wrinkled. This is the best method. A second method is to blanch the different vegetables in small batches for 5–10 seconds in a pot of boiling water. Take care to cool and air-dry the blanched vegetables. A third way is to spread the vegetables on rattan trays and dry under a fast-moving fan for 12 hours or till wrinkled.

8. Using a large clean dry spoon, stir cooled wrinkled vegetables into cold pickling mix. Stir in toasted pounded sesame seeds last.

9. Let stand for at least a week in the pot before dividing into clean glass jars. The vegetables will be ready for consumption at different times. Carrots and green beans are ready after a week followed by cucumbers, cauliflower and chillies. The garlic and shallots will take at least a month.

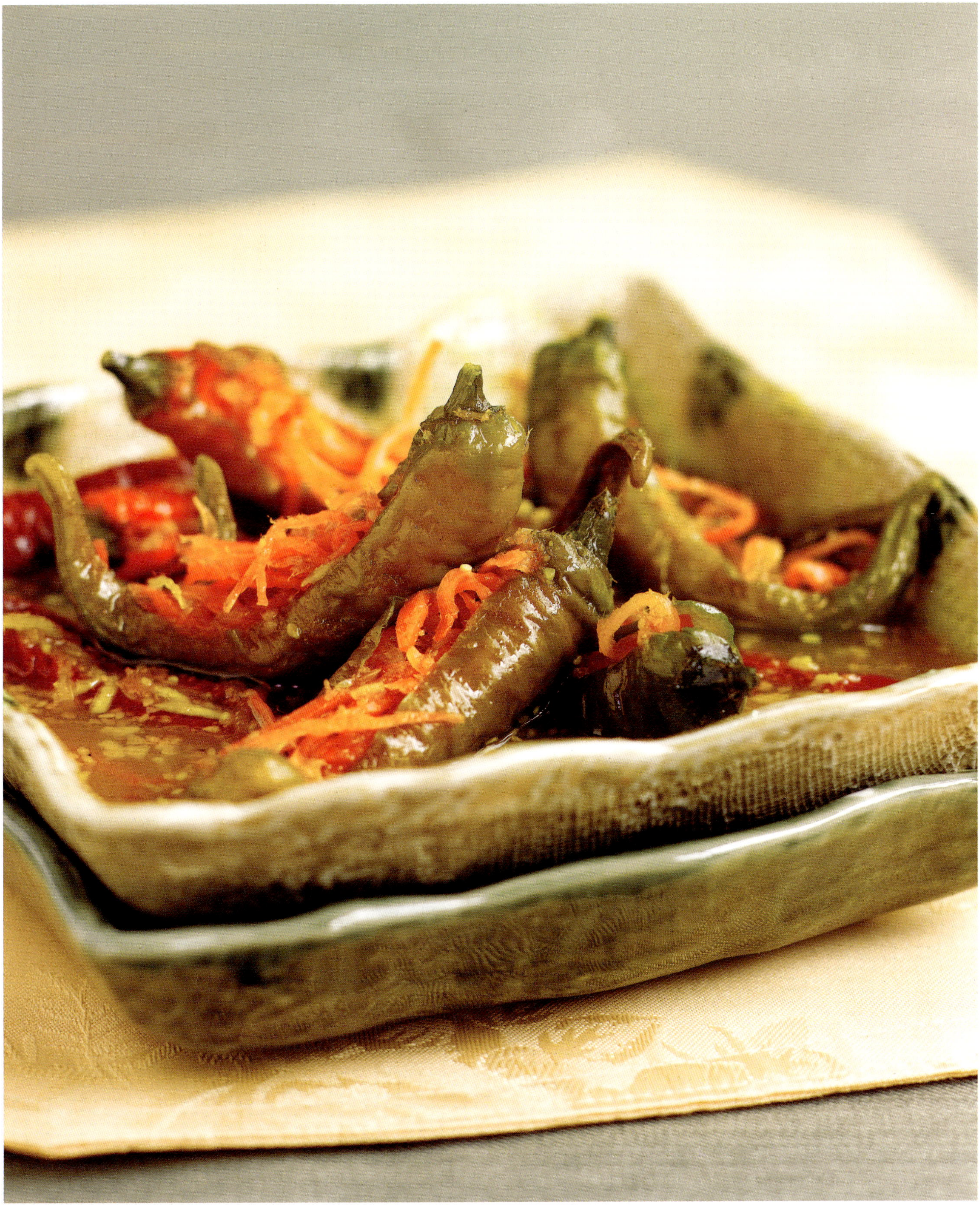

Petai and Prawn Kerabu

Preparation and cooking: 45 minutes

Petai aka stink beans aka twisted cluster beans are from a large tropical tree and once harvested in the wild. Apparently some are now farmed. If very young, *petai* may be eaten raw. If mature, a light blanching may be preferred. Mature beans have marked ridges, unlike young beans. Unless you don't mind the occasional extra protein, *petai* needs to be split before cooking because perfect-looking beans may harbour a little worm inside.

300 g (11 oz) *petai*

50 g (1¾ oz) onion, peeled and finely sliced

1 Tbsp *sambal belacan* (page 14)

1 tsp salt

2½ Tbsp lime juice

400 g (14 oz) prawns, boiled, shelled and deveined

Choice of fresh herbs (kaffir lime leaves/laksa leaves/spring onion/coriander leaves/turmeric leaves), finely shredded or chopped

50 g (1¾ oz / ¼ cup) fresh grated coconut/*serunding* (page 18) (optional)

1. Rinse, then blanch *petai* for 20 seconds in a pot of boiling water. Drain.

2. Mix onion with *sambal belacan*, salt and lime juice. Stir in prawns and *petai,* followed by herbs and grated coconut.

3. Serve with rice.

Bean Sprout Kerabu

Preparation and cooking: 10 minutes

This is another common Straits-born dish given the ubiquity of bean sprouts. Recently, I found a way to keep bean sprouts fresh for up to a week or more: pluck the tails and wrap the unwashed sprouts in paper towels or sheet of newspaper to absorb the moisture that turns the bean sprouts slimy if just kept in a plastic bag. This method works well for bean sprouts that are sold dry as in Singapore. In some places, such as Kuala Lumpur, stallholders keep bean sprouts in tubs of water!

300 g (11 oz) bean sprouts, tails removed, rinsed

25 g ($^4/_5$ oz) shallots, peeled and thinly sliced

2 Tbsp lime juice

1 Tbsp *sambal belacan* (page 14)

$^1/_2$ tsp salt

3 Tbsp toasted grated coconut (*serunding*) (page 18)

Chopped spring onion/coriander leaves/chives/laksa leaves

1. Bring a pot of water to the boil and blanch bean sprouts for 10 seconds Remove and drain well.

2. In a serving bowl, mix shallots, lime juice, *sambal belacan* and salt. Stir in blanched bean sprouts.

3. Top with toasted grated coconut and choice of chopped herbs. Serve with rice.

Kerabu Kay (Chicken Salad)

Preparation and cooking: 30 minutes

You can tell just by the name of the dish that this is a Penang recipe. It was from my aunt in Alor Star, but where she got it from is unknown. What makes the *kerabu* special is the flavour of fresh coconut cream combined with crunchy boiled cabbage and a tart spicy dressing. Packaged coconut cream just won't do. So when I prepared coconut cream for this picture, I had three great meals with this *kerabu*. Home-made coconut cream keeps fresh even without salt added.

300 g (11 oz) chicken, boiled

400 g (14 oz) cabbage, cut into bite-size pieces

50 g (1³/₄ oz) shallots, peeled and finely sliced

2 Tbsp *sambal belacan* (page 14)

2 Tbsp lime juice

1 tsp salt

250 ml (8 fl oz / 1 cup) fresh coconut cream (page 20)

1. Remove and set aside bones from boiled chicken and cut meat into bite-size pieces. Return bones to boil further and save the chicken stock for something else.

2. Bring a pot of water to the boil and blanch cabbage for 10 seconds. Remove and drain well, then cool and squeeze out water from the cabbage. This makes the cabbage more crunchy.

3. Mix shallots slices, *sambal belacan*, lime juice and salt together. Stir in chicken and cabbage.

4. Top with coconut cream and adjust seasoning to taste.

5. Serve with rice.

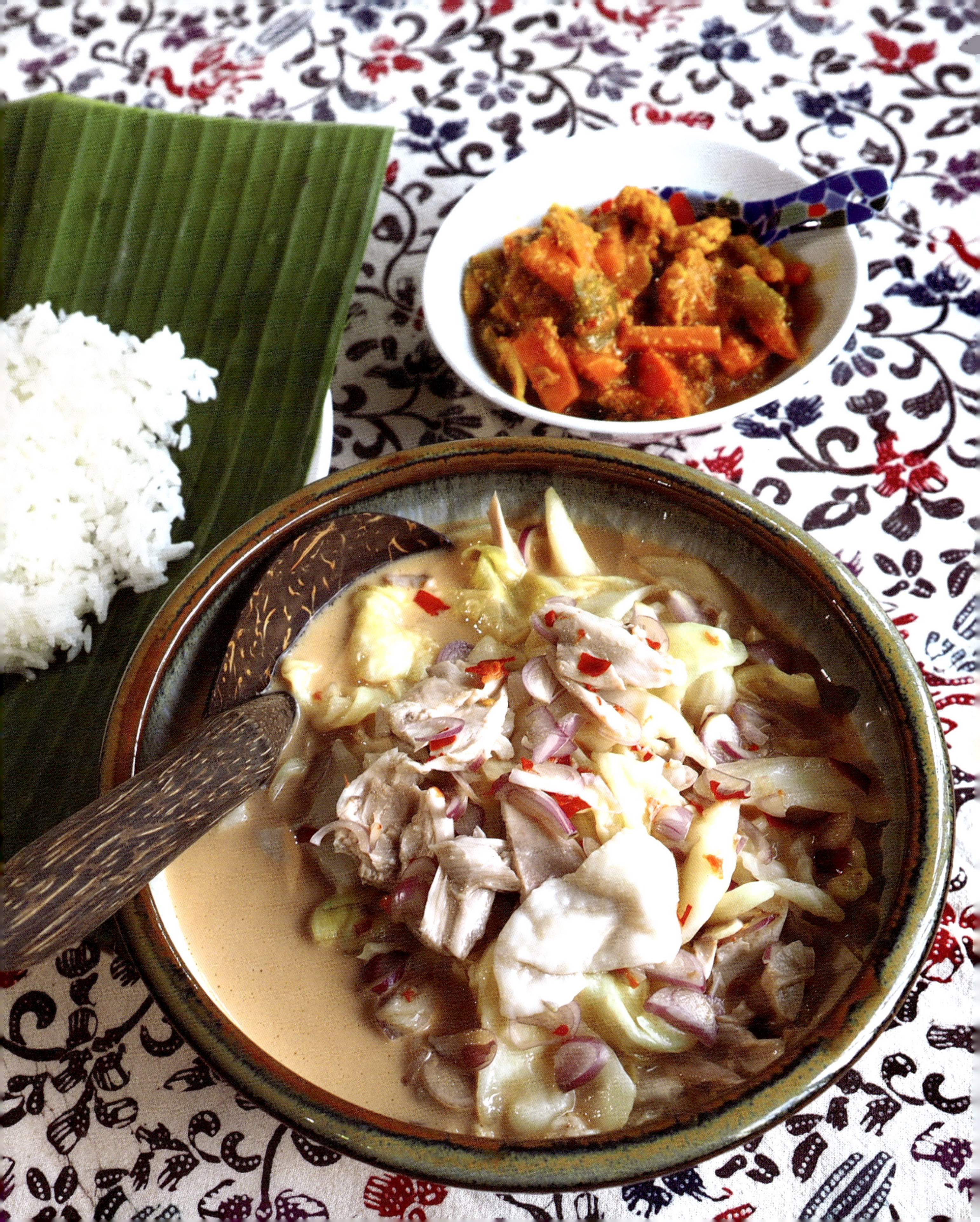

Cabbage, Wood Ear and Prawn Kerabu

Preparation and cooking: 30 minutes

The chopped torch ginger bud makes this a very fragrant *kerabu*. It is also full of different textures because of the wood ear fungus, cabbage and boiled prawns. If torch ginger bud is not available, substitute with whatever fragrant herb is available, for example, laksa leaf or mint. The black wood ear fungus may also be substituted with a head of white fungus.

$^1/_4$ cup wood ear fungus, softened in water

200 g (7 oz) cabbage, finely shredded

300 g (11 oz) prawns, boiled and peeled

2 spring onions, chopped

1 bunch coriander leaves, chopped

Dressing

50 g ($1^3/_4$ oz) shallots, peeled and thinly sliced

$1^1/_2$ Tbsp *sambal belacan* (page 14)

2 Tbsp lime juice

1 tsp salt

1 tsp sugar

1 Tbsp finely sliced torch ginger bud

4 Tbsp fresh coconut cream (page 20) (optional)

1. To clean softened wood ear fungus, trim off any woody, sandy bits. Cut into bite-size pieces/strips. If using white fungus, trim off woody bits and cut into bite-size pieces.

2. Bring pot of water to the boil and blanch cabbage for 5 seconds. Drain well and leave to cool before squeezing out excess moisture.

3. Return the pot of water to the boil and blanch wood ear fungus for 2 minutes. Drain well and leave to cool.

4. To make dressing, combine all the ingredients and mix well.

5. Stir wood ear fungus, cabbage, prawns and herbs into dressing. Mix well and adjust seasoning to taste.

6. Serve at once with rice.

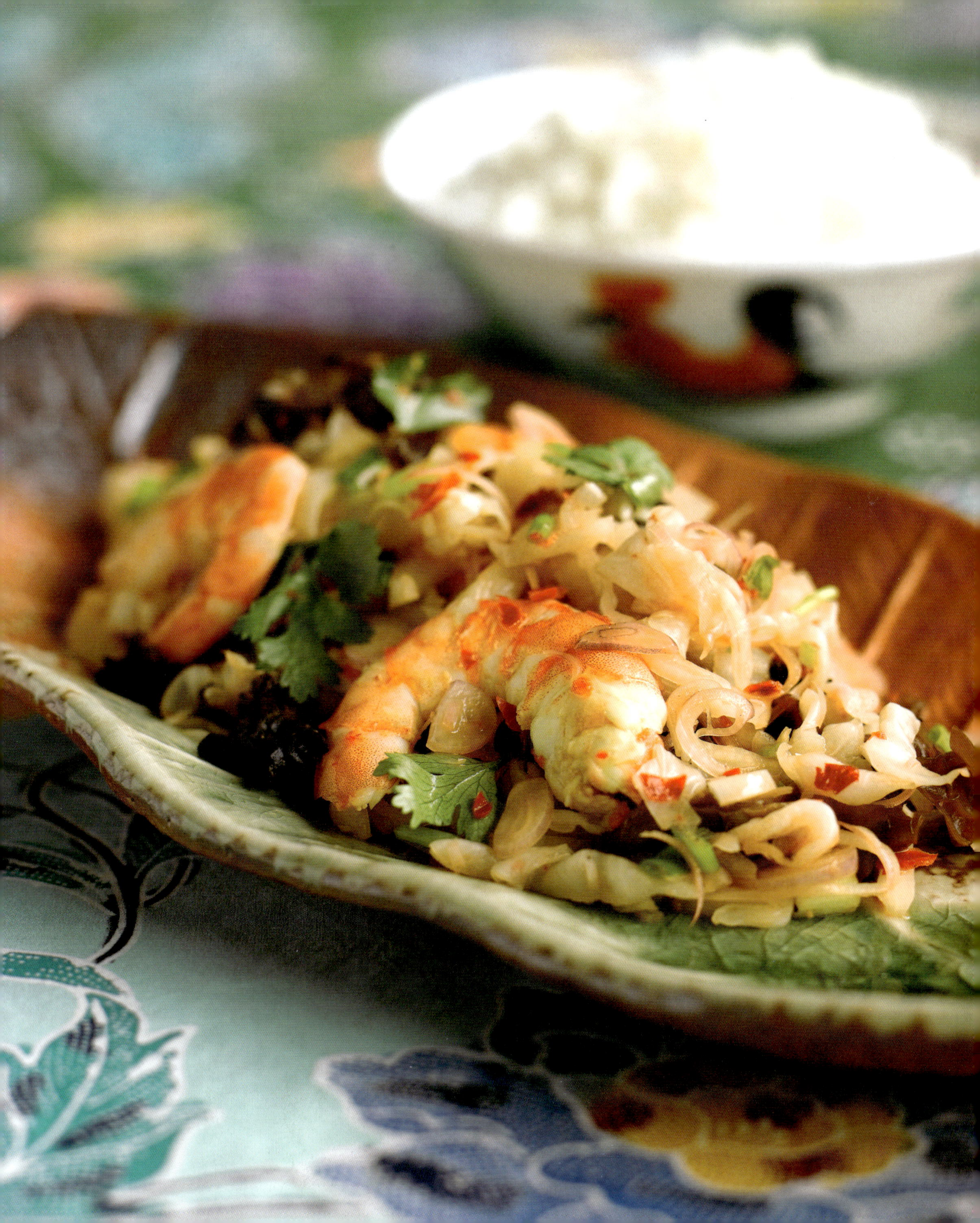

Belimbing and Prawn Kerabu

Preparation and cooking: 20 minutes

The belimbing tree is a pretty tree and once popular in Straits-born gardens. The fruit is easily picked and used in salads, chutneys and stews. As belimbing is always sour, no lime or tamarind juice is required in salads and stews made with belimbing. This fruit is never available commercially, but its relative, carambola or star fruit, is seasonally available. Green star fruit is tart, but when ripe, may be sweet-sour. This salad can also be made with *buah binjai* (another super sour tropical fruit related to the mango, but which is also usually home-grown) or the more easily available pineapple or green mangoes.

200 g (7 oz) belimbing/star fruit/ pineapple/green mangoes

1 Tbsp *sambal belacan* (page 14)

50 g (1³/₄ oz) onion, peeled and thinly sliced

¹/₂ tsp salt

200 g (7 oz) grilled or fried fish, meat or boiled prawns, flaked or sliced

2 spring onions, chopped

1 bunch coriander leaves, chopped

¹/₄ cup *serunding*, *krisek* (page 18) or toasted and ground peanuts

1. Cut the belimbing or fruit of choice to your preferred shape and taste. Slice into small wedges, sliver thinly or chop coarsely. If using star fruit, the tough seed sacs in the centre must be pulled out before slicing. Chopping coarsely spreads the tartness.

2. Mix the *sambal belacan* with the onion and salt and stir in the sour fruit. Add the flaked fish, meat slices or prawns.

3. Top with *serunding*, *krisek* or peanuts.

4. Serve with rice.

Green Mango Kerabu

Preparation and cooking: 30 minutes

When picking green mangoes for a *kerabu*, note that not all green mangoes are sour, although for many varieties, green or even yellowing mangoes tend to be more sour than sweet.

1 Tbsp *sambal belacan* (page 14)

1–2 tsp sugar

¹/₂ tsp salt

4 shallots, peeled and finely sliced

3 small green mangoes, peeled and shredded

1 tomato, seeded and cubed

Kaffir lime leaf, finely sliced (optional)

1. Mix *sambal belacan*, sugar, salt and shallots together.

2. Stir in shredded mangoes and adjust seasoning to taste. Stir in cubed tomato.

3. Top with finely sliced kaffir lime leaf.

Pictured: Belimbing and Prawn Kerabu

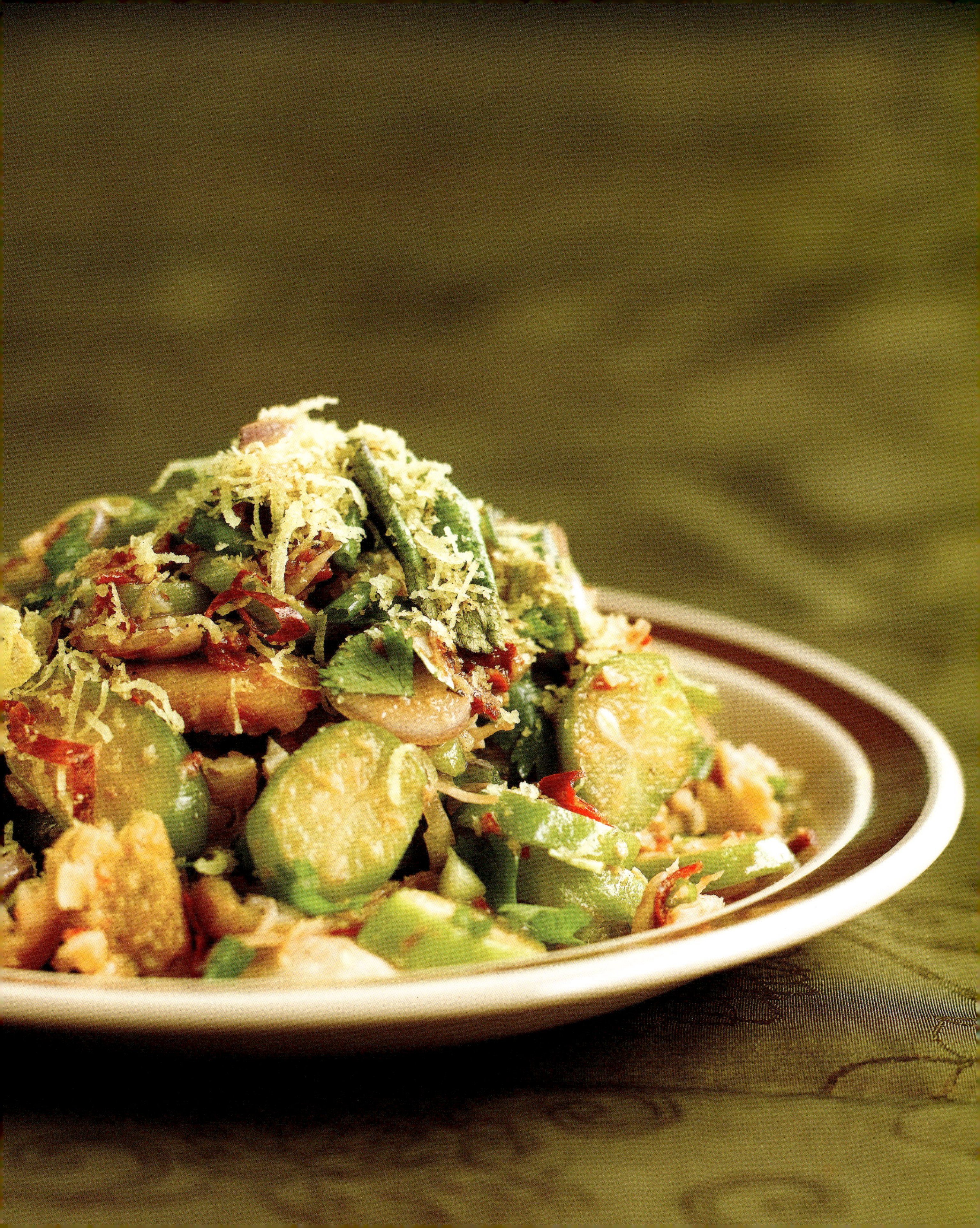

Penang Fruit Rojak

Preparation and cooking: 1 hour

Rojak appears to have been a 10th century Javanese dish, but the dressing would have been very different. Penang *rojak* has remained a fruit salad while Singapore *rojak* is a mix of *taupok* (fried tofu puffs), *eu char kuay* (fried dough crullers), cucumber, *bangkuang* (yam bean), pineapple, blanched bean sprouts and *kangkong* (water convolvulus). These days, most *rojak* stalls in Singapore skip the bean sprouts and *kangkong*, and you may find green apple or green mango instead. Some stalls have taken it upmarket by adding pickled jellyfish, grilled dried squid or century egg.

Any combination of the fruits below to make up 800 g ($1^3/_4$ lb):
- Green mango
- *Buah kedondong* (ambala)
- Pineapple
- *Jambu ayer* (water apple)
- Guava
- Green apple

Dressing

3 Tbsp sugar

2 Tbsp lime juice

1 Tbsp tamarind juice

75 g ($2^2/_3$ oz / $^1/_3$ cup) *hae ko* (black prawn paste)

1 Tbsp cooked dried red chilli paste (page 17)

1 Tbsp chopped torch ginger bud (optional)

250 g (9 oz) peanuts, toasted and ground

1. Peel mango, *buah kedondong* and pineapple and cut into bite-size pieces. For the other fruits that do not require peeling, cut them into bite-size pieces as well.

2. In a large mixing bowl, mix together sugar, lime juice and tamarind juice, black prawn paste and red chilli paste. Dilute with more tamarind juice/water, if needed to get a dressing that will drip slowly off a spoon.

3. Add torch ginger bud, if using, cut fruits and peanuts. Mix well and adjust seasoning to taste.

4. Serve at once.

Hae Ko Dip

Preparation time: 10 minutes

This is a quick dip for assorted vegetable sticks like cucumber or carrots, and also tart fruit like pineapple or mango. Adjust the proportions to suit your taste and what you are eating it with.

Hae ko (black prawn paste)

Fresh sliced red chilli/ *sambal belacan* (page 14)

Sugar

Lime juice (optional)

1. Combine the ingredients in a bowl and adjust according to taste.

2. Serve with cucumber or carrot sticks, slices of green mango, star fruit, pineapple, guava, *jambu ayer* (water apple) and/or *buah kedondong* (ambala).

Pictured: Penang Fruit Rojak

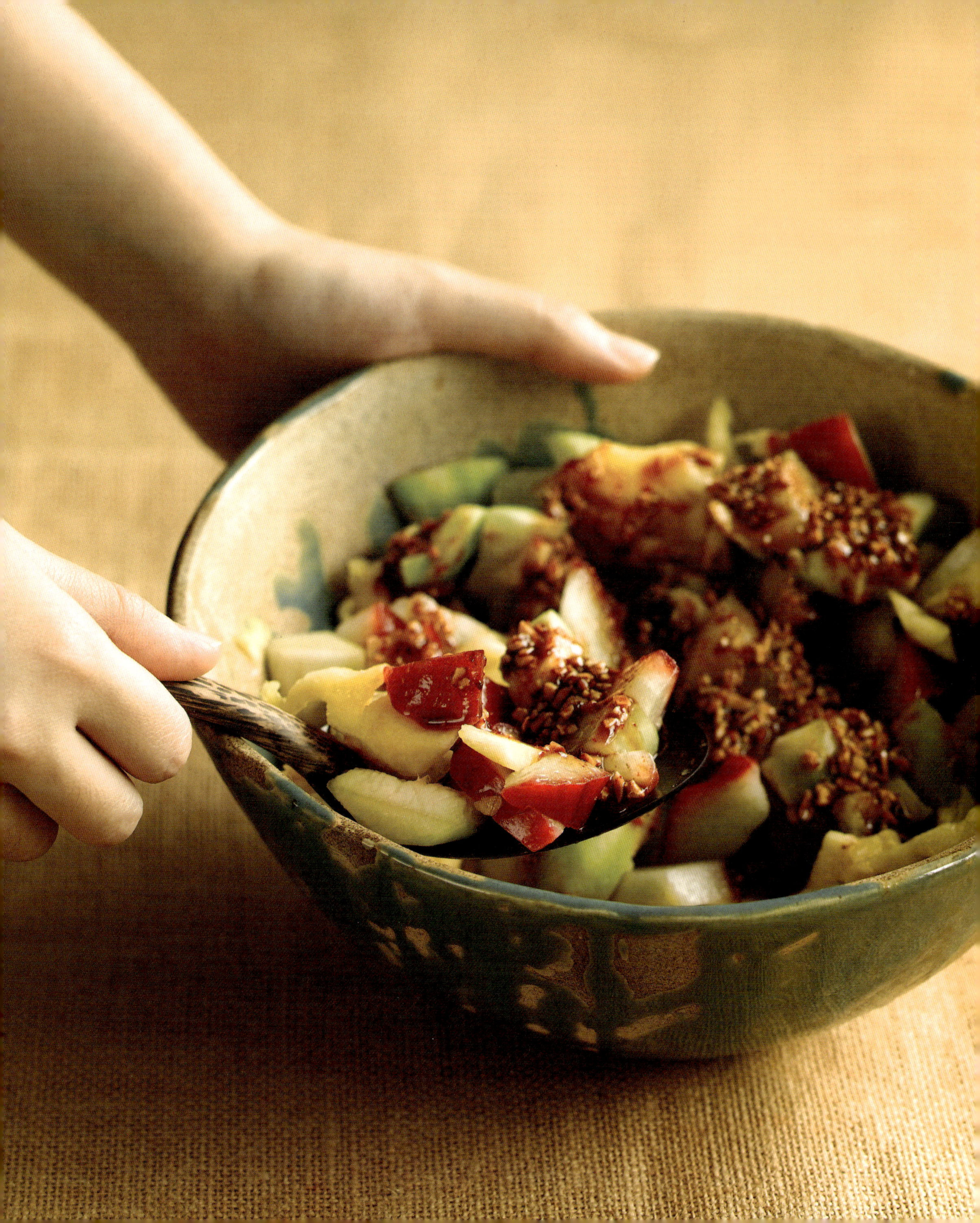

Chetti Urap Timun (Cucumber Salad)

Preparation and cooking: 30 minutes

This is a salad that is soured with belimbing, if you can get it. Otherwise, use lime juice or any sour fruit such as pineapple or green mango. The dried seafood can be substituted with boiled prawns or meat.

30 g (1 oz / 1 Tbsp) dried shrimps (*hae bee*)/dried krill (*grago*)/boiled prawns

250 g (9 oz) cucumber

150 g (5$^1/_3$ oz) tomatoes

100 g (3$^1/_2$ oz) onion, peeled and finely sliced

2–3 fresh red chillies, seeded and thinly sliced

1–2 tsp *sambal belacan* (page 14) (optional)

1 tsp sugar

$^1/_2$ tsp salt

4 belimbing, cut into thin strips/ 1 green mango, cut into thin strips/2–3 limes, juice extracted

1. If using dried shrimps, rinse and pick through to remove any bits of shells. Chop or pound if dried shrimps are large. If using dried krill, rinse clean and pick through any debris. Dry-fry or grill the dried seafood over low heat until fragrant. If using boiled prawns, shell them.

2. Discard the core of the cucumber, then shred the cucumber. Soak the shreds in water for 10 minutes to crisp them. Spin-dry the cucumber strips if you have a salad spinner.

3. Halve the tomatoes and remove the watery seeds, if desired. Cut into strips or pieces to preferred thickness.

4. Mix sliced onion, chillies, *sambal belacan* if using, sugar and salt together. Add belimbing/green mango/lime juice. Stir in the chosen seafood, cucumber and tomatoes.

5. Serve with rice.

Pineapple Kerabu

Preparation and cooking: 30 minutes

A 1-kg (2 lb 3-oz) pineapple will give about 500 g (1 lb 1½ oz) of fresh pineapple — about half the weight is in the skin and the core. Make an appetiser with half and turn the other half into a chutney for another meal.

250 g (9 oz) fresh pineapple, peeled and coarsely shredded

2 shallots, peeled and finely sliced

½ Tbsp *sambal belacan* (page 14)

½ Tbsp dark soy sauce

2 tsp sugar

1. Combine all the ingredients and mix well. Adjust seasoning to taste.

2. Serve with rice.

Pineapple Chutney

Preparation and cooking: 30 minutes

2 Tbsp cooking oil

25 g (⅘ oz) onion, peeled and cut into rings

2 fresh red chillies, seeded and cut into strips

250 g (9 oz) pineapple, cut into wedges

2 tsp sugar

1 tsp dark soy sauce

1 tsp light soy sauce

2 Tbsp water

1. Heat oil in a frying pan and fry onion and chilli until onion is transparent. Add pineapple wedges followed by sugar and soy sauces, and finally water.

2. Turn down the heat and simmer until pineapple is limp and transparent. Add more water if necessary. Adjust seasoning to taste.

3. Serve with rice or cool and refrigerate for another meal.

Mushroom Stem Sambal

Preparation and cooking: 1 hour

This unusual dish was the creation of a Hakka nun in this Buddhist temple in Singapore that my Hakka grandmother supported. In the 1950s and 1960s, the nuns would prepare a vegetarian lunch for the iconic Buddhist festivals when devotees showed up for prayers at the temple. Decades later, I was swept by a sudden nostalgia for a taste of this *sambal*. Racking my taste memory banks and after several experiments, I hit upon the recipe. This all-vegetarian *sambal* has plenty of umami because mushrooms and tomatoes are rich in glutamic acid. What's more, it is an expression of the "waste not, want not" mantra that prevails in many cultures. Shiitake stems are usually discarded in other dishes, but not here: they are essential.

65 g (2 oz) dried shiitake mushroom stems

30 g (1 oz) dried shiitake mushrooms

375 ml (12 fl oz / 1½ cups) water

2 Tbsp tamarind paste

3 Tbsp cooking oil

1 tsp salt

200 g (7 oz) ripe tomatoes, cubed

100 g (3½ oz) cauliflower, cut into florets

100 g (3½ oz) *petai*, split open

100 g (3½ oz) green beans/sugar snap peas

3 green chillies, seeded and coarsely sliced

Spice Paste

5-cm (2-inch) galangal, scraped and sliced

10-cm (4-inch) old turmeric, scraped and sliced

4 Tbsp water

2–3 Tbsp dried red chilli paste (page 17)

1. Make the spice paste by blending together the galangal, turmeric and water to a fine paste. Add the dried red chilli paste. Set aside.

2. Rinse the mushroom stems and caps clean. Soak in 250 ml (8 fl oz / 1 cup) water until soft. Trim any stems from the caps. The caps can be kept whole or sliced if preferred. Save the mushroom water for cooking.

3. Trim any woody bits from the mushroom stems and squeeze them dry. Pound the stems in a mortar and pestle until shredded. (Note that chopping in a food processor will give an entirely different texture to the *sambal*.)

4. Measure out 125 ml (4 fl oz / ½ cup) of the mushroom water and mix with the tamarind paste. Strain and discard the solids.

5. Heat oil in a saucepan or wok and fry the spice paste until oil rises to the top. Add the pounded mushroom stems, mushroom caps, salt and tamarind juice. Mix well and bring to the boil. Turn down heat and simmer for 10 minutes.

6. Add the tomatoes, cauliflower and *petai* and simmer for 10 minutes or until tomatoes are soft. Add beans/sugar snap peas. Cook until beans/peas change colour before adding green chillies. If too dry, add more mushroom water/tamarind juice.

7. Add green chillies, stir well and adjust seasoning to taste. Turn off heat when green chillies change colour. The *sambal* should be moist and the vegetables still be firm.

8. Let stand for at least half a day before serving with rice.

9. This *sambal* keeps well for weeks in the fridge. Reheat only serving quantities or serve at room temperature.

Sambal Belacan Vegetables

Preparation and cooking: 20 minutes

A vegetable stir-fried with *sambal belacan* is one of the usual Straits-born ways of serving vegetables and the vegetable most commonly cooked this way is the much-loved *kangkong* (water convolvulus). Firm vegetables such as French beans and long beans may even be turned into a quick stir-fried *sambal* with the addition of fresh prawns and cubes of fried tofu or tempeh.

2 Tbsp cooking oil

2 cloves garlic, peeled and finely chopped

1 Tbsp *sambal belacan* (page 14)

$^1/_2$ tsp salt

2 Tbsp water

Choice of Vegetable
- *Kangkong* (water convolvulus)
- Local spinach (*bayam*)
- Ladies' fingers (okra)
- Long beans
- French beans (green beans)
- Drumsticks (*moroongakai*)

Optional Ingredients
- 2 tsp brown miso/1 Tbsp fermented soy beans (*taucheo*), rinsed and mashed
- 2 shallots, peeled and pounded
- 1 Tbsp dried shrimps (*hae bee*), softened in water, then drained and pounded
- 100 g (31/2 oz) prawns, shelled

1. Heat oil in a wok, add garlic and *sambal belacan* and stir-fry until fragrant. Add salt, followed by the vegetable of choice.

2. If using any of the optional ingredients, add miso/fermented soy beans/shallots/dried shrimps at the same time as the garlic and *sambal belacan*.

3. If using fresh prawns, add them when the *sambal belacan* smells fragrant. When the prawns change colour, add the vegetable and 2 Tbsp water.

4. If frying ladies' fingers, add them at the same time as the prawns. Ladies' fingers take at least 5 minutes to become pleasantly crunchy and double that to soften. Add more water, if needed. This is one vegetable that tastes better on the soft side.

5. Serve with rice.

Rich Spicy Kangkong with Sweet Potato

Preparation and cooking: 30 minutes

Sweet potatoes originated in Central and South America and were introduced to South East Asia by either the Portuguese or Spanish. Sweet potatoes can be used in desserts or savoury dishes. This dish pairs sweet potato with *kangkong* (water convolvulus), but the spice paste here can be used for cooking other vegetables too. I have provided a list of vegetable substitutes below. The dried shrimps in it give the dish its umami. You could also substitute the sweet potato with a potato or two. Potatoes were also originally from South America. In South East Asia, potatoes and sweet potatoes are treated as vegetables and eaten with rice.

200 g (7 oz) sweet potato, peeled and cut into chunks

1 kg (2 lb 3 oz) *kangkong* (water convolvulus), rinsed

2 Tbsp cooking oil

500 ml (16 fl oz / 2 cups) water

125 ml (4 fl oz / ¹⁄₂ cup) packaged coconut cream

1 tsp salt

200 g (7 oz) small prawns, shelled (optional)

Spice Paste

50 g (1³⁄₄ oz / ¹⁄₂ cup) shallots, peeled and coarsely sliced

2 Tbsp dried shrimps (*hae bee*), softened in water

50 g (1³⁄₄ oz) fresh red chillies, seeded

1 tsp dried shrimp paste (*belacan*)

1. Make the spice paste by blending together all the ingredients in a food processor to a smooth paste.

2. Peel the sweet potato and cut into small chunks.

3. Clean and rinse *kangkong*, then trim roots and split large stems in half. Cut *kangkong* into finger-lengths, keeping leaf and stem together.

4. Heat oil in a saucepan and fry spice paste until fragrant and oil rises to the top.

5. Add water, salt, half the coconut cream and sweet potato. Simmer until sweet potato is tender. Depending on the size of the chunks, it should take about 20 minutes. It should be soft, but not disintegrating.

6. Stir in prawns if using and when prawns change colour, add the remaining coconut cream and *kangkong*. Cook until the leaves change colour. This should take only about 1 minute. The leaves should be dark green and the stems still crunchy.

7. Serve with rice.

Vegetable Substitutes
• Cabbage and carrot
• Long beans/French beans
• Brinjals
• Local spinach (bayam)
• Ladies' fingers (okra)
• Squash and petai

Note
The pink milk drink in the picture is *susu bandung*. It is simply rose syrup in milk and ice water. It is a classic Straits-born drink.

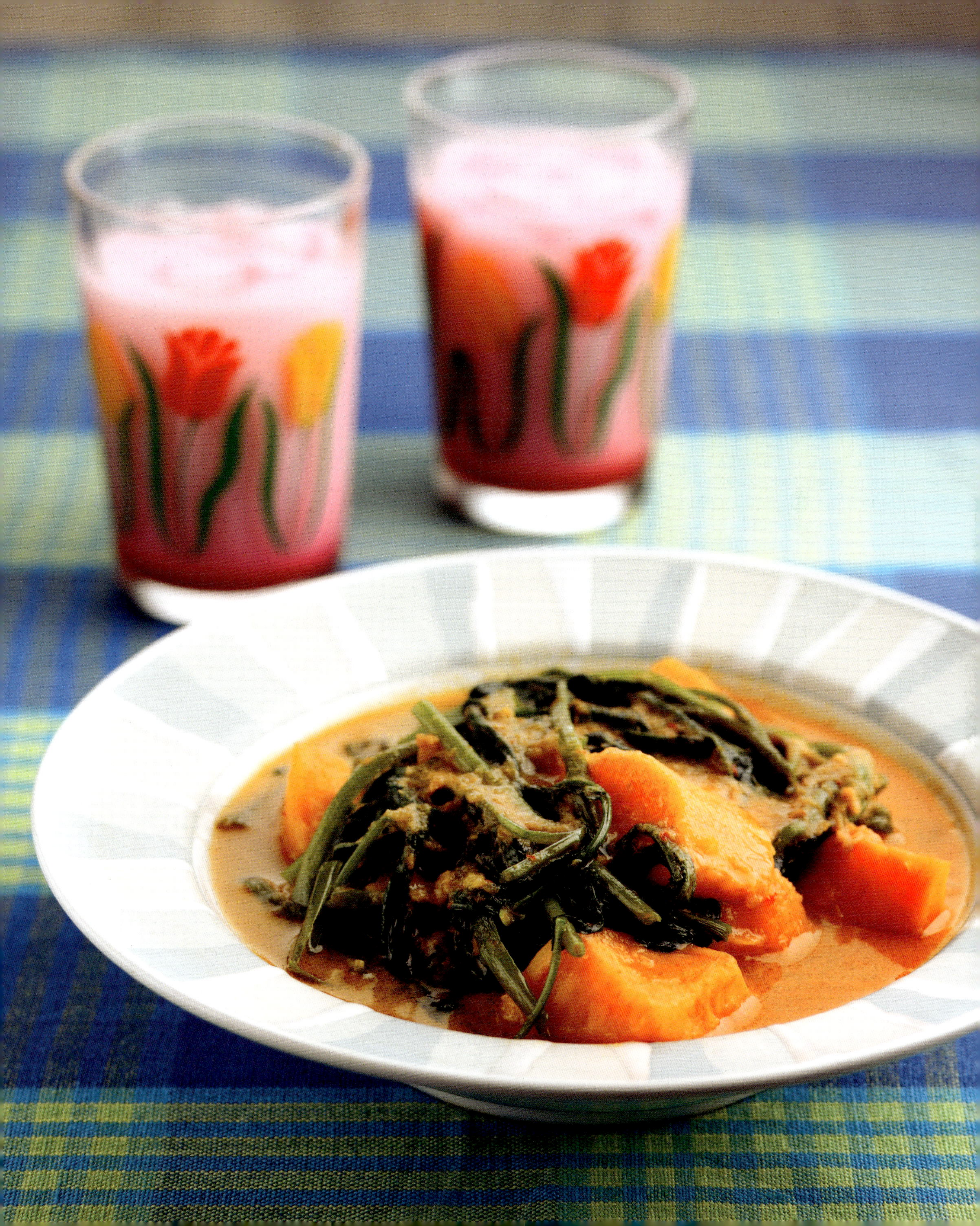

Rice and Noodles

Rice is Nice and Noodles Too

Traditional rice-eating cultures cover a huge geographical swathe stretching from India to Japan. All of South East Asia is rice-eating and where there is land for agriculture, rice is grown — even when terraces have to be cut into mountain slopes for rice to be grown. While the hill tribes of South East Asia grow short-grain rice (*Oryza sativa japonica*), the common rice everywhere else is long-grain rice (*Oryza sativa indica*). This is the rice that takes pride of place in daily and celebratory meals in all the Straits-born communities. Plain boiled rice is eaten with elaborate or simple side dishes. For special occasions, the rice may be dressed up in a variety of ways to get a biryani or the simple yet delicious coconut rice or *nasi lemak* (rich rice). Rice-eating cultures even eat rice for breakfast and little packets of *nasi lemak* wrapped in banana leaf are still sold for breakfast in coffee shops throughout Singapore. *Nasi lemak* makes a great weekend special and is even enjoyed stone-cold. Sides are easily increased for unexpected guests and leftovers are just as delicious. Grandma and Grandpa can sit down to mahjong sessions and the younger folks can take care of themselves meal-wise. Other flavoured rice such as biryanis and *nasi kebuli* are also good for family gatherings and celebrations. Who would say "no" to prawn biryani, fond as so many of us are of prawns? And glutinous rice plays a prominent role in not only Straits-born desserts and treats.

Where there is rice, there are rice noodles and more than that. The Straits-born have elaborate rice noodle dishes — laksa, *mee siam*, *kerabu bee hoon* (a very Penang way of preparing rice noodles), fried noodles and noodle soups. Wheat noodles are just as popular. Hokkien mee aka yellow mee aka cooked wheat noodles get star status for Chinese New Year and birthday celebrations in many families. In mine, every Chinese New Year's Eve dinner had to include a plate of fried Hokkien mee which everyone had to eat, even if only a token mouthful. "*Jia tng mia*" my mother would say in Hokkien, meaning "eat for a long life", although I have never needed any persuading to eat noodles of any kind.

Straits-born celebrations were always with a huge spread of soups, curries, salads, *acar*, chutneys, rice and noodles laid out on a long table literally, a *tok panjang*. The term came to be synonymous for celebratory meals. Guests took turns to enjoy the food which would be constantly replenished. The seniors and men were the first to sit at the table, followed by the women and children. The communal nature of all Straits-born dishes made them perfect for a *tok panjang*.

Nasi Lemak (Rich Rice)

Preparation and cooking: 2¹/₂ hours

Rice cooked with coconut milk is a dish prepared wherever there are coconut trees in a rice-eating culture. Besides coconut milk, herbs such as pandan leaves, lemongrass or cloves are added for fragrance. The side dishes that my mother prepared for *nasi lemak* meals were always seafood: fried fish, *otak-otak*, *asam* prawns, fish cake, *ikan bilis* with peanuts and sometimes, onion omelette. *Nasi lemak* with fried chicken is a popular combination today. The Chettis used to make *nasi lemak* by steaming rice with coconut milk. The modern rice cooker simplifies things. Getting the right flavour for the *sambal* is critical to a good *nasi lemak*. Avoid over-sweetening it. A tangy *sambal* is a better foil for the richness of *nasi lemak*.

400 g (14 oz / 2 cups) long-grain rice

250 ml (8 fl oz / 1 cup) packaged coconut cream

500 ml (16 fl oz / 2 cups) water

1¹/₂ tsp salt

3 pandan leaves, rinsed and knotted

Nasi Lemak Chilli Sambal

250 ml (8 fl oz / 1 cup) water

50 g (1³/₄ oz) tamarind paste

65 g (2¹/₃ oz) shallots, peeled

30 g (1 oz) dried red chillies, seeded and softened in water

1 tsp salt

1 Tbsp sugar

4 Tbsp cooking oil

Optional Side Dishes
Otak-otak (page 84)

Fried fish (page 73)

Fish cake (page 74)

Onion omelette

Asam prawns

Cucumber slices

1. Rinse rice in a sieve, then soak in water for an hour and drain.

2. Combine coconut cream and water to get coconut milk. Put rice, coconut milk, salt and knotted pandan leaves in the rice cooker. and cook.

3. When rice is ready, fluff it to mix in the coconut cream that has risen to the top. Alternatively, stir rice and coconut milk half-way through the cooking. Serve with *nasi lemak* chilli *sambal*, cucumber slices and an assortment of side dishes of your preference.

4. Make *nasi lemak* chilli *sambal*. Mix water with tamarind paste and strain away the solids. Blend the tamarind juice with the shallots and softened dried chillies.

5. In a microwave-safe dish, stir together the shallot-chilli mixture with the remaining ingredients. Cook in the microwave oven on High for about 20 minutes, stirring after every 5 minutes, until the mixture has a thick, glossy consistency. Add more tamarind juice if it dries out too much. Overcooking is better than undercooking.

6. Leave to cool before serving with *nasi lemak*. Store excess in a glass jar and freeze.

Asam Prawns

1 kg (2 lb 3 oz) prawns

1 tsp salt

1 Tbsp water

50 g ($1^3/_4$ oz) tamarind paste

3 Tbsp cooking oil

Asam Prawns

1. To prepare prawns, cut prawn eyes, tips and whiskers off with a pair of scissors. Shell trimmed prawns until just before the tail. Split prawns along the top to remove the dirt thread.

2. Mix salt, water and tamarind paste together and rub into prepared prawns, taking care not to get stabbed by the prawn tails. Leave prawns to marinate, preferably overnight or even for a couple of days.

3. Before frying prawns, scrape off tamarind paste marinade and rinse in water. Pat prawns dry with paper towels.

4. Heat oil in a wok until smoking hot and add prawns. If prawns exude a lot of liquid, scoop them out once they have changed colour and let the liquid boil off. This often happens with prawns that are not super fresh, or which have been frozen for too long. Once all the liquid has boiled off, return prawns to the pan to brown nicely. Add more oil, if needed.

Onion Omelette

2 Tbsp cooking oil

100 g ($3^1/_2$ oz) onion, peeled, halved and sliced

4 eggs, beaten

$^1/_2$ tsp salt

1 fresh red chilli, seeded and sliced

Onion Omelette

1. Heat 1 Tbsp oil in a frying pan and sauté sliced onion until transparent. Remove from the pan and mix into beaten eggs. Add salt and sliced chilli.

2. Heat $^1/_2$ Tbsp oil in the frying pan and pour in half the egg mixture. When the bottom of the omelette begins to firm up, fold it over and continue cooking until eggs are cooked through. Remove from the pan onto a serving plate.

3. Repeat to use up the remaining oil and egg mixture.

4. To serve, cut omelette into slices.

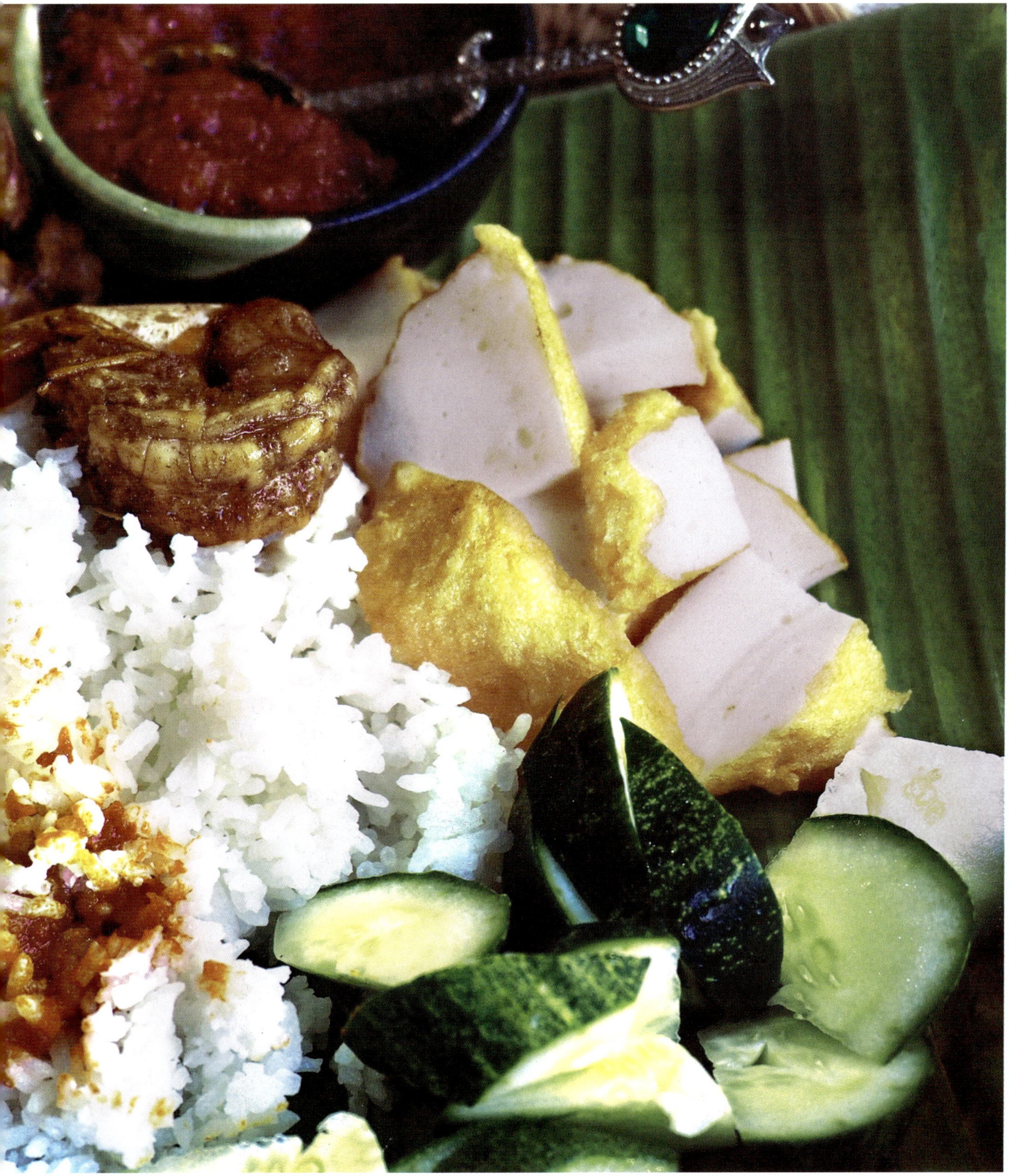

Nasi Ulam (Rice Salad)

Preparation and cooking: 2 hours

This very flexible rice salad can be found in one form or another on the table of many Straits-born families from Penang to Indonesia and its flavour depends very much on what there is to go into the rice. The fresh herbs and raw vegetables, as well as the fried fish, may be varied according to taste. In Penang, salted threadfin (*tanau kiam hoo*) is the salted fish of choice. If sliced thinly, salted threadfin can be fried to a crisp, then pounded for mixing into the *nasi ulam*. If sliced thickly, it can be fried and shredded instead. To make the *nasi ulam* a treat for the eyes too, colour the rice blue with butterfly pea flower.

1 Tbsp dried butterfly pea flower (optional)

1 Tbsp warm water (optional)

100 g (3¹/₂ oz) cabbage, finely shredded

1 cup cored and cubed/shredded cucumber

1 cup shredded carrot

Any preferred fresh herb (spring onion/mint/coriander/turmeric leaf/kaffir lime leaf/basil/*chekok* leaf/ginger leaf/*krachai* leaf), finely shredded

200 g (7 oz) fried fish, flaked

50 g (1³/₄ oz) salted threadfin, fried and shredded

6 cups cooked long-grain rice, kept hot

Dressing

100 g (3¹/₂ oz) shallots, peeled and finely sliced

2 Tbsp *sambal belacan* (page 14)

2 Tbsp lime juice

1 tsp salt

1. Soak dried butterfly pea flowers in warm water while you prepare the other ingredients.

2. In a large mixing bowl, mix together all the ingredients for the dressing. Add shredded vegetables and herbs and mix well.

3. Stir in flaked fish and shredded salted fish.

4. Stir strained butterfly pea water into the rice to get streaks of colour, then add the vegetable-fish mixture. Mix well.

5. Serve warm.

Chetti Nasi Kebuli

Preparation and cooking: 2 hours

Biryani arrived in India in the 16th century with the Mughals, who brought with them the courtly Persian treatment of rice. *Nasi kebuli* is a biryani cooked with coconut milk instead of yoghurt. Mughal biryani typically combined dried fruit and nuts with meat and rice, also typical in Iranian cooking today. The Straits-born touch is in the use of pandan leaves, kaffir lime leaves, lemongrass and coconut milk instead of yoghurt.

400 g (14 oz / 2 cups) basmati/ long-grain rice

1 litre (32 fl oz / 4 cups) water

2 tsp tamarind paste

4 Tbsp cooking oil

4-cm (1¹/₂-in) cinnamon stick

800 g (1³/₄ lb) mutton/lamb, cut into pieces

1 stalk lemongrass, smashed

2 tsp salt

250 ml (8 fl oz / 1 cup) packaged coconut cream

3 pandan leaves, rinsed and knotted/4–5 kaffir lime leaves

A handful of raisins

A handful of roasted cashew nuts/ almonds

Spice Paste

100 g (3¹/₂ oz) shallots, peeled and chopped

2 cloves garlic, peeled

25 g (⁴/₅ oz) turmeric, scraped

50 g (1³/₄ oz) ginger, scraped and sliced

4 Tbsp water

2 tsp ground coriander

¹/₂ tsp ground cumin

¹/₂ tsp ground fennel

2 tsp red chilli powder

¹/₂ tsp ground white pepper

¹/₄ tsp ground cloves

¹/₄ tsp ground cardamoms

1. Make spice paste by blending the shallots, garlic, turmeric and ginger with water in a food processor until smooth. Mix the ground spices with the wet paste. Set aside.

2. Rinse the rice, then soak in water for an hour.

3. Make tamarind juice with the tamarind paste and 250 ml (8 fl oz / 1 cup) water. Strain and discard the solids. Use this tamarind juice to boil the meat.

4. Heat cooking oil in a saucepan and fry the cinnamon stick for a few minutes. Add the spice paste and fry until oil rises to the top. Add the meat with the tamarind juice, lemongrass and 1 tsp salt. Bring to the boil. Turn down heat and simmer until meat is tender. Add a little more water if necessary, but the meat should be dry when tender. Near the end of cooking, stir to prevent spice paste from burning or catching at the bottom. Set aside 3 Tbsp gravy for flavouring the rice.

5. To cook the rice, drain off the soaking water. Add enough water to the coconut cream to make 750 ml (24 fl oz / 3 cups) coconut milk. Combine rice, coconut milk, salt, 3 Tbsp meat gravy, knotted pandan/kaffir lime leaves and raisins in a rice cooker. Stir to mix well.

6. When rice is done, stir the meat and gravy into the rice.

7. Garnish rice with some cashew nuts/almonds. Serve hot with rasam (page 52), *acar* (page 26), a *kerabu* and/or *urap timun* (page 38).

Prawn Biryani

Preparation and cooking: 1½ hours

This is a Straits-born creation inspired by how the Straits-born communities in Singapore love these crustaceans, in part because they get them very fresh and in different sizes. This is another no-fuss biryani that can be prepared in a rice cooker. If, like me, you stock up on prawn shells and heads whenever you cook prawns, there will always be enough in the freezer to make the stock for this biryani.

400 g (14 oz / 2 cups) basmati/
  long-grain rice

2 tsp red chilli powder

1 tsp ground coriander

2 tsp salt

1 kg (2 lb 3 oz) shelled prawns
  with tails on

4 Tbsp + 2 Tbsp cooking oil

200 g (7 oz) onions, peeled and
  finely sliced

1 packed cup coriander leaves,
  chopped

Prawn Stock

Prawn shells and heads

1 litre (32 fl oz / 4 cups) water

2 stalks lemongrass, bruised

1 tsp cracked white pepper

6-cm (2½-in) stick cinnamon

6 green cardamom pods

Rasam

½-1 Tbsp tamarind paste

500 ml (16 fl oz / 2 cups) water

1 Tbsp cooking oil

¼ tsp ground black/white pepper

½ tsp ground cumin

3 dried red chillies, seeded and
  cut into several pieces

¾ tsp salt

2 ripe tomatoes, cubed

1. Make prawn stock ahead of time so it can cool before using. Combine all the ingredients in a stockpot and bring to the boil. Turn down the heat and simmer for 15 minutes. Scoop out the solids and discard. Set stock aside to cool.

2. Rinse the rice, then soak in water for an hour.

3. Mix chilli powder, ground coriander and 1 tsp salt with shelled prawns and marinate in the fridge for at least 30 minutes or overnight.

4. Heat 4 Tbsp cooking oil in a frying pan and fry sliced onions until nicely browned. Remove half the browned onions from the frying pan and set aside.

5. Add marinated prawns to the pan and sauté until prawns begin to change colour. Turn off the heat.

6. Put drained rice into the rice cooker, add 750 ml (24 fl oz / 3 cups) cooled prawn stock, the remaining 2 Tbsp cooking oil and 1 tsp salt.

7. When rice is ready, stir in prawns, chopped coriander leaves and reserved browned onions. Cover and rest rice for 30 minutes or so.

8. Serve hot with rasam, *urap timun* (page 38), *acar* (page 26) and/ or pineapple *kerabu* (page 39).

9. To prepare rasam, mix tamarind paste with water and strain. Combine tamarind juice with the remaining ingredients in a saucepan and bring to the boil. Simmer until tomatoes are very soft. Adjust the seasoning to taste.

Nasi Kunyit (Yellow Rice)

Preparation: Overnight soaking   Cooking: 30 minutes

Nasi kunyit or *nasi kuning* (both terms mean yellow rice and the former also means turmeric rice) used to have ritual significance in the Straits-born communities, and maybe it still does for some families. Amongst the Straits Chinese, *nasi kunyit* made with glutinous rice was part of the celebrations for introducing a one-month-old baby to the extended family. For the Chetti Melaka, *nasi kunyit* was prepared as ritual offerings for ancestor worship. In other parts of South East Asia, yellow is linked to the colour of gold and associated with special occasions. For instance, in various parts of Indonesia, rice cooked for special occasions is often coloured yellow. Whether prepared with glutinous or long-grain rice, *nasi kunyit* is always enriched with coconut cream. It is basically a variation of *nasi lemak*.

600 g (1 lb 5^1/$_3$ oz / 3 cups)
  glutinous rice

2 tsp ground turmeric

1 tsp salt

125 ml (4 fl oz / ^{1}/$_2$ cup) packaged
  coconut cream

3 pandan leaves, knotted/
  6 kaffir lime leaves, crushed/
  6 cloves (optional)

Note
If making this dish with long-grain rice, soak rice for 1 hour only. Add ^{1}/$_2$ tsp ground turmeric to the coconut cream. Use the proportions for *nasi lemak* (page 46).

1. Rinse rice, then soak in some water mixed with ground turmeric. Leave overnight.

2. The following day, drain rice, then steam for 10 minutes.

3. Mix together salt and coconut cream and stir into the half-cooked rice. Mix well.

4. Add pandan leaves/kaffir lime leaves/cloves and continue steaming rice for another 20 minutes, turning rice over after 10 minutes so that the grains at the top are cooked properly.

5. Continue steaming rice to your preferred texture. The rice should be al dente rather than soft. Remove the pandan leaves/kaffir lime leaves/cloves before serving.

6. Serve with chicken curry (page 106), *asam* prawns (page 47), *acar* (page 26) and a *kerabu*.

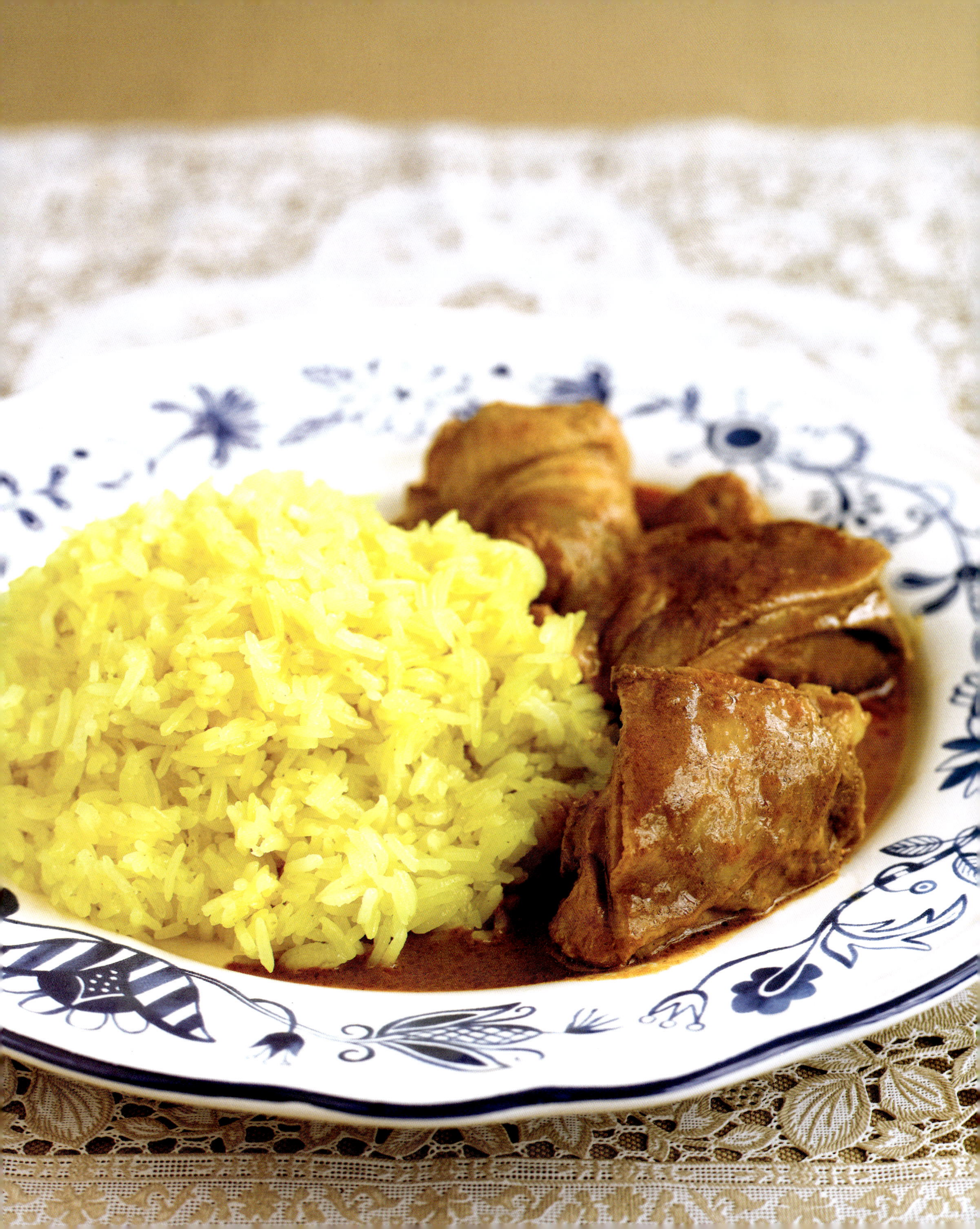

Penang Laksa (Rice Noodles in Fish Soup)

Preparation and cooking: 2 hours

Unlike Singapore laksa, which is rich, Penang laksa is tangy. The fish gravy has not only tamarind juice, but also dried sour fruit slices (*asam gelugor*) and limes. Penang laksa must be a relative of traditional Burmese *mohinga* (Burmese laksa) which is also fish-based. Myanmar is Penang's close neighbour, as is Thailand. The fish of choice for Penang laksa used to be wolf herring aka *dorab* aka *ikan parang*, a fish with great umami, but also numerous small bones. Overfishing has probably led to the disappearance of *ikan parang*, but *ikan selar* and *ikan kembong* are decent substitutes and easier to flake. Many cooks add mashed canned sardines sans the tomato sauce to further enrich the gravy.

600 g (1 lb 5^1/$_3$ oz) fresh medium-coarse rice vermicelli

Spice Paste

150 g (5^1/$_3$ oz) shallots, peeled

1 tsp dried shrimp paste (*belacan*)

100 g (3^1/$_2$ oz) fresh red chillies, seeded

2 Tbsp water

Gravy

100 g (3^1/$_2$ oz) tamarind paste

1.25 litres (40 fl oz / 5 cups) water

600 g (1 lb 5^1/$_3$ oz) fish

4–6 large dried sour fruit slices (*asam gelugor*)

1 torch ginger bud, split

1 bunch laksa leaves (*daun kesom*)

Garnishing

^{1}/$_2$ tsp salt

1 large onion, peeled and thinly sliced

200 g (7 oz) cucumber, cored and shredded

1 cup mint leaves, stems discarded

1 torch ginger bud, finely chopped

1 small head lettuce, shredded

1 small pineapple, peeled, cored and coarsely chopped

Limes, halved

2 Tbsp *hae ko* (black prawn paste)

3 Tbsp hot water

1. Make spice paste by blending all the ingredients together in a food processor until smooth. Set aside.

2. Make gravy by mixing tamarind paste with water. Strain tamarind juice into a large pot. Discard solids.

3. Add fish and sour fruit slices to the pot and bring to the boil. Simmer gently until fish is cooked.

4. Remove fish and leave to cool. Flake fish and set aside. Return fish bones and skin to pot and boil for another 10 minutes. Using a wire sieve with a long handle, scoop out solids from stock and discard.

5. Add spice paste to the pot and return to the boil.

6. Trim torch ginger bud and halve it. Add to the boiling gravy. Rinse laksa leaves and add to the gravy. Add flaked fish and bring gravy to the boil.

7. Prepare garnishing. Mix salt with sliced onion and stand for 5 minutes. Squeeze juices from onion, then arrange slices on a serving dish together with the cucumber, mint leaves, ginger bud, lettuce, pineapple and limes.

8. Dilute *hae ko* with hot water and stir to get an even pouring liquid. Pour diluted paste into a small bowl for drizzling over noodles.

9. To serve noodles, bring a large pot of water to the boil. Blanch noodles until tender but still al dente. (If you get laksa noodles from Penang, the cooking time should be reduced. Singapore laksa noodles are not as tender.) Divide into serving bowls. Ladle hot gravy over noodles and garnish as desired.

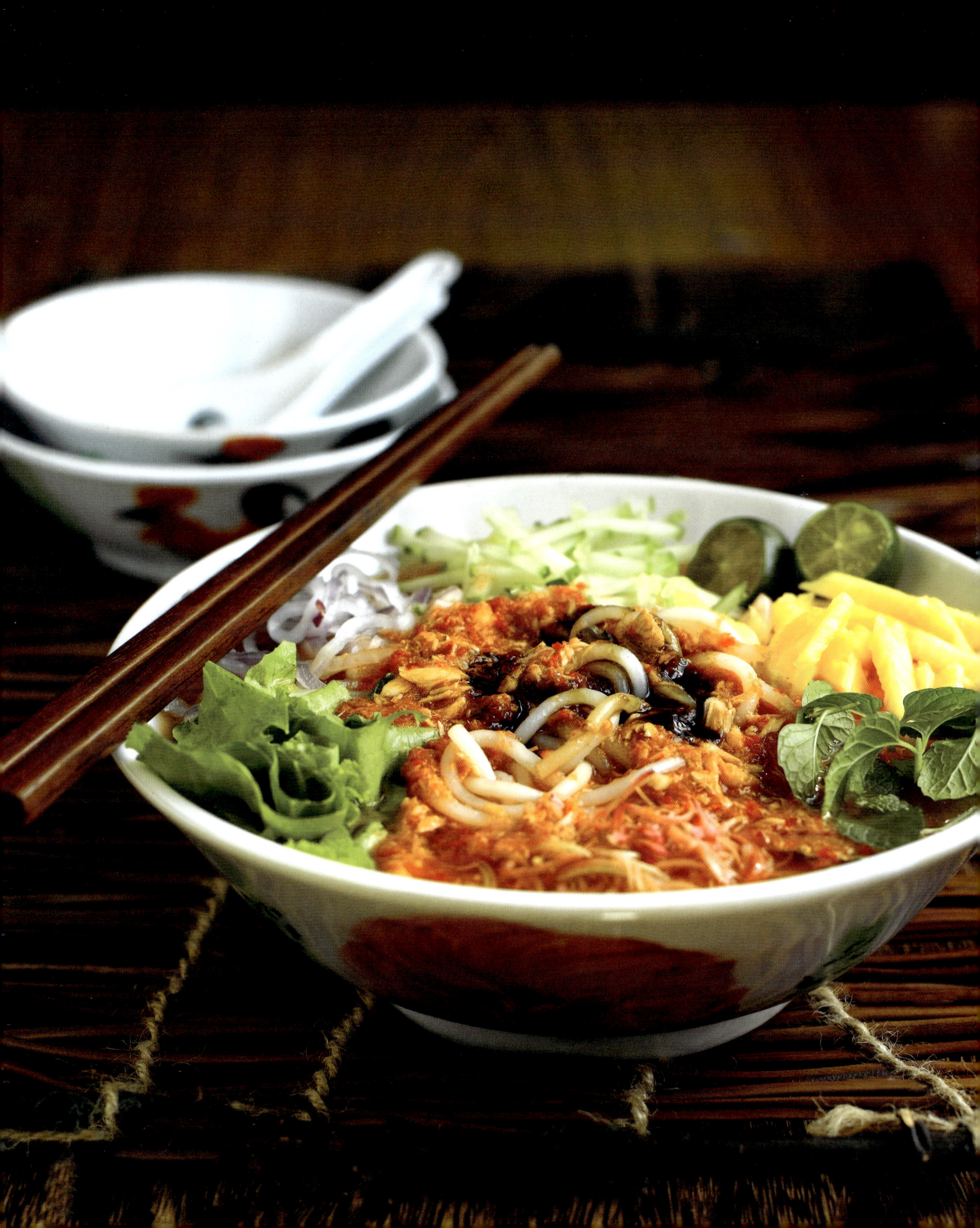

Penang/Singapore Char Kuay Tiao

Preparation and cooking: 30 minutes

Penang *char kuay tiao* (flat rice noodles) is more savoury than sweet because it does not have sweet black sauce in it, unlike Singapore *char kuay tiao*. Penang *char kuay tiao* also has chives, unlike Singapore *char kuay tiao*. The flavour is better if this quantity of noodles is fried in two or three batches rather than one large batch. For a more traditional flavour, use lard.

5 Tbsp cooking oil/lard (page 23)

2 Tbsp chopped garlic

600 g (1 lb 5¹/₃ oz) fresh flat rice noodles

2 Tbsp light soy sauce

1 tsp dark soy sauce

1 tsp salt

300 g (11 oz) bean sprouts

1 Tbsp cooked dried red chilli paste (page 17)

3 eggs

2–3 Tbsp water

100 g (3¹/₂ oz) prawns, boiled and shelled, leaving tails intact

100 g (3¹/₂ oz) fish cake, sliced

200 g (7 oz) blood cockles (*haam*), blanched and shelled (optional)

1 bunch Chinese chives, cut into finger-lengths (optional)

1. Heat 4 Tbsp oil/lard in a wok and fry garlic until fragrant. Add noodles, light and dark soy sauces and salt. Mix well.

2. Stir in bean sprouts and cooked dried red chilli paste. Mix well and stir-fry for a minute or so.

3. Make a well in the centre of noodles and add remaining spoonful of oil/lard as well as eggs. Cover with hot noodles and let eggs cook for a couple of minutes before stirring up the noodles.

4. Sprinkle water over noodles and stir in boiled prawns, fish cake slices, blood cockles, if using, and chives. Mix well. Remove from heat when chives wilt.

5. Serve hot.

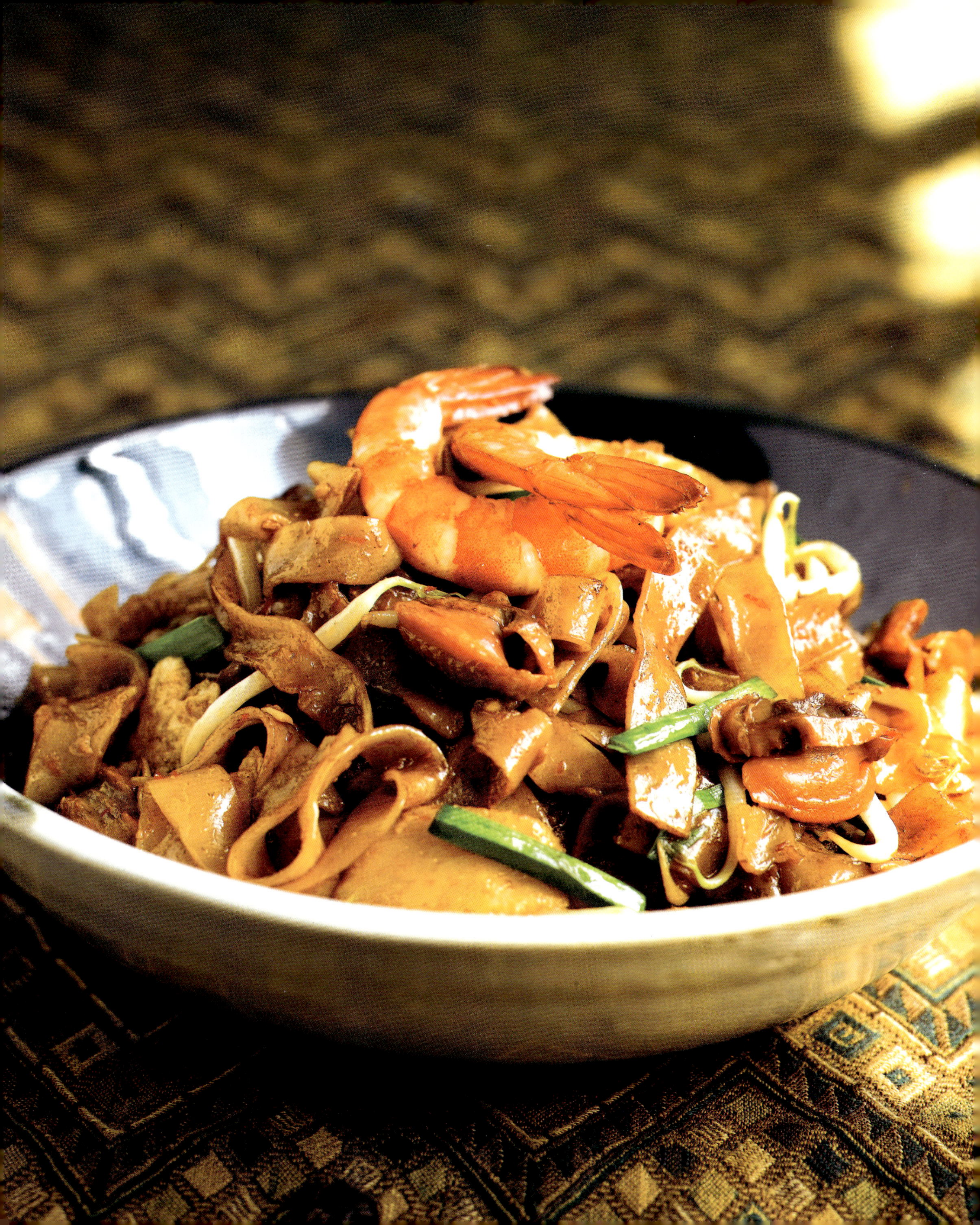

Penang Prawn Mee Soup

Preparation and cooking: 2 hours

Penang-style prawn mee soup starts out spicy and comes with more pork than Singapore-style prawn mee. However, like Singapore-style prawn mee, Hokkien mee can be combined or substituted with *bee hoon* (fine rice vermicelli).

Prawn Stock

500 ml (16 fl oz / 2 cups) water

500 g (1 lb 1½ oz) prawn shells

Pork Stock

1.5 litres (48 fl oz / 6 cups) water

1 kg (2 lb 3 oz) pork ribs with meat

1 pig's tail, chopped into short lengths

1½ tsp salt

1 Tbsp fried shallots

1 Tbsp cooked dried red chilli paste (page 17)

1 tsp dried shrimp paste (*belacan*)

300 g (11 oz) prawns, boiled and shelled

400 g (14 oz) *kangkong* (water convolvulus)

200 g (7 oz) bean sprouts

600 g (1 lb 5⅓ oz) Hokkien mee or substitute half with fine rice vermicelli (*bee hoon*)

Garnishing

Fried shallots

Cooked dried red chilli paste (page 17)

1. To prepare prawn stock, combine water and prawn shells in a saucepan and boil for 10 minutes. Strain and discard shells. Use this prawn stock to boil prawns for garnishing.

2. To prepare pork stock, combine water, pork ribs and pig's tail in a large saucepan and bring to the boil. Skim off any scum that rises to the top before adding the other ingredients.

3. Continue simmering over low heat until ribs and pig's tail are tender. This should take about 1½ hours.

4. Stir in prawn stock and adjust the seasoning to taste. Return to the boil and keep stock hot.

5. Prepare *kangkong* by trimming off the last 6 cm (2.5 in) of its stalk from the root up. Split thick stems down the middle and cut each stalk into finger lengths. Rinse clean in several changes of fresh water.

6. Bring a pot of water to the boil and blanch *kangkong* until the colour changes. Scoop out and drain.

7. Return water to the boil and blanch bean sprouts for 10 seconds. Drain and divide bean spouts, along with *kangkong*, into serving bowls.

8. Return water to the boil and cook noodles until tender but al dente. If combining Hokkien mee with rice vermicelli, cook both types of noodles separately.

9. Portion noodles into serving bowls and ladle hot soup over noodles. Top with pieces of pork ribs, pig's tail and prawns.

10. Garnish with fried shallots and serve hot with dried red chilli paste on the side.

Singapore Hokkien Mee Soup

The soup is dark coloured and does not have dried shrimp paste (*belacan*) nor chillies in it. Cut chillies or chilli paste *sambal* is served on the side. To get its dark colour, raw and cooked prawn shells are dry-fried in a stockpot to achieve a layer of blackened prawn juices. The frying also reduces the prawn shells' fishy odour. Add 2 litres (64 fl oz / 8 cups) water and bring prawn shells to the boil. Turn down the heat and simmer for 15 minutes. Scoop out and discard prawn shells. Put in pork bones and another 500 ml (8 fl oz / 2 cups) water, then return stock to the boil. Turn down the heat and simmer for another hour or more to get a good stock.

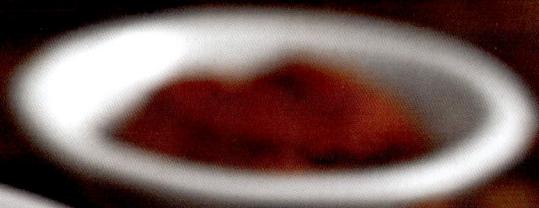

Penang Birthday Mee Sua

Preparation and cooking: 1 hour   Makes 1 serving

I always got this for dinner, with a boiled egg that had been dyed red, on my birthday when I was a child. I used to look forward to this treat because I liked the texture of kidney and that of the slippery, soft wheat thread noodles! Kidneys need to be well-cleaned to taste good.

1 pig's kidney

$^1/_2$ tsp cornflour

1 Tbsp minced pork

$^1/_4$ tsp ground white pepper

$^1/_2$ tsp salt

$^1/_2$ Tbsp cooking oil

2 tsp chopped garlic

$^1/_2$ Tbsp shredded ginger

500 ml (16 fl oz / 2 cups) pork/ chicken stock

30 g (1 oz) pig's liver, sliced

50 g (1$^3/_4$ oz) fine wheat vermicelli (*mee sua*)

Chopped spring onions

1 hard-boiled egg, dyed red with food colouring

Dark soy sauce (optional)

Sliced fresh red chilli (optional)

1. To clean kidney, split in half crosswise and trim off smelly, white renal tubes. Using a sharp knife, score outside of kidney with a criss-cross pattern, then cut into bite-size rectangles. Soak in several changes of water until kidney no longer smells of urine.

2. Mix cornflour into minced pork with a touch of pepper and a pinch of salt. Form mixture into 2 or 3 small balls.

3. Heat oil in a saucepan and sauté garlic and ginger until garlic begins to brown. Add stock and season with salt and pepper. Bring to the boil.

4. Add pork balls, liver and kidney slices and cook for a minute.

5. Rinse noodles quickly under running water before adding to pot. Bring to the boil and cook for a minute.

6. Garnish with spring onions and serve immediately with a dyed hard-boiled egg and a dip of dark soy sauce, with slices of fresh red chilli, if preferred.

Sweet Birthday Noodles

Preparation and cooking: 30 minutes   Makes 1 serving

A friend with a Penang-born mother always got this sweet *mee sua* soup as a birthday breakfast when she was young. Lilian Lane in *Malayan Cookery Recipes* gave a similar recipe also as a birthday treat, but with two hard-boiled eggs in the soup.

1 Tbsp sugar

1 pandan leaf, knotted

250 ml (8 fl oz / 1 cup) water

25 g (1 oz) fine wheat vermicelli (*mee sua*)

1 egg

1. Put sugar, pandan leaf and water in a pot and bring to the boil. When sugar has dissolved, remove and discard pandan leaf.

2. Rinse noodles quickly under running water, then add to boiling syrup. Return syrup to the boil, lower heat and simmer for a minute.

3. Crack egg into pot and turn off heat. Cover pot to semi-cook egg for 2 minutes, until white begins to solidify. Dish out into a bowl without breaking egg yolk and serve immediately.

Pictured: Penang Birthday Mee Sua

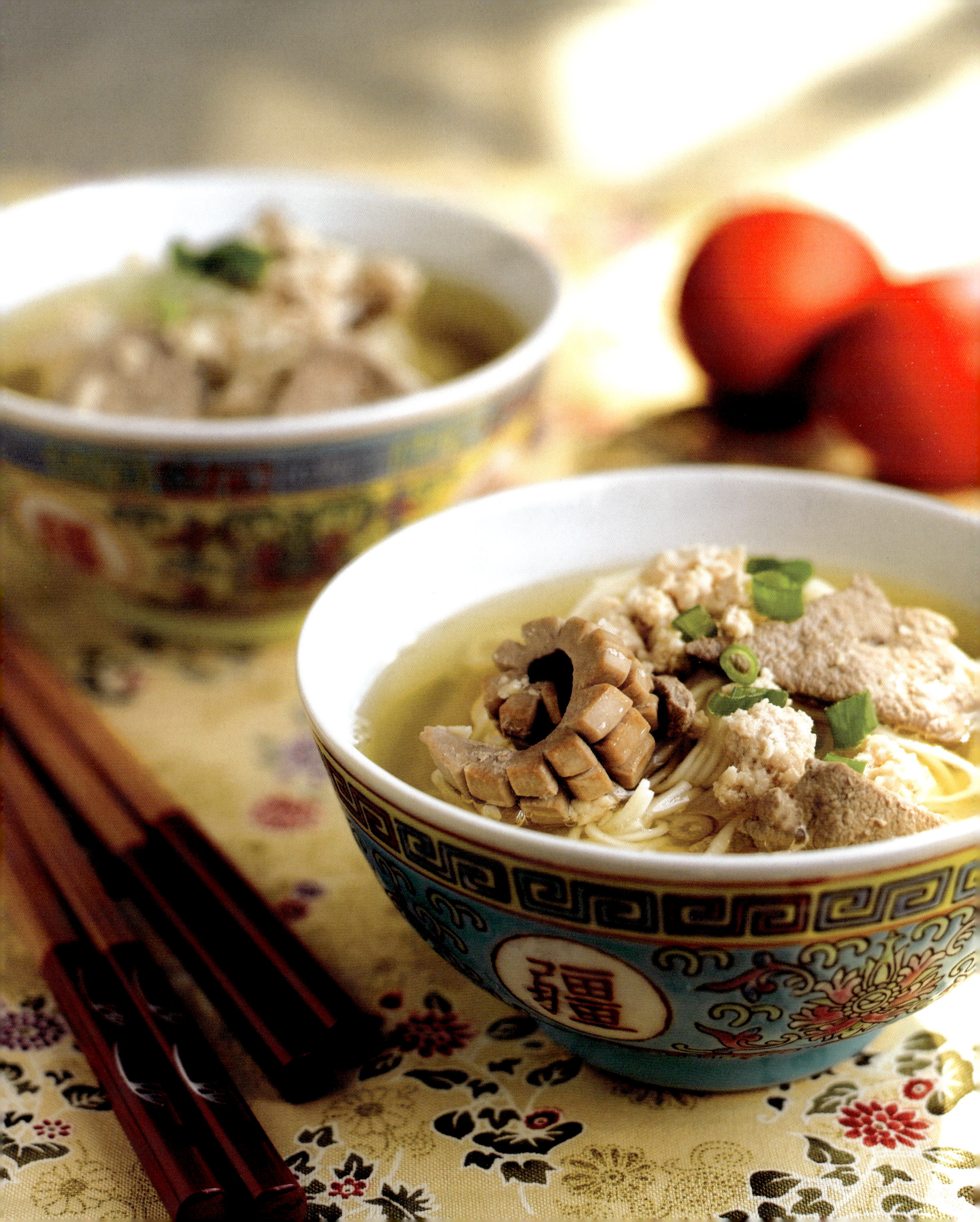
疆

Seafood
The Bounty of the Seas

The Straits-born originated from Island South East Asia, a region with long coastlines. In the 14th and 15th centuries, Singapore's original inhabitants, the *Orang Laut* (Malay for Sea People) even lived on water either in boats or water villages in sheltered coves. The seas have always been the region's source of sustenance and riches. The seas not only connected the peoples of the region to the world, but they were also responsible for creating the Straits-born communities. Since ancient times, explorers, traders and colonisers came on sailing ships in search of trade and economic opportunities. The ports of South East Asia were the crossroads for international traders east and west of the region.

The bounty of the seas set up some of the earliest links between the region, South India and China. From ancient times, Chinese traders came south in search of tropical trade goods such as dried sea cucumbers (*tripang*), salted fish and shrimps, seaweed aka agar-agar as well as fragrant woods and exotic animals. Fishing in the region has long been a very traditional industry, as is the drying and salting of fish, shrimps and squid. *Grago* (krill) would be made into *belacan* or fermented into *cinchalok*.

The seas brought new spices, herbs, fruits, vegetables to the bazaars of South East Asian ports. Cooks experimented with and made them essential in their cooking. Imported seed spices were combined with introduced and native roots and gingers with chillies and seafood. The cooking of seafood developed into a fine art, but many also made much of the freshness of the seafood by simply grilling or steaming. Straits-born seafood cooking ranges from simple boiled or grilled prawns and crabs, to fried, stuffed and grilled fish, curries, soups, stews and *sambals*, pickled and fermented seafood, condiments from seafood, and garnishes for rice and noodles. Prawns are stars in celebratory meals. The rich variety of fish yielded numerous fish-based dishes whether simple or complex. All are delicious.

If there is one thing that characterises Straits-born cuisine, it is this prominence given to seafood in small and big ways. One small but critical way is in the role of *belacan* in Straits-born cooking. The big way is in the huge array of fresh seafood available to a Straits-born cook in Singapore every day. Even in the time of Covid-19, it has been possible to buy fresh clams, crabs, prawns, shrimps and fish by the kilos. Straits-born cooks are united by the loving attention given to seafood and the wide array of similar yet different culinary creations that have emerged over time in the communities.

Squid Asam Pedas with Pineapple

Preparation and cooking: 30 minutes

This *asam pedas* (spicy sour) dish may be equally called a *lauk pindang* by a Chetti or *singgang* by a Portuguese Eurasian. All three are a sour-sweet seafood dish prepared with fish, but it is also good with any seafood. It is common to add a sour fruit such as belimbing or pineapple to the dish.

4 Tbsp *asam pedas* paste (page 16)

2 Tbsp lime/lemon juice

500 ml (16 fl oz / 2 cups) water (omit if using tamarind juice)

1 tsp salt

1 tsp sugar

200 g (7 oz) pineapple wedges

600 g (1 lb 5^1/$_3$ oz) squid/fish/ prawns

2 green chillies, seeded (optional)

1 torch ginger bud, halved (optional)

Some sprigs of laksa leaves (*daun kesom)* (optional)

Tamarind Juice

2 Tbsp tamarind paste

500 ml (16 fl oz / 2 cups) water

> **Note**
> This dish may also be prepared with an alternative spice paste. Blend 50 g (1^3/$_4$ oz) peeled shallots, 2 peeled garlic gloves, 3 candlenuts (*buah keras*), 5-cm (2-in) knob scraped turmeric and 2 Tbsp water in a food processor to a smooth paste. Add 1 Tbsp dried red chilli paste (page 17). Use in place of the *asam pedas* paste.

1. If using tamarind juice, mix the tamarind paste with 500 ml (16 fl oz / 2 cups) water. Strain the juice and discard the solids.

2. In a saucepan, combine the *asam pedas* spice paste with the tamarind juice or lime/lemon juice, water (omit if using tamarind juice), salt, sugar and pineapple wedges, and bring to the boil. Simmer for about 10 minutes until pineapple wedges look transparent.

3. Add squid/fish/prawns and bring to the boil. If cooking squid, turn off the heat as soon as the squid changes colour. Overcooking makes squid tough. If this happens, continue boiling and squid will soften again. If cooking fish/prawns, also turn off heat once the seafood changes colour. Cover the pot and the residual heat will finish the cooking.

4. If using any of the optional aromatics, add in the last minutes of cooking. Serve as a side dish with rice.

Lauk Pindang

Preparation and cooking: 15 minutes

Chetti *lauk pindang* needs very fresh seafood, usually fish. This simple *lauk pindang* may also be enriched with coconut milk, in which case, halve the water. When enriched with coconut milk, the taste will be rather like a fish version of northern Thai *thom kha gai* (spicy coconut soup with chicken). Again, this dish will pack more punch with fresh coconut milk rather than packaged coconut milk. The simplicity of the seasoning emphasises the freshness of the ingredients.

2 Tbsp tamarind paste

500 ml (16 fl oz / 2 cups) water/ fresh coconut milk

2 stalks lemongrass, split bottom half lengthwise, but keep it whole

3-cm (1½-in) knob galangal, sliced/smashed

20 g (⅔ oz) shallots, peeled and chopped

3–4 fresh red/green chillies, seeded

1 torch ginger bud, kept whole, but slit four-ways

1 tsp salt

1 tsp sugar

600 g (1 lb 5⅓ oz) fish

1. Combine tamarind paste with water. Strain away the solids. If adding coconut milk, make tamarind juice with just 250 ml (8 fl oz / 1 cup) water. Prepare 250 ml (8 fl oz / 1 cup) coconut milk for topping up gravy.

2. In a saucepan, combine tamarind juice with all the other ingredients, except the fish and coconut milk, if using. Bring to the boil and simmer for 15 minutes.

3. If using coconut milk, add milk and fish to the pot and return to the boil. Remove from heat when fish changes colour.

4. Serve hot as a side dish with rice.

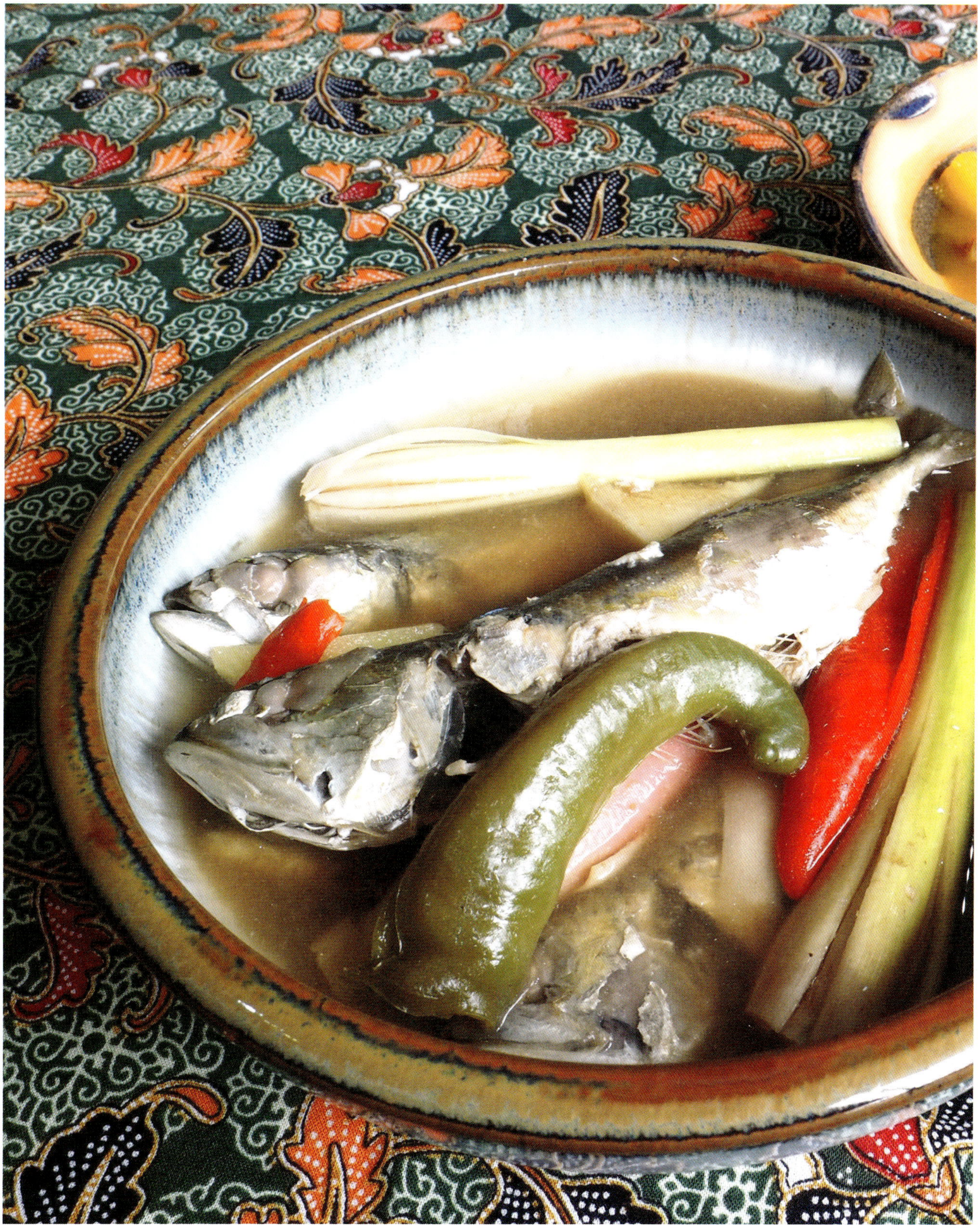

Asam Fish Head Curry

Preparation and cooking: 1 hour

The head of any large fish usually has the best-tasting morsels of flesh — and there are people who love the gelatinous eyeballs, too. With such a well-developed community taste for fish head, getting one at the market means going early or asking your favourite fishmonger to keep one for you. If a fish head is not available, use the spice paste to cook fish steaks or small whole fish, but halve the quantity of the spice paste or make and freeze.

100 g (3¹/₂ oz) tamarind paste

1.25 litres (40 fl oz / 5 cups water)

4 Tbsp cooking oil

1¹/₂ tsp salt

500 g (1 lb 1¹/₂ oz) aubergines (eggplants), trimmed and thickly sliced

1–1.5 kg (2 lb 3 oz–3 lb 5 oz) fish head

2 torch ginger buds, halved/ 10 large sprigs laksa leaves (*daun kesom*)

Spice Paste

200 g (7 oz) shallots, peeled

2 cloves garlic, peeled

5-cm (2-in) knob turmeric, scraped and sliced

2 large stalks lemongrass, peeled and thinly sliced

2.5-cm (1-in) knob galangal, scraped and thinly sliced

6 candlenuts (*buah keras*), broken up coarsely

1¹/₂ tsp dried shrimp paste (*belacan*)

30 g (1 oz) dried red chillies, seeded and softened in water

4 Tbsp water

1. Make spice paste by blending all the ingredients together in a food processor until fine. Set aside.

2. Mix tamarind paste with 1.25 litres (40 fl oz / 5 cups) water and strain away the solids. Set aside tamarind juice.

3. Heat oil in a large saucepan and fry spice paste until fragrant and oil rises to the top.

4. Add tamarind juice, salt and aubergines and bring to the boil. Turn down the heat and simmer until vegetables are soft or to your preferred texture.

5. Place fish head in the pot with the eye facing upwards if it is half a fish head. Add torch ginger halves/laksa leaves and bring to the boil. Turn down the heat and simmer for 10–15 minutes or until the eye begins to turn opaque. Carefully transfer fish head to a microwave-safe serving bowl. Cover the bowl and cook in the microwave oven for 2 minutes on High. Keep it covered. Residual heat will continue to cook the fish.

6. Meanwhile, adjust the seasoning in the curry to taste. Boil for 1–2 minutes, then turn off the heat and rest the curry gravy for several hours before serving.

7. To serve, boil up curry gravy. Reheat fish head for 2–3 minutes in the microwave oven on High. Pour gravy and vegetables over the fish head.

8. Serve with rice.

Note
Any leftover gravy may be poured over blanched bean sprouts and boiled Hokkien mee/fine rice vermicelli (*bee hoon*) for a special treat.

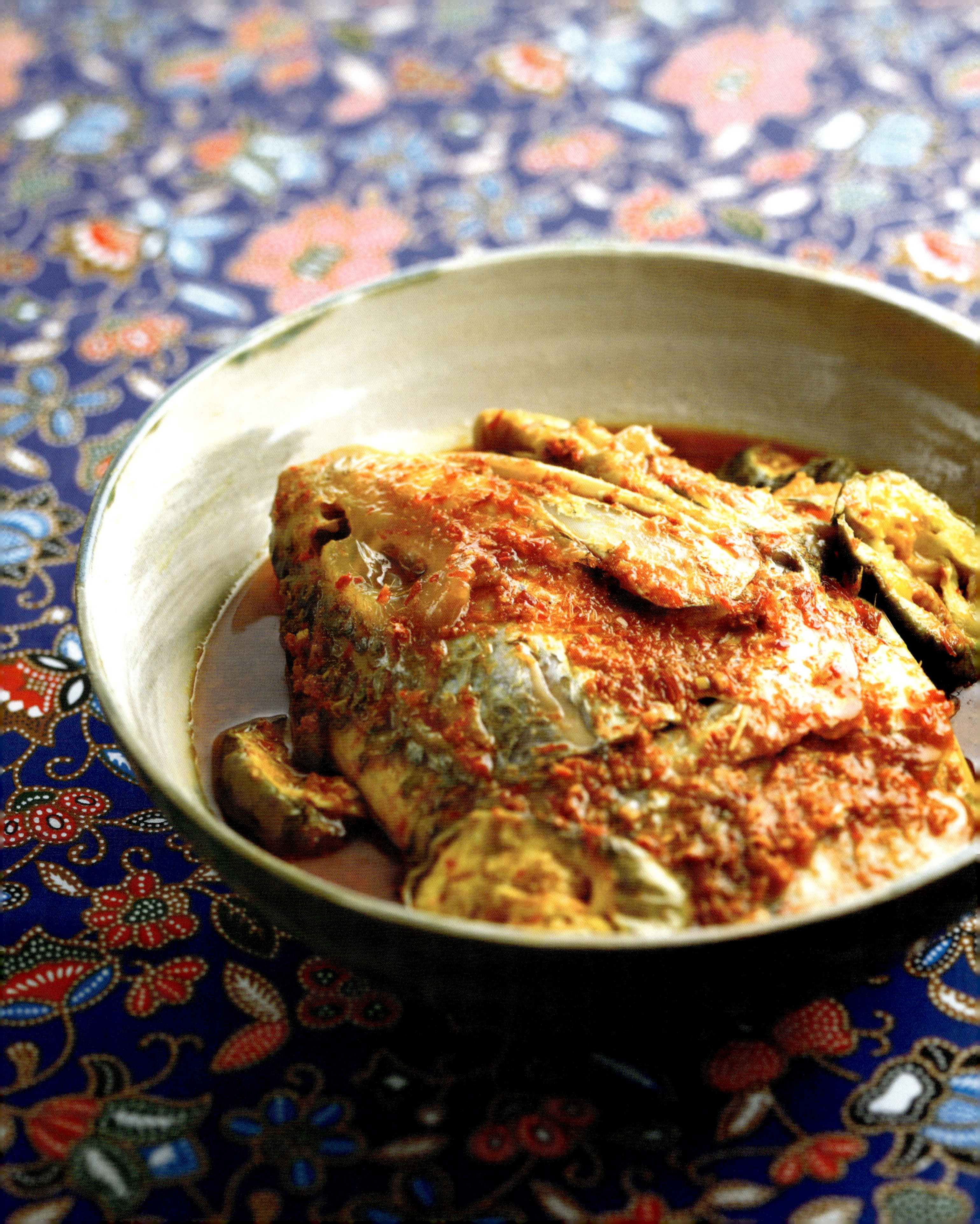

Sambal Ikan Bilis (Spicy Anchovies)

Preparation and cooking: 10 minutes

Ikan bilis (dried anchovies) are a good standby in the larder. They will make a tasty stock or a quick *sambal,* and can be fried with peanuts for a snack or as a side dish with *nasi lemak. Ikan bilis* is sold cleaned or whole. However, even when cleaned, you still need to pick through them to remove any bits of bitter innards or gritty heads. The backbones are fine, because they are a good source of calcium.

2 Tbsp cooking oil

4 Tbsp *asam rempah* (page 16)

150 g (5^1/$_3$ oz) *ikan bilis*, cleaned and rinsed

125 ml (4 fl oz / ^{1}/$_2$ cup) coconut milk/1 Tbsp coconut oil lees (page 20)

1. Heat oil in a pot and fry spice paste until oil rises to the top. Add *ikan bilis* and coconut milk and bring to the boil. Rest for an hour before serving.

2. Serve with crusty bread rolls or rice.

Variations
Halve the amount of *ikan bilis* and substitute with one or a combination of these:
• Firm tofu, cubed and fried
• French beans, sliced
• Long beans, cut into finger-lengths
• *Petai*
• Tempeh, cubed and fried
• Pineapple, cubed
• Ripe tomato, cubed

Fried Fish with Dark Soy Sauce

Preparation and cooking: 20 minutes

This is a fried fish with a very simple sauce. It is one of my top comfort foods to be enjoyed with a bowl of hot rice porridge. The sauce can be made for any fish, but usually one with minimal bones.

3 Tbsp cooking oil

250 g (9 oz) threadfin (*ikan kurau*)

1 fresh red chilli, seeded and sliced

30 g (1 oz) onion, peeled and cut into half-rings

1 Tbsp dark soy sauce

4 Tbsp water

^{1}/$_2$ tsp sugar

^{1}/$_2$ tsp ground white pepper

1. In very hot oil, brown both sides of the fish. It does not have to be cooked through. Add onion rings and sliced chilli to the oil and sauté for 2 minutes.

2. Add dark soy sauce followed by water, sugar and pepper. Mix well. Simmer until fish is cooked through. Turn the fish halfway through to coat the other side with sauce.

3. Serve with rice or rice porridge.

Fried Fish

Preparation and cooking: 20 minutes

- Any fried fish is divine with *sambal belacan* and hot rice. It's even better when the *sambal belacan* has a dash of lime juice in it.

- The simplest dressing for fried fish is salt.

- Ring in a change in flavour by rubbing the fish with tamarind paste. However, before frying the fish, the tamarind paste must first be rinsed off and the fish patted dry with paper towels.

- Another option is salt and ground turmeric. No rinsing is needed with this seasoning.

- A variation of turmeric seasoning involves pounding fresh turmeric with some candlenuts to make a paste that is spread over the fish. Season the paste with salt before using.

- To fry a fish without its skin coming off, make sure the oil is very hot. This forms the crust quickly so that when you turn the fish over, its skin is less likely to come loose or stick to the pan.

- Do not flip the fish over before the skin has crusted.

Fish Balls and Fish Cakes

Preparation: Half-day   Cooking: 30 minutes

Before fish balls and fish cakes were commercially available, my mother used to her own fish paste with the redbelly yellowtail fusilier (*Caseio cuning*) aka *huan cho hoo* (sweet potato fish). This fish is now rarely available, but any kind of fish can be used so long as it does not have small bones and the flesh is easily scraped off the bone. (Boil the fish trimmings with ginger and fried garlic to make a tasty fish soup.) A powerful food processor speeds up developing the bounce in the fish paste. Alternatively, pound the flesh in a mortar and pestle or slam the fish paste hard against the mixing bowl repeatedly.

500 g (1 lb 1¹/₂ oz) boneless fish meat

2 Tbsp tapioca starch

³/₄ tsp salt

125 ml (4 fl oz / ¹/₂ cup) water

A bowl of water with 1 tsp salt added

Cooking oil for greasing and frying

Chilli Fish Balls

Chopped coriander leaves

Chopped spring onion

2 fresh red chillies, seeded and chopped

1. In a food processor, blend fish meat, tapioca starch and salt. Add water in small amounts as fish paste stiffens. Process for 5–10 minutes to a smooth shiny paste. The length of time depends on the power of your food processor. If making chilli fish balls, mix chopped coriander, spring onion and seeded chillies into fish paste before forming the balls.

2. Have ready a bowl of cold salted water.

3. To make fish balls, mash paste together with one hand to make sure paste is smooth. Holding a handful of fish paste, form a circle with your thumb and forefinger. Have the circle small for a small ball and bigger for a big ball. Use your other fingers to slowly push fish paste up through the circle and squeeze out a round ball of fish paste, making sure that there are no cracks or grooves in the ball. (See picture.) If there is, start over. Do not wet the hand holding the fish paste in the water or the fish paste will separate into little lumps.

4. Using a metal tablespoon dipped in the cold salted water, scoop out the fish ball and drop it into the bowl of salted water. Repeat until the fish paste is used up.

5. To cook fish balls, bring a pot of water to the boil and drop in the fish balls. The fish balls are cooked when they float up.

Fish Cakes

1. To make fish cakes, grease a plate and rub some oil on your hands. Divide the fish paste into roughly equal amounts and shape the fish paste into patties/tubes.

2. Fill a frying pan with about 2 cm (1 in) depth of oil. To check that the oil has reached the right temperature, press the tips of wooden chopsticks against the bottom of the frying pan. The tips should send out tiny air bubbles, not vigorous ones. The oil should be hot enough to quickly form a crust, but not so hot that the outsides brown too quickly before cooking through. Lower fish cakes into the oil.

3. When cakes puff up, they are cooked through.

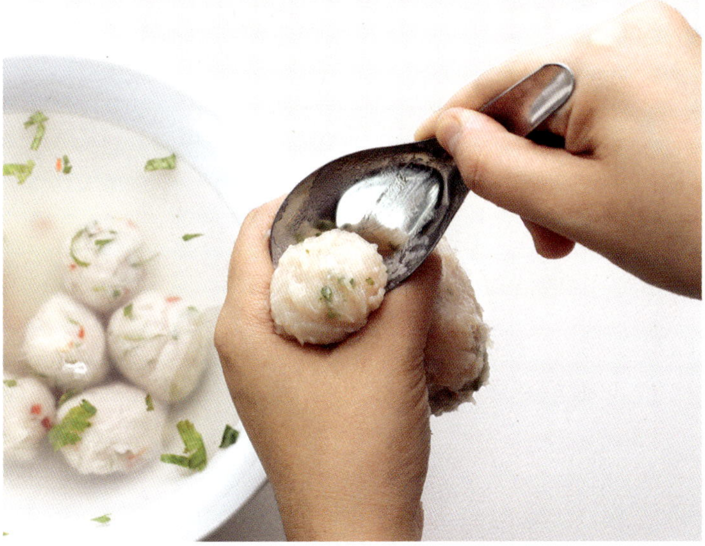
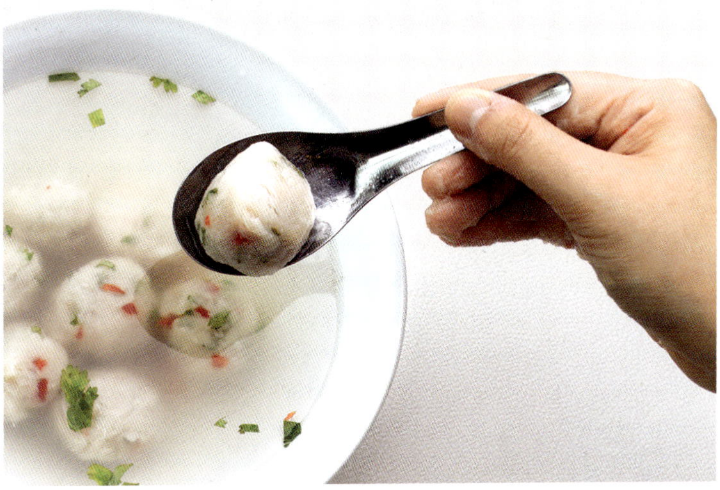
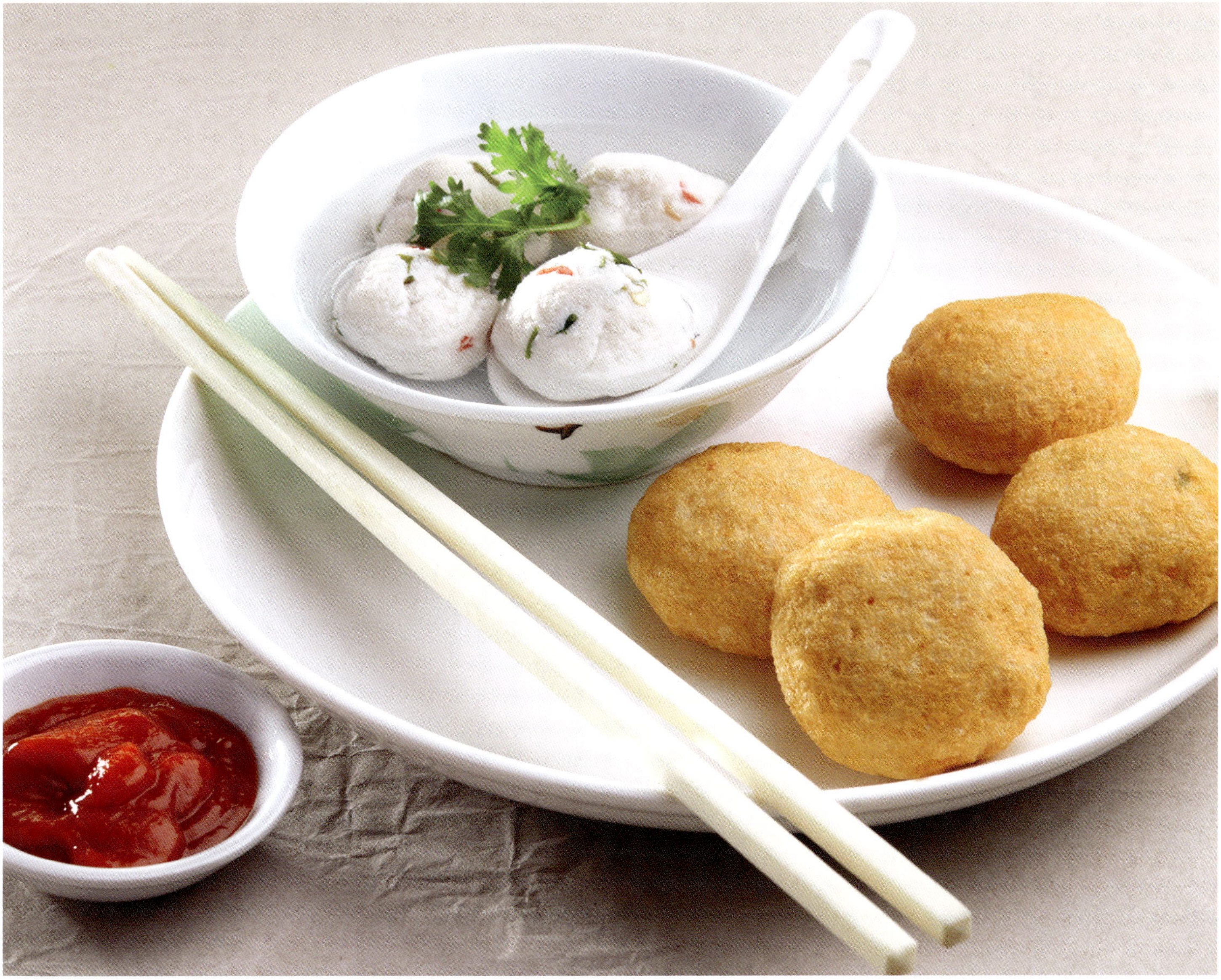

Grilled Fish in Banana Leaf

Preparation and cooking: 40 minutes

Grilled seafood with *sambal belacan* is a common Straits-born dish. My top choice for grilled fish is *ikan terubok* (*toli* shad, *lupea toli*) although any kind of fish, with or without scales, may be grilled. *Ikan terubok,* also known as Chinese herring or sablefish, is especially flavourful grilled because it is an oily fish with very tender flesh and great umami. The fish may be grilled or steamed, but in either method, leave the thick scales on. They keep the flesh moist and are easily lifted up whole once cooked. If the fish has been scaled, slash the thickest parts to even out the grilling time. The condiment of *sambal belacan* with shallots is good with any steamed or boiled seafood.

600 g (1 lb 5$^1/_3$ oz) *ikan terubok* (*toli* shad) with scales

1 large banana leaf, leaf stem discarded

Sambal Belacan with Shallots

1 Tbsp *sambal belacan* (page 14)

50 g (1$^3/_4$ oz) shallots, peeled and finely sliced

1 Tbsp dark soy sauce

2 tsp sugar

1$^1/_2$ Tbsp lime juice

1. Wrap fish in several layers of banana leaf and grill at 190°C (375°F) for about 20 minutes on each side. The time depends on the thickness of the fish.

2. Unwrap, and using the tip of a blunt knife, ease the scales away from the flesh.

3. Serve hot with rice and *sambal belacan* with shallots.

4. To prepare *sambal belacan* with shallots, mix all the ingredients together. Adjust sugar and lime juice to taste.

Steamed/Boiled/Grilled Prawns

Preparation and cooking: 15–30 minutes

When prawns are super fresh, steam, boil or grill them and enjoy with a *sambal belacan* dip as well as rice if you prefer. Boiled prawns are also staples in many Straits-born salads and are also garnishes for noodle dishes. To keep prawns juicy, grill, steam or boil them in their shells. They should not be overcooked, or the flesh toughens. However, there are some variety of prawns that are enjoyed for their firm, almost crunchy texture. If you boil prawns, save the resulting stock for another dish. The length of cooking time depends on the size of the prawns, but generally, prawns are done when the colour changes and deepens.

1. Rinse prawns in their shells clean.

2. To steam prawns, place them in a steamer and steam for 5–7 minutes until the colour changes. The time needed depends on the size of the prawns.

3. To boil prawns, bring a pot of water to the boil before adding prawns. When prawns change colour, turn off the heat and let them cook for another minute or so in the residual heat.

4. To grill prawns, set the temperature to 190°C (375°F). Brush some oil on the prawns and grill for 10–20 minutes on each side, depending on the size of the prawns.

5. Serve hot with *sambal belacan* with shallots (page 76).

Steamed/Boiled Crabs

Preparation and cooking: 15–30 minutes

Live crabs should be prepared as soon as possible on coming back from the market. They should not be allowed to die on their own before cooking. When that happens, the flesh turns mushy. This is why the texture of blue swimmer crabs, which are never alive when you buy them, can be chancy. Even if you intend to make chilli crabs for dinner, steam them first.

1. Kill live crabs by stabbing between the eyes. Clean them by brushing the shells under running water. Brush under the carapace and in between the legs under running water.

2. Steam or boil crabs in their shells whole. Crabs are done when the shells change colour.

3. Serve hot with a *sambal belacan* dip, turn them into chilli crabs (page 78) or shell them for crab cakes (page 80).

Chilli Crabs

Preparation and cooking: 45 minutes

Crabs — you either love them or hate them. It's not so much the flavour, but rather the big bother of having to shell them before you can enjoy them. Straits-born cooks love crabs, and they were my favourite seafood even as a child. My mother's idea of a Sunday lunch treat was chilli crabs and hers was superior to the sweet tomato-based chilli crabs that I came across in seafood restaurants years later. Since she was from Penang, the debates between Malaysian and Singaporean foodies over the origins of chilli crab have always been intriguing if insoluble. My mother never went for any cooking classes or read a cookbook in her life, and the family rarely ate out. So where did her chilli crab recipe come from? Sadly, I never thought to ask her when I could. Perhaps from my aunt in Alor Star?

1.5 kg (3 lb 5 oz) crabs

3 Tbsp cooking oil

Sauce

1 Tbsp cornflour/tapioca starch

375 ml (12 fl oz / $1^1/_2$ cups) water

4 Tbsp white rice vinegar

$1^1/_2$ Tbsp sugar

1 egg

Spice Paste

100 g ($3^1/_2$ oz) fresh red chillies, seeded

5-cm (2-in) knob ginger, scraped and thinly sliced

6 cloves garlic, peeled

2 Tbsp water

Garnishing

Coriander leaves

1. Make spice paste by blending all the ingredients together in a food processor until smooth.

2. Prepare crabs by stabbing them between the eyes with a knife. When crabs are no longer moving, carefully brush the shells clean before pulling the carapace off to rinse the underneath of the flaps. Quarter the crab if it is large, half it if it is small. (*See also* Steamed/Boiled Crabs, page 77.)

3. Combine all the ingredients for the sauce in a bowl. Set aside.

4. Heat oil in a wok and sauté spice paste until fragrant. Stir in prepared crabs and mix well. Continue stir-frying until crabs begin to change colour.

5. Give the sauce mixture a stir before adding to the wok. Mix well. Cover, turn down the heat and continue cooking for another 5 minutes. Stir well and dish out. Garnish with coriander leaves.

6. Serve hot with rice or bread.

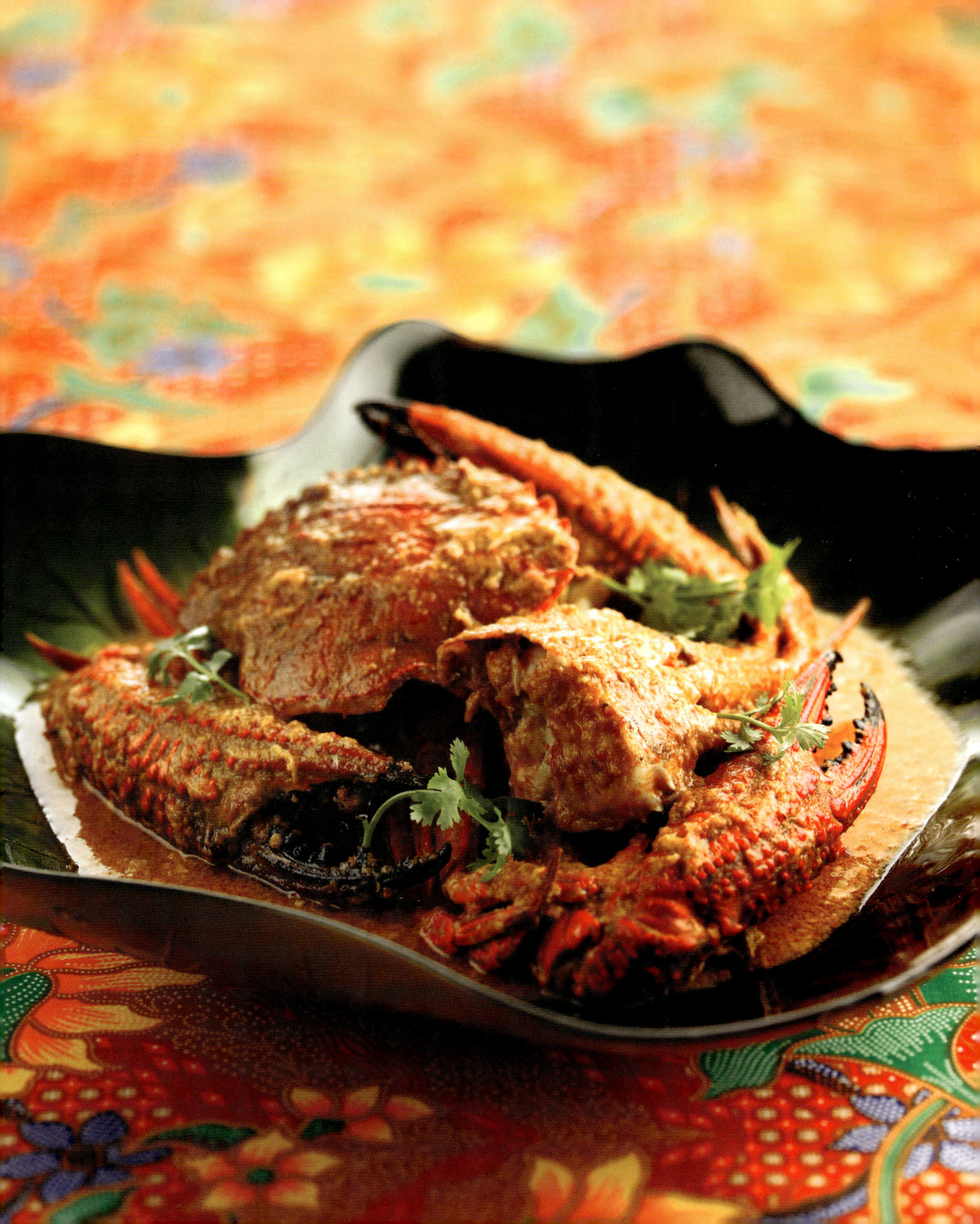

Crab Cakes

Preparation and cooking: 45 minutes

There should be plenty of fat in the minced pork to keep these crab cakes juicy. Adding water chestnuts gives the crab cakes crunch, while bamboo shoot gives it a distinct Straits-born flavour. My mother would plan ahead and save empty crab shells to use as casings for the crab cakes, or she would cook chilli crabs so that there would be enough shells for crab cakes. The shells must be thoroughly cleaned and patted dry with paper towels before they are filled.

150 g (5^1/$_3$ oz) crab meat

3 water chestnuts, peeled and chopped

150 g (5^1/$_3$ oz) minced pork

150 g (5^1/$_3$ oz) shelled prawns

^{1}/$_2$ cup finely shredded bamboo shoot

^{1}/$_2$ tsp salt

^{1}/$_2$ tsp ground white pepper

2 tsp cornflour

Cooking oil for frying

1. Mix all the ingredients, except the cooking oil, together. Divide into 12 portions and form each one into a patty. If crab shells are available, stuff the mixture into the shells. Any leftover mixture can be made into patties. Alternatively, turn leftovers into meatballs and cook in pot of chicken stock to make *bakwan kepiting* (crab and pork ball soup).

2. Heat 3-cm (1^1/$_2$-in) depth of oil in a frying pan. The patties need to be more than half-covered in oil so that the sides are browned nicely. If frying stuffed crab shells, fry the shells meat-side down until nicely brown before frying shell-side down to cook the filling inside.

3. Serve as a side dish with rice or as a main course with a dressed salad on the side.

Mum's Sambal Hae Bee (Dried Prawn Sambal)

Preparation and cooking: 1 hour

This dried prawn *sambal* was my mother's emergency side dish to eat with rice porridge, or mix into hot rice. It's another comfort food. Strangely, however, it is not so tasty with bread. For sandwiches, the *hae bee hiam* sandwich filling (page 122) is much better.

100 g (3½ oz) dried shrimps (*hae bee*), softened in water

1 tsp sugar

Juice of 2–3 limes

3 Tbsp vegetable oil

Spice Paste

4 candlenuts (*buah keras*)

30 g (1 oz) fresh red chillies, seeded and chopped

50 g (1¾ oz) shallots, peeled and chopped

1 tsp dried shrimp paste (*belacan*)

1. It is better to pound this spice paste using a mortar and pestle because any addition of water to blend the ingredients will make for a very damp mixture, and lengthen the frying process.

2. Pound softened dried shrimps until fine. Remove and set aside.

3. Place the ingredients for the spice paste in the mortar and pound until fine. Mix the spice paste into the pounded dried shrimps together with the lime juice and sugar.

4. Heat oil in a wok. Add dried shrimp mixture and stir-fry over low heat until mixture is fragrant and dry. While frying, do a taste test and adjust seasoning to taste.

5. Leave to cool in the wok, then bottle and refrigerate.

6. Serve as a side dish with rice or rice porridge.

Grilled Stringray

Preparation and cooking: 1 hour

Slathering a *sambal* dressing on fish, wrapping it in banana leaf and grilling it is classic Straits-born, especially in the days when many cooked with open coal or wood fires. These days, you would pop it under a grill in the oven. If you cook with gas and don't have an oven, you can create a make-shift grill if you have an Indian *tawa* (the flat cast iron plate that is used for cooking *roti paratha*) or a heat diffuser plate (a layered piece of metal used in West Asia to control diffusion of heat). Japanese household shops or supermarkets often have a contraption for grilling over gas. These come in several sizes. Place a rack on the *tawa* or diffuser plate and the wrapped fish on the rack. Stingray (*ikan pari*) is common in South East Asian waters. It is also tasty made into *asam pedas/ lauk pindang* (page 16).

500 g (1 lb 1¹/₂ oz) stingray, cleaned

1 tsp salt

¹/₂ Tbsp tamarind paste

2 Tbsp water

2 Tbsp oil

2 tsp sugar

Banana leaves, cleaned

Lime halves for garnishing

Spice Paste

3 fresh red chillies, large, seeded and chopped

50 g (1³/₄ oz) shallots, peeled and chopped

6 slices galangal

1–2 stalks lemongrass, thinly sliced

¹/₂ tsp dried shrimp paste (*belacan*)

4 Tbsp water

1¹/₂ Tbsp dried red chilli paste (page 17)

1. Make spice paste by blending all the ingredients, except the dried red chilli paste, together in a food processor until smooth. Add the dried red chilli paste last.

2. Rub stingray with ¹/₂ tsp salt and set aside.

3. Make tamarind juice by mixing tamarind paste and 2 Tbsp water. Strain the juice and discard the solids.

4. Heat oil in a small saucepan and fry spice paste until oil rises to the top. Add tamarind juice, sugar and remaining ¹/₂ tsp salt. Simmer for 1–2 minutes until oil rises to the top. Leave spice paste to cool if preparing the stingray overnight for grilling.

5. Spread out several layers of cleaned banana leaves. Spoon a layer of cooked spice paste on the leaves, lay the stingray on top of the spice paste, then spread the remaining spice paste over the stingray. Wrap banana leaves over the stingray and skewer the ends together with a sharp toothpick.

6. Heat the grill to 190°C (375°F). Grill the wrapped stingray for about 20 minutes on each side. The time needed depends on the thickness of the stingray.

7. To serve, squeeze wedges of lime over the stingray. Serve with rice or as it is.

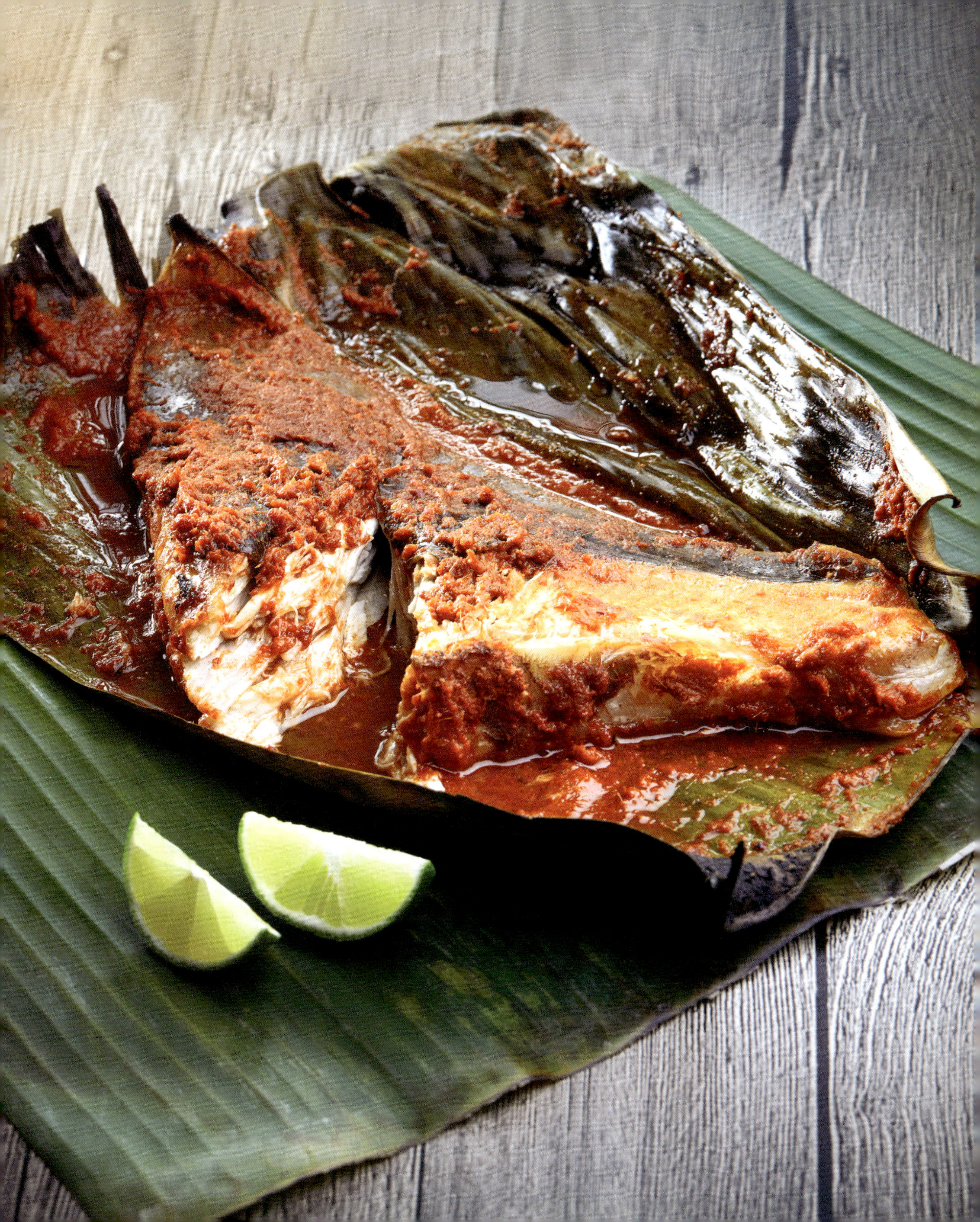

Otak-Otak Panggang (Grilled Spicy Fish Custard)

Preparation and cooking: 2 hours

Otak-otak was one of the few dishes to be identified as Straits Chinese in the very first pre-war Malayan-Singapore cookbook. *Otak* is the Malay word for "brains". Brains are soft and custard-like which may be how *otak-otak* got its name since it can be custard-like especially when cooks add beaten eggs to the spicy mixture. Penang and Singapore *otak-otak* differ in that the former is steam-cooked in a banana leaf cup lined with *daun kadok* (wild pepper leaf), and the latter grilled either wrapped in banana leaf or coconut leaves. Susie Hing's recipe in *In a Malayan Kitchen* was an interesting combination of an *otak* mixture wrapped in banana leaf and steamed. My mother's grilled *otak-otak* was always scented with kaffir lime leaves, which was also used in some of the old recipes.

500 g (1 lb 1¹/₂ oz) boneless fish meat

¹/₂ tsp salt

125 ml (4 fl oz / ¹/₂ cup) packaged coconut cream

8 kaffir lime leaves, shredded finely

Banana leaves, cut into 12 square sheets, each about 20 x 20-cm (8 x 8-in) for wrapping

24 bamboo skewers with sharp ends, each 3-cm (1¹/₂-in) long

Spice Paste

5-cm (2-in) knob turmeric, scraped

10 g (¹/₃ oz) *krisek* (page 18)

4 candlenuts (*buah keras*)

30 g (1 oz) fresh red chillies, seeded and chopped

2 Tbsp dried red chilli paste (page 17)

¹/₂ tsp dried shrimp paste (*belacan*)

150 g (5¹/₃ oz) shallots, peeled and chopped

125 ml (4 fl oz / ¹/₂ cup) packaged coconut cream

1. Set aside 100 g (3¹/₂ oz) of fish meat and chop coarsely.

2. Make spice paste by blending all the ingredients together in a food processor until smooth. Remove and set aside.

3. Combine remaining fish meat, salt and coconut cream in the food processor and blend for 3 minutes. Remove and stir into the ground spice paste.

4. Stir the coarsely chopped fish meat into the *otak-otak* paste and divide into 12 portions.

5. Clean the banana leaves, then blanch in boiling water to make them pliable. Wipe dry with paper towels.

6. Spoon a portion of fish paste on a banana leaf and bring two opposite sides up over the paste to enclose. Secure the two ends with bamboo toothpicks. Repeat for the other portions.

7. Grill *otak-otak* for 10 minutes, turning them over after 5 minutes. The time required depends on the thickness of the *otak-otak*. The *otak-otak* may also be baked in a 180°C (356°C) oven.

8. Serve *otak-otak* as a snack, as a sandwich filling or as a side dish with *nasi lemak* (page 46).

> **Note**
> *Otak-otak* paste can also be spooned into individual ramekins, then baked covered with foil or steamed uncovered, Penang-style. If taking the *otak-otak* to a potluck meal, line the bottom of a couple of foil containers with banana leaves cut to fit, then fill with the mixture. Bake/steam until cooked through.

Variations
- Fish head *otak-otak*: Chop some pieces of fish head and mix with the *otak-otak* paste at step 4.
- Fish cake *otak-otak*: Add slices of fish cake to the *otak-otak* paste at step 4.
- Prawn *otak-otak*: Add whole shelled prawns to the *otak-otak* paste at step 4.
- Crab meat *otak-otak:* Add 100 g ($3^1/_2$ oz) crabmeat to the *otak-otak* paste at step 4.

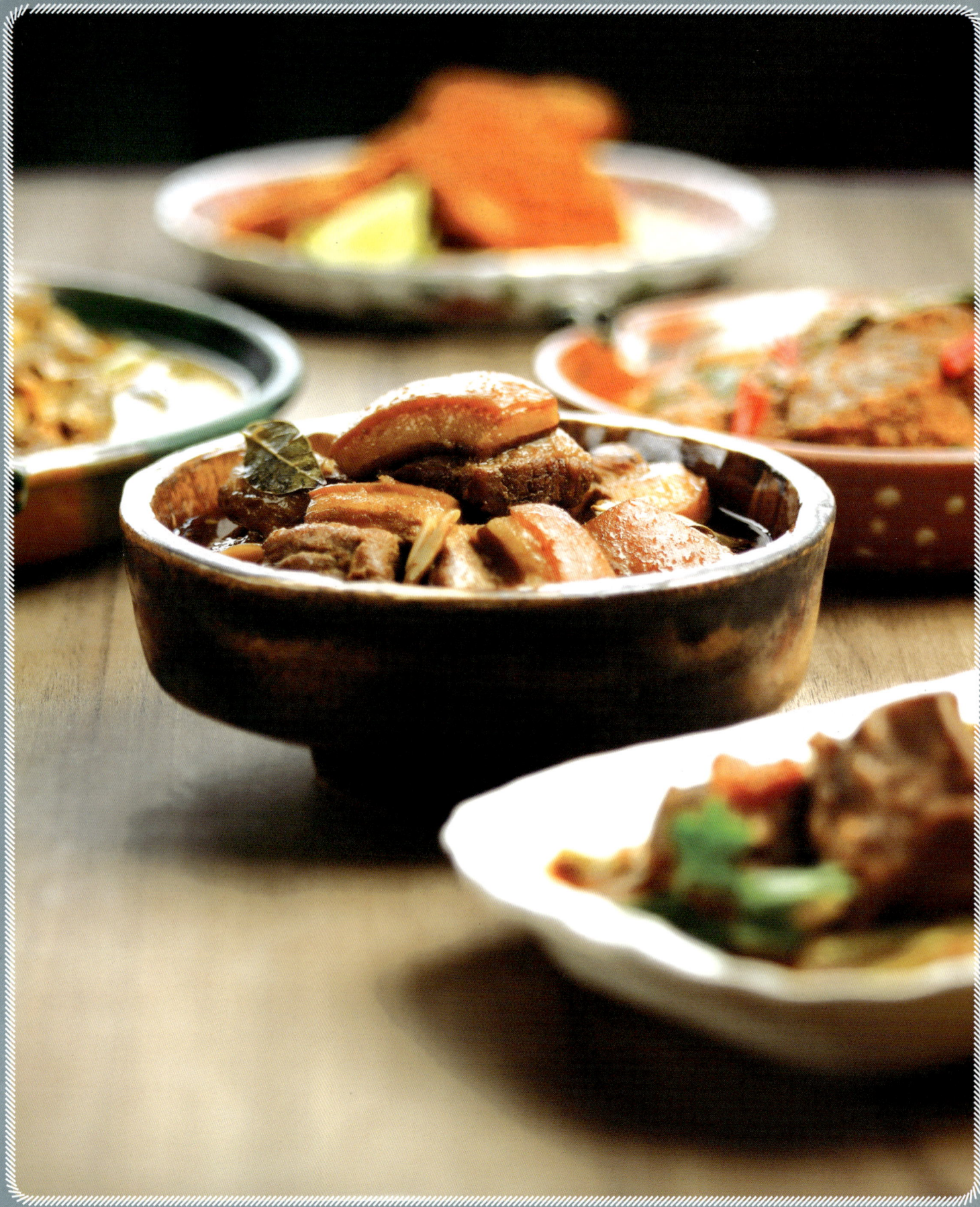

Meats

Comforting Curries, Stews and Soups

While there are very few differences in the ways that Straits-born communities prepare seafood, there are significant differences in their approach to meats, starting with what they do and do not eat. The Hindu Chettis, for instance, do not eat beef and if they have intermarried into the Indian Muslim or the Malay Muslim communities, then they do not eat pork. Straits Chinese Buddhists traditionally do not eat beef, also for religious reasons, and many may not eat mutton because this meat is thought of as "strongly flavoured". The Christian Straits-born, which include the Eurasians and Chinese, go with personal preference, and roasts and grills are popular. The one meat that all except vegetarians will eat is chicken. Pork, too, is popular with the non-Muslim Straits-born and practically a must for most Straits Chinese.

Meat dishes in Straits-born cooking are often tweaks of ethnic recipes. Take a Straits Chinese classic like *babi pong teh*. This pork dish probably evolved from Hokkien *tau eu bak* (braised pork in dark soy sauce), but the addition of spices has turned it into quite a different dish. Portuguese Eurasian vindaloo, too, originated from a combination of New World chillies with a traditional South Indian dish to give another Straits-born classic. *Ayam buah keluak* is another iconic Straits-born dish that is a clear example of how geographical proximity contributed to the development of the dish. *Buah keluak* come mostly from East Java where *nasi rawon* is a culinary specialty of beef braised with *buah keluak*. *Mrs Lee's Cookbook* (1974) even has a *nasi rawon* recipe. Both dishes may look somewhat alike, but they taste different. Eurasian and Singapore Straits Chinese cookbooks usually have a *keluak* recipe, but Penang heritage cookbooks do not.

Straits-born meat dishes are characterised by the fact that practically all benefit from being given a rest period after cooking for the flavours to mature. Even some of the soups taste better if left to sit for a while. Many of the stews and curries freeze very well and almost all may be left outside the fridge overnight without spoiling if handled correctly.

The celebration of festivals are times when this old trick that my mother taught me may come in useful. To store outside the fridge, bring the pot of food to the boil, stir it and keep it on the boil for about 5 minutes before turning off the heat. Cover the pot and leave it alone and the food will not go off for up to 24 hours. Before serving, return it to the boil. This process may be repeated for a number of days, but of course, the meat will begin to fall apart with repeated boiling. So keep the meat firm if you think you might have to resort to this method of storage. Alternatively, store in two smaller pots to minimise repeated boiling.

Buah Keluak Curry with Chicken and Pork

Advance preparation and cooking: several days

Although I never ate *buah keluak* when I was young because my mother never cooked it, I acquired a taste for it as well as this recipe through trial and error. I had been challenged by a friend who asked how come a Nonya like me did not eat it. This dish has since become one of my signature dishes at Chinese New Year for friends who love it. It also became one of my father's Chinese New Year favourites although my mother never acquired a taste for it. The stew freezes well and is an excellent standby for unexpected guests. While some recipes do not mix ground pork with the pounded nut meat, note that the nut meat will dissolve into the curry without minced pork as the binder. In Singapore and Indonesia, shelled *buah keluak* may be sold by weight and the empty shells sold separately by the piece. Whole unshelled nuts may be sold by weight or by the piece. While shelled *buah keluak* can be seen to be good at a glance, some whole nuts may be off, so it is worth getting extra.

150 g (5$^1/_3$ oz) tamarind paste

1 litre (32 fl oz / 4 cups) water

4 Tbsp cooking oil

25–30 *buah keluak*

2 tsp salt

500 g (1 lb 1$^1/_2$ oz) pork ribs, cleaned

500 g (1 lb 1$^1/_2$ oz) chicken, cleaned

Buah Keluak Stuffing

$^1/_2$ tsp salt

200 g (7 oz) minced pork

Spice Paste

300 g (11 oz) shallots, peeled

3 cloves garlic, peeled

4 stalks lemongrass, finely sliced

5-cm (2-in) knob galangal, scraped and sliced

1$^1/_2$ tsp dried shrimp paste (*belacan*)

30 g (1 oz) dried red chillies, seeded and softened in water

2 Tbsp water

1. Clean and prepare *buah keulak* (pages 90–91).

2. Prepare *buah keluak* stuffing. Mix salt with minced pork and pounded nut meat. Stuff mixture into shells.

3. Make spice paste by blending all the ingredients together in a food processor until fine.

4. Mix tamarind paste with 1 litre (32 fl oz / 4 cups) water. Strain and discard the solids. Set aside tamarind juice.

5. Heat oil in a large pot and fry spice paste until fragrant and oil rises to the top.

6. Add stuffed *buah keluak*, tamarind juice, salt and pork ribs. Bring to the boil, turn down the heat and simmer for 30 minutes.

7. Adjust seasoning with tamarind juice and salt to taste before adding the chicken. The curry should taste slightly tart and there should be at least 2 cups of gravy. Return to the boil, turn down the heat and simmer for another 30 minutes or until the chicken is cooked. Neither the pork ribs nor chicken should be so tender that the meat is falling off the bone. The curry will darken once the *buah keluak* has steeped in the gravy for at least half a day.

8. Reheat before serving. Enjoy with rice.

> **Note**
> When this curry is frozen, the *buah keluak* changes the acidity and spiciness of the stew. Adjust with more tamarind juice, cooked dried red chilli paste (page 17) or a few spoonfuls of cooked spice paste when reheating.

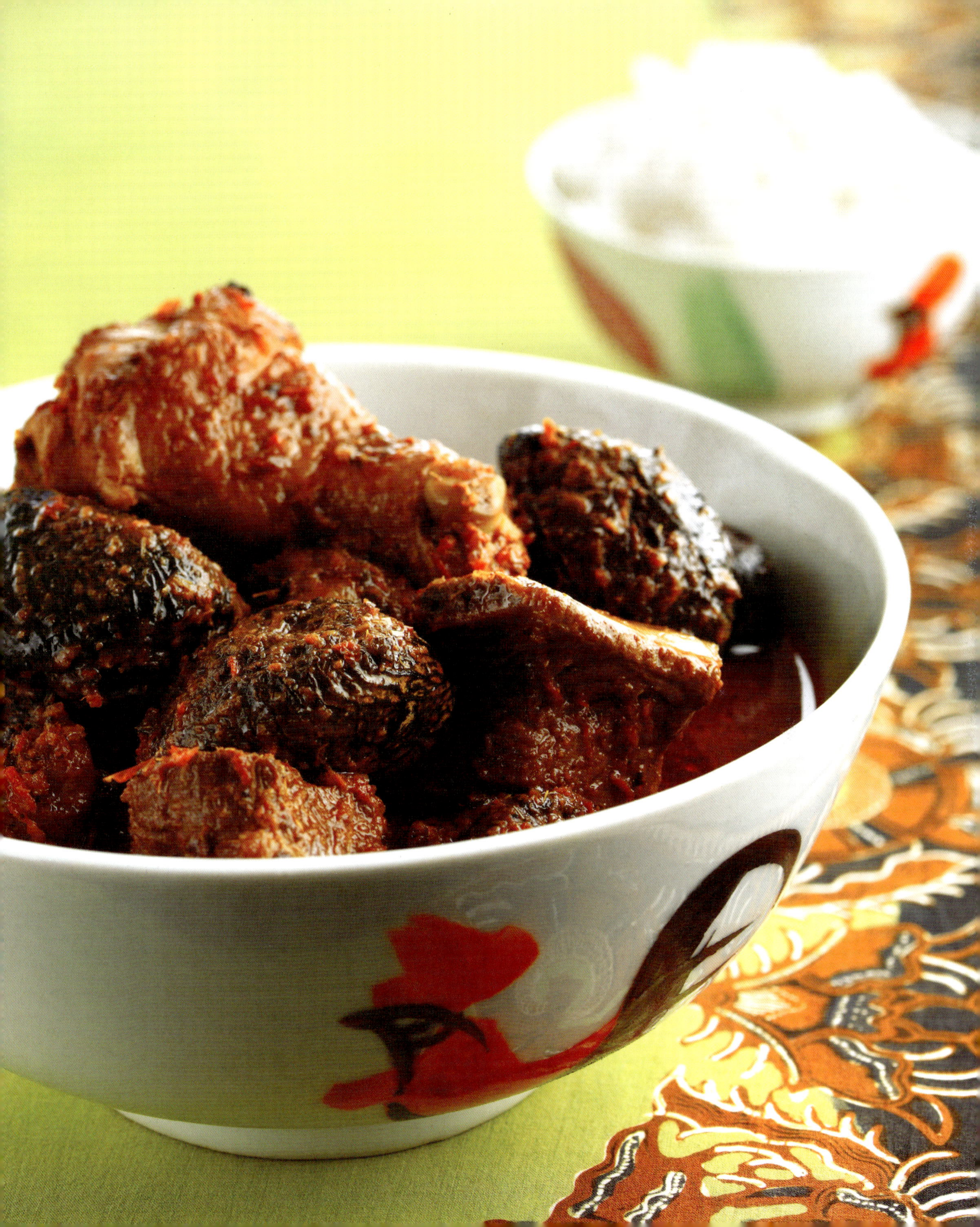

Preparing and Stuffing Buah Keluak

Buah keluak is the nut of the *kepayang* (*Pangium edule*), a tree native to Indonesia, Malaysia and Papua New Guinea. Called *keluak* in Malay and *kluwak* in Indonesia, the soft, oily meat has to be extracted from the very hard shells before cooking. But before they are even sold, *buah keluak* has to be processed by boiling and pickling in ashes due to the poisonous hydrocyanic acid in the nuts. If looking for *buah keluak* outside of this region, try Asian stores patronised by Indonesian Chinese. Preparing *buah keluak* for cooking is a time-consuming process. The method shown here is the only safe way I know for cracking open the nut while leaving the shells intact for stuffing. I keep a pair of blunt-nose pliers and a short-handled broad-edge screwdriver reserved for preparing the annual *buah keluak* curry.

1. *Buah keluak* must be soaked in daily changes of water for at least 3 days and not more than 6. Before soaking, place the *buah keluak* in a basin of water and using a firm brush, scrub and wash the nuts clean.

2. Place cleaned *buah keluak* in a basin of water. To immerse the nuts properly and stop them from bobbing up, cover with a plate large enough to push the nuts down. Top with a waterproof weight such as a mortar pestle. Change the water daily.

3. Place a chopping board on some newspapers on a good working surface and where you can also sit down. The job will take about an hour or less, depending on how adroit you are and how many nuts you are cracking. I take about an hour to do 50 nuts.

4. Hold the screwdriver and nut together in one hand. The flat tip of the screwdriver should be pressed firmly into the thin groove at the broader end of the nut. Hold the stone or pestle with the other hand and strike the top of the screwdriver hard enough to puncture the nut. This groove is the weakest part of the nut.

5. Prise out the little plate-like part of the nut with the tip of your screwdriver. The nut meat can now be extracted with a teaspoon. Take care to remove any shell shards.

6. If the opening is rather small, which is usually the case, widen it with the blunt-nosed pliers by nipping off the edge of the opening. It should be large enough to allow the insertion of a teaspoon into the nut, but not so large that the filling will drop out.

7. Discard any nut that is not black, smells off or is watery. Good *buah keluak* is thick, soft, black and smells fragrant. Black but dry nut meat that does not smell bad may be soaked in some water and prepared with the good soft nut meat.

8. The nut meat must be pounded to a smooth paste. Take care to remove any shell shards before or during pounding.

9. Mix in salt and minced pork with the pounded nut meat. Stuff the mixture back into the shells right up to the top. Smooth the paste so that it is level with the end of the shell. Set aside until ready to cook.

Press the flat tip of the screwdriver into the thin groove at the top of the nut.

Widen the opening with the blunt-nose pliers by nipping off the edge of the nut.

Extract the nut meat with a teaspoon.

Ayam Tempra (Chicken in Sour-Spicy Sauce)

Preparation and cooking: 45 minutes

My mother called this simple dish *char sui kay*. In Hokkien, it means chicken in sour sauce. A closer look at Lloyd Matthew Tan's *Daily Nonya Dishes* showed that the main seasoning ingredients in his *tempra* recipes were the same as my mother's *char sui kay*! So were those for *babi tempra* in *Mrs Lee's Cookbook*. "*Tempra*" is a corruption of the Portuguese "*tempero*" meaning seasoning. Interestingly, the word "tempering" is found in some old Indian cookbooks to describe the process of frying spices and curry powder in oil to develop the seasoning. Mary Gomes in *The Eurasian Cookbook* translated "*tempra*" as *rempah* or spice paste. Her *tempra* recipes were without lime juice, but had sugar in them.

2 Tbsp cooking oil

5-cm (2-in) knob ginger, scraped and cut into thin strips

25 g ($^4/_5$ oz) shallots, peeled and thinly sliced

3 cloves garlic, peeled and thinly sliced

2 fresh red chillies, seeded and sliced

800 g ($1^3/_4$ lb) chicken, cleaned and cut into small pieces

4 Tbsp lime juice

1 Tbsp sugar

2 Tbsp light soy sauce

1 tsp dark soy sauce

250 ml (8 fl oz / 1 cup) water

1. Heat oil in a saucepan and fry ginger, shallots, garlic and chilli until shallots are soft.

2. Add chicken and the rest of the ingredients and bring to the boil. Turn down the heat and simmer for about 20 minutes or until chicken is cooked. The time depends on the size of the pieces.

3. Adjust seasoning to taste and leave to rest for at least an hour before serving with rice.

4. This dish keeps very well.

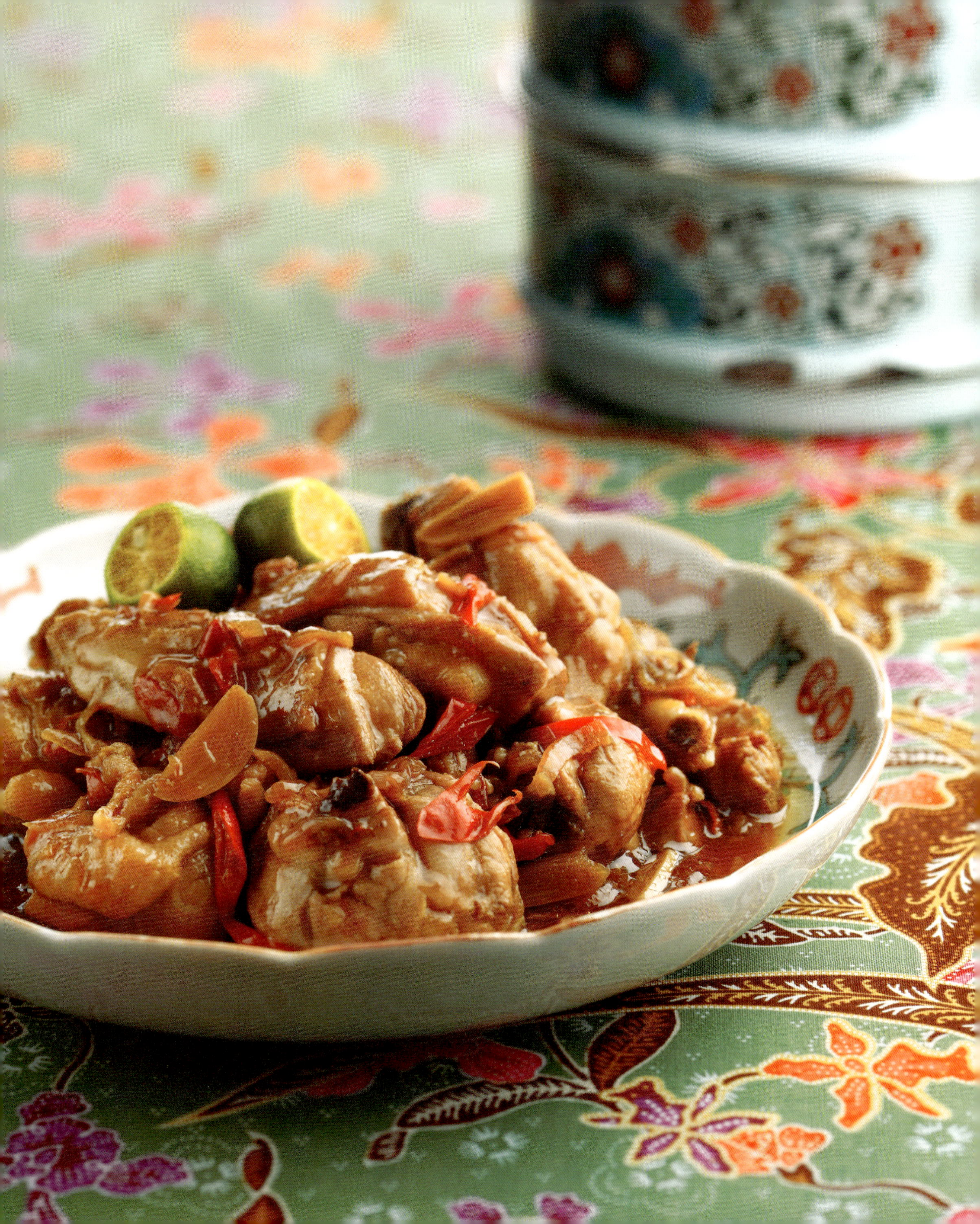

Too Kah Sui (Sour Pork Trotters)

Preparation and cooking: 2¹/₂ hours

Pineapple (*Ananas comosus*) was introduced to South East Asia by the Portuguese and Spanish in the 16th century and the Straits-born ended up cooking the fruit with meat, seafood and even salted fish bones besides making salads and chutneys with it. A tropical South American native, pineapple found a ready home in colonial Malaya and Singapore. In the 19th century, Singapore even exported canned pineapples. This stew keeps well, tastes great, is easy to prepare and has a festive flavour because of the trotters. No wonder it was one of my mother's favourite dishes.

1 kg (2 lb 3 oz) pork foreleg, chopped into serving-size pieces

1 fresh pineapple, about 500 g (1 lb 1¹/₂ oz)

25 g (⁴/₅ oz) tamarind paste

1 litre (32 fl oz / 4 cups) water

2 Tbsp cooking oil

6 dried sour fruit slices (*asam gelugor*)

1 tsp salt

1 Tbsp light soy sauce

A few drops dark soy sauce

3 tsp sugar

Spice Paste

30 g (1 oz) dried red chillies, seeded and softened in water/ 2 Tbsp dried red chilli paste (page 17)

100 g (3¹/₂ oz) shallots, peeled and chopped

4 cloves garlic, peeled and chopped

4 Tbsp water

1. When using pork foreleg, wash meat carefully to remove any bone fragments: feel each piece of meat carefully.

2. Make spice paste by blending softened dried chillies, shallots and garlic with water in a food processor until fine. If using chilli paste, blend shallots and garlic first before adding it.

3. Trim pineapple, removing the skin, eyes and core. Cut pineapple into 2.5-cm (1-in) thick wedges. Set aside.

4. Mix tamarind paste with 1 litre (32 fl oz / 4 cups) water and strain away the solids. Set aside tamarind juice.

5. Heat oil in a large saucepan and sauté spice paste until oil rises to the top.

6. Add pork, tamarind juice and the rest of the ingredients and bring to the boil. Turn down the heat and simmer for about 2 hours or until meat is tender. Adjust seasoning to taste. If pineapple is sweet, add more tamarind juice. Add more water for stewing, if needed. There should be a thick gravy.

7. Rest the stew for several hours before serving. Reheat and serve with rice.

Note

Why is the foreleg or *chiew* (Hokkien, meaning "hand") used in this dish? Because it is believed that the meat here takes less time to cook till tender compared to the hind leg. So said my mother. I have not checked out the truth of this belief.

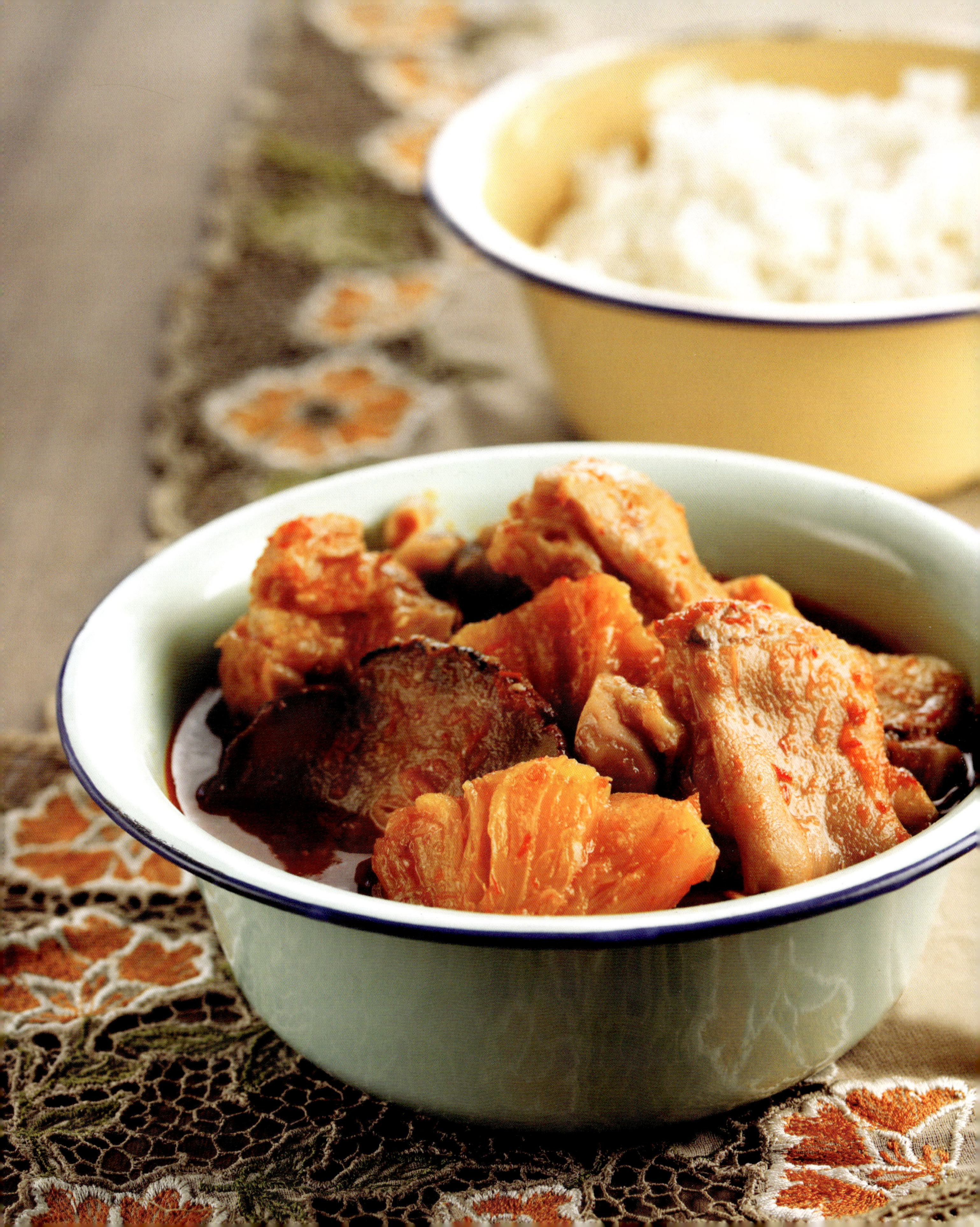

Babi Pong Teh (Pork and Cinnamon Stew)

Preparation and cooking: 2 hours

The Straits Chinese added fried shallots, spices and bamboo shoots to Hokkien *tau eu bak* (pork braised in dark soy sauce) to get *babi pong teh*. In the Philippines, which also has a lot of Hokkien migrants, cooks turned the same dish into *adobo* by adding vinegar to it. "*Adobo*" comes from a Spanish word that means marinade or seasoning. Amongst the Straits Chinese, this dish may have had ritual significance in the days when many practised ancestor worship. Like the Chettis, certain dishes would be cooked as prayer offerings and the dark-coloured *babi pong teh* was considered more appropriate than brightly coloured curries.

1 kg (2 lb 3 oz) pork foreleg/belly pork, chopped into pieces

3 Tbsp cooking oil

200 g (7 oz) shallots, peeled and thinly sliced

30 g (1 oz) garlic, peeled and thinly sliced

2 Tbsp fermented soy beans (*taucheo*), rinsed and mashed/ brown miso

1 tsp dark soy sauce

3 Tbsp light soy sauce

1 tsp salt

2 tsp sugar

$^1/_2$ tsp ground white pepper

1 tsp ground cinnamon

6-cm (3-inch) cinnamon stick

552 g (1 lb 4 oz) canned bamboo shoots, cut into chunks/thick slices

1 litre (32 fl oz / 4 cups) water

Condiments

Green chillies

Sambal belacan with or without lime juice

> **Note**
> Japanese brown miso has the same flavour as Chinese fermented soy beans (*taucheo*), but it is easier to use and comes ready-mashed.

1. If using pork foreleg, wash meat carefully to remove any bone fragments: feel each piece of meat carefully. Clean belly pork.

2. Heat oil in a saucepan and fry shallots until they begin to turn golden. At this point, add garlic slices and continue frying until shallots are brown. Keep a close eye on the shallots as they brown fairly quickly once they turn golden.

3. Stir in miso/mashed fermented soy beans and fry for 10 seconds. Add pork and the rest of the ingredients and bring to the boil.

4. Turn down the heat and simmer for about 1$^1/_2$ hours until the meat is tender, but not falling off the bone. Adjust the seasoning to taste and add a few more drops of dark soy sauce if the stew is too pale. There should be about 500 ml (16 fl oz / 2 cups) of gravy left.

5. Rest the stew for half a day before serving. Reheat and serve with crusty French bread or rice and fresh green chillies and *sambal belacan* as condiments.

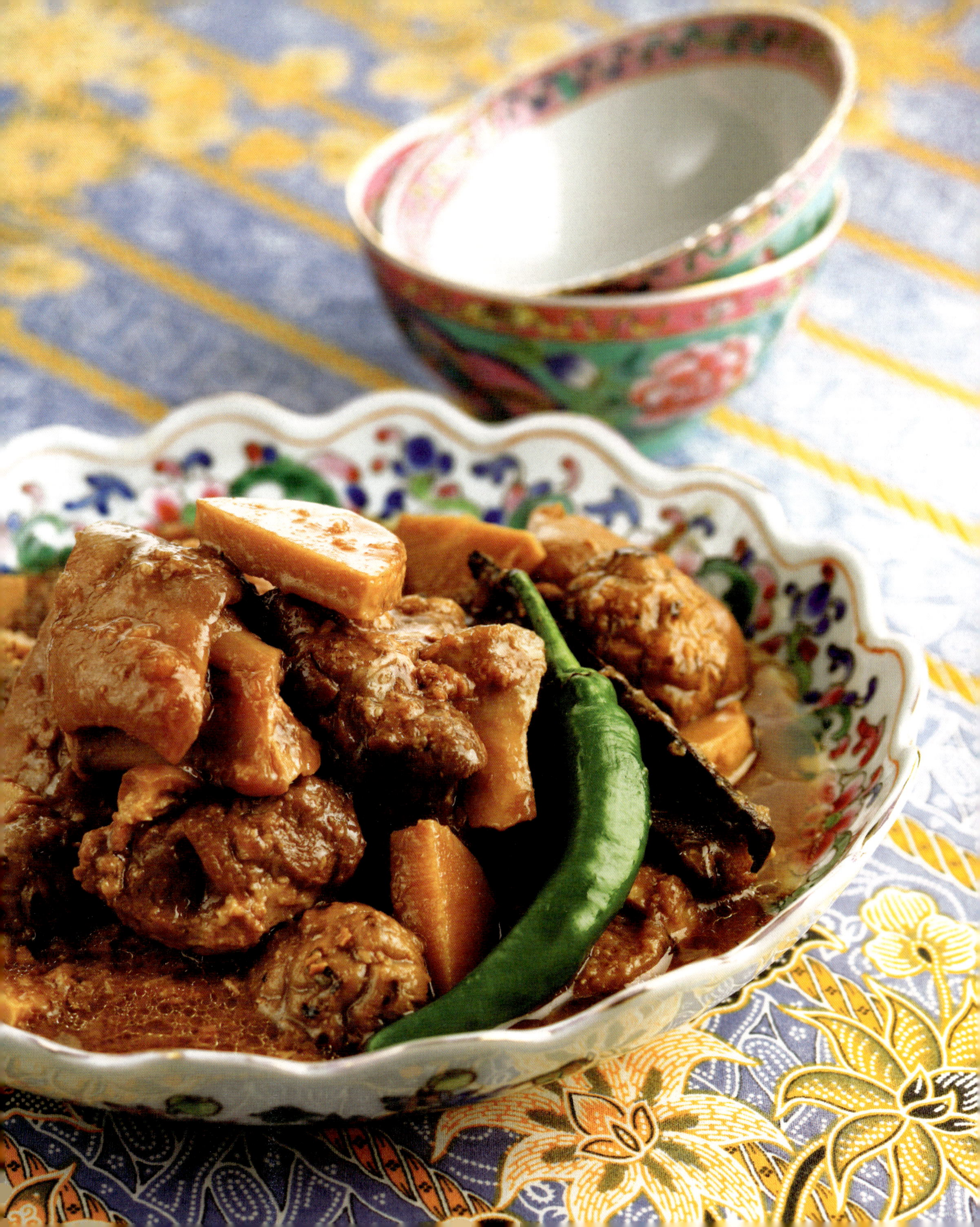

Devil Curry

Preparation and cooking: 1 hour

This Eurasian curry gets its name from its fiery look and spiciness. It's a must-have in many Eurasian homes as a festive dish, but it is one of those dishes where no one does it quite like Mum! This is to be expected as it is one of those stews where family tradition and tastes dictate what goes into it. I have heard that some cooks add *bak kuah* (Chinese barbecue pork jerky) to it, while others add *lap cheong* (Chinese sausage). Feel free to fiddle with the ingredients, as well as the amount and mix of vegetables. If using Spam/luncheon meat, cut into cubes. If using bacon, dice it. If using ham hock/knuckles, just add to the stew.

200 g (7 oz) chicken, cleaned and cut into pieces

1 tsp dark soy sauce

1¼ tsp salt

3 Tbsp cooking oil

200 g (7 oz) Spam/luncheon meat/bacon/cleaned ham hock/cleaned knuckles/Chinese sausage (*lap cheong*)

250 g (9 oz) potatoes, peeled and quartered

65 g (2 oz) onion, peeled and quartered

3–4 Tbsp white vinegar

2 tsp sugar

1 litre (32 fl oz / 4 cups) water

150 g (5⅓ oz) carrots, peeled and cut into chunks

200 g (7 oz) long beans, cut into finger lengths

400 g (14 oz) cabbage, cut into bite-size pieces

Spice Paste

200 g (7 oz) onion, peeled and chopped

3 cloves garlic, peeled

5-cm (2-in) knob ginger, scraped

1 tsp dried shrimp paste (*belacan*)

125 ml (4 fl oz / ½ cup) water

½ cup vindaloo spice mix (page 14)

1. Make spice paste by blending onion, garlic, ginger and dried shrimp paste with water in a food processor until smooth. Stir in vindaloo spice mix.

2. Rub chicken with soy sauce and ½ tsp salt and set aside for 30 minutes.

3. Heat oil in a saucepan and fry spice paste until oil rises to the top. Except for the carrots, long beans and cabbage, add all the other ingredients to the saucepan and bring to the boil. Turn down the heat and simmer for about 30 minutes. If using ham hock/knuckles, the meat should be getting tender before you add the vegetables.

4. Add carrots, long beans and cabbage. Time the addition of vegetables to how tender you like them. Adjust seasoning to taste during simmering. There should be a fairly thick gravy in the curry.

5. Serve with rice or crusty French bread.

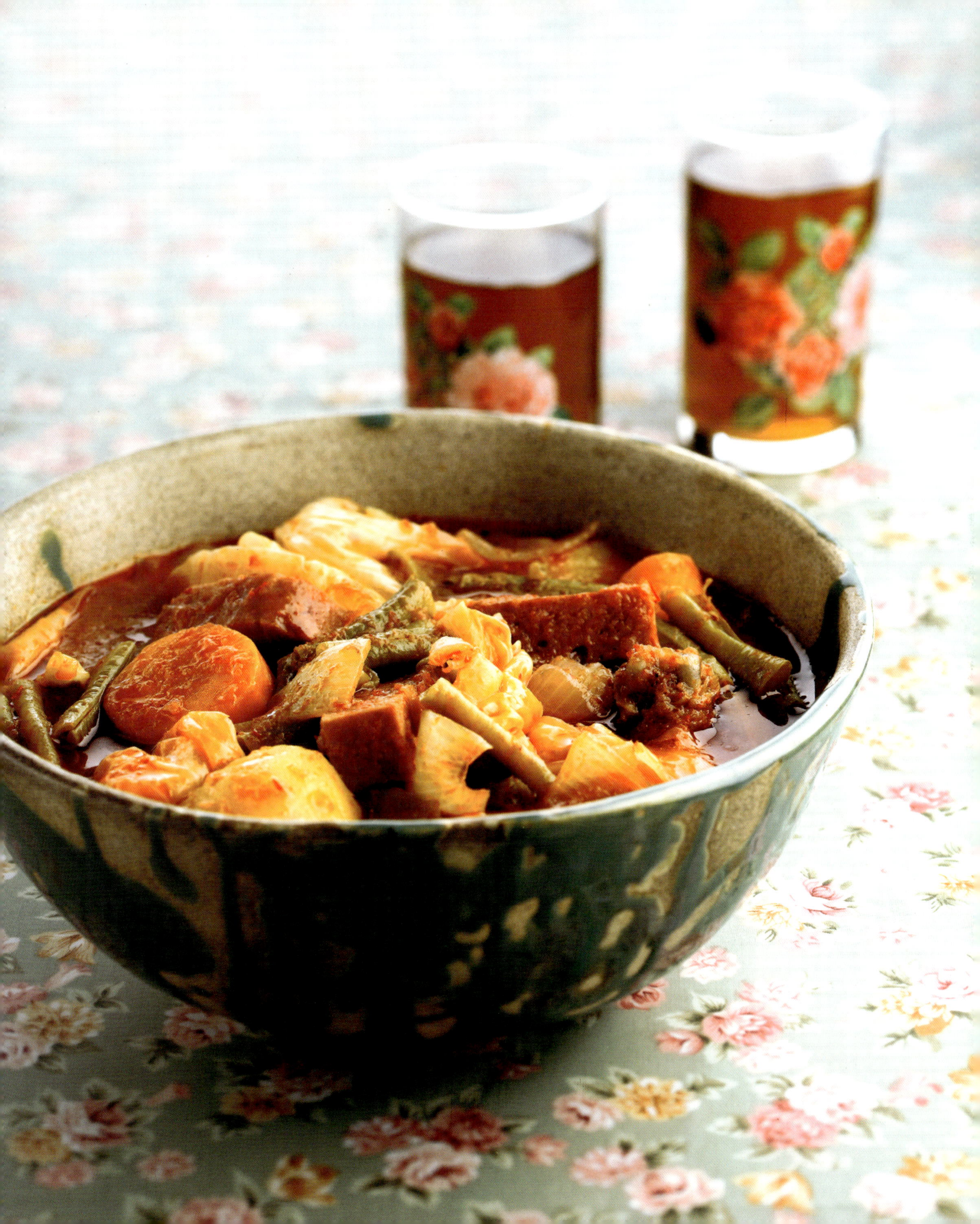

Too Torh Tng (Pig's Stomach Soup)

Preparation and cooking: 3 hours

This is another example of the "waste not, want not" philosophy of the traditional Chinese. This soup is Hokkien in origin and is common in Penang, too. The soup is enjoyed for the slightly chewy texture of the meat even when it is tender. As my grandfather liked the soup, my mother would always cook it for Chinese New Year. When it came to writing up this recipe, luckily for me, my mother was still alive and I could turn to her for instructions on how to clean pig's stomach. That was the first time I had cooked this dish. To be tasty, the meat has to be well-cleaned.

1 pig's stomach, about 500 g (1 lb 2 oz)

Coarse salt for cleaning

2 litres (64 fl oz / 8 cups) water

$\frac{1}{2}$ Tbsp cracked white/black peppercorns

150 g ($5\frac{1}{3}$ oz) onion, peeled and quartered

200 g (7 oz) potatoes, peeled and quartered

Garnishing
Chopped spring onions

Chopped coriander leaves

Condiment
Dark soy sauce with cut fresh red chillies

1. To clean the pig's stomach, place it in a pot, cover with water and bring to the boil. Cook for 10 minutes. Turn off the heat and let the stomach cool until it can be handled. Split cooled stomach open into two flat pieces. Rub coarse salt on the surfaces to clean it well. Rinse off salt and repeat.

2. If you have a gas stove, each piece can be held over the flames with a pair of tongs to singe it and burn off any odours. If you use an electric stove, heat a wok or frying pan and use a spatula to press each piece down on the hot pan, searing both sides.

3. Cut stomach into pieces, each about 1 x 2-cm ($\frac{1}{3}$ x $\frac{3}{4}$-in).

4. Combine 2 litres (64 fl oz / 8 cups) water, cleaned stomach, peppercorns, onion and potatoes in a pot and simmer until the meat is fork-tender. Adjust seasoning to taste. Alternatively, place ingredients in a slow-cooker and cook overnight, or place in a covered casserole and cook overnight in an oven set to 90°C (195°F). It is better for the meat to be too tender rather than not tender enough.

5. Serve with rice and a condiment of dark soy sauce and cut red chillies.

Chap Chye (Mixed Vegetable Stew)

Preparation and cooking: 3 hours

This classic stew that's popular with Straits Chinese and Eurasians is a misnomer as there are only two vegetables in it. Everything else in this rich stew is actually dried Chinese vegetarian ingredients such as bean sticks, glass noodles, dried mushrooms and wood ears. This stew has a way of expanding beyond expectations because all manner of dried ingredients may be added to it. These are first softened in water and added in stages, depending on how quickly they cook. Take your pick. The basics are the stock, prawns and the condiment of *sambal belacan*.

Basic Stock

3 litres (96 fl oz / 12 cups) water

1 kg (2 lb 3 oz) pork ribs and bones

2 Tbsp cooking oil

4 cloves garlic, peeled and finely chopped

2 Tbsp fermented soy beans (*taucheo*), rinsed and mashed/brown miso

$1^1/_2$ tsp salt

250 g (9 oz) prawns in their shells, whiskers trimmed

Choice of Ingredients

100 g ($3^1/_2$ oz) *bangkuang* (yam bean), peeled and cut into sticks or slices

4–8 dried shiitake mushrooms, softened in water, kept whole or sliced depending on size

20 g ($^2/_3$ oz) dried lily buds, softened in water, tough stems trimmed and buds knotted

20 g ($^2/_3$ oz) wood ear fungus (fresh or dried), trimmed and cut into bite-size pieces

50 g ($1^3/_4$ oz) gingko nuts, shelled and skinned, germ removed

50 g ($1^3/_4$ oz) Chinese red dates, rinsed

50 g ($1^3/_4$ oz) dried bean sticks, softened in water and cut into 1-cm ($^1/_3$-in) pieces

50 g ($1^3/_4$ oz) glass noodles, softened in water

100 g ($3^1/_2$ oz) cabbage, cut into 3-cm ($1^1/_2$-in) squares

30 g (1 oz) lotus seeds, softened in water, germ removed

Condiment

Sambal belacan (page 14)

1. Combine water, pork ribs and bones in a large stockpot and bring to the boil. Turn down the heat and simmer for 2 hours to get a rich stock. Discard the bones, but keep some of the more meaty ribs.

2. While the stock is boiling, prepare the dried ingredients.

3. In a frying pan, heat oil and fry chopped garlic until golden. Stir in mashed fermented soy beans/brown miso and fry for 30 seconds. Tip the contents of the frying pan into the pork stock and bring the stock back to the boil.

4. Except for the glass noodles, prawns and cabbage, add the rest of the ingredients to the stew and continue simmering until *bangkuang* is tender. This should take about 30 minutes unless the *bangkuang* has been cut too thickly.

5. Add glass noodles, cabbage and prawns and continue simmering until the prawns change colour. Turn off the heat, cover the pot and rest the stew for at least an hour before serving.

6. Reheat and serve with *sambal belacan*. Rice is optional.

Pictured: Chap Chye (Mixed Vegetable Stew)

Stir-fried Chap Chye

Preparation and cooking: 30 minutes

This is a dry version of *chap chye* with fewer ingredients. If desired, other quick-cooking ingredients may be added.

4 Tbsp water/pork stock

1 tsp salt

A few drops dark soy sauce

2 Tbsp cooking oil

1 Tbsp chopped garlic

125 g (4^1/$_2$ oz) wood ear fungus, softened, trimmed and cut into bite-size pieces

A few dried shiitake mushrooms, softened in water and sliced

30 g (1 oz) glass noodles, softened in water

400 g (14 oz) cabbage, shredded

Condiment
Sambal belacan (page 14)

1. Mix water/pork stock, salt and dark soy sauce together. Set aside.

2. Heat oil in a wok and fry garlic until fragrant. Add wood ear fungus, mushrooms, glass noodles and soy sauce mixture. Bring to the boil, turn down heat and simmer for 5 minutes.

3. Stir in shredded cabbage and continue stir-frying for another 5 minutes. Adjust seasoning to taste.

4. Serve with rice and *sambal belacan*.

Kiam Chye Arh Tng
(Duck Soup with Salted Mustard Greens)

Preparation and cooking: 4 hours

In Penang, this soup is called *kiam chye arh tng*. In Singapore, it is called *itek tim* by Straits Chinese and Eurasians. It is a good festive dish as it requires very little attention. It was one of my mother's backyard soups prepared on a charcoal stove. Today, a slow cooker makes it even easier. Alternatively, combine everything in a large covered casserole and leave it overnight in an oven set to 90°C (195°F). The Penang version has a cracked nutmeg in the soup. On the topic of nutmegs, here's an interesting bit of history. After the second Anglo-Dutch War in 1667, the Dutch exchanged Manhattan Island on which New York City sits for Pulo Ran in the Moluccas because they wanted the nutmeg trees there. Early British colonial planters tried growing nutmeg in Singapore but unsuccessfully. However, they succeeded in Penang and today, the former Prince of Wales Island still produces nutmegs.

500 g (1 lb 1¹/₂ oz) salted mustard greens (*kiam chye*)

Coarse salt for cleaning duck

1 duck, cut into 8 pieces or kept whole

4 salted plums

400 g (14 oz) tomatoes, halved

100 (3¹/₂ oz) old ginger, scraped and smashed

Salt to taste

4 litres (128 fl oz / 16 cups) water

2 tsp cracked white peppercorns

2–4 Tbsp brandy

1 nutmeg, broken into pieces

Finely sliced spring onion

Condiments
Sambal belacan (page 14)

Dark soy sauce with cut fresh red chillies

1. Rinse salted mustard greens and cut into large pieces.

2. To clean duck, rub coarse salt all over duck, taking care to remove any tiny feathers or feather shafts. Rinse well with water.

3. Combine all the ingredients in a large stockpot and bring to the boil. Skim off any scum that rises to the top. Turn down the heat and simmer until the duck is tender. Adjust seasoning to taste. If the duck has not been cut up, use a fork to break the duck into pieces for serving.

4. Rest the soup for at least half a day before serving hot with/ without rice. Garnish with finely sliced spring onion.

5. The tender meat is best enjoyed with a condiment of either *sambal belacan* or dark soy sauce with cut fresh chillies.

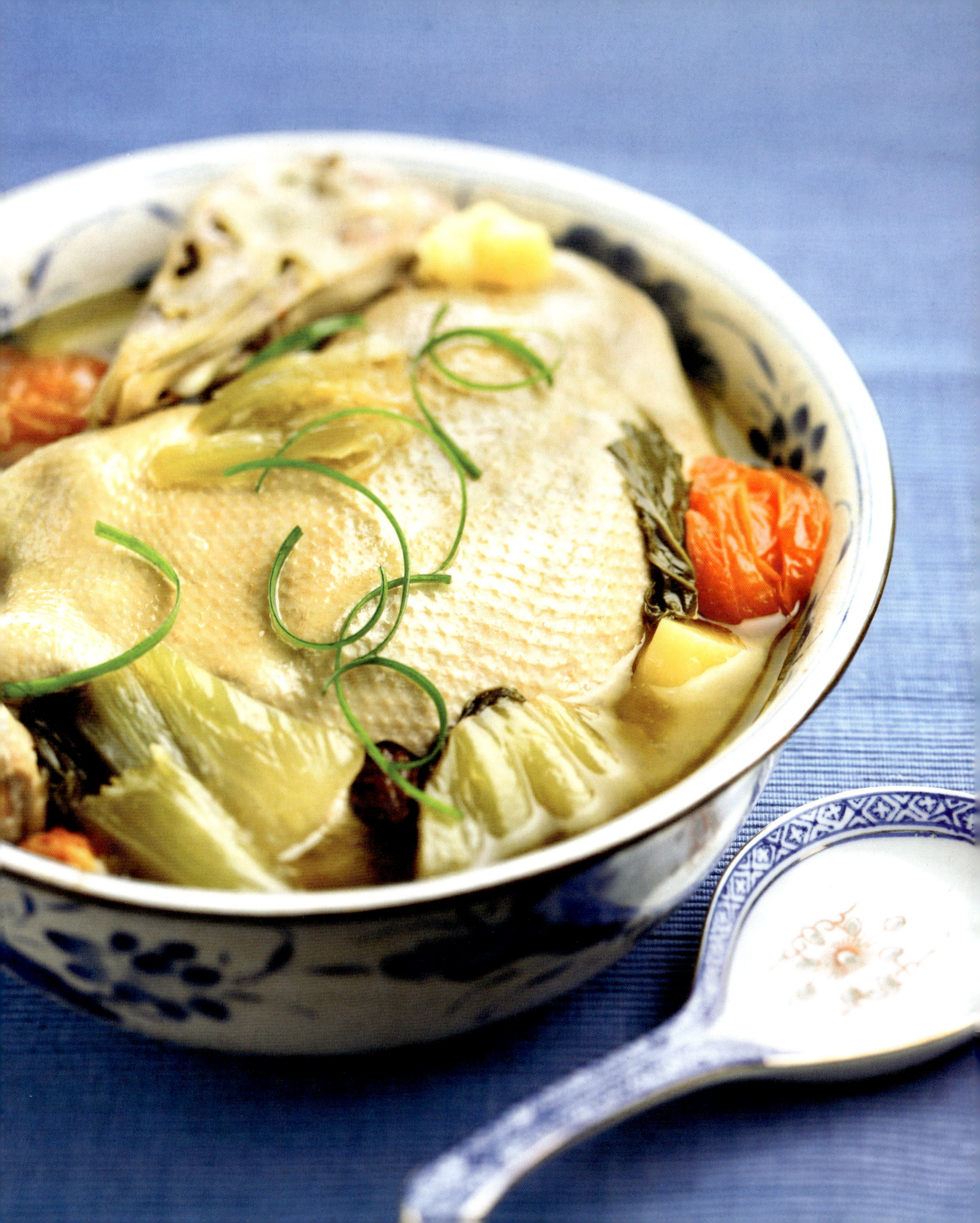

Chicken Curry with Potatoes

Preparation and cooking: 1 hour

Whenever the idea of spending a Sunday at Changi Beach entered my father's head, my mother would prepare this curry for our picnic lunch. It was eaten with crusty bread which she called *bangali roti*. *Bangali* was a word used to describe Sikhs who came from the Punjab area, except that for some reason, the word came to be associated with a particular type of bread sold by the Indian bread peddler. These itinerant hawkers were not Sikhs, and the chewy loaves that had the appearance of French baguettes were baked by Hainanese bakers. The gravy from this curry, if made with meat curry powder, makes an excellent topping for boiled Hokkien mee/*bee hoon/kuay tiao*. By using a *kurmah* spice mix instead of the meat curry powder, you get an entirely different dish.

100 g (3¹/₂ oz) shallots, peeled and chopped

4 cloves garlic, peeled and chopped

1 Tbsp tamarind paste

250 ml (8 fl oz / 1 cup) water

4 Tbsp cooking oil

300 g (11 oz) potatoes, peeled and quartered

4-cm (1¹/₂- in) cinnamon stick

3 cloves

3 cardamoms

1 kg (2 lb 3 oz) chicken, cleaned and cut into 16 pieces

500 ml (16 oz / 2 cups) coconut milk

1¹/₂ tsp salt

4 Tbsp packaged coconut cream

Spice Paste
8 Tbsp meat curry powder/ *kurmah* spice mix (both page 15)

4 Tbsp water

1. Make spice paste by mixing meat curry powder/*kurmah* spice mix with 4 Tbsp water. Set aside.

2. Blend shallots and garlic with 2 Tbsp water in a food processor until smooth .

3. Mix tamarind paste with 250 ml (8 fl oz / 1 cup) water. Strain away the solids. Leave out the tamarind if making chicken *kurmah*.

4. Heat oil in a large pot and brown quartered potatoes. They need not be cooked through. Set aside.

5. Into the hot oil, add the cinnamon stick, cloves and cardamoms and fry for a couple of minutes. Add the shallot-garlic mixture and fry until fragrant. Stir in the spice paste and continue frying until oil rises to the top.

6. Add chicken, browned potatoes, coconut milk, tamarind juice and salt. Stir well and bring to the boil. Turn down the heat and simmer for about 45 minutes or until the meat and potatoes are tender.

7. Adjust seasoning to taste, add the coconut cream and return to the boil. Turn off the heat and rest the curry for at least a couple of hours before serving.

8. Reheat and serve with rice, baguette or *roti jala* (page 124).

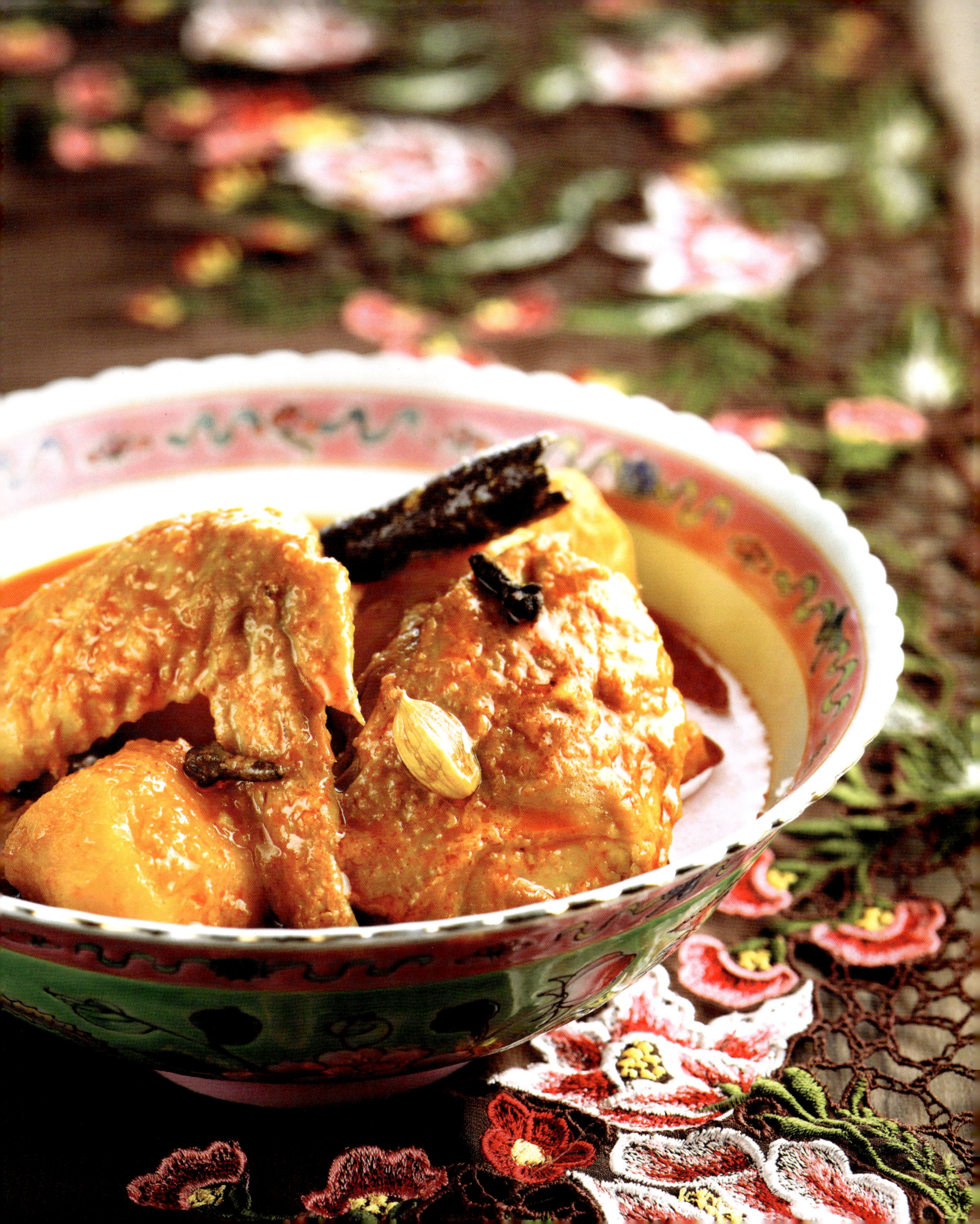

Inchee Kebin (Penang Nonya Fried Chicken)

Preparation and cooking: 1 hour

This recipe is from my aunt in Alor Star, and it's better than any *inchee kebin* that I have had in Penang where restaurants still serve it with the traditional dip of Lea & Perrins Worcestershire sauce. Mrs Leong Yee Soo's *Singaporean Cooking* had a very different recipe for it — the chicken was marinated in a number of ingredients different from this recipe and the chicken was fried whole. The skin of the chicken is supposed to crisp up, but crisp or not, this chicken is very flavourful. Keeping the fried chicken warm in a 50°C- (120°F-) oven will prevent it from turning soggy.

$1^1/_2$ tsp ground white pepper

$1^1/_2$ Tbsp ground coriander

$2^1/_2$ tsp five-spice powder

1 tsp salt

1.5 kg (3 lb 5 oz) chicken, cleaned and cut into 2 x 2-cm ($^3/_4$ x $^3/_4$-in) pieces

Juice of 1 large lemon

125 ml (4 fl oz / $^1/_2$ cup) packaged coconut cream

A few drops dark soy sauce

Peanut oil for deep-frying

1. Mix ground spices together with salt, then add to chicken pieces together with lemon juice and mix well. Marinate overnight. The marinated chicken can be kept in the fridge and fried in batches over several days.

2. Fill a quarter of a medium saucepan with peanut oil and place over medium heat. The oil needs room to bubble up when the chicken is added to it. Check if oil is hot enough by pressing the tips of wooden chopsticks to the bottom of the pot. If you get vigorous bubbles, the oil is hot enough.

3. Place marinated chicken, a few pieces at a time, into the hot oil and deep-fry until nicely browned. Remove and drain well.

4. Serve hot with Lea & Perrins Worcestershire sauce or chilli sauce.

Penang-Style Roast Chicken

Preparation and cooking: 3 hours

This was my mother's recipe for roasting chicken.

1.5 kg (3 lb 5 oz) whole chicken, cleaned

Seasoning

1 tsp salt

$^1/_2$ tsp dark soy sauce

1 tsp light soy sauce

2 tsp five-spice powder

1–2 Tbsp sugar/honey

1. Rub seasoning ingredients all over the inside and outside of chicken. Tuck feet into body cavity and fold wing tips under bird. Cover with plastic wrap and stand overnight in the fridge.

2. Preheat oven to 180°C (350°F). Place chicken breast-side down in a roasting pan and roast for 30 minutes to brown skin. Turn chicken breast-side up and roast until skin begins to brown or about 30 minutes.

3. Turn oven temperature down to 160°C (325°F) and continue roasting for another 1 hour.

4. Rest roast for 10 minutes before serving with rice or a salad.

Mutton Kurmah

Preparation and cooking: 2 hours

Originally a South Indian dish, *kurmah* is popular beyond the Straits-born community. As a South Indian dish, it would be cooked with yoghurt. As a Straits-born dish, it would be enriched with coconut milk and a stalk or two of lemongrass added to it. *Kurmah* can be quick-cooking or slow, depending on the choice of meat. Beef or chicken may substitute for mutton. I have never tried cooking *kurmah* with pork, but it should be tasty too.

4 Tbsp cooking oil

2 stalks lemongrass, smashed

800 g (1¾ lb) boneless mutton, cleaned and cubed

750 ml (24 fl oz / 3 cups) water

250 ml (8 fl oz / 1 cup) packaged coconut cream

1 tsp salt

2–3 green chillies, seeded (optional)

Spice Paste

90 g (3 oz) shallots or onion, peeled and chopped

3 cloves garlic, peeled

5-cm (2-in) knob ginger, scraped

2 Tbsp water

90 g (3 oz) *kurmah* spice mix (page 15)

1. Make spice paste by blending shallots or onion, garlic and ginger with 2 Tbsp water in a food processor until fine. Stir in *kurmah* spice paste.

2. Heat the cooking oil in a saucepan. Add the spice paste and smashed lemongrass and stir-fry for 2 minutes or until oil rises to the top.

3. Add the mutton, water and half the coconut cream and bring to the boil. Turn down heat and simmer for an hour or so until the meat is tender. Add more water and the remaining coconut cream as needed. The gravy should be thick.

4. Add salt to taste when meat is nearly tender, followed by the green chillies, if using, in the last 10 minutes of cooking.

5. Rest the *kurmah* for at least an hour before serving with rice or *nasi kebuli* (page 51).

Beef Rendang

Preparation and cooking: 3½ hours

This classic Malay and Indonesian dish has become part of Straits-born cooking, and I associate it with Hari Raya Puasa in the 1950s. This was when my family would visit our Malay neighbours to extend festive greetings and enjoy a *rendang* feast.

1 Tbsp tamarind paste

4 Tbsp cooking oil

1 kg (2 lb 3 oz) beef cubes/ shin beef

5-cm (2-in) knob galangal, scraped and smashed

4 stalks lemongrass, smashed

1 litre (32 fl oz / 4 cups) water

250 ml (8 fl oz / 1 cup) packaged coconut cream

1½ tsp salt

3 tsp sugar

6 kaffir lime leaves, crushed (optional)

Spice Paste

65 g (2⅓ oz) shallots, peeled and chopped

30 g (1 oz) *krisek* (page 17)

2.5-cm (1-in) knob old ginger, scraped

25 g (⅘ oz) dried red chillies, softened in water

125 ml (4 fl oz / ½ cup) water

1½ Tbsp ground coriander

½ Tbsp ground fennel

1. Make spice paste by blending shallots, toasted grated coconut, ginger and softened chillies with water in a food processor until fine. Stir in ground spices.

2. Make tamarind juice by combining tamarind paste with some water. Strain away the solids.

3. Combine all the ingredients in a pot and bring it to the boil. Turn down the heat and simmer gently for about 2–3 hours or until beef is tender. When the rendang begins to look dry, stir continuously to prevent the meat from sticking to the bottom of the pot. If beef is still not tender, add enough water to continue simmering until meat is tender. Do not add too much water as this is a dry dish. If using kaffir lime leaves, add at the end of cooking.

4. Rest the rendang for at least an hour before serving. Reheat and serve with rice, bread or rice cakes (*ketupat*).

Pig's Lungs with Pineapple

Preparation and cooking: 1 hour

I was reminded of this dish when a newspaper clipping dropped out of a book I was looking at. It featured my mother and her recipe for pig's lung. Published some decades ago, that was probably when I last ate this dish! I can still remember the texture of the strips of spongy lung and the gingery pineapple flavour, but it looks unlikely that I will ever be able to cook this dish again. When I asked the young butcher at the market for pig's lung, he was astonished and said nobody ate lung. So this recipe is for the record and also in the hope that somebody somewhere more traditional may need a recipe for pig's lungs. It is a tasty dish.

1 pair pig's lungs

3 Tbsp cooking oil

5-cm (2-in) knob ginger, scraped and finely shredded

4 cloves garlic, peeled and chopped

$^1/_2$ Tbsp light soy sauce

1 tsp dark soy sauce

$^1/_2$–1 tsp salt

1 small ripe pineapple, skinned, cored and coarsely shredded

1–2 tsp sugar (optional)

1. Boil the lungs in a large saucepan of water for about 15 minutes to get rid of the blood and foam. Cool and rinse well under tap water, then slice lungs into strips.

2. Heat oil in a wok and fry ginger and garlic until garlic begins to brown.

3. Add prepared lungs, soy sauces and salt and mix well.

4. Add pineapple and stir-fry until most of the pineapple juices have been absorbed. The dish should be quite dry. Adjust seasoning to taste. Add sugar if pineapple is sour.

5. Serve hot with rice.

Pork Vindaloo

Preparation and cooking: 1 hour

The kind of vinegar that you cook vindaloo with will affect its taste. Some vinegars are more acidic, while others tend to be fruitier. This vindaloo combines a fruity vinegar with tamarind juice for mellow sourness. Taste-test and adjust the seasoning to preferred taste. Vindaloos must be allowed to mature. They keep well for days in the fridge.

50 g (1³/₄ oz / ³/₄ cup) vindaloo spice mix (page 14)

85 ml (2 fl oz / ¹/₃ cup) apple cider vinegar

1 Tbsp tamarind paste

500 ml (16 fl oz / 2 cups) water

3 Tbsp cooking oil

200 g (7 oz) onion, peeled and finely sliced

4-cm (1¹/₂-in) knob ginger, scraped and cut into fine strips

1 kg (2 lb 3 oz) pork trotters/ belly pork, cut into pieces

2 tsp sugar

1 tsp salt

6 green chillies, seeded and halved

1. Stir vindaloo spice mix into the vinegar and set aside.

2. Mix tamarind paste with 250 ml (8 fl oz / 1 cup) water. Strain and discard solids. Set aside.

3. Heat oil in a saucepan and fry onion and ginger until fragrant. Add spice paste and sauté for a few minutes until oil rises to the top.

4. Add pork, tamarind juice, remaining cup of water, sugar and salt. Mix well and bring to the boil. Turn down heat and simmer for about 45 minutes or until meat is tender, but not falling apart. If curry is dry and meat is still not tender, add water in small amounts to continue simmering. The curry should be dry at the end of cooking.

5. When meat is tender, mix in green chillies and simmer until chillies change colour.

6. Rest vindaloo for several hours or overnight before serving. Reheat and enjoy with rice.

Duck Vindaloo

Preparation and cooking: 2–3 hours

As a Portuguese Eurasian dish, vindaloos originated from Goa in South India. Vindaloos should have as little gravy as possible. If using a slow cooker, add only a little water. If cooking in a covered casserole in a 90°C- (195°F-) oven, do not add any water as there is virtually no evaporation. One way to get a drier vindaloo is to remove the meat once it is tender and boil down the gravy to the preferred consistency before returning the meat to the gravy. Chinese white rice vinegar can be strong; substitute with a milder, more fruity vinegar if preferred.

1 kg (2 lb 3 oz) duck/pork trotters/belly pork

2 Tbsp tamarind paste

250 ml (8 fl oz / 1 cup) water

1 tsp salt

2–3 tsp sugar

125 ml (4 fl oz / $^1/_2$ cup) white rice vinegar

3 Tbsp cooking oil

Spice Paste

75 g (2$^2/_3$ oz) onion, peeled and coarsely chopped

2 cloves garlic, peeled

3 Tbsp dried red chilli paste (page 17)

1 Tbsp ground coriander

1 tsp ground cumin

2 tsp ground turmeric

$^1/_2$ tsp ground black pepper

$^1/_2$ tsp mustard powder

1. If cooking a whole duck, clean well, taking care to remove any tiny feathers or feather shafts. Cut into 8 pieces. If cooking pork trotters, rinse, then carefully feel the meat to remove any bone fragments. Rinse belly pork.

2. Make tamarind juice by mixing tamarind paste with 125 ml (4 fl oz / $^1/_2$ cup) water. Strain and discard the solids.

3. Make spice paste by blending ginger, onion and garlic together with 125 ml (4 fl oz / $^1/_2$ cup) tamarind juice in a food processor until smooth. In a large bowl, mix ginger-onion-garlic paste with chilli paste, ground spices, salt, sugar, vinegar and cooking oil.

4. Mix meat and spice paste together. Place in a container and leave to marinate in the fridge for 24 hours.

5. To cook, transfer meat and marinade to a saucepan and bring to the boil. Turn down heat and simmer until meat is tender. Adjust amount of water, vinegar or tamarind juice in vindaloo so that meat gets tender, but curry stays fairly dry. Adjust vinegar to taste. This is a dry curry with a sourish spicy flavour. Duck and pork trotters take about 2 hours to become tender, while belly pork takes about 45 minutes.

6. Rest for several hours before serving. Reheat and serve with rice.

Light Bites and Sweet Treats

For all Occasions

When it comes to light bites and sweet treats, the Straits-born share many identical recipes that arose from communal proximity and simply good neighbourliness. In the days before the proliferation of cookbooks, cooking shows and the Internet, good recipes spread by word of mouth. If you enjoyed something very much, you asked for the recipe. If you found a good recipe or were given a good recipe, you shared it with whoever asked you for it. This was how I accumulated many of my favourite festive treats.

While each community may have had recipes that were once special to them, many have been happily adopted by others. Practically everyone enjoys *sugee* cake or Christmas cake, both Eurasian specialties. My recipe for Christmas cake given here goes way back to a recipe for an English fruit cake. It was the first kind of cake that I tried baking in my teens.

Traditional Straits-born desserts were made with rice/rice flour, tapioca/bean flour (*tepung hoen kwe*, a Dutch-Indonesian product). They still are. Think *cendol*, *kueh bangket* or *kueh salat*. The use of wheat flour and butter for cakes, pineapple tarts and biscuits, and even love letters were the result of colonisation. Butter and wheat flour were introduced by the Europeans. All the three earliest local cookbooks —*The YWCA International Cookery Book* (1931), *My Favourite Recipes* (1952) and *In a Malayan Kitchen* (1954) — had numerous recipes for very colonial era cakes and biscuits as well as recognisably Straits-born favourites. *In a Malayan Kitchen* had a Dutch slant because Mrs Susie Hing was Indonesian Chinese. And hers was the only one of the three with a recipe for pineapple tarts which looked similar to any of today's pineapple tart recipes. So did pineapple tarts originate in Dutch Indonesia? While the first clearly labelled Nonya cookbook, *Mrs Lee's Cookbook* (1974), did not include desserts, Mrs Leong Yee Soo's *Singaporean Cooking* (1976) had recipes for both open pineapple tarts as well as pineapple-shaped tarts, like Mrs Hing's, but this was some 20 years later than Mrs Hing's 1954 recipe. My recipe for pineapple tarts dates back to the 1950s and came from my Malay neighbours. It was given to them by an aunt who had learnt how to make the tarts at a baking class. Wherever that recipe came from, to me, it is one of the best, if not *the* best, pineapple tart recipe ever.

Straits-born treats are not all sweet treats. A number are savouries and are often prepared for special occasions. *Kueh pie tee* is one such festive treat and actually made for large crowds. So is *roti jala*. Others like *sambal lengkong* are more homely but special because it is time-consuming. If making something time-consuming is a way to show affection for family and friends, then Straits-born treats are certainly expressions of affection. Enjoy!

Kueh Pie Tee

Preparation and cooking time: 1 day

This Nonya festive treat shows up in the run-up to Chinese New Year when the shops that specialise in Straits-born treats will also offer *pie tee* shells for sale. Some 30 years ago, my huge disappointment with one of these commercial preparations prompted me to experiment and quantify a vague recipe that my mother gave me together with her *pie tee* irons. Since then, making *pie tee* shells has become an annual all-day, pre-Chinese New Year party for myself and a few friends. Note that the number of shells you get from this quantity of batter depends on how much wastage there is as well as how you thin the batter. There was no wastage when I worked out the number of shells from this batch of batter. There are two recipes for fillings here. The bamboo shoot filling is a traditional Nonya filling also used for Nonya *popiah*. The other is a simple yam bean filing that my mother preferred.

Pie Tee Shells

Makes about 150 small shells

180 g (6¹⁄₃ oz) plain (all-purpose) flour

65 g (2 oz) rice flour

¹⁄₂ tsp salt

2 eggs

400 ml (13 fl oz / 1²⁄₃ cups) water + more as needed

2 litres (64 fl oz / 8 cups) peanut oil

2 *pie tee* irons

1. Sift plain flour, rice flour and salt together. In a large mixing bowl, beat eggs lightly and stir in the water. Using a wire whisk, add the sifted flour mixture.

2. Pour the batter through a large sieve and rub the lumps through the sieve. A wire sieve works best.

3. To adjust the batter to the right thickness, first fry a test shell once the *pie tee* irons are hot. Thick batter gives you fewer and thicker shells. If the shell is too thick, dilute the batter with some water to your preferred thickness.

4. Heat the peanut oil in a pot together with the *pie tee* irons. The pot should be at least half-full of oil. When coated with batter, the irons must not touch the bottom of the pot, but be held suspended in the hot oil.

5. To check the temperature of the oil, press the tips of wooden chopsticks against the bottom of the pot. You should see vigorous bubbles when the temperature is right. If using an electric fryer, set the temperature to 180°C (350°F).

6. Stir up the batter and ladle enough batter to fill a large mug that is kept near the pot of hot oil. Cover the bowl of batter with a damp cloth to prevent drying out.

7. Dip one of the hot irons halfway up into the batter. If the iron is hot enough, the batter will sizzle. Return the coated iron to the hot oil and fry until the shell is firm, but not yet brown. Using the tips of the chopsticks, push the shell out of the iron and continue frying to brown the shell.

8. Repeat until the batter is used up.

Bamboo Shoot Filling

1 kg (2 lb 3 oz) pork bones

3 litres (45 fl oz / 12 cups) water

4 Tbsp cooking oil

1 Tbsp chopped garlic

2 Tbsp fermented soy beans (*taucheo*), rinsed and mashed/ 1 Tbsp brown miso

1 kg (2 lb 3 oz) bamboo shoots, finely shredded

1 Tbsp light soy sauce

2 tsp dark soy sauce

1 tsp salt

Bangkuang (Yam Bean) Filling

500 g (1 lb 1½ oz) prawn shells and heads

250 ml (8 fl oz / 1 cup) water

4 Tbsp cooking oil

1 Tbsp chopped garlic

1 kg (2 lb 3 oz) *bangkuang* (yam bean), peeled and finely shredded

1 carrot, peeled and shredded

1 Tbsp light soy sauce

1 tsp dark soy sauce

1 tsp salt

Chilli Sauce

200 g (7 oz) fresh red chillies, seeded

2 Tbsp water

2 Tbsp white rice vinegar

½ tsp salt

1 tsp sugar

Topping

Boiled prawns or crabmeat (page 77)

Coriander leaves

Bamboo Shoot Filling

1. Make pork stock. Combine pork bones with water and bring to the boil. Skim off the scum that rises to the top. Simmer for an hour until liquid is reduced by half.

2. Heat oil in a pot and sauté chopped garlic until it begins to turn yellow. Add fermented soy beans/brown miso. Mix well.

3. Add bamboo shoots and pork stock without the meat. Bring to the boil and simmer for an hour or until filling is moist, but not watery. Adjust seasoning to taste.

Bangkuang (Yam Bean) Filling

1. Make prawn stock. Combine prawn shells and heads with water and bring to the boil. Simmer for 15 minutes. Strain and discard solids.

2. Heat oil in pot and sauté chopped garlic until it begins to turn yellow. Add shredded *bangkuang* and carrot, soy sauces, salt and prawn stock. Mix well and bring to the boil.

3. Turn down heat, stir-fry and cook for 15 minutes or until *bangkuang* turns very limp and transparent. Filling should be moist.

Chilli Sauce

1. Make chilli sauce. Blend all the ingredients together in a food processor until smooth. Adjust seasoning to taste.

1. To serve, fill a pie tee shell with either filling and top with boiled prawns or crabmeat, coriander leaves and chilli sauce.

Tips on Frying Pie Tee Shells

- A seasoned iron fries best as the shell will slip off easily. The first few times with a new iron can be frustrating as the shell tends to stick.
- The iron must be hot before you dip it into the batter.
- Before dipping the iron into the batter, first wipe the oil off the bottom of the iron by pressing it against a piece of folded paper towel. If there is too much oil on the iron, the bottom of the shells will develop holes.
- Do not let the shell brown on the iron or it will break when you try to slip it out. On the other hand, do not let it slip out of the iron before it is fairly firm or the shell will collapse.
- To make "top hats", coat the iron with batter almost up to the top edge, not over the top. When dipping iron back in the hot oil, dunk it up and down to flip out the edge to form the lip. It does not always flip out though. (Top hats are inconvenient to eat, but look interesting. I prefer small shells that you can pop whole into your mouth as in the picture.)
- To keep the iron well-seasoned, never wash a well-used *pie tee* iron with soap and water. To clean, unscrew it from the handle and rub off the oil with several paper towels. To drain the oil in the screw holes, set the irons upside down on a paper towel overnight.
- Wrap the irons in paper towels and keep them tied up with the handles in a plastic bag until needed.
- Peanut oil handles deep-frying best and the shells stay fresh for longer.
- Make sure that the bottoms of the shells are nicely browned. If not, the shells will soften more quickly during storage.
- Pie tee shells well-fried in peanut oil will keep fresh-tasting for up to a month, longer if refrigerated.

Hae Bee Hiam (Spicy Dried Shrimp)

Preparation and cooking time: 45 minutes

This is a recipe that dates back to the 1950s and 1960s when mission schools held funfairs and sold home-made goodies to raise money for the school building fund. One day, a classmate brought sandwiches with this dried shrimp filling. I liked it so much that I asked for her mother's recipe which she shared with me. It's a gift I continue to enjoy decades later. The saltiness of dried shrimps varies, so soak them for 10–15 minutes, then do a taste-test. Too bland is better than too salty.

200 g (7 oz) dried shrimps (*hae bee*), softened in water

65 g (2 oz) shallots, peeled and thinly sliced

3 Tbsp dried red chilli paste (page 17)

3 tsp sugar

3 Tbsp lime juice

4 Tbsp cooking oil

1. Pound dried shrimps and shallots until fine, then mix in chilli paste, sugar and lime juice.

2. Heat oil in a wok and stir-fry mixture over low heat until mixture is crumbly and fairly dry. Adjust seasoning to taste.

3. Cool, bottle and refrigerate or freeze.

4. To make sandwiches, butter slices of bread generously, and sprinkle a tablespoon of filling in between two slices of bread. If preferred, top the filling with shredded cucumber.

Sambal Lengkong (Spicy Fish Floss)

Preparation and cooking time: 2 hours

Making *sambal lengkong* is a labour of love, but well worth the time. The best fish for this treat is wolf herring (*Chironcentrus dorab*), aka *ikan parang* or dorab, because its flesh is very fine and it crisps up nicely. The fish has a lot of bones though, so keep an eye out for any bones you might have missed out while stir-frying. As this fish is hard to get these days, you can substitute with mackerels.

1 kg (2 lb 3 oz) wolf herring/mackerel (*ikan selar*)/Indian mackerel (*ikan kembong*)

3 tsp sugar

1 tsp salt

125 ml (4 fl oz / ½ cup) packaged coconut cream

10 kaffir lime leaves, crushed

Spice Paste

300 g (11 oz) shallots, peeled and chopped

10 candlenuts (*buah keras*)

4 stalks lemongrass, thinly sliced

4 slices galangal, chopped

6 fresh red chillies, seeded and chopped

4 Tbsp water

1. Make spice paste by blending all the ingredients together in a food processor until smooth.

2. Steam/poach fish until cooked through. When cool enough to handle, flake fish very finely, taking care to remove all the bones. If using either mackerel, discard any hard bits that won't flake. The discarded bits of fish may be boiled in a pot of water to make fish stock for some other dish.

3. Mix fish with spice paste, sugar, salt and coconut cream. Place in a dry wok over low heat and dry-fry with kaffir lime leaves until the floss is very dry and golden brown. Stir and scrape the bottom of the wok constantly to prevent burning.

4. Leave to cool, then bottle and store in the fridge to extend shelf life.

5. Sprinkle generously on buttered bread as a sandwich filling or serve on rice porridge or steamed rice.

Roti Jala (Lacy Pancakes)

Preparation and cooking time: 1 hour

In the 1980s, I came across *roti jala* at a Kelantan foodfest where the pancakes were coloured yellow with food colouring or powdered turmeric. *Roti jala* probably originated as a Malay dish that was adopted by some Straits-born cooks and it came to be considered as Nonya. *Mrs Lee's Cookbook* (1974) has a *roti jala* recipe that uses milk instead of coconut milk, unlike mine. In Malay, *jala* means net and *roti* means bread. It is made using a special cup with holes, but you can improvise with a small tin, punched with small holes. Other alternatives are a plastic water bottle pierced with holes or a squeeze bottle with a nozzle.

360 g (13 oz) plain (all-purpose) flour

¾ tsp salt

3 eggs, lightly beaten

500 ml (16 fl oz / 2 cups) water

250 ml (8 fl oz / 1 cup) packaged coconut cream

Cooking oil for greasing pan

1. Sift flour and salt into a mixing bowl. In another bowl, mix beaten eggs, water and coconut cream. Using a wire whisk, stir liquid into flour mixture. Pass batter through a sieve and rub any lumps down with a metal spoon. There must be no lumps.

2. Heat a non-stick flat frying pan or an Indian *tawa* until a drop of water dances on the hot surface. Grease the pan lightly. The first few pancakes will need a greased pan, but after a while, the pan will be greasy enough for the pancake to slip easily off the pan.

3. Stir the batter and scoop some into the *roti jala* cup. Swirl it over the hot pan to form the pancake. The speed of the swirling to get the strands depends on the flow rate of your batter. If the batter is too thick, you get little drops instead of a steady stream of thin strands. If the batter is too thin, the rapid flow gives you thick, spread-out strands. Do a test pancake and adjust the batter cautiously.

4. When the thin lines of batter start to curl at the edges., the pancake is done. Fold the pancake in half, then fold again into a quarter. Arrange on a serving plate.

5. *Roti jala* keeps well packed into an airtight container in the fridge for a couple of days. The pancakes do not need to be reheated, but bring them to room temperature before serving.

6. Serve with chicken curry (page 106) or rendang (page 110).

Pulot Rempah Udang
(Glutinous Rice Rolls with Spiced Prawns)

Preparation and cooking time: Half a day; advance preparations required

Once upon a time, this *kueh* would be made as part of the buffet of a Penang Straits Chinese wedding, but my mother made it often as a treat for us. It was always eagerly looked forward to once I saw the preparations get underway. This rather substantial snack may be called *pulot rempah udang* (glutinous rice rolls with spiced prawns) or *pulot panggang* (grilled glutinous rice rolls). Make sure that you have extra banana leaves on hand so that torn leaves can be repaired by covering it with more leaf.

600 g (1 lb 5$^{1}/_3$ oz / 3 cups) white glutinous rice, soaked overnight

1$^{1}/_2$ tsp salt

4 Tbsp water

4 Tbsp packaged coconut cream

Filling

200 g (7 oz / 2 cups) *krisek* (page 17)

1 Tbsp ground coriander

2 tsp ground white pepper

2 Tbsp cooking oil

400 g (14 oz) shelled prawns, cleaned and chopped

$^{3}/_4$ tsp salt

3 tsp sugar

125 ml (4 fl oz / $^{1}/_2$ cup) water

For Wrapping

Banana leaves, cut into 12 pieces, each 20 x 30-cm (8 x 12-in)/ 24 pieces, each 10 x 30-cm (4 x 12-in)

A bowl of cold water

Oil for greasing leaves

24/48, bamboo skewers with sharp ends, each 3-cm (1$^{1}/_2$-in) long

1. Drain rice and steam for 10 minutes. Mix salt, water and coconut cream together and stir into rice. Continue steaming for another 30 minutes until rice is cooked through.

2. Prepare filling. Blend *krisek* in a food processor or pound in a mortar and pestle until fine. Mix ground coriander and pepper with a little water to form a thick paste. Heat oil in a pan and fry the paste until fragrant, taking care not to burn it. Add prawns, salt and sugar and stir-fry until prawns change colour. Add pounded *krisek* and water and cook for 5 minutes. The filling should be just moist. You should get about 500 g (16 oz) filling.

3. Clean the banana leaves, then blanch in boiling water to make them pliable. Wipe dry with paper towels and cut to size.

4. Have handy a bowl of cold water for moistening hands, a bowl of cooking oil with brush for greasing leaves, and the bamboo skewers.

5. Dampen hands in cold water and divide steamed rice into the preferred portions. Do the same for the filling.

6. Spread each portion of rice into a rectangle on a piece of greased banana leaf. Spread filling on the rice and shape the rice into a roll to enclose the filling. Roll the banana leaf round the rice. Fold the ends together and fasten with bamboo skewers. Trim off excess leaf from ends.

7. Grill rolls under a 220°C (425°F) grill or on a wire rack placed over a gas burner for 5 minutes to warm through. If not consumed on the same day, refrigerate rolls. Toast or microwave just before serving. Do not serve cold.

Nonya Rice Dumplings

Preparation and cooking time: Half a day; advance preparations required

Making rice dumplings was an annual ritual that my mother practised. Sometime during the fifth Chinese lunar month, the kitchen would be filled with the scent of boiling bamboo leaves and soon a frenzy of activities would erupt, from knotting together a bundle of straw strings to the chopping and cutting of boiled pork and mushrooms. My mother's dumplings were different from the Chinese ones. Hers were what are called Nonya dumplings today. These dumplings are so popular that they are available year-round today.

600 g (1 lb 5^1/$_3$ oz / 3 cups) white glutinous rice, soaked overnight

2 tsp salt

125 ml (4 fl oz / ^{1}/$_2$ cup) water

2 Tbsp cooking oil

3 Tbsp dried butterfly pea flower (*Clitoria ternatea*)

Filling

300 g (11 oz) lean pork

30 g (1 oz) dried shiitake mushrooms, soaked for 30 minutes in 250 ml (8 fl oz / 1 cup) water

1^1/$_2$ Tbsp ground coriander

1 tsp ground white pepper

2 Tbsp cooking oil

85 g (3 oz) shallots, peeled and thinly sliced

100 g (3^1/$_2$ oz) sugared winter melon, finely chopped

1 Tbsp dark soy sauce

For Wrapping

40 dried bamboo leaves

^{1}/$_2$ tsp alkali salts/bicarbonate of soda

Straw twine/raffia string

A small bowl of oil for greasing leaves

A bowl of water for dampening hands

1. Prepare filling and bamboo leaves a day ahead. When buying bamboo leaves, try to get larger leaves, which make for easier wrapping. If small, estimate two leaves per dumpling with some extras in case of tears.

2. To prepare bamboo leaves, heat a wok of water with ^{1}/$_2$ tsp alkali salts/bicarbonate of soda and boil leaves for 10 minutes. Boiling makes the leaves more pliable. Wipe dry as leaves are sometimes dusty, then keep the leaves soaking in water until needed.

3. Cut straw twine/raffia string into 16 pieces, each about 60-cm (24-in) long. Bundle them together and knot one end into a loop so that the strings can be hung up. If using straw twine, soak in water for 10 minutes to make it more pliable.

4. To prepare filling, boil pork in a pot of water for 10–15 minutes until cooked through. Drain and cool, then cut into very small cubes. Trim stems off dried mushrooms and slice caps into small pieces. Keep the soaking water for cooking filling.

5. Mix ground coriander and pepper with 1 Tbsp water into a paste.

6. Heat oil in a pot and fry shallots until golden brown, taking care not to burn them. Scoop out and set aside. Fry coriander and pepper paste until fragrant, then stir in pork, mushrooms, winter melon, dark soy sauce and mushroom water. Mix well and stir-fry until all the liquid is almost evaporated and the filling is just moist. Stir in fried shallots.

7. When ready to make dumplings, start by preparing rice. Steam soaked rice for 10 minutes. Mix salt, water and oil together, then stir into steamed rice. Leave rice covered until cool enough to handle.

8. If colouring rice, soak dried butterfly pea flowers in 1–2 Tbsp water for 30 minutes. Use the liquid to streak the rice blue.

9. Divide rice and filling into 12–16 portions depending on the size of your bamboo leaves. Wrap dumplings (see below).

10. The wrapped dumplings can be either boiled or steamed. The latter reduces the risk of soggy corners or lost dumplings if they haven't been well wrapped.

11. To steam dumplings, place a Chinese steamer with at least 2 layers on a large wok filled with water. Arrange the dumplings loosely on the tray, if possible. Steam for an hour, adding water to the wok as needed. Remove dumplings and hang them up to drain.

12. To boil, fill with water a pot large enough to contain all the dumplings and bring to the boil. Place dumplings in and return to the boil. Keep it on the boil for an hour, adding more water to the pot as needed. Remove dumplings and hang them up to drain.

13. Rice dumplings should be kept well aired by hanging them up in a cool spot, and never packed inside a plastic bag. Kept well aired, they do not need refrigeration. If not consumed after 3 days, boil them up again for 30 minutes.

How to Wrap Rice Dumplings

1. Assemble the rice, filling and leaves near you. Have a large bowl of water nearby for wetting your hands, and a bowl of oil with a brush for greasing the leaves.

2. Grease two leaves and layer them such that you get a larger surface area to work with. Fold the leaves in half, then fold one-third from the edge near you. Open out the leaves into a cone.

3. With dampened hands, put a portion of rice into the cone. Press the rice upwards as a thin layer round the cone and a little higher than the dumpling will be for covering up the filling. Keep the centre hollow for the filling.

4. Spoon filling into the hollow, pressing it to pack tightly. Turn the excess rice near the top to cover the filling. Add more rice, if needed.

5. Fold the ends of the leaves to cover the opening and shape the ends to form a neat four-cornered pyramid-shaped dumpling. Tie with twine in such a way that the ends are held tightly round the dumpling.

Kee Chang (Soda Dumplings)

Preparation and cooking time: 1 day; advance preparations required

These dumplings are Chinese in origin, but much loved by the Straits Chinese. Many families kept their ancient Chinese rituals even as their daily diet was evolving to accord with their environment. These dumplings were made as prayer offerings during the fifth lunar month which was also the season for savoury rice dumplings. Unlike Nonya *chang*, *kee chang* are still mostly home-made, and only for a few days during the festival when the market stalls display fresh bamboo leaves, straw twine and alkali salts. These dumplings are usually eaten with a sugar dip, but I love my aunt's combination of palm sugar syrup with white grated coconut. The flavour is rather like *kueh lopes*, a Malay *kueh*, but with a distinct alkali flavour to it. Note that too much alkali salt will turn the rice hard.

Dumplings

600 g (1 lb 5^1/$_3$ oz / 3 cups) white glutinous rice

3^1/$_4$ tsp alkali salts

40 bamboo leaves, fresh if possible

Straw twine/raffia string

Dressing

500 g (1 lb 1^1/$_2$ oz) palm sugar

200 ml (7 fl oz / ^{4}/$_5$ cup) water

Freshly grated coconut

1. Soak glutinous rice with 2 tsp alkali salts in a basin of water overnight. The rice will turn a pale yellow. Drain rice and spread on a steamer lined with muslin. Steam for 20 minutes. The rice does not have to be cooked through.

2. Soften bamboo leaves before using by soaking in water overnight or boiling them in a large pot of water with ^{1}/$_4$ tsp alkali salts for 10 minutes. Wipe dry with paper towels as leaves are sometimes dusty.

3. Cut straw twine/raffia string into 16 pieces, each about 60-cm (24-in) long. Bundle them together and knot one end into a loop so that the strings can be hung up. If using straw twine, soak in water for 10 minutes to make it more pliable.

4. With a bowl of water at hand, wrap dumplings (page 130). The wrapping is easier because there is no filling.

5. When dumplings are done, bring a large pot of water to the boil together with 1 tsp alkali salts. Lower in dumplings and boil for 2 hours.

6. Take dumplings out and hang them up to cool before serving.

7. The dumplings do not need to be refrigerated for several days. If refrigerated, boil for 15 minutes and cool before serving.

8. Make dressing by cooking palm sugar and water in a small pot to get a syrup. Strain to remove any impurities and bottle the syrup. Keep excess in the fridge.

9. To serve, unwrap a dumpling and cut into bite-size pieces. Spoon palm sugar syrup over dumpling and garnish with freshly grated coconut.

Pengat Pisang (Bananas in Coconut Milk)

Preparation and cooking time: 1¹/₂ hours

This is another very old Straits-born dessert that is ideally made with *pisang raja*. If you substitute *pisang raja* with another cooking banana, make sure the bananas are very ripe. *Pengat pisang* becomes plain *pengat* if you add cubes of steamed sweet potato and yam. Sago beads are another possible addition. In the old days, cooks would boil strips of coloured tapioca dough to make the dish more colourful. These coloured bits of tapioca dough are now available commercially.

500 g (1 lb 1¹/₂ oz) ripe *pisang raja*/ other cooking bananas

500 ml (16 fl oz / 2 cups) water

3 pandan leaves, rinsed and knotted

Sago beads/tapioca strands (optional)

100 g (3¹/₂ oz) palm sugar

1. Peel bananas, halve them lengthwise and slice about 1-cm (¹/₂-in) thick at a slant.

2. Bring the water to the boil with pandan leaves. If adding sago beads/tapioca strands, boil them at this point and keep the pot on a low boil until sago beads/tapioca strands become translucent. It takes about an hour even for small beads. Top up with more water, if needed.

3. Add banana slices and simmer for 15 minutes until bananas are soft. Stir in palm sugar to taste and bring it back to the boil. Turn off the heat and discard pandan leaves.

4. Serve warm or chilled if preferred.

Pengat Durian (Durian Cream)

Preparation and cooking time: 40 minutes

There are 10th century Javanese inscriptions that list durians as one of the fruits eaten. This dessert is incredibly rich and a little goes a long way. For the best durian *pengat*, make it with a soft, strong-smelling durian. If you have a stick blender, getting a smooth cream quickly is easy. If not, rub the durian meat through a sieve to get a smooth cream.

85 g (3 oz) palm sugar

500 ml (16 fl oz / 2 cups) water

450 g (1 lb) durian meat

250 ml (8 fl oz / 1 cup) packaged coconut cream

1. In a saucepan, combine palm sugar with 250 ml (8 fl oz / 1 cup) water to make a syrup. Strain syrup through a sieve to remove any impurities.

2. In a clean saucepan, combine palm sugar syrup, durian meat, the remaining water and coconut cream and bring to the boil. Turn off the heat.

3. If using a stick blender, blend *pengat* to a smooth cream. Alternatively, pour *pengat* into a large sieve and rub durian meat through. Discard the stringy bits.

4. Serve *pengat* warm.

Pictured: *Pengat Pisang* (Bananas in Coconut Milk)

Jemput-Jemput aka Kueh Kodok

Preparation and cooking time: 40 minutes

My mother's recipe for this *kueh* was just banana, flour and water mixed into a drop batter. It was strictly *agak-agak*. She didn't even add sugar as *pisang rajah* is usually very sweet. Then my aunt in Alor Star suggested adding grated coconut to it. The Malay word *kodok* means "toad", but why this *kueh* is so-called is a mystery. If there is no *pisang rajah*, any very ripe banana will do. The amount of sugar depends on the sweetness of the bananas. Keep the bananas lumpy. Some recipes suggest moistening with coconut cream instead of water. *Kueh kodok* are best eaten freshly fried and warm. Refrigerate leftovers and reheat briefly in the microwave oven before serving.

1–2 Tbsp sugar

100 g (3¹/₂ oz) plain (all-purpose) flour

100 g (3¹/₂ oz) white grated coconut

400 g (14 oz) very ripe bananas, peeled and mashed coarsely

65 ml (2 fl oz / ¹/₄ cup) water/ packaged coconut cream

Oil for deep-frying

1. Mix sugar and flour together to get rid of any lumps. Add grated coconut, mashed bananas and water or coconut cream to form a drop batter, that is, batter that will drop out of a spoon as a lump.

2. Fill one-quarter of a small saucepan with cooking oil and place over medium heat. Test the temperature by pressing the tips of wooden chopsticks against the bottom of the pan. When you see bubbles, the oil is ready.

3. Using a teaspoon, gently drop a heaped teaspoon of batter into the hot oil and deep-fry until the *kueh* is brown and puffed up. The first few may stick to the bottom of the pot, especially if there is a lot of banana in that spoonful and the pot is not a non-stick pot. Scrape it off with a metal spatula. Keep the heat at medium so that the centres will cook well.

4. Remove the cooked *kueh* and drain on paper towels.

5. Serve warm.

Banana Agar-agar

Preparation and cooking time: 30 minutes

Hundreds of years ago, agar-agar (*Spherococcus lichenoldes*) was one of the tropical products that Hokkien traders came in search of in Island South East Asia. The seaweed was made into a succulent jelly, but according to Hobson-Jobson, "[The Chinese] also employ it as a glue and apply it to silk and paper intended to be transported." This banana agar-agar recipe dates to the 1960s. It came from a Chinese newspaper clipping that my father brought home one day and it became a family favourite. If *pisang rajah* is not available, any very ripe cooking banana such as *pisang tandok* or *pisang mas* would be good too. Do not substitute agar-agar with gelatine.

500 g (1 lb 1^1/$_2$ oz) ripe *pisang raja/* other cooking bananas

35 g (1^1/$_3$ oz) agar-agar straws, rinsed

1.5 litres (48 fl oz / 6 cups) water

165 g (7 oz) white sugar

185 ml (6 fl oz / ^{3}/$_4$ cup) evaporated milk

^{1}/$_3$ tsp banana essence

1. Peel bananas and halve them lengthwise, then cut into chunks slantwise. Place in a large saucepan.

2. Add rinsed agar-agar straws, water and sugar to the saucepan and bring to the boil. Turn down the heat and simmer until sugar and agar-agar straws are dissolved and the bananas are soft.

3. Stir in evaporated milk and banana essence.

4. Pour hot mixture into a 22-cm (8-in) square pan/jelly mould. Set aside to cool before refrigerating until agar-agar hardens.

5. Cut into pieces and serve chilled.

Kueh Ee (Glutinous Rice Dumplings)

Preparation and cooking time: 40 minutes

This was a Straits Chinese ritual sweet that my mother prepared for the Chinese winter solstice on 21 December. In the 1950s before the easy availability of ground glutinous rice flour, she had to make her own. She had a stone rice mill in the backyard just for this task which was often passed on to the kids. Besides being a sweet treat for the family, it was said that eating these sweet rice balls ensured good health and prosperity for the family in the coming winter. Traditional *kueh ee* has merged with a Teochew dessert called *arh borh ling* (meaning mother duck's eggs) which are available frozen. The same glutinous rice flour dough is formed into big round balls and filled with ground black sesame seeds or ground roasted peanuts, and is also served in a sugar syrup like *kueh ee*.

12 pandan leaves, cleaned

250 ml (8 fl oz / 1 cup) water

300 g (11 oz) white glutinous rice flour

Food colouring

Sugar Syrup

150 g (5$^1/_3$ oz) white sugar

2 pandan leaves, cleaned and knotted/5-cm (2-in) knob ginger, scraped and smashed

500 ml (16 fl oz / 2 cups) water

1. Make sugar syrup by combining ingredients in a pot and melting the sugar. Boil for a few minutes to extract the flavour from the pandan leaves/ginger.

2. To make pandan-flavoured rice balls, cut up the pandan leaves and grind a small batch with 125 ml (4 fl oz / $^1/_2$ cup) water to pulp. Strain the liquid and use the pandan juice to grind more leaves to pulp. Repeat until all the leaves are ground up. Mix this pandan juice with part of the glutinous rice flour to form a dough stiff enough to roll into a ball that will hold its shape.

3. To make plain rice balls, combine the remaining glutinous rice flour and water to make a dough stiff enough to roll into a ball that holds its shape. Divide dough into two parts if making two colours. Colour one red with red food colouring and leave the other part white.

4. Divide the dough into small even portions and roll each portion into a ball between your palms.

5. Bring a pot of water to the boil and drop the balls into the boiling water. The balls are cooked when they float. Scoop out and drop into the pot of sugar syrup.

6. Serve *kueh ee* warm or at room temperature with a portion of the sugar syrup.

Sago Pudding

Preparation and cooking: 1 hour

This is another Straits-born classic that is in all the old cookbooks. Sago palms were once plentiful in Island South East Asia, but as sago starch is extracted by cutting down the palm, the disappearance of sago palms may be why "sago beads" today are made from tapioca. Genuine sago has a slight chewiness that tapioca flour does not have. Ellice Handy noted in *My Favourite Recipes* (1960) that "good sago [is] the kind that will not become starch during boiling". Unfortunately, there is no way to tell genuine sago from tapioca "sago". Sago pudding is another dessert that is enhanced by freshly squeezed coconut milk rather than packaged coconut cream.

180 g (6^1/$_3$ oz / 1 cup) sago beads

150 g (5^1/$_3$ oz) palm sugar

125 ml (4 fl oz / ^{1}/$_2$ cup) + 750 ml (24 fl oz / 3 cups) water

2 Tbsp sugar

250 ml (8 fl oz / 1 cup) fresh coconut cream (page 20), chilled and lightly salted

1. Rinse sago beads in a sieve, then soak in a bowl of water for an hour.

2. While sago is soaking, combine palm sugar with 125 ml (4 fl oz / ^{1}/$_2$ cup) water in a small saucepan and bring to the boil. Continue boiling until palm sugar is dissolved. Strain syrup through a sieve to remove any impurities. Bottle the syrup until needed.

3. Bring 750 ml (24 fl oz / 3 cups) water to the boil in a pot. Drain sago and add to the boiling water. Return water to the boil, then turn down the heat and continue boiling for about 10 minutes, stirring constantly.

4. Stir in sugar, cover pot and simmer over low heat until mixture thickens and sago beads are transparent. This will take about 30 minutes. Keep an eye on it.

5. Wet a 1-litre (32-fl oz / 4-cup) jelly mould and pour in the hot pudding. Cool, then refrigerate until the pudding is firm.

6. To serve, unmould the pudding into a deep dish, pour coconut cream and palm sugar syrup over the pudding. Serve cold.

Kueh Salat

Preparation and cooking: 1 hour

This *kueh* has several names and is another very old Straits-born dessert. It may be called *kueh salat*, *kueh serimuka* or *kueh serikaya*. Whatever its name, it is basically steamed glutinous rice with a layer of *kaya* (coconut egg custard) on top. Penangites make a similar dish known as *pulot tatai* which is steamed glutinous rice pressed into a flat cake that is cut into bite-size pieces and served with a bowl of *kaya*.

Rice Layer

300 g (11 oz / 1 cup) white glutinous rice, soaked overnight

2 Tbsp water

2 Tbsp packaged coconut cream

$^1/_2$ tsp salt

2 Tbsp dried butterfly pea flowers

Kaya Topping

150 g (5$^1/_3$ oz) sugar

4 Tbsp plain (all-purpose) flour

6 eggs

150 ml (5 fl oz / $^2/_3$ cup) packaged coconut cream

150 ml (5 fl oz / $^2/_3$ cup) pandan juice

Pandan Juice

12 pandan leaves, rinsed and chopped

250 ml (8 fl oz / 1 cup) water

> **Note**
> To make durian *kueh salat*, substitute the pandan juice with 1 cup durian puree when making the *kaya* topping.

1. Drain the rice and spread on a steamer lined with muslin. Steam for 10 minutes. Mix water, coconut cream and salt together, then stir into rice. Let rice continue steaming for another 30 minutes until tender.

2. If using dried butterfly pea flowers for colouring, soak dried flowers in 1 Tbsp water to extract the colour. When rice is done, streak colouring through the rice. (The blue in the picture is pale because I ran out of pea flowers.)

3. Line a 22-cm (8-in) square cake tin/glass dish with banana leaves, then pack in the hot rice. Press rice down with a spoon to get a level surface.

4. Prepare the *kaya* topping by first making pandan juice. Blend the chopped pandan leaves with water to pulp. Strain to obtain the juice and discard the fibre. Measure out 150 ml (5 fl oz / $^2/_3$ cup) pandan juice for flavouring the *kaya*.

5. To make the *kaya* topping, mix sugar and flour to dissolve any lumps of flour. Beat eggs, coconut cream and pandan juice into sugar mixture. Pour mixture into a small saucepan and cook over low heat, stirring constantly to thicken the custard. The mixture should be smooth. Be careful not to allow lumps to form. The *kaya* does not need to be fully cooked.

6. Spread the hot *kaya* mixture over the rice layer. Place cake tin/ glass dish in a steamer filled with water and steam for 30 minutes or until the *kaya* topping is cooked through.

7. To prevent drops of water from falling onto the custard and making pockmarks in it, tie a large piece of cloth or clean dish towel around the lid to soak up the water vapour.

8. Cool *kueh* thoroughly to allow the topping to set before cutting to serve.

Kaya (Coconut Egg Jam)

Preparation and cooking: 30 minutes   Makes 300 g (11 oz)

Kaya (coconut egg jam) on crisp toast with two soft-boiled eggs and a cup of coffee has long been a popular breakfast food in Singapore coffee shops. Nowadays, you can get this breakfast combination all day long at some chain outlets. My mother used to make *kaya* for us and it was always caramel-flavoured because without a food processor, getting pandan juice meant pounding the leaves. Besides *kaya* on toast, try it on cream crackers, soda crackers or saltines. Neither *kaya* keeps well because it has no preservatives, but the caramel *kaya* keeps longer because of the sugar.

Caramel-flavoured Kaya

100 g (3^1/$_2$ oz) white sugar

1^1/$_2$ Tbsp plain (all-purpose) flour

4 eggs

125 ml (4 fl oz / ^{1}/$_2$ cup) packaged coconut cream

1. Divide the sugar into two bowls.

2. Mix one bowl of sugar with the flour to break up any lumps. Add eggs and beat with a wire whisk. Set aside.

3. Caramelise the second bowl of sugar by melting it in a saucepan over low heat. Let the melted sugar turn brown, but not black. Remove from heat once it starts to brown. Residual heat will brown it further.

4. Warm the coconut cream for 30 seconds in the microwave oven, then add to the caramelised sugar. Mix well. Spoon some of the warm coconut mixture into the egg mixture to warm it up slowly. (This step prevents the eggs from curdling.) Once the eggs are warm, combine the egg mixture with the warm coconut mixture in the saucepan.

5. Stir the mixture continuously over low heat until it thickens. This will take about 10 minutes.

6. Let the mixture cool, before bottling and storing in the fridge. Consume within a week.

Pandan-flavoured Kaya

50 g (1^3/$_4$ oz) white sugar

3 Tbsp plain (all-purpose) flour

4 eggs

4 Tbsp packaged coconut cream

100 ml (3 fl oz / ^{3}/$_4$ cup) pandan juice

Pandan Juice

6 pandan leaves, rinsed and chopped

100 ml (3 fl oz / ^{3}/$_4$ cup) water

1. To make pandan-flavoured *kaya*, start by making pandan juice. Blend chopped pandan leaves with water and strain the juice. Measure out 100 ml (3 fl oz / ^{3}/$_4$ cup) juice.

2. In a saucepan, mix the sugar and flour together to break up any lumps. Add the pandan juice, eggs and coconut cream and beat with a wire whisk.

3. Place a heat diffuser plate or a frying pan filled with water over medium heat. Sit the saucepan on the diffuser plate or in the frying pan. Stir the *kaya* for about 10 minutes or until it thickens.

4. Let the mixture cool, before bottling and storing in the fridge. Consume within 4 days.

Coconut Candy

Preparation and cooking: 1 hour

In the 1950s, before the proliferation of imported chocolates and sweets, coconut candy was *the* treat. It was also a favourite item for sale at school funfairs and fundraisers for school building funds. The traditional way to make coconut candy is by boiling, but I find that making it in the microwave oven is faster and avoids painful splatters. Either way, take care not to overcook or the candy will become rock hard. Timing is important when it comes to sweet-making. To check if the candy is ready, use the ice water test for the soft-ball stage. With candy-making, the amount of sugar cannot be reduced. To get two colours, the candy has to be made in two batches as the colour is added during cooking.

1 Tbsp butter

50 g (1³/₄ oz) cocoa powder

600 g (1 lb 5¹/₃ oz) white sugar

200 ml (7 fl oz / ⁴/₅ cup) evaporated milk

400 g (14 oz) white grated coconut

1 tsp vanilla essence

Ice-cold water

Note
This recipe can also be cooked in a microwave oven as outlined in the Rose-flavoured Coconut Candy, but it will take longer to reach the soft-ball stage because of the evaporated milk.

1. Butter a 22-cm (8-in) square baking tin. Set aside.

2. In a thick-bottom saucepan, mix cocoa powder and sugar to break up any lumps of cocoa. Stir in evaporated milk and grated coconut.

3. Set the stove to medium heat and bring the mixture to the boil. If you have a heat diffuser plate, use it. When the mixture is very wet, give it an occasional stir every 5 minutes. Once it starts to dry on the sides of the pot, stir continuously to the soft-ball stage.

4. To test for the soft-ball stage, drop a small lump of the mixture into a cup of ice-cold water. The candy should still be soft and chewy, but with hard bits formed around the soft bit.

5. Add vanilla essence at this stage and continue cooking for another 2 minutes. Do the ice water test first before scraping the candy into the buttered tin.

6. Smooth the candy evenly into the baking tin. If you like a more "rocky" look, use a fork to rough the candy up. Using a buttered knife, mark the candy into cubes while it is still hot. Work quickly.

7. Let the candy cool, then break into pieces. Store in an airtight container and consume within a month. To keep candy fresh for several months, refrigerate it..

Rose-flavoured Coconut Candy

Preparation and cooking: 35 minutes

1 Tbsp butter

400 g (14 oz) white grated coconut

345 g (12 oz) white sugar

1 can (392 g / 14 oz) condensed milk

A few drops of rose essence

Food colouring

Ice-cold water

> **Note**
> Not all microwave ovens have the same wattage. Use the soft-ball test to check the timing.

1. Butter a 22-cm (8-in) square baking tin. Set aside.

2. Combine the grated coconut, sugar and condensed milk in a large microwaveable bowl. Cook the mixture in the microwave oven on High for 10 minutes, then remove and stir. Cook for another 5 minutes, then add rose essence and food colouring and mix well. Continue to cook for another 5 minutes, then stir and cook for 2 minutes. Test for the soft-ball stage at this point (see page 146, step 4).

3. When candy is ready, scoop it out into the buttered tin and spread out evenly. Using a buttered knife, mark the candy into cubes while it is still hot. Work quickly.

4. Let the candy cool, then break into pieces. Store in an airtight container and consume within a month. To keep candy fresh for several months, refrigerate it..

Pulot Hitam (Black Glutinous Rice Porridge)

Preparation and cooking: 3–4 hours

This Straits-born dessert is sometimes called *bubor pulot hitam* (black glutinous rice porridge) and it is best dressed with a dollop of fresh coconut cream just before serving. If fresh coconut cream is not available, cook the rice with packaged coconut cream and dispense with the topping. Cooking this dessert in a slow cooker overnight is fail-proof and no soaking is needed. Equally easy is slow-cooking it overnight in an oven set to 90°C (195°F). If cooked on the stove top, the porridge must be stirred frequently to prevent the rice from catching at the bottom. Sometimes black glutinous rice is not particularly glutinous, but you can't tell by reading the label. If so, you could cheat by mixing a spoonful of glutinous rice flour with water and boiling it with the rice. Good black glutinous rice has a special fragrance.

200 g (7 oz / 1 cup) black glutinous rice

4 pandan leaves, rinsed and knotted

1.5 litres (48 fl oz / 6 cups) water

500 ml (16 fl oz / 2 cups) second coconut milk (page 20)

A pinch of salt

150 g (5^1/$_3$ oz) sugar

375 ml (12 fl oz/ 1^1/$_2$ cups) fresh coconut cream (page 20)

On the stove top

1. Rinse rice, then leave to soak overnight.

2. Put soaked rice, pandan leaves and water into a large pot and bring to the boil. If you have a heat diffuser plate, use it. Skim off any scum, then turn down heat and simmer until rice is soft. Stir occasionally and start adding the second coconut milk to the rice after an hour or so. Even if using a heat diffuser plate, the rice needs to be stirred to prevent it from sticking to the bottom of the pot.

3. When the rice is soft enough, add salt and sugar to taste. Bring to the boil and adjust with more sugar and salt, if needed.

4. To serve, ladle hot porridge into individual serving bowls and drizzle generously with fresh coconut cream.

Using a slow cooker

1. Rinse rice, then place in a slow cooker with pandan leaves and 1 litre (32 fl oz / 4 cups) water. Leave rice to cook overnight.

2. When rice is a thick porridge, thin out with second coconut milk and return to the boil. Add sugar and salt to taste.

3. To serve, ladle hot porridge into individual serving bowls and drizzle generously with fresh coconut cream.

Pineapple Tarts

Preparation and baking: Half a day   Makes about 100 bite-size tarts

Once upon a time, before the appearance of special pineapple tart cutters with fluted edges, each tart had to be hand-pinched with special pincers. A little "leaf" would then be cut for the top of each tart. (The picture shows the original "pinched" tarts that my mother used to make as well as the stamped tarts that I make in the foreground.) This same pastry can be used to make "apples" which are balls filled with jam and a clove pressed into the top to simulate a stem. It can also be shaped into an oval to make "pineapples" with "spikes" made using a small pair of scissors. Both shapes should be brushed with beaten egg. Thank you, Zaiton's aunty, for one of the best tart recipes ever. When making the pineapple jam, the skins, eyes and cores will add up to about half the weight of a pineapple. This amount of fruit will give you about 1.6 kg ($3^1/_2$ lb) cooked pineapple filling and will be enough for the quantity of pastry in this recipe. Be generous with the balls of jam because they shrink after baking. If there is any jam left over, freeze until you want to make pineapple tarts again. The addition of butter helps to keep the tart filling slightly moist, as does not overcooking or over-sugaring the jam. The lime juice gives a different tartness.

Pastry
450 g (1 lb) plain (all-purpose) flour, sifted

2 tsp baking powder

A pinch of salt

$1^1/_2$ tsp castor sugar

2 egg yolks

250 g (9 oz) salted butter, softened

4 Tbsp water

Pineapple Jam
3 kg (6 lb 9 oz) whole pineapples

250–300 g (9–11 oz) sugar

5 cloves (optional)

1 tsp lime juice

30 g (1 oz) unsalted butter

1. Prepare pineapple jam. Cut off the skin and gouge out the eyes. Rinse clean. Quarter pineapples, then cut off and discard the hard core. Chop pineapple into very small pieces. Strain away some of the juice if the pineapple is very ripe. Reserve it until the jam is done.

2. In a stockpot, boil chopped pineapple with sugar and cloves, if using, until pineapple juice has evaporated and you have a thick jam that can be rolled into a ball. It should take about $1^1/_2$ hours on a slow boil. Adjust sugar to taste.

3. Stir in lime juice and butter, then remove the cloves. Cool and store in a box in the fridge until ready to use.

4. Prepare pastry. Sift flour and baking powder together twice.

5. On a flat surface such as a pastry board, silicon pastry sheet or clean table top, use your hand to mash together the salt, sugar and egg yolks.

6. Mix the butter into the egg yolk mixture with your hand. Gradually add the flour mixture and water to get a soft, but not sticky dough. Let dough rest for 30 minutes.

7. Roll the rested dough into a 0.5-cm ($^1/_5$-in) thick sheet. Use a pineapple pastry cutter to cut out tart bases. Arrange on a baking tray. No lining or greasing is necessary.

8. To make the process of filling the pastry easier, first roll the jam into small balls, each about 6 g ($^1/_5$ oz) before you start cutting out the pastry. Set aside.

9. Using your thumb, widen the dent in the centre of the bases and fill with pineapple jam. Smooth neatly. Decorate as desired.

10. Preheat the oven to 150°C (300°F) and bake tarts on the lower bottom shelf for 30–40 minutes. The time depends on size of the tarts. Browner tarts do not turn mouldy as quickly without refrigeration unlike paler tarts.

Java Jades

Preparation and baking: 1 hour   Makes about 100 biscuits

This recipe is in the fine old tradition of recipe-sharing, in this case between teacher and student. Some 60 years ago, I visited Miss Wee, one of my school teachers, for Chinese New Year. I was so taken with her coffee biscuits that I boldly asked her for the family recipe. It became my mother's top favourite biscuit and one that I had to bake every year for her. The biscuits may be shaped with a cookie press or piped through a star nozzle. Those in the picture were done with two very old and tiny different-sized star nozzles made of brass. A little more flour to the dough will give a more defined pattern, but harder biscuits. These are very light.

1$^1/_2$ Tbsp instant coffee granules

140 g (5 oz) castor sugar

1 egg

1 tsp vanilla essence

250 g (9 oz) salted butter, at room temperature

350 g (12$^1/_3$ oz) plain (all-purpose) flour, sifted

1. Mix instant coffee granules with sugar, pressing the granules into the sugar to grind them into powder.

2. Combine coffee mixture, egg, vanilla essence and butter in a mixing bowl and beat until light and fluffy.

3. Beat in flour to form a soft dough that can be piped through a cookie press or a piping bag fitted with a star nozzle.

4. Grease two baking trays with oil, then press greaseproof paper onto the trays. Turn paper over to line the trays and spread the oil over the paper.

5. Use a cookie press or piping bag to form dough shapes on the baking trays. If using a piping bag, hold the tube just 0.5-cm ($^1/_5$-in) from the tray and pipe. The biscuits may be piped close to each other because they hardly expand.

6. Preheat the oven to 160°C (320°F) and bake biscuits on the centre shelf for 15 minutes or until nicely brown.

7. Transfer biscuits onto a wire rack to cool. Store in an airtight jar.

Indonesian Kueh Lapis (Spekkoek)

Preparation and baking: 3–4 hours

This was originally a Dutch-Indonesian cake known as *spekkoek* which literally means "bacon cake" because of its streaky bacon look. The Indonesian name simply means "layered cake". Once sourced mainly from Indonesian Chinese housewives who baked the cake for festive sales, *kueh lapis* has since become one of the top festive treats in Singapore for all communities — in part because it is now easily available from some bakeries. The original recipe came from a women's magazine, with nearly twice the butter and sugar, and 30 egg yolks. It was richer and a little bigger. This tweaked recipe is more cake-like. This cake may get mouldy if left too long without refrigeration. It freezes well.

125 g (4 oz) plain flour

4 Tbsp mixed spice (page 158)

200 g (7 oz) castor sugar

15 egg yolks

2 Tbsp brandy/rum/whisky

360 g (13 oz) salted butter

1¹/₂ tsp vanilla essence

6 egg whites

1. You will need an oven with a top and bottom heating function as the first layer is baked with top and bottom heating, like a regular cake, but the second layer onwards is grilled.

2. Prepare a 19-cm (7-in) square baking tin. Cut out a piece of greasproof paper the size of the base of the tin. Oil the base and place the paper on the oil to keep the paper in place.

3. Sift flour and mixed spice together. Stir to mix well.

4. Divide sugar into two portions. Set one aside for beating egg whites. Beat other portion with egg yolks until thick and pale. Halfway through, add brandy/rum/whisky and continue beating.

5. Using an electric mixer, beat butter and vanilla essence until light and fluffy. With the mixer on low speed, add the beaten egg yolks in a slow steady stream. Add spoonfuls of flour while mixing. This will prevent the mixture from separating or curdling. Set aside.

6. Beat egg whites with remaining half of sugar to stiff peaks. Using a metal spoon, fold the egg whites into the butter mixture.

7. Preheat the oven to 170°C (340°F).

8. Ladle about 150 g (5¹/₃ oz) batter into the prepared tin and spread it out to the edges of the tin. Place the tin on the second bottom rack of the oven and bake for about 3 minutes.

9. Turn off the bottom heating function of the oven. Ladle an even spoonful of batter on top of the first layer and grill for 3 minutes.

10. Repeat until the batter is used up.

11. Cool the cake in the tin for 10 minutes. To turn the cake out, run a knife along the inside of the tin, then position a wire rack carefully over the tin to ensure the cake gets an even pattern. Turn cake out.

12. Cut into small, thin slices. Serve with coffee or tea.

Tips for Baking

- Check the temperature and time for the first few layers carefully so that the layers bake/grill evenly. Each layer should turn a nice dark brown in no more than 4 minutes. A too-low temperature is better than too high.

- Measure out equal ladlefuls for each layer so that the layers are of even thickness. Spread out the batter evenly.

- Sometimes air pockets form in part because of uneven heat and the top then colours unevenly. Prick the air bubble and press it down with a flat bottom cup or spatula. Change the position of the tin to optimise even heating. Or move the shelf lower down.

Almond and Cherry Cake

Preparation and baking: 1¹/₂ hours

This recipe was from a very old cookbook that I found in the National Library in the 1960s. But it was more than a decade later before I finally tried the recipe. It became a family favourite not just for its rich buttery flavour, but also for its excellent keeping qualities. This cake tastes better the day after, which was why this was the cake that I brought on camping trips. Even when left to sit in a box under the shade of a tree on a desert island improved its flavour.

125 g (4¹/₂ oz) glacé cherries

200 g (7 oz) plain (all-purpose) flour

¹/₂ tsp baking powder

125 g (4¹/₂ oz) ground almonds

250 g (9 oz) salted butter, at room temperature

140 g (5 oz) castor sugar

3 eggs

1 tsp vanilla essence

2 Tbsp milk

Note

If using an unlined tin, buttering is preferable to oiling the surface. Unlike oil, butter does not slide to the bottom of the tin.

1. Prepare a 1-litre (32-fl oz / 4-cup) baking tin. Grease the bottom and sides of the tin with butter and line with greaseproof paper. If using a fluted tin, butter sides generously. Preheat oven to 150°C (300°F).

2. Quarter glacé cherries and rinse in cold water to remove syrup. Pat dry with paper towels, then spread cherries out on more paper towels to dry them.

3. Sift together the flour and baking powder. Set aside 2 Tbsp flour mixture for dredging cherries.

4. Mix the rest of the flour mixture with the ground almonds.

5. Using an electric mixer, beat butter and sugar together for 10 minutes. Add eggs, one at a time, and continue beating until mixture turns pale, and is light and fluffy. Add vanilla essence and mix well.

6. Turn mixer speed to low, then add flour and ground almond mixture by the spoonful. Add milk halfway through.

7. Dredge cherries in reserved flour mixture and add to batter. Mix well.

8. Pour batter into prepared tin. Hollow out centre slightly so cake will level out when baking. Place in the centre of the oven and bake for 45 minutes to 1 hour, until a skewer inserted into the centre of the cake comes out clean.

9. Turn cake out onto a wire rack to cool. Remove the greaseproof paper lining, if used.

10. When cool, cut into slices and store in an airtight container at room temperature for up to a week. To keep for longer, freeze immediately.

Christmas Fruit Cake

Preparation and baking: 2 hours

I made my first fruit cake from a recipe taken from a National Library cookbook, and I kept tweaking the recipe. The original was practically all fruit with very little cake, so I adjusted it to have less fruit and more cake. I then changed almonds to walnuts because I thought walnuts kept their crunch better. In recent years, I added dried cranberries and currants because they are more tart. One early tweak was to increase the quantity as well as the quality of the alcohol that went into the cake because quality matters. Bake this cake at least a month ahead, preferably longer. (The cake in the picture was nearly a year old.) The cake freezes well.

430 g (16 oz) plain (all-purpose) flour

6 tsp mixed spice (recipe below)

2¼ tsp baking powder

375 g (13 oz) butter, at room temperature

200 g (7 oz) molasses sugar

2 Tbsp molasses

6 eggs

250 ml (8 fl oz / 1 cup) brandy/ rum/whisky

Dried Fruit and Nuts
65 g (20 oz) glacé cherries

300 g (11 oz) raisins

85 g (3 oz) golden raisins

200 g (7 oz) sultanas

50 g (1¾ oz) walnuts/almonds

50 g (1¾ oz) mixed peel

50 g (1¾ oz) cranberries

100 g (3½ oz) currants

Mixed Spice
2 tsp ground cinnamon

1 tsp ground ginger

1 tsp allspice

½ tsp ground cloves

½ tsp ground nutmeg

½ tsp mace

½ tsp cardamom

1. Grease and line two 18 x 25-cm (7 x 10-in) baking tins. Preheat the oven to 150°C (300°F).

2. Prepare dried fruit and nuts. Quarter glacé cherries and rinse in cold water to remove syrup. Pat dry with paper towels, then spread cherries out on more paper towels to dry them. Chop raisins and sultanas. Chop walnuts/almonds coarsely.

3. Combine spices for mixed spice, then sift together with flour and baking powder. Add dried fruit and nuts. Mix well and set aside.

4. Using an electric mixer, beat together butter, molasses sugar and molasses. Add eggs one at a time and continue beating until mixture is light and fluffy.

5. With mixer on low speed, add flour mixture. Mix until incorporated.

6. Divide batter evenly into prepared tins. Make a fairly deep well in the centre so cake will level out when baking.

7. Bake for 1 hour or until a skewer inserted into the centre of cake comes out clean. Remove from the oven and unmould cakes onto a wire rack. Peel away greaseproof paper and leave cake to cool thoroughly before the next step.

8. Prepare 2 sheets of greaseproof paper large enough to wrap cakes in. Spoon brandy/rum/whisky all over top, bottom and sides of cakes and wrap with greaseproof paper. Store in an airtight container or plastic bag at room temperature for up to a month. Refrigerate if not served.

9. If kept refrigerated for more than 6 months, repeat the alcohol bath with 125 ml (4 fl oz / ½ cup) brandy/rum/whisky about 10–14 days before cutting cake.

10. Before cutting the cake, refrigerate it overnight. Cut the cold cake into thick pieces with a sharp knife. Store in a covered container at room temperature for serving.

A Select Glossary of Ingredients

Alkali salts (*kee chui, ayer abok, ayer kapok,* lye water)

Kee chui is a Hokkien term meaning lye water. *Ayer abok* is Malay meaning ash water and *ayer kapok* is Malay meaning limestone water. Alkali salts are added to rice or flour to give the dessert or noodles a springy texture. A good example is *kee chang* (soda dumplings). Alkali salts is sold as a liquid, a paste or as orange or pale yellow crystals. Substitute with bicarbonate of soda.

Ambarella (*Spondias cytherea*)

Called *buah long long* by the Chinese in Singapore, *kedondong* by the Malays and *ambla* by the Penang Chinese, this tropical fruit is rarely seen commercially in Singapore. It is common in Penang where the young fruit is made into a drink flavoured with salted sour plum called *ambla chui*. The green fruit is an ingredient in Penang *rojak*, eaten with a *hae ko* (black prawn paste) dip or as part of a Straits salad. The fruit turns golden and soft when ripe and the pit becomes very fibrous. The fruit is more tart when ripe.

Anchovies (*Engraulidae*)

There are various types of small fish called anchovies. The best known in Straits-born cooking is the Indian anchovy called *ikan bilis* in Malay. These are usually sold salted and dried although fresh anchovies may sometimes be seen in wet markets. Anchovies and other similar small fish are fermented into fish sauce as well as made into bouillon cubes.

Anise, Aniseed (*Pimpinella anisum*) *see* Fennel; *see also* Star Anise

Asam *see* Tamarind

Asam Gelugor (*Garcinia atroviridis*)

Translated into English as dried sour fruit slices, it is better known to Straits-born cooks as *asam gelugor, asam keping, asam puay* (Hokkien meaning *asam* skin), *asam Jawa* and sometimes tamarind slices. However, *asam gelugor* is not related to the tamarind (*Tamarindus indica*). It is the fruit of a tropical tree believed to be native to Peninsular Malaya. The fruit is sold in dried slices and used in Straits-born cooking to give the dish a tang either with or without tamarind juice. *Asam gelugor* is in Penang laksa and may be used in Chetti *lauk pindang* or any dish where some fruity sourness is desired.

Asam Jawa *see* Asam Gelugor

Bananas (*Musa*)

Originating in South East Asia, there are some 1,000 types of bananas and banana hybrids today. Markets in Singapore display a large variety of bananas, both the eating variety and the cooking variety. Cooking bananas or plantains are less sweet. A number of classic Straits-born desserts have bananas in them and the most popular banana for desserts is *pisang rajah* which in Malay translates as king banana. It is an excellent eating banana as well. There are two types of *pisang rajah*. One kind has a slightly orange-coloured flesh, and the other has a paler more white flesh. The former is often sweeter than the latter. To be cooked, bananas must be very ripe. Otherwise, it can be "sappy".

Bangkuang *see* Yam Bean

Bay Leaves (*Laurus nobilis, Cinnamonum tejpata, Eugenia polyantha Wight*)

There are several kinds of bay leaves. The bottled bay leaves sold in western supermarkets is *Laurus nobilis* also called Turkish bay leaf. Indian bay leaf is *Cinnamonum tejpata* and sold loose in Indian grocery stores. Indian bay leaf, called *tejpata* in Hindi, was known to Roman cooks as *malabathrum/malobathrum*. Indonesian bay leaf (*Eugenia polyantha Wight*) or *daun salam* is mostly used fresh in Malay and Indonesian cooking, but also dried. *Daun salam* is sometimes mistakenly called Indian bay leaf.

Belacan *see also* Krill

This is one of the essentials of Island South East Asian cooking. Called *belacan* in Malay, *terasi* in Indonesia and *bagoong* in the Philippines, it is sometimes translated as "dried prawn paste". *Belacan* is more accurately translated as "dried shrimp paste" as it is made from krill which are tiny shrimps and shrimps are defined as very small prawns. *Belacan* comes in colours that range from pink to a dark brown. The texture may be soft and liquid enough to be bottled or as a block that is cut up. *Belacan* must not to be mistaken for *hae ko* (black prawn paste). Unlike *hae ko*, *belacan* is never eaten as it is, but always as part of a spice paste or made into *sambal belacan*.

Belimbing *see* Camias

Black Shrimp Paste *see Hae Ko*

Buah belimbing *see* Camias

Buah kedondong *see* Ambarella

Buah keras see Candlenuts

Camias (*Averrhoa bilimbi*)

A relative of star fruit, this juicy fruit is invariably sour and is best cooked into chutney or in a seafood curry for some tang. Chopped up, it is added in small quantities to a salad as the souring agent. It is not available commercially, but was once common in Straits-born homes. Unripe star fruit is a substitute.

Candlenuts (*Aleurites moluccanus*)

Commonly called *buah keras* (Malay meaning hard fruit or nut) or *kemiri* in Indonesia, the oily soft nut of a tropical tree is used as a thickener in Straits-born spice pastes and never eaten any other way. The tree is native to Maluku in Indonesia, Malaysia and the South Pacific. The closest substitute for candlenuts is the much more expensive macadamia nuts. The oily nuts were once made into candles.

Cassia, Chinese (*Cinnamonum cassia (L.) Presl*)

Originally from southern China, this fragrant bark comes in pieces of thick grey-brown bark with patches of pale grey, not neatly curled into quills, unlike Ceylon or Indonesian cinnamon. Cassia is also vaguely bitter, unlike Ceylon cinnamon. Cassia is one of the key ingredients in Chinese five-spice mixes as well as in the spice mixes of Central and West Asia, regions that were once part of the ancient Silk Road. Vietnamese cinnamon is related to Chinese cassia but is a thinner bark that looks like its Chinese cousin.

Cekur see Ginger

Chillies (*Capsicum annuum L. and Capsicum frutescens L.*)

Originally a native of Central America, its introduction by the Portuguese to their colonies in South Asia and Melaka in the 16th century changed totally the character of South and South East Asian cuisines. Distinguish between fresh and dried, green and red, large and tiny chillies. They all give different flavours and different degrees of heat. The smaller ones are more deadly: Scotch bonnet, bird's eye and the very small Thai *prik kee nu*. Mexican chillies also come smoked.

Chinese Celery (*Apium graveolens var. secalinum*)

Chinese celery is a relative of western celery and sometimes called leaf celery. Unlike Western celery, Chinese celery is green, leafy and with round hollow stems. It has a much stronger flavour than western celery and one that resembles flat-leaf (Italian) parsley. It is sometimes called Chinese parsley. It is popularly added to soups, hence its Malay name *daun sop* (soup leaves). In Straits-born cuisines, it may be chopped up alone or together with spring onions and coriander leaves to give a herby accent to salads, soups and some stews. The leaves of western celery, parsley or Italian flat-leaf parsley are good substitutes for Chinese celery.

Chinese Chives (*Allium tuberosum*)

A member of the onion family, Chinese chives are sometimes called garlic chives or flat chives. They come as young flat leaves or mature stems with flowers. The mature chives are cooked as a vegetable, the young leaves are chopped and an essential garnish for Straits-born *mee siam*, Penang *char kuay tiao*

and Singapore fried Hokkien mee. Chinese chives also come a pale yellow colour, the result of growing in deep shade. Yellow chives are very mild compared to the dark green ones.

Chinese Keys see Ginger

Cilantro see Coriander

Cinnamon, Ceylon see Cinnamon, Indonesian

Cinnamon, Chinese see Cassia

Cinnamon, Indonesian (*Cinnamonum burmannii* [Nees & T. Nees] *Blume*)

Compared to Ceylon cinnamon (*Cinnamomum verum*), Indonesian cinnamon bark is thicker and harder to break into smaller pieces. It is the more common of the two cinnamons in Singapore markets and used as a whole spice in curries. Ceylon cinnamon has a more delicate and sweeter fragrance.

Cinnamon, Vietnamese see Cassia

Coconut (*Cocos nucifera*)

Coconut palms are native to Island South East Asia and the nuts play a significant role in Straits-born cooking. Coconut cream (*santan*) and coconut milk are added to the cooking as well as enjoyed raw. Grated coconut is used both raw and dry-fried. Palm sugar is extracted from the inflorescence as is the sap for making palm wine and arack. Numerous Strait-born dishes from savoury to sweet depend on the richness of coconuts. Although packaged coconut milk and coconut cream are available worldwide, these heavily processed products cannot match the flavour of fresh coconut milk in certain savouries and desserts.

Coriander (*Coriandrum sativum*)
Originally from the southern Mediterranean and West Asia regions, coriander is eaten both as a herb and as a seed spice in Straits-born cooking. As leaves (called cilantro in the US), coriander is both a herb as well as a garnish. The seed spice is coriander or *ketumbar* in Malay and it is an important ingredient in curry powders. Note that there are people who are genetically programmed to taste coriander leaves as soapy or "buggy".

Cumin (*Cuminum cyminum*)
Originating from the Eastern Mediterranean and stretching across West Asia into India, cumin aka *jintan puteh* in Malay is an essential ingredient in curry powders, but used in smaller quantities than coriander.

***Daun Kesom* (*Pericaria adorata, syn.Polygonum adoratum*)**
Often called laksa leaf perhaps because it is essential in both Penang and Singapore laksa, it is also known as Vietnamese mint and is a common herb used in salads as well as seafood curries.

***Daun Salam* see Bay Leaves**

***Daun Sop* see Chinese Celery**

Dried Sour Fruit Slices see *Asam Gelugor*

Fennel (*Foeniculum vulgare*)
Fennel and anise or aniseed are often used interchangeably, but are two different things. Originally from southern Europe, fennel is consumed as a vegetable while anise (*Pimpinella anisum*)/aniseed is used as a spice. Called *jintan manis* in Malay, the spice is labelled as "fennel" in Singapore and Malaysia.

Fenugreek (*Trigonella foenum-gracum*)
Originally native to south-eastern Europe and West Asia, fenugreek is a Latin word meaning Greek hay. Called *methi* in Hindi and *venthiyam* in Tamil, the little squarish seeds are used in small quantities in some Straits-born dishes of South Indian origins.

Fried Tofu Puffs see Soy Beans

Galangal, Greater (*Alpinia galanga*)
Believed to be of South East Asian origin, this tough fibrous rhizome with an orange-red skin is another essential in Straits-born cooking. A member of the ginger family, it is the largest of the several South East Asian native gingers. Young rhizomes are paler, less fibrous and less well flavoured. Galangal is used fresh, ground up as well as whole when it is smashed or sliced for flavouring. Powdered galangal is now to be found in Asian stores.

***Gandum* see Wheat**

Ginger (*Zingiber officinale*)
Native through a wide swathe of South Asia through South East Asia and into southern China, this common ginger is eaten either young or old, the older having the sharper flavour. It is common in Straits-born cooking. Two other gingers native to South East Asia are sometimes used in Straits-born cooking, particularly recipes of Javanese or Thai origins. One is *kencur* (*Kaempferia galanga*) aka *cekur* in Malay. The leaves of *kencur* are used as a herb in salads and *nasi ulam*. The other is *krachai* (*Boesenbergia rotunda*) aka Chinese keys aka lesser galangal aka finger root. The leaves are also fragrant and are used as a herb in salads. Like the greater galangal, the rhizomes of both gingers are used in small quantities and ground as part of spice pastes.

***Grago* see Krill**

***Hae Bee* see Shrimps and Prawns**

Hae Ko
A Hokkien term usually translated as "black prawn paste", this is a Penang speciality made by cooking together a very strongly flavoured prawn stock with salt, palm sugar and flour. It is called *petis udang* in Malay. Online sources say it is also made in Indonesia. Black shrimp paste must not be confused with dried shrimp paste (*belacan*).

***Halba* see Fenugreek**

***Herbs* see**
• Bay Leaves
• Chinese Celery
• Chinese Chives
• *Daun kesom*
• Ginger
• Limes
• Mint
• Torch ginger bud
• Turmeric

***Ikan Bilis* see Anchovies**

***Jicama* see Yam Bean**

***Jintan Manis* see Fennel**

***Jintan Puteh* see Cumin**

Kaffir Lime Leaves see Limes

***Kayu Manis* see Cinnamon, Indonesian**

Kedondong see Ambarella

Kencur see Ginger

Ketumbar see Coriander

Kin Chye see Chinese Celery

Koo Chye see Chinese Chives

Krachai see Ginger

Krill (*Eusausiacea*)
These are very small crustaceans or shrimps and known locally as *grago*. This is the Kristang word for krill. (Kristang is old Melaka Portuguese.) Found in all the world's oceans, they are eaten in South East Asia in *belacan*, fermented into *cinchalok* and fried with egg in omelettes. Krill may also be salted and dried and used as a garnish in salads and soups. It is sometimes found in wet markets as fresh or boiled. Krill is an essential food source for bigger marine creatures such as whales and dolphins.

Lemongrass (*Cymbopogan citratus*)
Native to South and South East Asia, lemongrass is *serai* in Malay. It is one of the essential ingredients in most Straits-born spice pastes for *sambals* and curries. It may also be bruised and used whole in certain recipes. Only the 10 cm (4 in) up from the bulbous end of the stalk is used and before processing, the stalk must be sliced thinly. If not, the spice paste will be unpleasantly full of fine lengths of lemongrass fibres. The citronella oil used as a mosquito repellent is distilled from a different lemongrass (*Cymbopogan nardus*) which looks and even smells like *serai*, but is supposed to be inedible. Edible lemongrass is believed to be

also useful as a mosquito repellent if crushed and put near doors and windows.

Lengkuas see Galangal, Greater

Limes (*Citrofortunella macrocarpa, Citrus aurantifolia,Citrus hystrix*)
There are several types of limes popular in South East Asia, but the commonest one in Straits-born cooking is calamansi (*Citrofortunella macrocarpa*), originally a Philippines native. Also called *limau kasturi* in Malay, calamansi is a small round juicy lime with thin skin. The flesh of calamansi is yellow. This is unlike the white coloured flesh of *limau nipis* (*Citrus aurantifolia*) which is a larger lime with a thicker paler green skin and juice with a different stronger fragrance. *Limau nipis* is also known as key lime and is a popular lime for Thai salads because of the fragrance. The other important lime in Straits-born cooking is kaffir lime (*Citrus hystrix*) aka makrut lime or *limau perut* in Malay. A South East Asian native, the knobbly lime is often used as a prayer offering, but the peel and leaves may also be added to certain types of curries, *sambal belacan*, salads and *nasi ulam* for that herby accent. It is probably the most fragrant of all the limes.

Mint (*Mentha piperita*)
There are several varieties of mint, of which the most popular are spearmint and peppermint. Fresh mint is used as herb in salads and some *sambals*, and essential in Penang laksa.

Miso see Soy Beans

Palm Sugar
This is the sugar from the sap of a variety of palm trees native to the

region. The sap is boiled down to a heavy sugar that has a fragrance all its own. Sometimes called coconut sugar, palm sugar is dark brown if it comes from West Malaysia where it is called *gula Melaka* because the best was supposed to have come from this west coast town. Palm sugar from Indonesia may be dark brown or golden like the palm sugar from Indochina. Palm sugar is the main sweetener in many Straits-born desserts such as *bubor terigu* or *pengat*. Good palm sugar is soft rather than hard. If hard, it's a sign that the palm sugar has been cooked with cane sugar. Another sign is when the palm sugar is super sweet. Genuine palm sugar has a lighter sweetness than cane sugar.

Pandan (*Pandanus amaryllifolius Roxb.*)
Native to South East Asia and also known as screwpine, pandan leaves give a lovely fragrance to numerous Straits-born desserts as well as special Straits-born rice preparations like *nasi lemak* or *nasi kebuli*. The leaves are knotted and used whole when cooking rice or a porridge-like dessert. The juice extracted from the leaves is also used to colour certain desserts and cakes green as well as give fragrance.

Pepper (*Piper nigrum*)
In the 15th and 16th centuries, wars were fought over pepper. Originally a native of India, pepper was distributed in Island South East Asia by Indian traders long before that. The Portuguese and Spanish landed in Central and South America in their search for pepper. The Portuguese, Spanish, Dutch and English came to South and South East Asia also in of pepper. It was once worth its weight in gold and was even used as currency in some

places. Black and white peppercorns are from the same pepper vine. White is more expensive than black because the green skin of the pepper seeds has to be stripped off to get beige-coloured white peppercorns. Black pepper is dried with the green skin on. Pepper is *lada* in Malay.

Petai (*Parkia speciosa*)

This tropical tree is native to South East Asia and the beans are popular throughout. It is sometimes called twisted cluster beans because the long bean pod is somewhat twisted, or stink bean because several hours after eating *petai* your urine will smell of the bean. *Petai* is believed to be good for the kidneys. Young *petai* may be eaten raw in a salad. More mature beans are usually cooked into a *sambal*. Before cooking, the thin membrane covering the bean should be peeled off and the bean split to check for worms in the centre – unless you don't mind the extra protein. The worms are harmless.

Prawn Paste, Black *see Hae Ko*

Prawns *see* Shrimps and Prawns

Screwpine *see* Pandan

Semolina (*Triticum durum*) *see also* Wheat

Semolina is made from durum wheat, the same hard wheat as that for making Italian pastas. It is a high-protein granular flour best enjoyed amongst the Straits-born in *sugee* cake aka *suji* cake. Because semolina is also widely used in Indian sweets, it is available in Indian grocery stores.

Serai see Lemongrass

Shrimp Paste, Dried *see Belacan*

Shrimps and Prawns (*Decapods*)

Linguistically, the word "shrimps" refers to small prawns. There are several types of shrimps and prawns with slightly different textures ranging from firm to tender. In South East Asia, wet markets have both shrimps and prawns and krill too. Shrimps may be salted and dried to make *hae bee* (Hokkien) or *udang kering* (Malay). Dried shrimps are also used in some Straits-born spice pastes. The smallest shrimps are called krill or *grago*. *See also* Krill

Shrimp Paste, Dried *see Belacan*

Shrimps, Dried (*Hae Bee*) *see* Shrimps and Prawns

Soy Beans (*Glycine max.*)

Cultivated for thousands of years in China originally, soy beans are made into numerous products popular in Asian cooking. The most widely used are the soy sauces, light and dark, the former for salting and the latter for colouring as well. Other soy bean products used in Straits-born cooking are tofu, both soft and firm, bean sticks (*taukee* in Hokkien), fried tofu puffs or pockets (*taupok* in Hokkien), tofu skins (*taukee puay* in Hokkien), sweet bean curd and fermented soy beans called *taucheo* in Hokkien. Japanese brown miso has the same flavour as *taucheo*. Note that Japanese miso ranges from pale white to near-black with differences in the intensity of the flavour.

Spices *see*
- Anise *see* Fennel
- Aniseed *see* Fennel
- Cassia (Chinese cinnamon)
- Cinnamon, Indonesian (*kayu manis*)
- Coriander (*ketumbar*)
- Cumin (*jintan puteh*)
- Fennel (*jintan manis*)
- Fenugreek (*halba*)
- Pepper (*lada*)
- Turmeric (*kunyit*)

Star Anise (*Illicium verum Hooker fill.*)

Originally from southern China, star anise takes its name from its shape. In Chinese, the name means "eight corners". Star anise is one of the spices in Chinese five-spice powder. It is also used whole with strong-tasting meat such as duck or beef. It has strong anise-citrus-clove fragrances.

Star Fruit *see* Camias

Sugee see Semolina

Sugi see Semolina

Tamarind (*Tamarindus indica*)

Originally native to India, tamarind is eaten all over South and South East Asia in various forms. Known also as *asam*, tamarind is an essential souring agent in Straits-born curries, *sambals*, and even soups. The fleshy ripe fruit is shelled, salted and mashed into a block of paste that includes seeds and bits of the brittle shell. The paste is added to water to get tamarind juice and the seeds and other bits discarded. Outside South and South East Asia, look for tamarind as dried pods or in powdered form. Tamarind paste may be kept for a long time although it will dry out. To revive, soak in a covered glass jar of water and refrigerate until needed. If in dried pods, shell and soak the flesh in water. Tamarind paste may also be available as paste only without the seeds and the tub misleadingly labelled as *asam Jawa*, a term also used for *asam gelugor* in Indonesia. Tamarind is not *asam gelugor*.

Terigu see Wheat

Torch Ginger Bud (*Etlingera elatior, Nicolaia elatior, Phaeomeria speciosa, Nicolaia speciosa, Alpinia nutans*)
Another South East Asian native ginger and also called *bunga kantan* and *bunga siantan* in Malay, the flower bud may be split and put into a fish soup such as Penang laksa gravy or chopped and used as a herb in salads and *rojak*. It is also a fragrant addition to seafood curries and sambals.

Turmeric (*Curcuma longa*)
A member of the ginger family, turmeric is native to India. Known as *kunyit* in Malay, meaning "yellow" and *ooi keorh* in Hokkien meaning "yellow ginger", the very name indicates its potential as a natural yellow dye. Turmeric comes in whole pieces, fresh or dried, as well as in powder form. Straits-born cooks use both fresh turmeric as well as turmeric powder. Fresh turmeric in a spice paste will not give the same taste as turmeric powder. There is old as well as young turmeric, the older having a stronger colour and therefore perfect for vinegar *acar*. The younger turmeric is more common. Scraping the thin skin off the rhizome will stain your fingers yellow. Wear disposable gloves. Processing a spice paste with fresh turmeric in a food processor may stain a plastic bowl yellow too. Soak the bowl in bleach diluted with water. Diluted bleach is the only way to remove turmeric stains from tablecloths. Turmeric leaves are used as a herb in Straits-born salads.

Wheat (*Triticum*)
Wheat flour was introduced into South East Asia in the colonial era. Besides plain aka all-purpose flour or self-raising flour used in cakes and biscuits, there is also pearl wheat aka *biji gandum* aka *terigu* which is boiled into a porridge (*bubor* in Malay) dessert popular all over Island South East Asia. Enriched with coconut milk and sweetened with palm sugar, *bubor terigu* or *bubor biji gandum* is popular even for special occasions such as wedding feasts. The kernels of the same soft wheat as that in plain flour, *terigu* is rich in fibre and protein. It is easiest cooked in a slow cooker.

Yam Bean (*Pachyrhizus erosus*)
A native of Mexico where it is eaten as a fruit with a sprinkling of chilli powder and sugar, this tuberous root known as *bangkuang* in Hokkien and Malay is now essential in several iconic Straits-born dishes such as *popiah*, Penang *joo hoo char* and Singapore *rojak*. It was first introduced to South East Asia by the Spaniards via the Philippines. Crisp, bland and juicy, yam bean keeps its shape despite long cooking and has become the cheap substitute for bamboo shoots or paired with bamboo shoots to tone down the stronger flavour of the latter.

Ingredients on Front Cover
Clockwise from bottom left: lemongrass, turmeric, kaffir lime leaves, knotted pandan leaves, bunch of kaffir lime leaves, turmeric leaves (on the left), candlenuts in pounding stone, green and red bird's eye chillies, galangal, belimbing, pepper leaves, fresh green peppercorns, calamansi limes.

Bibliography

Deutrom, Hilda. *Ceylon Daily News Cookery Book.* [Colombo], Stamford Lake, 2013. 17th reprint ed of 5th ed. 1964. Originally published 1929.

Fernandes, Jennifer. *100 Easy-to-Make Goan Dishes.* New Delhi, Vikas Publishing House, 1993. 22nd reprint. First published 1977.

Gomes, Mary. *The Eurasian Cookbook.* Singapore, Horizon Books, 2001.

Hing, Mrs [Susie]. *In A Malayan Kitchen.* Singapore, [The Author, 1954?] Printed by Mun Seong Press.

Hobson-Jobson: A Glossary of Coloquial Anglo-Indian Words and Phrases, and of Kindred Terms, Etymyological, Historical, Geographical and Discursive. See Yule, Henry, Col. and Burnell, AC.

Kwa Chong Guan and Kua Bak Lim, eds. *A General History of the Chinese in Singapore.* Singapore, Singapore Federation of Chinese Clan Associations and World Scientific, 2019.

Kwa Chong Guan, Derek Heng, Peter Borschberg, and Tan Tai Yong. *Seven Hundred Years: A History of Singapore.* Singapore, Published by National Library Board Singapore in partnership with Marshall Cavendish International (Asia), 2019. Supported by Singapore Bicentennial Office.

Lane, Lilian. *Malayan Cookery Recipes.* Singapore, Eastern Universities Press, 1964.

Lee, Mrs Chin Koon. *Mrs Lee's Cookbook: Nonya Recipes and Other Favourite Recipes.* Singapore, The Author, 1974.

Leong, Mrs Yee Soo. *The Best of Singapore Cooking.* Singapore, Times Books International, 1988. Reprint ed 1992.

Leong, Mrs Yee Soo. *Singaporean Cooking.* Singapore, Eastern Universities Press, 1976.

Matsuyama, Akira. *Traditional Dietary Culture of Southeast Asia: Its Formation and Pedigree.* London, Kegan Paul, 2003. Translated by Atsunobu Tomomatsu.

Ong, Jin Teong. *Penang Heritage Food: Yesterday's Recipes for Today's Cook.* Singapore, Landmark Books, [2010]

Owen, Sri. *The Rice Book.* New York, St Martin's Press, 1993. First published by Doubleday.

[Sanmugam, Devagi]. *Banana Leaf Temptations: The Complete Collection of Mouth-Watering South Indian Recipes.* Reprint ed. Singapore, VJ Times, 2000. First published 1997.

Singapore National Bibliography, 1967-1991. CD Rom. First issued in 1967.

Tan, Cecilia. *Penang Nonya Cooking: Foods of My Childhood.* Petaling Jaya, Eastern Universities Press, 1983.

Tan, Lloyd Matthew. *Daily Nonya Dishes. (Laok Hari Hari: Heritage Recipes for Everyday Meals).* Singapore, Landmark Books, 2010.

Veerasawmy, EP. *Indian Cookery For Use In All Countries.* 5th reprint ed. Bombay, Jaico Publishing House, 1963. First published 1956.

Wee Eng Hwa. *Cooking for the President: Reflections and Recipes of Mrs Wee Kim Wee.* Singapore, The Author, 2010.

Wee, Sharon. *Growing Up in a Nonya Kitchen.* Singapore, Marshall Cavendish, 2012.

The Y.W.C.A. of Malaya Cookery Book: A Book of Culinary Information and Recipes Compiled in Malaya. Ed.: Mrs AE Llewellyn. 9th ed. Kuala Lumpur, Persatuan Wanita Keristian di-Malaya (The Y.W.C.A of Malaya), 1962. First published 1931.

Yeap Joo Kim. *The Penang Palate.* Penang, The Author, 1990.

Yeo, Elaine. *Irene's Peranakan Recipes.* Singapore, Epigram, c. 2006.

Yu, Yoon Gee. *Nonya Cake Recipes.* [Singapore, The Author, c.1970]

Yule, Henry, Col. and Burnell, AC. *Hobson-Jobson: A Glossary of Colloquial Anglo-Indian Words and Phrases, and of Kindred Terms, Etymological, Historical, Geographical, and Discursive.* Calcutta, Rupa, 1986. 3rd reprint. First published 1886. The Bengal Chamber Edition on the Occasion of the Tercentenary of Calcutta.

Internet Sources
Chetti Melaka. https://www.roots.gov.sg/learn/stories/chetti-melaka#chetti-creole>. Retrieved on 28 September 2020.

Google Search
Chetti Peranakan-influenced family recipes for Deepavali from octogenarian home cook Sushila Nadarajah in https://www.thestar.com.my/starplus. Retrieved on 4 October 2020.

Chetty Peranakan Culinary Staples by Abirami Durai. In https://www.thestar.com.my/lifestyle/family/2019/10/04/chetti-culinary-staples. Retrieved on 4 October 2020.

Nasi Kembuli (Chitty) https://www.kuali.com/recipe/nasi-kembuli-chitty/. Retrieved on 4 October 2020.

Dedication

This book is dedicated to all cooks and cookbook writers who have inspired me with their recipes, shared their knowledge and contributed in ways big and small to everything I know today about cooking. The two biggest inspirations were my late mother, Goh Eng Thye, and my late aunt, Lee Lian Kim, both of whose recipes I enjoy to this day and whose recipes I pass on so that more may enjoy the pleasure of great home cooking. This book is also dedicated to all wannabe cooks — my daughters Shakun and Savitri, and grandsons Arjuna and Rama too — to help them keep alive a most delectable cuisine with an incredible history.

Acknowledgements

Friendship in the time of Covid-19! Social distancing notwithstanding, I was able to get additional photographs for this cookbook thanks to good friends. My many thanks to Low Lih Jeng for some of the photography, Tan Khim Yong for styling some of the pictures, especially the cover, and to Tan Hwa Luck for giving me most of the green stuff for the shoot seen on the cover. He is the only person I know in Singapore with a pepper vine in his kitchen garden. Some of the photographs were done with my iPhone 11 Pro and boosted with Snapseed, a free app in the Apple App Store. Thanks to Emil Parkalis of the iPhone Editing Academy (https://iphonephotographyschool.com) for pointing me to this app. And special thanks to Kwa Chong Guan for the loan of Akira Matsuyama's *The Traditional Dietary Culture of South East Asia: Its Formation and Pedigree*. It was the essential peg on which to hang together my ideas about Straits-born food. He not only brought my attention to his aunt's *In A Malayan Kitchen*, but even gave me a photocopy of the cookbook borrowed from a cousin. This cookbook contributed enormously to a fuller picture of 1950s Straits-born cooking and its Indonesian Chinese links.

Weights and Measures

Quantities for this book are given in imperial, metric and American spoon and cup measures. Standard spoon and cup measurements used are: 1 teaspoon = 5 ml, 1 tablespoon = 15 ml. All measures are level unless otherwise stated.

LIQUID AND VOLUME MEASURES

Metric	Imperial	American
5 ml	$1/6$ fl oz	1 teaspoon
10 ml	$1/3$ fl oz	1 dessertspoon
15 ml	$1/2$ fl oz	1 tablespoon
60 ml	2 fl oz	$1/4$ cup (4 tablespoons)
85 ml	$2^1/2$ fl oz	$1/3$ cup
90 ml	3 fl oz	$3/8$ cup (6 tablespoons)
125 ml	4 fl oz	$1/2$ cup
180 ml	6 fl oz	$3/4$ cup
250 ml	8 fl oz	1 cup
300 ml	10 fl oz ($1/2$ pint)	$1^1/4$ cups
375 ml	12 fl oz	$1^1/2$ cups
435 ml	14 fl oz	$1^3/4$ cups
500 ml	16 fl oz	2 cups
625 ml	20 fl oz (1 pint)	$2^1/2$ cups
750 ml	24 fl oz ($1^1/5$ pints)	3 cups
1 litre	32 fl oz ($1^3/5$ pints)	4 cups
1.25 litres	40 fl oz (2 pints)	5 cups
1.5 litres	48 fl oz ($2^2/5$ pints)	6 cups
2.5 litres	80 fl oz (4 pints)	10 cups

DRY MEASURES

Metric	Imperial
30 grams	1 ounce
45 grams	$1^1/2$ ounces
55 grams	2 ounces
70 grams	$2^1/2$ ounces
85 grams	3 ounces
100 grams	$3^1/2$ ounces
110 grams	4 ounces
125 grams	$4^1/2$ ounces
140 grams	5 ounces
280 grams	10 ounces
450 grams	16 ounces (1 pound)
500 grams	1 pound, $1^1/2$ ounces
700 grams	$1^1/2$ pounds
800 grams	$1^3/4$ pounds
1 kilogram	2 pounds, 3 ounces
1.5 kilograms	3 pounds, $4^1/2$ ounces
2 kilograms	4 pounds, 6 ounces

LENGTH

Metric	Imperial
0.5 cm	$1/4$ inch
1 cm	$1/2$ inch
1.5 cm	$3/4$ inch
2.5 cm	1 inch

OVEN TEMPERATURE

	°C	°F	Gas Regulo
Very slow	120	250	1
Slow	150	300	2
Moderately slow	160	325	3
Moderate	180	350	4
Moderately hot	190/200	370/400	5/6
Hot	210/220	410/440	6/7
Very hot	230	450	8
Super hot	250/290	475/550	9/10

ABBREVIATION

tsp	teaspoon
Tbsp	tablespoon
g	gram
kg	kilogram
ml	millilitre